Transactions
of the
American Philosophical Society
Held at Philadelphia
For Promoting Useful Knowledge
Volume 96 Parts 2 & 3

ALHACEN ON THE PRINCIPLES OF REFLECTION

A Critical Edition, with English Translation and Commentary, of Books 4 and 5 of Alhacen's *De aspectibus*

VOLUME ONE
Introduction and Latin Text

VOLUME TWO
English Translation

A. Mark Smith

American Philosophical Society
Independence Square • Philadelphia
2006

ISBN-10: 0-87169-962-1
ISBN-13: 978-0-87169-962-6
US ISSN: 0065-9746

COVER ILLUSTRATION: Paris, Bibliothèque Nationale MS Lat. 7319, f114v.
The author wishes to express his deep gratitude for permission to use the
many figures from this manuscript that appear in this book.

Library of Congress Cataloging-in-Publication Data

Alhazen, 965-1039.
[Manazir Book 4-5. English & Latin]
Alhacen on the principles of reflection: a critical edition, with English translation and commentary, of books 4 and 5 of Alhacen's De aspectibus, the medieval Latin version of Ibn al-Haytham's Kitab al-Manazir/[edited by] A. Mark Smith.
p. cm. — (Transactions of the American Philosophical Society, v. 96, pts. 2 & 3)
Includes bibliographical references (p.) and indexes.
Contents: v. 1. Introduction and Latin text — v. 2. English translation.
ISBN 0-87169-962-1 (pbk.)
1. Optics—Early works to 1800. I. Smith, A. Mark, II. Title. III. Transactions of the American Philosophical Society; v. 96, pt. 2-3.
QC353
[.A32313 2001]
535'.09'021—dc21 2001041227

TRANSACTIONS

of the

American Philosophical Society

Held at Philadelphia for Promoting Useful Knowledge

VOLUME 96, Part 2

Alhacen on the Principles of Reflection:

A Critical Edition, with English Translation and Commentary, of Books 4 and 5 of Alhacen's *De aspectibus*

VOLUME ONE

To Lois, my sine qua non

CONTENTS

VOLUME I

VOLUME II

PREFACE

The passage from books 1-3 to books 4-6 of the *De aspectibus* is marked by an abrupt shift in topical focus. In the first three books, Alhacen was at pains to show that, under the right circumstances, the visual system yields a true and accurate perception of external reality. If the light is adequate, the eye healthy, the intervening medium appropriately transparent, and so forth, we see things as they actually present themselves to us in physical space. But only when the radial links between object and eye are uninterrupted. Under no circumstances is sight veridical when those links are broken. No matter how good the light, how healthy the eye, or how clear the air, things are never perceived correctly in mirrors or refracting media because, at best, they appear displaced and, at worst, distorted in shape, size, and distance. Reflection and refraction, in short, are by their very nature sources of visual illusion.

Alhacen's purpose in books 4-6 is to explain precisely how and why reflection deceives the eye. To that end, he establishes the basic principles of reflection in books 4 and 5 (the focus of the current edition) and then goes on to show in book 6 how those principles can be applied to the problem of image-distortion in plane, convex, and concave mirrors. Yet, however straightforward this goal, its accomplishment is anything but. On the contrary, as becomes clear in the course of Alhacen's analysis, an adequate explanation of reflection, as well as of image-formation and distortion, requires a great deal of sophistication in terms of both empirical verification and mathematical demonstration.

As a result, Alhacen's study of reflection seems more "scientific" and modern in its approach than his study of immediate, or direct, vision. No longer is common experience sufficient for empirical verification; controlled experiments are increasingly necessary. Admittedly, most of these experiments are primitive and loosely organized, but, as we shall see, one in particular calls for extremely tight controls. Likewise, the focus on mathematics is far more intense in books 4-6 than in books 1-3. And much of that focus is on the surprisingly complex problem of finding the exact spot, or spots, on a convex or concave mirror where a given ray of light will reflect to a given point on the other side of that spot.

It is easy to be misled by these changes in topical and analytic focus into thinking that Alhacen's primary concern in books 4-6 has shifted from the physiology and psychology of sight to the physics of light. After all, the way Alhacen uses ray-geometry in these books makes his approach look much like that of modern physical optics. It is crucial to realize, however, that throughout his analysis of reflection Alhacen's ultimate goal is to explain how and why things appear as they do in mirrors, not how light or illuminated color interacts physically with reflecting surfaces. No matter how far removed from visual issues the problem may appear to be at first glance, those issues are always lurking in the wings.

How Alhacen deals with the problem of finding precisely where reflection occurs on convex or concave spherical mirrors exemplifies this point. The givens of his analysis are 1) the great circle on the mirror defined by the plane of reflection, 2) the centerpoint of that circle, 3) the equal-angles law of reflection, 4) a point-source of radiation, and 5) a point to which that radiation will reach after reflection. But—and here is where Alhacen's true intentions are revealed—this latter point is not just any random point in space. It is *always* a center of sight. This is not to say that Alhacen was unconcerned with how light acts in its own right, but of far greater concern to him was how it acts with respect to specific viewpoints. This holds true whether the view is uninterrupted or interrupted by reflecting or refracting surfaces: the ultimate reference-point for *all* optical analysis is the center of sight. The apparent similarity between Alhacenian and modern physical optics is therefore just that: apparent. For, unlike its modern counterpart, Alhacenian optics is subjective, not objective in its analytic orientation. To lose sight of this fact is to miss the real point of his analysis of reflection, a point that becomes eminently clear in book 6, when Alhacen turns his attention to image-distortion in convex and concave mirrors.

Thematically bound by their focus on reflection and mirror-images, books 4-6 thus form a discrete topical unit. Nevertheless, my decision to separate books 4 and 5 from book 6 for this edition makes sense for at least two reasons. The first is practical. Taken as a whole, books 4-6 constitute nearly 40% of the entire text of the *De aspectibus*, almost as much as is occupied by the first three books. Given the size of this segment of text, and given my own time-constraints, I wanted to reduce the task of editing it to manageable proportions. The second reason has to do with topical organization. The purpose of books 4 and 5 is to establish and validate the rules of image-location in plane, convex, and concave mirrors on the basis of point-objects and point-images. The purpose of book 6, is to apply these rules to the analysis of two-dimensional images in order to explain precisely how the formation of such images determines the ways in which they appear deformed to the onlooker. There is thus a clear distinction between the section consisting of books 4 and 5, where the theoretical foundations are laid, and the section consisting of book 6, where those foundations are built upon.

Reduced though it was, the task of editing books 4 and 5 has been a long and complicated one during which I have accumulated many significant debts that demand acknowledgement. To start with, I wish to thank the NSF for its generous support during 2001-2003 (SES 0080445). Thanks are also due to the MU Research Council, the UM Research Board, and the American Philosophical Society for funding.

At a more personal level, I want to thank the librarians and archivists in charge of the manuscript collections I consulted for allowing me access to the necessary manuscripts and for the various courtesies they extended to me. For permission to reproduce pages from manuscripts in their collections, I am grateful to: The Master and Fellows of Trinity College, Cambridge; the Bibliothèque de l'agglomération de Saint-Omer; the Biblioteca Nazionale Centrale of Florence, by concession of the Ministero per i Beni e le Attività Culturali della Repubblica Italiana; the President and Fellows of Corpus Christi College, Oxford; the Crawford Collection of the Royal Observatory, Edinburgh; the Royal College of Physicians, London; and the Bibliothèque Nationale de France.

At the most personal level, I need to acknowledge my debt to the following people: Matt Shaw, whose talents as an editorial jack-of-all-trades lightened my authorial burden considerably; Mark Singer, whose knowledge of computer technology was invaluable in the production of the figures; Lindsey O'Donnell, whose uncanny arithmetical abilities went into the formatting of this edition; and Melinda Lockwood, without whose help with PageMaker I would have been lost in cyberspace. I also wish to thank Sandy Kietzman and Mona Burkett for their services in proofreading, and to John Frymire and Sabetai Unguru I owe a debt of gratitude for serving as sounding boards. Sabetai's critique of my introductory account of mathematical development during the Middle Ages and Renaissance was not only helpful, but also extraordinarily generous, given that my interpretation of Witelo is somewhat at odds with his. Thanks also to Abdullahi Ibrahim for his help with Arabic terms and concepts. And, of course, I cannot adequately thank my wife, Lois Huneycutt, both for her encouragement and for her willingness to take on so many family responsibilities in order to provide me the free time to work on this project.

INTRODUCTION

1. *Alhacen's Analysis of Reflection: An Overview*

Since the relevant background to both Alhacen and the *De aspectibus* has already been given in the preface and introduction to the edition of books 1-3, there is no need to rehearse the details here. Let us therefore turn directly to an examination of the form and content of books 4 and 5.

Alhacen's overall intent in these books is to determine as precisely as possible where the image of any given object-point seen in a mirror will lie with respect to a particular viewpoint, as defined by the center of sight at the vertex of the cone of visibility.[1] That location, according to Alhacen, is specified by the so-called cathetus-rule, which states that, for any given center of sight, the image of any object-point will lie where the normal, or cathetus, dropped from that object-point to the reflecting surface meets the straight line extending from the center of sight through the point of reflection. This line coincides with the light-ray reflected from the mirror, or, as Alhacen denominates it, the "line of reflection" (*linea reflexionis*). Thus, as illustrated in figure 1, p. 521, if center of sight A and object-point B face reflecting surface EF—be it plane, convex spherical, or concave spherical—and if ray BD reflects to A along ray DA, then, from the perspective of viewpoint A, the image of B will appear at I, where cathetus BI intersects the extension of AD. In plane mirrors, the cathetus is perpendicular to the reflecting surface itself, whereas in the two types of curved mirrors, it is perpendicular to the plane tangent to the reflecting surface where it intersects that surface. In those two types of mirrors, therefore, the cathetus passes through center of curvature G.

With this rule firmly established by the end of book 5, Alhacen turns in book 6 to the analysis of image-formation and image-distortion for object-surfaces, which, unlike points, have physical dimensions and are thus actually visible.[2] This analysis is based on reducing such surfaces to a representative sample of constituent points, locating the images of those points, and then extrapolating the composite image from them. How this works in practice is most easily illustrated in the case of plane mirrors. Let A in the top diagram of figure 2, p. 522, be the center of sight, BS the cross-section of some object-surface, D the point where incident ray BD reflects to A, and EF the common section of the surface of a plane mirror and the plane containing A, BS, and D. If BS is resolved into constituent points B, P, Q, and S, the incident ray from each point will reflect to A from a specific point of reflection on the mirror's surface. Accordingly, image B′ of B will lie where cathetus BB′ intersects line of sight ADB′, image P′ of P where cathetus PP′ intersects line of sight AD_1P', and so forth for points Q and S. The resulting composite image B′S′ will thus be the same shape and size as BS and will lie the same distance below the reflecting surface as BS lies above it. So, apart from apparent left-right reversal, image B′S′ will be an exact replica of BS.

The same kind of analysis can be applied to the situation illustrated in the bottom diagram of figure 2, where center of sight A and cross-section BS face a convex spherical mirror centered on G and containing arc EF. Let incident ray BD reflect to A from D, and let EF be the common section of the mirror's surface and the plane containing A, BS, and D. When the image-locations of constituent object-points B, P, Q, and S on BS are determined, the resulting composite image B′S′ turns out to be smaller than BS and somewhat bowed according to the curvature of arc EF. Moreover, B′S′ does not lie the same distance below the reflecting surface as BS lies above it. Far from being an exact replica, then, the image of BS is distorted in several ways that depend upon the mirror's curvature.

The fundamental principles underlying these two point-analyses, as well as the cathetus-rule upon which they are based, are few and fairly obvious. The first such principle is that the center of sight, the object-point, the point of reflection, and the image all lie in a plane normal to the mirror at the point of reflection. This plane constitutes what Alhacen calls the "plane of reflection" (*superficies reflexionis*). The second principle is that the angle of incidence, as measured according to the normal at the point of reflection (i.e., angle BDN in figure 2), is equal to the angle of reflection, as measured according to the same normal (i.e., angle ADN). The third is that the image lies on the straight line extending through the point of reflection from the center of sight. And the fourth is that the image also lies on the cathetus dropped from the object-point to the mirror. A considerable portion of Alhacen's analysis from the beginning of book 4 through the first quarter or so of book 5 is devoted to establishing these four principles as clearly and incontrovertibly as possible.

Neither the cathetus-rule nor the principles underlying it were new to Alhacen. All of these principles are at least implicit in Euclid's analysis of image-formation in the *Catoptrics* (c. 300 B.C.). In fact, Euclid went so far as to offer a mathematical demonstration of the equal-angles principle in the first proposition of that work.[3] The same fundamental principles are also implicit in Hero of Alexandria's *Catoptrics* (c. 50 A.D.). Like Euclid, moreover, Hero essayed a mathematical demonstration of the equal-angles principle.[4] But the clear articulation of all four principles had to await Ptolemy, who laid them out explicitly in the third book of his *Optics* (c. 160 A.D.).[5] Furthermore, Ptolemy departed from both Euclid and Hero by basing his analysis of reflection not just on mathematical principles but on empirical observation as well. This is most evident in his notorious experimental confirmation of the equal-angles principle, which is based on careful, measured observation of the angles of incidence and reflection for plane, convex cylindrical, and concave cylindrical mirrors.[6]

Not surprisingly, Alhacen follows Ptolemy's lead in his analysis of reflection, balancing empirical observation with mathematical demonstration in much the same way as Ptolemy.[7] Yet, despite obvious similarities, there are crucial differences between Alhacen's and Ptolemy's approaches. The most obvious lies in the fact that Alhacen bases his analysis on light-rays rather than on visual rays. A second and more significant difference is that Alhacen's analysis is far more rigorous and comprehensive than Ptolemy's. Take, for instance, Alhacen's experimental confirmation of the equal-angles principle in chapter three of book 4 (see pp. 300-312 below). Clearly modeled on Ptolemy's, Alhacen's experiment is nonetheless light-years beyond Ptolemy's in its instrumental and conceptual sophistication. Furthermore, unlike Ptolemy, Alhacen goes to great lengths to detail the mathematical implications of his empirical observations. In this way, for instance, he draws on the experimental confirmation of the equal-angles principle to verify that the plane of reflection is always normal to the mirror at the point of reflection. In addition, the rigor and sophistication of Alhacen's approach extends beyond the empirical to the mathematical basis of his analysis. Alhacen's proofs are incomparably more rigorous, sophisticated, and elegant than Ptolemy's. Thus, although there are traces of Ptolemy's influence throughout Alhacen's account of reflection, those traces dwindle to insignificance in the light of Alhacen's many original contributions to that account.

The third and most significant difference between Alhacen's and Ptolemy's analyses of reflection is that Alhacen undertakes to determine precisely where on the surface of a convex or concave spherical mirror the radiation from a given object-point will reflect to a given center of sight. Ignored entirely by Ptolemy—and for good reason, given its complexity—this problem, or, rather, its definitive solution, constitutes the *pièce de résistance* of Alhacen's analysis of reflection in book 5 of the *De aspectibus*. It also caps Alhacen's effort to hone the cathetus-rule to perfect sharpness for subsequent application in book 6. With the object-point, the cathetus, the center of sight, and the point of reflection all given and perfectly determined, the image-location can at last be found with absolute and unvarying precision.

The following overview of Alhacen's reflection-analysis is organized according to the structure of books 4 and 5. I will therefore proceed through those books in step-by-step, chapter-by-chapter order. I have chosen this approach not simply because it is straightforward. I have also chosen it because, in following the actual flow of Alhacen's analysis as it builds from empirical foundations to the elaborate mathematical structure erected on them in book 5, this approach reveals the underlying comprehensiveness, elegance, and logical seamlessness of that analysis.

Book Four: Alhacen opens his study of reflection in chapter 2 of book 4 by describing a set of simple experiments designed to show that the reflection of light is regular, that this regularity extends to every kind of light or luminous color, whether primary or secondary,[8] and that reflection weakens the resulting light or luminous color. The experimental apparatus consists of a room with one small opening through which sunlight or daylight streams in a relatively narrow shaft. An iron or silver mirror can then be placed toward the incoming light so as to reflect it to various places in the room. In this way it can be shown that, when the mirror is rotated, the spot of reflected light will follow the arc of rotation (4, 2.2-2.5, p. 296). It can also be shown that reflected light or illuminated color is less intense than the light or illuminated color at the source (4, 2.7-2.13, pp. 297-298) and that, if the same light or illuminated color shining on the mirror also shines directly on a white background, the reflected light or illuminated color will be less intense than that shining directly on the background (4, 2.15-2.20, pp. 298-299). On the basis of such experiments, therefore, we can conclude with Alhacen that "the forms of lights and colors are reflected from polished bodies and [are thereby] weakened" and that a "reflected form is brighter than a secondary one when they share the same source" (4, 2.21, p. 299).

Having set the stage with these rough, preliminary observations, Alhacen turns in chapter 3 to an experimental confirmation of the equal-angles principle. Before setting up the actual experiment, he establishes a few guidlines. First, the reflectivity of any polished body is due to the physical smoothness of its surface (4, 3.1, p. 300). Second, no matter the shape of the reflecting surface, reflection can occur from every point on it. In such reflection, moreover, the lines of incidence and reflection, as well as the normal to the point of reflection, will all lie in a single plane normal to the reflecting surface at the point of reflection, and within that plane the lines of incidence and reflection "will maintain an equivalent situation with respect to [that] normal and will form equal angles [with it]" (*tenebunt . . . eundem situm respectu perpendicularis et equalitatem angulorum*—4, 3.2, p. 300). Third, if light reaches a mirror along the normal, it reflects back along the normal. In plane mirrors, this normal is perpendicular to the reflecting surface itself; in cylindrical and conical mirrors, whether convex or concave, it is perpendicular to a plane tangent to the reflecting surface along a line of longitude; and in spherical convex and concave mirrors, it is perpendicular to a plane tangent to the reflecting surface at a point only (4, 3.3, p. 300).

In the next 38 paragraphs (4, 3.4-3.41, pp. 300-308), Alhacen describes the apparatus to be used for the experiment. This apparatus consists of three main components, the first of which is the register described in 4, 3.3-3.8, pp. 300-301. It consists of a thin bronze plaque formed from a semicircular section six digits in radius.[9] When finished, it takes the form illustrated in figure 4.3.2, p. 193, with its outer arc OFV centered on E. From that same centerpoint an arc five digits in radius is scribed on its surface to pass through point G. A selection of lines, such as HE and AE on the right side of EF, is also scribed on its surface. These lines are matched on the left side of EF by lines forming the same angle with it. Thus, angle HEF = angle MEF, and so forth. The bottom edge of the register is cut along the outermost of these converging lines, as well as along line VO, so that what is left at the base is a small triangular section with centerpoint E of the register at its vertex.

The second component consists of a wooden ring fourteen digits in diameter and seven digits high, its wall being two digits thick. Its inner hollow is therefore ten digits in diameter. The construction of this ring is described in 4, 3.9-3.22, pp. 301-304. On its top surface, lines are scribed to correspond precisely with those scribed on the face of the register. Thus, as illustrated in figures 4.3.3 and 4.3.4, pp. 193-194, lines AE, BE, CE, and DE, which converge at centerpoint E of the top base of the ring, form angles BEA, CEA, and DEA equal, respectively, to angles HEF, AEF, and MEF on the face of the register. From the inner and outer endpoints of these lines, as represented in figure 4.3.4, perpendicular lines are drawn along the inner and outer surfaces of the ring to form lines of longitude on both surfaces. Hence, when the cylinder is placed upright with its top and bottom bases perfectly horizontal, those lines will be perfectly vertical, as illustrated from an inside view in figure 4.3.6, p. 195.

A circle is then drawn on the inner surface of the ring parallel to its base at a height of two digits minus half a grain of barley above that base.[10] A notch is scooped along this circle to a depth of one digit and no thicker than the bronze register so that the register can be inserted snugly into the notch with its upper face parallel to the base of the ring at a distance of two digits minus half a grain of barley above that base (see figure 4.3.6, p. 195, for a cutaway diagram of the ring with its notch). When the register is properly inserted into this notch, the ring's axis will pass through centerpoint E of the register, and the inner arc passing through point G on the register will coincide with the inner surface of the ring. Also, line EF on the register will pass through point B on the ring's inner surface, line EH through point C, and so forth. Each line scribed on the register's face will thus intersect a corresponding vertical line on the inner surface of the ring along the orthogonal.

At a height of precisely two digits above the base on the outer surface of the ring, a circle is drawn parallel to the ring's base. Where that circle intersects each of the lines drawn vertically along the ring's outer surface, a hole one grain of barley in diameter is drilled through the wall with its axis parallel to the corresponding line inscribed on the upper face of the register. Accordingly, the line of longitude on the ring's outer surface and the corresponding line of longitude on its inner surface will bisect the outer and inner bases of the hole. Likewise each line scribed on the face of the register will line up perfectly with its appropriate hole, as represented in figure 4.3.7, p. 195. According to that alignment, the line on the register's face will coincide with the line of longitude at the bottom of the hole at the point where the vertical line on the inner surface of the ring intersects it.

For the final step of this phase of the construction, a square block of wood fourteen digits on a side and over one digit thick is formed.[11] At the very center of the top surface of this block, a square four digits to a side is drawn so that its center coincides with the center of the block. That square is excavated to a depth of one digit so that its bottom is perfectly flat and therefore parallel to the top surface of the block, as represented in figures 4.3.8 and 4.3.9, p. 196. The base of the ring is attached to this block so that the center of the circle at the base of the ring's hollow coincides with the center of the square on the top surface of the block. The axis of the ring will therefore pass through the center of the square at the bottom of the cavity in the block, the center of the square at the top, and point E of the register inserted into the ring above it, as illustrated from a bird's-eye view in figure 4.3.10, p. 197.

The third and final component consists of seven panels, each with a specific type of mirror inserted into it. Distinguished by the shape of their reflecting surfaces, these seven mirrors range from plane, through convex spherical, cylindrical, and conical, to concave spherical, cylindrical, and conical. Alhacen's choice of these particular mirrors was dictated by the relative simplicity of their shape and his capacity to explain their relevant properties with the mathematical tools available to him.[12] Accordingly, within the limits of his ability to analyze them mathematically, the six types of curved mirror selected by Alhacen are as representative as possible of all curved mirrors, whatever the complexity of their curvature.

The construction of the seven selected mirrors, all of which are iron, is described in 4, 3.23-3.41, pp. 304-308. The plane mirror consists of a fairly thin, flat disk three digits in diameter, as represented in the upper left-hand diagram of figure 4.3.11, p. 198.[13] The convex and concave cylindrical mirrors are formed from a hollow cylinder three digits high and six digits in diameter at the bases. Two sets of parallel chords three digits in length are measured off on the top and bottom bases of the cylinder, and the resulting sections are cut off to leave two cylindrical segments three digits high and three digits wide along the chords. The construction of one of these mirrors is represented in the upper right-hand diagram of figure 4.3.11. Then follows the construction of the convex and concave conical mirrors from a hollow right cone whose base is six digits in diameter and whose lines of longitude are four-and-a-half digits from vertex to base. When each mirror is properly excised from the cone, it will be four-and-a-half digits high and three digits across at the base, as represented in the lower right-hand diagram of figure 4.3.11. Finally, the spherical convex and concave mirrors are formed from a hollow iron sphere six digits in diameter. When properly excised, both mirrors will be three digits in diameter at the base, as represented in the lower left-hand diagram of figure 4.3.11.

For each mirror a rectangular wooden panel six digits high, four digits wide, and at least thick enough to stand firmly upright on its own is formed.[14] The plane mirror is inserted into its panel so that the reflecting surface is perfectly flush with the panel's face and so that its centerpoint coincides with the centerpoint of the panel's face–i.e., where the midlines along the length and width of the panel's face intersect. Accordingly, as represented in the upper right-hand diagram of figure 4.3.13, p. 199, the centerpoint of the plane mirror lies on the midline of the panel's face at a distance of precisely three digits above the panel's base.

The insertion of the convex cylindrical mirror is represented in the lower left-hand diagram of figure 4.3.13. It fits into its panel so that the midline along its length is perfectly flush with the face of the panel and thus coincides with the midline along the length of the panel's face. Likewise, the midline along the width of the panel's face bisects the mirror so that the centerpoint of the mirror's surface lies precisely three digits above the panel's base.

As represented in the upper left-hand diagram of figure 4.3.14, p. 200, the convex conical mirror is set into its panel so that line of longitude VF bisecting its surface is perfectly flush with the panel's face, its vertex lying at the very top of the midline along the length of the panel's face. Thus, the midline along the width of the panel's face intersects the mirror at a distance of precisely three digits above the panel's base along its midline of longitude. In this case, therefore, the midpoint on the panel's face will not coincide with the midpoint on the line of longitude bisecting the mirror's surface. But all that really matters is that the mirror's base lies below the midpoint of the panel.

The insertion of the convex spherical mirror, finally, is illustrated in the lower left-hand diagram of figure 4.3.14. It must be set into its panel so that the centerpoint of its surface coincides with the centerpoint of the panel's face and thus lies precisely three digits above the base of the panel on the midline along the length of the panel's face. In this case, placing the centerpoint of the mirror's surface precisely at the panel's centerpoint is crucial.

The three concave mirrors, on the other hand, must be inserted into their panels so that the chord (or chords) at their bases is perfectly flush with the panel's face or, in the case of the concave spherical mirror, so that the entire base is perfectly flush with that surface. Accordingly, as represented in the lower right-hand diagram of figure 4.3.13, p. 199, the concave cylindrical mirror fits into its panel with the chords at both ends flush with the panel's face. Hence, the line of longitude bisecting the mirror's surface will lie directly below, and parallel to the midline along the length of the panel's face, and the midpoint on its surface will lie precisely three digits above the panel's base.

Like its convex counterpart, the concave conical mirror is set into its panel so that its vertex touches the panel's top. However, as represented in the upper right-hand diagram of figure 4.3.14, p. 200, it must be set into its panel so that line of longitude VF bisecting its surface lies directly below, and parallel to the midline along the length of the panel's face. As before, although the midpoint of the panel's face will not be in line with the midpoint on line of longitude VF that bisects the mirror's surface, all that matters is that the mirror's base lie below the midpoint of the panel.

Finally, the spherical concave mirror is set into its panel so that its base is perfectly flush with the panel's face and the centerpoint of its surface lies directly below the centerpoint of the panel's face, as represented in the lower right-hand diagram of figure 4.3.14. Hence, the mirror's centerpoint lies precisely three digits above the panel's base. Here, too, the placement must be absolutely precise.

With their mirrors properly inserted, the panels are ready to be placed one at a time into the ring and stood perfectly upright on the floor of the square cavity at its base so that line FGE of the inserted register is perpendicular to the face of the panel containing the inserted mirror as well as to the midline along its length. So oriented, the panel is then slid toward point E of the register until that point touches the mirror. In the case of the plane, cylindrical, and conical mirrors, point E will touch the line of longitude bisecting the mirror's surface. As we have seen, this line either coincides with, or is parallel to and directly below the midline along the length of the panel's face. Thus, the line connecting point E and the midpoint of the mirror will either coincide with, or be parallel to the midline along the length of the panel's face. Moreover, by construction, the top face of the register lies half a grain of barley below the midpoint of the panel, so, in the case of the plane, cylindrical, and conical mirrors, point E will touch the mirror's midline of longitude half a grain of barley below the level of the panel's midpoint.[15]

Such is not quite the case for the spherical mirrors. Because of their curvature, point E at the vertex of the register will touch their surfaces either in front of or behind the midline passing through their centerpoints, so E will not be properly aligned with respect to those centerpoints.[16] In order to insure proper alignment, the panel containing the convex mirror must be pulled back slightly from the point of contact, whereas a hole must be drilled into the concave mirror at the point of contact so that point E of the register can be pushed ever so slightly into the panel beyond the mirror's surface. In both cases the required accommodation entails an adjustment of around .36 mm.[17]

Given this description of the apparatus and its appropriate setup, the purpose and outcome of the experiment are virtually self-evident. No matter which of the mirrors in its panel is properly disposed against point E of the register, when all but one of the holes in the ring's wall is blocked, and when the open hole is exposed to light, the beam of light passing through that hole should reach the mirror along the line matched to the hole on the face of the register. Likewise, after reflecting, the light-beam should follow the corresponding line on the face of the register to the corresponding hole on the other side. To establish that this is in fact the case, Alhacen proposes a variety of tests in 4, 3.43-3.64, pp. 309-312, for all seven mirrors on the basis of both primary light (i.e., sunlight) and secondary light (i.e., daylight). The test for illuminated color he defers until the very end of the chapter (4, 3.107-3.108, pp. 323-324). Given the simplicity of both the tests and their results, nothing more need be said about them than that they all confirm the equal-angles principle in one way or another. They also confirm that the plane of reflection is indeed normal to the mirror at the point of reflection.

With the empirical confirmation of the equal-angles principle complete, Alhacen goes on in the next section (4, 3.65-3.87, pp. 313-317) to explain the implications of the results. Although verbal, this explanation amounts to a geometrical demonstration based on the structure of the apparatus and the disposition of the mirrors in it. For one thing, Alhacen notes, the ring is formed so that the plane of the floor of the square cavity upon which the panels stand, the plane of the ring's bottom base, the plane of the top surface of the register, and the plane of the axes passing through the holes in the ring's wall are all horizontal and therefore parallel. For another, the plane of the register's upper face cuts the face of the panel at a level three digits minus half a grain of barley above the floor of the square cavity upon which the panel stands. Since the axes of the holes lie precisely half a digit above that level, the plane of the axes cuts the face of the panel at the midline along its width, which lies precisely three digits above its base and passes through or right in line with the centerpoints of the spherical and cylindrical mirrors.

Likewise, the vertical lines passing through the centerpoints of their respective holes along the inner and outer surfaces of the ring are all parallel to one another, as well as to the midline along the length of the panel's face, which coincides with or is parallel to the midline passing through the seven mirrors. So all those lines are perpendicular to the aforementioned planes. Finally, midline EGF of the register is perpendicular to the face of the panel.[18] Given, therefore, that light propagates in straight lines, and given the geometry of the apparatus in its proper disposition, it follows that, no matter the angle at which the incoming light-beam strikes the mirror, that light-beam will reflect at precisely the same angle from the mirror, and throughout the process, it will follow the appropriate lines scribed on the register's face. The axial rays of the incident and reflected beams will thus be perfectly parallel to their corresponding lines on the register at a distance of half a grain of barley above them, and they will be precisely the same length as those lines. So it follows that in all cases the plane of reflection, which is defined by the axial rays coinciding with the axes of their respective holes, is normal to the mirror at its centerpoint, because it is perpendicular both to the face of the panel and to the midline along its breadth.

From this it should be clear why Alhacen insists on such painstaking exactitude in the construction of the experimental apparatus, even to the point of adjusting the placement of the spherical mirrors by the negligible amount of .36 mm. The slightest variation in the parallelism of planes and lines, or in the measurement of angles and distances, or in the placement of the mirrors within their panels can skew the experimental results badly enough to disconfirm—or at least fail to confirm—the principle it is designed to validate. After all, the equal-angles principle mandates a specific and exact relationship between the angles of incidence and reflection, so any results that fall short of revealing precisely that relationship represent either failure of the experiment or failure of the principle.

That Alhacen's apparatus can be constructed more or less according to the exacting standards needed to yield the appropriate results was demonstrated in 1977 by Saleh Omar.[19] But Omar had the benefit of modern technology and machining. Alhacen did not. So the question inevitably arises: could Alhacen have possibly produced the apparatus he describes within the extraordinarily narrow tolerances required? To ask this question is to raise the issue of whether he really did construct the apparatus, or whether he imagined it, knowing full well that, if it were ever built to his specifications, it would work precisely as planned.

There are good arguments for both alternatives. On the one hand, the punctiliousness with which Alhacen describes various procedures in the construction of the apparatus suggests that he actually followed them. In 4, 3.10, pp. 301-302, for instance, he gives fairly clear instructions about how to set up a lathe for drilling out the central hollow in the ring. And in 4, 3.12-3.13, p. 302, he details two methods for producing the vertical lines on the outer and inner walls of the ring.[20] Furthermore, the apparatus as described is small enough that very slight errors in its construction can be tolerated without skewing the results beyond at least close approximation.

On the other hand, given the technological limits of Alhacen's time, there is ample reason to doubt that he could have constructed the apparatus as described, much less gotten the perfect experimental results he details. It is difficult to believe, for instance, that Alhacen could have drilled the holes in the ring's wall to the degree of accuracy required by the experiment, since the axis of every hole must be perfectly parallel both to the appropriate line on the face of the register and to the plane of the ring's base, which itself must be perfectly parallel to the top face of the register. It is also difficult to believe that he could have produced circles precisely half a grain of barley in radius (i.e., c. .21 cm.) with a compass, as he instructs us to do in 4, 3.17, p. 303. The adjustment of .36 mm. for the two spherical mirrors is also problematic. Failure to take it into account would have had a nugatory effect on the experimental results, so why bother to make it when the effort involved so clearly outweighs the benefit? Worse yet, it is difficult to believe that Alhacen could have formed the iron mirrors used in the experiment. For a start, casting the hollow iron sphere, cylinder, and cone from which the curved mirrors are supposedly formed would have been virtually impossible with the smelting technology available in Alhacen's day, so those forms must have been wrought or ground.[21] But cold iron is extraordinarily difficult to work because of its hardness, so, even if those hollow mirrors were somehow produced from blanks, the process of grinding and polishing them would have been inordinately time-consuming and expensive without the aid of advanced power technology, A far more practical, and equally effective, alternative would have been to form the requisite mirrors from bronze, which is considerably more tractable than iron.[22] There is no doubt that Alhacen was aware of this fact, so his failure to choose that alternative is puzzling. Whatever the case, until we know more about the technology available to Alhacen, the question of whether he conducted his reflection-experiments in physical or mental space must remain open.

Done with the reflection-experiments and their geometrical explanation, Alhacen attempts in the concluding section of chapter 3 to justify the equal-angles principle on theoretical and physical grounds (4, 3.88-3.106, pp. 317-323). He begins in 4, 3.88-3.92, pp. 317-319, by establishing that each point on a luminous surface facing a mirror forms the vertex of a cone of radiation with its base on the reflecting surface. By the same token, each point on the reflecting surface forms the vertex of a cone of radiation with its base on the luminous surface, the light from every point on this surface converging at that spot on the mirror. Furthermore, every point between the luminous and reflecting surfaces constitutes the vertex of two intersecting cones of radiation, one with its base on the luminous surface, the other with its base on the reflecting surface. All these cones taken together form a shaft of light whose outer edges are defined by the peripheries of the luminous and reflecting surfaces.

To illustrate, let AC in the top diagram of figure 3, p. 523, represent the cross-section of a luminous surface facing cross-section DD_4 of a plane mirror within the plane of the page. BD_1D_3 constitutes one of an infinite number of cones of radiation emanating from point B. D_2AC constitutes one of an infinite number of cones of radiation whose light emanates from every point on the luminous surface to converge at a single point on the reflecting surface. Point X, where the edges of these two cones intersect, constitutes the vertex of two opposite cones AXB and XD_1D_2 with bases AB and D_1D_2 on the luminous and reflecting surfaces, respectively. All such cones taken together will form truncated conical shaft ADD_4C of radiation.

Now, as Alhacen has established earlier on empirical grounds in 4, 3.46-3.47, pp. 309-310, when that conical shaft of light is reflected from the mirror it continues to spread out commensurately as the light within it radiates away from the mirror. Hence, the shaft of light after reflection is continuous with the incident shaft in the same way it would be if the light in the incident shaft had proceeded the same distance without being interrupted by the mirror. Suppose, therefore, that opaque surface EF in the top diagram of figure 3 intercepts the light from AC after reflection from base DD_4 on the mirror. That light will be projected to base HK on EF to create truncated conical shaft HDD_4K, which is continuous with incident shaft ADD_4C.

Like that incident shaft, this one is also composed of an infinite number of intersecting cones based on the reflecting and illuminated surfaces.

From every point on luminous surface AC, therefore, an infinite number of cones of radiation can be formed with their bases on reflecting surface DD_4. For each such point on AC, one of the cones emanating from it will have the entire reflecting surface as its base; the rest will stand on smaller segments of that surface. For every point on luminous surface AC, moreover, there will be one particular cone whose continuation after reflection has its base on segment HK of opaque surface EF. Thus, as illustrated in the middle diagram of figure 3, the light within cones ADD_3, BD_1D_4, and CD_2D_5 is reflected to HK from bases DD_3, D_1D_4, and D_2D_5 on the mirror along truncated conical shafts $DHKD_3$, D_1HKD_4, and D_2HKD_5, respectively. For every ray within the incident cone there is within the reflected shaft a corresponding ray, which, had it not been interrupted by the reflecting surface, would have continued along the line of incidence.

By symmetry, then, if HK is taken to represent the luminous source and AC the opaque surface upon which HK's light is reflected, that light will reach points A, B, and C from the same points on the mirror along the same rays as the light from points A, B, and C reached HK when AC was the luminous source. For each ray within incident shafts HKD_3D, HKD_4D_1, and HKD_5D_2, there will be a corresponding ray within the reflected cones D_3DA, D_4D_1B, and D_5D_2C, according to the continuity of reflection. So, if a center of sight is placed at point B, as represented in the bottom diagram of figure 3, the light from HK will radiate to the mirror along an infinite number of rays, such as HD, LD_1, MD_2, and KD_3, emanating from an infinite number of points on HK that, after reflecting, will converge on B along corresponding rays DB, D_1B. D_2B. and D_3B. This correspondence, of course, lies in the symmetry with which the ray-pairs are inclined to the reflecting surface—i.e., at equal angles. The same holds for any point on AC; if a center of sight is placed there, the entire form of HK will converge on it. Accordingly, the form of HK can reflect to an infinite number of centers of sight from the mirror, and the form of each point on HK will reflect to that center of sight in such a way that the incident and reflected rays will form equal angles with the normal dropped to the point of reflection.

In addition to using this model to justify the equal-angles principle on theoretical grounds, Alhacen draws upon it to explain the effect of reflection on light-intensity. As he points out in 4, 3.93, p. 319, light can be weakened by reflection in three ways: through increasing distance as it draws away from its source, through the deadening effect of reflection itself, and through the dispersal of the light as it spreads out. Accordingly, in the top diagram of figure 3, the light at HK is less intense than the light at source

AC because all three of these factors come into play. On the other hand, in that same diagram, the light from AC converges on point D2 of the mirror, so it is strengthened by concentration at that point. This intensification also applies to points H and K in the middle diagram of figure 3, the appropriate rays from every point on AC being concentrated at those two points after reflection—and indeed at every other point on HK. Altogether, then, the intensity of light at HK depends on how the weakening due to distance and reflection (as well as to dispersal) balances out against the strengthening due to concentration. If the weakening factors outweigh the strengthening factor, the resulting light will be less intense than, the light at the source. If not, it will be just as intense as, or more intense than, the source-light. Here Alhacen presumably has in mind the case of concave spherical or parabolic mirrors at or near whose focal points incoming parallel light-rays are concentrated with extraordinary intensity.[23]

Moving from the geometry to the physics of reflection, Alhacen draws on an analogy between projectile motion and light-radiation in order to explain reflection in terms of physical rebound. He lays the theoretical groundwork for this analogy by resolving the surface of any given luminous source into quanta of "least light" (*lux minima*) that lie at the threshold of effective luminosity. Nothing below that threshold—i.e., no spot of light smaller than this quantum—constitutes actual light because it will have dwindled to invisibility (4, 3.97, p. 320). Ill-defined though it may be (see note 71, pp. 359-360, for elaboration), this notion of minimal light nonetheless serves to reduce the luminous surface to such tiny spots that the beams of light emanating from them have almost no breadth. Virtually equivalent to mathematical rays, therefore, they can be understood as trajectories followed by the light-quanta as they fly outward from their source (4, 3.98, pp. 320-321).

How such light-quanta interact with a given physical surface is determined by the structure of that surface. When it is rough or porous, the light-quanta striking it are dispersed haphazardly because of its unevenness, some of the impinging quanta being scattered outward in random directions, others being trapped in the pores and thereby absorbed.[24] The less rough the surface, the less haphazard the dispersion, until eventually the surface becomes smooth enough to cause some reflection, however imperfect. The more perfect the reflection, the less haphazard the dispersal, since reflection is regulated by the equal-angles principle. Perfect smoothness therefore yields perfect reflection, which is perfectly regulated by that principle (4, 3.99-3.100, p. 321).

Now, if we think of these light-quanta as tiny projectiles shot at great speed from luminous sources, the analogy between reflection and physical rebound becomes obvious. Accordingly, light-quanta striking a mirror along the normal and reflecting back along that normal act like hard spheres dropped to an unyielding, horizontal surface and bouncing straight back along the original line of descent (4, 3.100-102, pp. 321-322). The smoother and more reflective the mirror's surface, the stronger the rebound as measured by the intensity of the reflected light. By the same token, the more unyielding the surface struck by the sphere, the stronger the rebound. So the polish of the mirror is analogous to the physical hardness of the surface from which the sphere rebounds.[25] However perfect the reflection or rebound, though, the light or sphere loses some of its original impulse after impact with the resisting surface—hence, the deadening effect reflection has on light.

Like reflection or physical rebound along the normal, reflection or physical rebound along the oblique follows the same dynamic principles. We can thus draw an analogy between light striking a mirror at an angle and a hard sphere projected obliquely at great velocity against an unyielding surface (4, 3.102, pp. 321-321). If the light or sphere were allowed to penetrate the given surface unhampered, it would continue along its original path. But such penetration is balked by the surface, which poses the same resistance it did when the light or sphere struck it orthogonally. Being directly opposed to the vertical impulse of the light or sphere, this resistance forces both to reverse direction along the orthogonal component of the path they followed in their incidence. But the surface poses no resistance whatever along the horizontal, so, when the light or sphere strikes it obliquely, its motion in that direction remains unaffected. Consequently, the motion is fully reversed along the vertical, while it remains unchanged along the horizontal. Altogether, then, the incidence and rebound of the obliquely striking light or sphere are perfectly symmetrical with respect to the point of impact, which is to say that they occur at equal angles with respect to the normal erected at that point (4, 3.103-106, pp. 322-323).

The idea of likening reflection to physical rebound is hardly original with Alhacen. Both Hero of Alexandria and Ptolemy apply the same analogy to the analysis of reflection, and Ptolemy extends that analogy to refraction (as in fact Alhacen does later on in book 7 of the *De aspectibus*). However, Alhacen's dynamic account of reflection according to the composition of vertical and horizontal motions and forces does seem to be original to him, and it has significant ramifications for the development of optics in Europe from the thirteenth to the seventeenth century. Clear traces of that account can be seen in Descartes's vector-analysis of reflection and refraction in the *Dioptrique* of 1637 (see note 79, p. 361, for elaboration on this point).

Bringing chapter 3 to a close with his long-deferred experimental confirmation of the equal-angles principle for illuminated color (4, 3.107-3.108, pp. 323-324), Alhacen undertakes in the next chapter to refute both the visual-ray account of image-formation in reflection and the account of those who "claim that the form of the object is impressed upon a facing mirror [so as to be] seen in the mirror the same way that natural forms of objects are perceived in objects" (4, 4.1, p. 324). In fact, Alhacen does not address the visual-ray account until the fifth chapter (4, 5.3, p. 326), so the whole of chapter 4 is devoted to undermining the impression-theory of image-formation. The arguments Alhacen adduces against this theory in 4, 4.2-4.6, pp. 324-325, are obvious enough that they warrant neither retailing nor discussion here. Suffice to say, they center on the fact that, contrary to what would be expected if they were actually impressed on the mirror's surface, images seen in reflection shift their location and disposition on the mirror's surface as we look at them from changing perspectives.

Alhacen's purpose in the fifth and final chapter of book 4 is to establish that, for any given viewpoint, every point on the exposed portion of a mirror's surface constitutes a point of reflection. To that end, he deals in order with plane mirrors (4, 5.1-5.10, pp. 325-329), convex spherical mirrors (4, 5.11-5.14, pp. 329-331), convex cylindrical mirrors (4, 5.15-5.26, pp. 331-335), convex conical mirrors (4, 5.27-5.5.45, pp. 335-342), concave spherical mirrors (4, 5.46-5.50, pp. 342-343), concave cylindrical mirrors (4, 5.51-5.56, pp. 343-344), and concave conical mirrors (4, 5.57-5.60, pp. 344-345). For each mirror, Alhacen determines how much of its surface is exposed to a given viewpoint, how any given plane of reflection passing through this viewpoint will cut that surface, and how many possible reflection-points there are within each plane.

Under normal circumstances, Alhacen begins, the entire surface of a plane mirror is exposed to any facing viewpoint. Any plane of reflection passing through that viewpoint will cut the reflecting surface along a straight line, and every point on this line can serve as a point of reflection for that viewpoint (4, 5.9, pp. 328-329). Since an infinite number of such planes can be passed through the given viewpoint along the normal dropped from it to the mirror, reflection can occur to that viewpoint from every point on the mirror (4, 5.10, p. 329).

In convex spherical mirrors, on the other hand, less than half the surface is exposed to any facing viewpoint (4, 5.11, p. 329). Every plane of reflection passing through that viewpoint will form a great circle on the mirror's surface, and every point on every great circle within the exposed portion of the mirror can serve as a point of reflection for that viewpoint.

Since an infinite number of such planes can be passed through any given viewpoint along the normal dropped from it to the mirror's surface, every point on that surface can serve as a point of reflection for that viewpoint (4, 5.13, p. 330).

As with convex spherical mirrors, so with convex cylindrical mirrors, less than half the surface is exposed to any facing viewpoint (4, 5.14-5,16, pp. 330-331). Planes of reflection passing through that viewpoint can cut the mirror's surface in three ways (4, 5.21-5.22, p. 333). If the cut is parallel to the cylinder's axis, the plane will form a line of longitude on the exposed surface, and every point on that line can serve as a point of reflection for the given viewpoint (4, 5.23, pp. 333-334). If the cut is perpendicular to the axis, the plane will form a circular section on the mirror's surface, and every point on that circle within the exposed portion of the mirror can serve as a point of reflection (4, 5.24, p. 334). But if the cut is oblique to the axis, the plane will form an elliptical section (or, as Alhacen calls it, a "cylindric section" [*sectio columpnaris*]) on the mirror's surface. Only one point on that section within the exposed portion of the mirror can serve as a point of reflection, and it lies where the minor axis of the ellipse intersects the mirror's surface (4, 5.25-5.26, pp. 334-335). Since an infinite number of planes can be passed through the given viewpoint along the normal dropped from it to the mirror's surface, an infinite number of elliptical sections can be formed on the exposed portion of the mirror. For each elliptical section, there is a unique minor axis intersecting the mirror's surface at a unique pair of points. Such points cover the entire exposed surface, so each and every one of them can serve as a point of reflection for the given viewpoint.

In convex conical mirrors, the amount of surface exposed to any viewpoint depends on where that viewpoint lies with respect to the mirror's vertex. If the line of sight extending from it to the mirror's vertex is perpendicular to the mirror's axis, then precisely half the mirror's surface is exposed to view (4, 5.30, p. 336). If that line of sight forms an acute angle with the axis on the side of the viewpoint, less than half the mirror's surface is exposed to view (4, 5.29, p. 336). And if that line of sight forms an obtuse angle with the axis on the side of the viewpoint, more than half the mirror's surface is exposed to view (4, 5.31, pp. 336-337). All, or virtually all, of that surface is exposed to view when the line of sight coincides with the edge of the cone or penetrates the cone through the vertex (4, 5.32-5.34, pp. 337-338).

Again, depending on where the viewpoint lies with respect to the mirror's vertex, any given plane passing through that viewpoint will form a line of longitude, a circle, or a conic section on the exposed portion of the reflecting surface (4, 5.40-5.41, p. 340). Under no circumstances will a plane forming a circle on this surface constitute a plane of reflection, because in a right cone, the plane of any circular section is oblique to its edge (4, 5.42, pp. 340-341), so the diameters of that section will be oblique to its edge. Since none of those diameters is normal to the reflecting surface at the points where they intersect it, none of those intersection-points can serve as a point of reflection. Consequently, in conical mirrors reflection is limited to planes that cut lines of longitude or conic sections on the mirror's surface.

When the viewpoint lies directly above the mirror's vertex along the axis, every plane passing through it along that axis will form a line of longitude on the exposed surface, and every point on that line can serve as a point of reflection. Since there is an infinite number of such planes, every point on the mirror's surface can serve as a point of reflection for that viewpoint (4, 5.40, p. 340). On the other hand, if the viewpoint does not coincide with the mirror's axis, then, no matter where it is located with respect to the vertex, only one plane of reflection passing through it will form a line of longitude on the mirror's surface.[26] The rest will form conic sections.

When a given plane of reflection cuts a conic section on the mirror's surface, there can be one or two but no more than two points of reflection within that plane (4, 5.43, p. 341). If the plane is normal to the mirror's surface, there can only be one point of reflection, and it lies where the axis (or, in the case of an ellipse, the major axis) of the conic section intersects the mirror's surface (4, 5.44, p. 341). If the axis or major axis of the conic section is not normal to the mirror, there can be two points of reflection, because in that case two lines within the given plane can be drawn normal to the mirror's surface from the point at which the cutting plane intersects the cone's axis, i.e., the section's focus, or nearer focus for an elliptical cut. For each such section, there is a unique pair of such lines intersecting the mirror's surface at a unique pair of points (4, 5.45, p. 341-342). Therefore, since an infinite number of cutting planes can be passed through the viewpoint along the normal dropped from it to the mirror, an infinite number of conic sections can be formed on the mirror's surface. Within each plane, the particular conic section will have a unique pair of lines extending from its focus to intersect the mirror along the normal. There will thus be an infinite number of such intersection-points on the exposed portion of the mirror's surface, each serving as a point of reflection for the given viewpoint.

In concave spherical mirrors, the entire surface is exposed to a viewpoint lying inside the mirror (4, 5.46, p. 342). Otherwise, no matter how far outside the mirror it lies, more than half the surface will be exposed to that viewpoint. Every plane passing through the given viewpoint along the normal dropped from it to the reflecting surface will cut a great circle on that surface, and every point on this circle within the exposed portion can serve as a point of reflection. Since an infinite number of planes can be passed through the viewpoint to form great circles on the exposed surface, every point on that surface can serve as a reflection-point (4, 5.48-5.50, p. 342-343). When the viewpoint lies at the very center of curvature, the lines extending from it will all be normal to the mirror. Since light radiating along a normal reflects back along it, and since there is an infinite number of normals reaching the mirror in all possible directions from the viewpoint, every point on the exposed surface constitutes a point of reflection for that particular viewpoint (4, 5.47, p. 342). In all cases, then, any point on the mirror's surface can serve as a point of reflection for any given viewpoint.

In concave cylindrical mirrors, when the viewpoint lies inside the mirror, its entire surface is exposed to view. Wherever else it lies outside the mirror, more than half the reflecting surface will be exposed to view (4, 5.51, p. 343). Like its convex counterpart, a concave cylindrical mirror can be cut in three ways by any given plane of reflection: i.e., along a line of longitude, along a circle, or along an elliptical section. When the cut occurs along a line of longitude or a circle, every point on that line or circle can serve as a point of reflection (4, 5.54-5.55, pp. 343-344). When the cut forms an ellipse, there will be only two points of reflection within the plane, those points lying where the minor axis of the ellipse intersects the mirror's surface (4, 5.56, p. 344). The same reasoning therefore applies to this mirror as to its convex counterpart. An infinite number of planes can be passed through the viewpoint along the normal dropped from it to the mirror so as to form elliptical sections on the exposed portion of the mirror. For each elliptical section, there is a unique minor axis intersecting the mirror's surface along the orthogonal at a unique pair of points. Such points cover the entire exposed surface of the mirror, so each of them can serve as a point of reflection for the given viewpoint.

In concave conical mirrors, finally, if the line of sight between the viewpoint and the mirror's vertex is perpendicular to the axis, precisely half the reflecting surface will be exposed to view. If, on the other hand, the line of sight forms an acute angle with the axis, more than half the reflecting surface will be exposed to view. The entire surface will be exposed to view if the viewpoint lies below the mirror's base and the line of sight coincides with the edge of the mirror or passes through the mirror's base to its vertex. By the same token, if the line of sight forms an obtuse angle with the axis, less than half the reflecting surface will be exposed to view, and none of it will be visible when the line of sight coincides with the edge of the cone or passes into the mirror through its vertex (4, 5.57, p. 344).

As for the selection of possible reflection-points, the same reasoning that applies to convex conical mirrors applies here. That is, if the plane passing through the viewpoint cuts a circle on the mirror's surface, there will be no point of reflection within that plane. If it cuts a line of longitude on the mirror's surface, every point on that line can serve as a point of reflection.[27] And if it cuts a conic section on the mirror's surface, there will be at least one and at most two points of reflection in that plane (4, 5.58, p. 344). Since an infinite number of planes of reflection can be passed through the viewpoint along the normal dropped from it to the mirror, and since each of them contains a unique set of reflection points, those points taken as a whole cover the entire reflecting surface. Every point on the exposed portion of that surface can therefore serve as a point of reflection for any given viewpoint.

Although descriptive rather than theorematic, Alhacen's analysis in chapter 5 involves complex geometrical reasoning based on mental visualization rather than on actual diagrams. At times, therefore his train of description is exceedingly difficult to follow. This is especially the case with his analysis of reflection-points in cylindrical and conical mirrors, because in those mirrors most of the planes of reflection form conic sections on the reflecting surface. The determination of reflection-points within such sections can be extraordinarily hard to visualize, particularly when it must be done in three dimensions. All of this effort seems wasted in view of Alhacen's ostensible purpose in chapter 5, which is to establish the intuitively and empirically obvious fact that reflection can occur to any facing viewpoint from every point on a reflecting surface. But within that context, Alhacen has a deeper purpose that betrays both the rigor and comprehensiveness of his approach. He means to show unequivocally that what holds for any given plane of reflection in any mirror necessarily holds for all such planes in that mirror. In other words, mathematical conclusions drawn on the basis of one, or one kind of plane of reflection will apply universally to every other plane of that kind. Accordingly, Alhacen has been at pains throughout chapter 5 to justify as fully as possible the detailed mathematical analysis of reflection to be developed in book 5 on the basis of single planes of reflection within each of the seven mirrors.

Book Five: By the end of book 4, Alhacen has established three of the four principles underlying the cathetus-rule: namely, that the plane containing the center of sight, the object-point, the reflection-point, and the image must be normal to the reflecting surface; that within this plane the angles of incidence and reflection are invariably equal; and that the image lies on the line of reflection. The first part of chapter 2 of book 5 is devoted to verifying the fourth and final principle, which puts the image on the cathetus dropped from the object-point to the reflecting surface. As with the other three principles, so with this one, the verification is empirical. Accordingly, for each mirror Alhacen proposes a variety of simple tests based on posing thin rods, cones, or needles normal to the reflecting surface, sighting along them in various ways, and observing that their images, or at least the image of their farther endpoints, lie in a direct line with the center of curvature. These tests, which pretty much speak for themselves, are described in 5, 2.1-2.5, pp. 385-387 for plane mirrors; in 5, 2.6-2.9, pp. 387-388, for convex spherical mirrors; in 5, 2.10-2.19, pp. 388-391, for convex cylindrical mirrors; in 5, 2.20, p. 391, for convex conical mirrors; in 5, 2.21-2.26, pp. 391-393, for concave spherical mirrors; in 5, 2.27-2.31, pp. 393-394, for concave cylindrical mirrors; and in 5, 2.32, p. 394, for concave conical mirrors.

Alhacen follows this empirical verification with a theoretical explanation of why the image appears on the cathetus rather than on some other line dropped from the object-point to the mirror. The gist of his argument is that, since images in plane mirrors are perfect replicas of their objects in size and distance from the reflecting surface, they must lie on the cathetus, because if they did not, they would appear smaller or larger than they should (5, 2.35-2.36, pp. 395-396). This argument is based on the assumption that we perceive images as if they were actual objects in front of us and judge their size and distance accordingly (5, 2.33-2.34, pp. 394-395). Thus, if the image-location were to lie on OI′ outside normal OAI in the top left-hand diagram of figure 5.5, p. 218, it would appear farther away and smaller than it should from viewpoint E, whereas if it were to lie inside that normal along OI″, it would appear closer and larger than it should. Alhacen then extends this argument to convex spherical mirrors in 5, 2.37-2.40, pp. 396-397, the point being to show that, in curved mirrors, the image, however distorted it may be, would not appear as it does were it to lie on a line other than the cathetus.

In the next section, Alhacen gives a brief description of the cathetus-rule and some of its implications for image-formation in plane mirrors (5, 2.42, p. 397), convex spherical mirrors (5, 2.43, pp. 397-398), convex cylindrical mirrors (5, 2.44, p. 398), concave spherical mirrors (5, 2.45, p. 398), and concave cylindrical and conical mirrors (5, 2.46, pp. 398-399). Alhacen's purpose here is to lay out various points to be dealt with later in his mathematical analysis of reflection. Most important among these is that in plane and convex mirrors there can only be one point of reflection for any given viewpoint and object-point, whereas in concave mirrors there can be as many as four.

The remainder, and by far the lion's share of book 5 consists of 54 propositions divided into seven unequal sets according to the type of mirror analyzed. Within each set Alhacen addresses three basic problems: how to determine image-location when the object-point, the center of sight, and the point of reflection are given; how to determine the reflection-point(s) when the center of sight and the object-point are given; and how to determine the number of reflection-points for any given center of sight and object-point. Propositions 1-4, pp. 399-403, deal with plane mirrors; propositions 5-25, pp. 403-432, with convex spherical mirrors; propositions 26-29, pp. 432-438 with convex cylindrical mirrors; propositions 30-31, pp. 438-446, with convex conical mirrors; propositions 32-49, pp. 446-475 with concave spherical mirrors; propositions 50-52, pp. 475-482, with concave cylindrical mirrors; and propositions 53-54, pp. 482-485, with concave conical mirrors.

The first 18 propositions of book 5, which deal with various aspects of reflection from plane and convex spherical mirrors, are fairly straightforward and unremarkable. Starting with proposition 19, pp. 415-419, however, there follows a set of six lemmas that are instrumental in one way or another to the subsequent determination of reflection-points in convex and concave mirrors. Of the six, however, only four—i.e., lemmas 3-6—are directly implicated in those determinations; the other two play ancillary yet critical roles in laying the requisite groundwork for them.

In the first lemma, which is dealt with in proposition 19, Alhacen shows how to generate a line extending from some randomly chosen point on a circle to the extension of its diameter such that the segment of this line between where it intersects the circle and where it intersects the diameter is equal to some randomly chosen line. Thus, as illustrated in figure 4, p. 524, a line must be extended from point A on the circle to meet the extension of diameter BG at point D in such a way that the segment between where it intersects the circle and point D is equal to randomly chosen line QE.

Alhacen subdivides this problem into four cases. In the first case, as illustrated in the upper left-hand diagram of figure 4, A is the point on the circle where lines AG and AB intersect so as to be equal. The objective here is to generate line AD from A to the extension of diameter BG such that HD = QE. The second case actually consists of three subcases according to the inequality of AG and AB. In the first subcase, which is illustrated in the upper right-hand diagram of figure 4, AG < BG in such a way that AD will be tangent to the circle. In that case, AD itself is to be produced equal to QE. The construction and proof in this particular subcase is noteworthy, because it depends on certain properties of hyperbolic sections (see esp. 2.147, p. 417). In the last two subcases, AG is either greater than or less than BG in such a way that AD intersects the circle. In the former case, illustrated in the lower left-hand diagram of figure 4, AD passes through point H on the circle such that HD = QE. In the latter case, illustrated in the lower right-hand diagram of figure 4, AD is extended to point H opposite point D such that DH = QE. This lemma comes into play only once in subsequent analysis (proposition 21, lemma 3, pp. 420-422—see esp. 2.169, p. 421).

Proposition 20, lemma 2, pp. 419-420, shows how to drop a line from some point on the circumference of a circle through the diameter such that the segment of this line from where it intersects the diameter to where it intersects the opposite arc on the circle is equal to some randomly chosen line. Thus, as illustrated in figure 5, p. 525, the objective is to drop line AD from randomly chosen point A through diameter GB of the circle so that ED equals randomly chosen line HZ. In all but one case, two such lines, i.e., AD and AD′ in the top diagram, will meet the specified condition, both DE and D′E′ being equal to HZ. The exception is illustrated in the lower diagram of figure 5, where AD passes orthogonal to diameter GB through the circle's center and therefore forms a diameter in that circle. Since no other line dropped from A through diameter GB can be as long as AD, no segment on that other line can be as long as ED. ED is therefore unique. Like the previous lemma, this one also depends on certain properties of hyperbolic sections. Unlike that lemma, this one is applied directy more than once to subsequent analysis (proposition 23, lemma 5, pp. 425-426—see esp. 2.187, p. 425—and proposition 24, lemma 6, pp. 426-427—see esp. 2.194 and 2.197, p. 427). It is also applied indirectly, but in a crucial way, in proposition 46, pp. 467-470.[28]

The point of proposition 21, lemma 3, pp. 420-422, is best explained in the context of figure 6, p. 526. Take right triangle ABG, and choose some point D on either of the legs forming right angle ABG. Let BG be the selected leg. As is evident from the figure, D can lie on the leg itself (as in the left-hand diagram at the top) or outside the triangle on the extension of the leg (as in the remaining two diagrams). From point D a line is to be drawn to or through point T on hypotenuse AG so as to pass to or through point Q on the opposite leg of right angle ABG in such a way that TQ is to TG as some randomly chosen line E is to some other randomly chosen line Z—i.e., TQ:TG = E:Z. Alhacen applies this lemma twice in subsequent analysis (proposition 22, lemma 4, pp. 422-425—see esp. 2.175, p. 422—and proposition 38, pp. 458-459—see esp. 2.375, p. 458).

In proposition 22, lemma 4, pp. 422-425, the problem is as follows: given two randomly chosen points, such as E and D in figure 7, p. 527, and given a circle, to find point A on that circle where the line tangent to that point bisects the angle formed by the lines extending from the given points to that point. Thus, as illustrated in the figure, the problem posed in this lemma is to find point A at which tangent AH bisects angle EAD. This lemma is used only once in subsequent analysis (proposition 31, pp. 441-446—see esp. 2.295, p. 445).

The penultimate lemma, proposition 23, lemma 5, pp. 425-426, entails dropping a line from outside a circle to a radius in that circle such that the segment of that line between where it intersects the circle and where it intersects the radius is equal to the segment of the radius between this latter point of intersection and the circle's center. Given the circle with radius GB in figure 8, p. 527, and given point E outside it, the construction calls for extending line EZ from E through point D on the circle to point Z on the radius such that DZ = GZ. Like the previous lemma, this one comes into play only once in subsequent analysis (proposition 31, pp. 441-446—see esp. 2.290, p. 444).

The sixth and final lemma, proposition 24, pp. 426-427, is really just a special case of proposition 21, lemma 3. As in that case, some point D in figure 9, p. 528, is taken on leg BG of the right angle in triangle ABG. From it a line is dropped to point T on the other leg of the right angle, and that line is extended in the opposite direction to point Q on hypotenuse AG so that TQ:QG = E:Z, E and Z being randomly chosen lines. As is shown in the figure, two such lines, TDQ and T'DQ', can be extended from point D to fulfill the requisite proportionality. This lemma is applied twice in subsequent analysis (proposition 25, pp. 427-432—see esp. 2.200 and 2.201, pp. 427-428—and proposition 47, pp. 470-471—see esp. 2.464, p. 471).

Immediately following this set of six lemmas, the problem of finding the point of reflection on a convex spherical mirror faced by a center of sight and an object-point that are chosen at random is taken up in proposition 25, pp. 427-432. This same problem is addressed in proposition 29, pp. 437-438, for convex cylindrical mirrors and in proposition 31, pp. 441-446 for convex conical mirrors. In the case of the three concave mirrors, the problem is considerably more complicated, because in those mirrors there can be as many as four points of reflection, depending on how and where the center of sight and the object-point are disposed with respect to one another, as well as to the reflecting surface. Accordingly, the determination of all possible reflection-points for concave spherical mirrors requires four separate solutions, which are given in propositions 36, 37, 38 and 47, pp. 452-459 and 470-471. On the basis of these solutions, the points of reflection are determined for concave cylindrical mirrors in proposition 52, pp. 478-481, and for concave conical mirrors in proposition 54, pp. 482-485. These propositions lie at the very heart of Alhacen's account of reflection, those devoted specifically to convex and concave spherical mirrors forming the core of what has come to be known as "Alhazen's Problem." Because of their complexity, I have remanded the detailed analysis and explanation of all of them to a separate section (pp. xlvi-lxvi below) in order not to disrupt the narrative flow of this section unduly.

Aside from 25, 29 and 31 (see pp. xlvii-lvi below for a detailed analysis), the remaining set of seven propositions devoted to convex mirrors—i.e., 25-31—is straightforward enough to warrant no discussion (see pp. xlvii-lii below for a detailed analysis of proposition 25). Starting with proposition 32, however, the set of 18 theorems devoted to concave spherical mirrors does merit discussion. As a whole, these theorems specify the conditions under which reflection will occur from one, two, three, or four points in a given plane of reflection within the mirror. As mentioned earlier, this specification will depend on how the center of sight and the object-point are disposed with respect both to one another and to the reflecting surface.

As far as their disposition with respect to one another is concerned, three factors come into play. First, the center of sight and object-point may or may not lie on the same normal. Second, if they do, they may or may not be equidistant from that centerpoint. And third, they may or may not lie on opposite sides of the mirror's centerpoint. As far as their disposition with respect to the reflecting surface is concerned, that depends on whether they lie on different normals. If they do, they face opposite arcs on the mirror according to the intersection of the normals at the mirror's centerpoint. The various dispositions just specified are illustrated in figure 10, p. 528, where B and B′ represent object-points, A a center of sight, and G the center of the mirror. In the left-hand diagram B, B′ and A lie on the same normal on opposite sides of G, and BG = AG, whereas B′G ≠ AG. In the right-hand diagram, the two object-points and the center of sight lie on different normals LF and KH, which necessarily intersect at G. As before, BG = AG, and B′G ≠ AG. In this case, reflection can occur from B to A, or from B′ to A from arc KDL facing angle KGL, or from arc LD′K facing angle LGK.

Alhacen deals with both these cases in proposition 34, pp. 450-451. For the situation in which B, B′, and A lie on the same normal, as represented in the left-hand diagram of figure 10, he demonstrates that within the plane of the circle (which constitutes a great circle on the sphere from which the mirror is formed) reflection can occur from B or B′ to A at two corresponding points D and D′ on the mirror within respective arcs of the circle. Overall, then, reflection can occur from every point in the circle generated on the mirror's surface by the rotation of D and D′ about normal AGB as axis.

Things are somewhat different when B and B′ do not lie on the same normal with A, as represented in the right-hand diagram of figure 10. Although in that situation there will be reflection from B or B′ to A from some points D and D′ on opposite arcs of the circle, points D and D′ for B′ and A do not lie at corresponding positions within their respective arcs. Hence, D and D′ for B′ and A will be specific to the plane of reflection, which is to say that reflection will not occur from a circle generated on the mirror's surface by the rotation of D and D′ about an axis. Furthemore, if B and A lie on the same normal and on the same side of G, as represented in the left-hand diagram of figure 11, p. 529, then no reflection can occur, because the normal dropped from centerpoint G to the supposed reflection-point D will not bisect angle BDA, leaving angle of incidence BDG ≠ angle of reflection ADG. By the same token, when A and B lie on different normals, as represented in the right-hand diagram of figure 11, reflection cannot occur from arcs LH or KF for the same reason. Hence, in that case, reflection is restricted to arcs KL and HF.

Once the method for determining points D and D′ for any center of sight A and any object-point B on the same normal is given in proposition 36, pp. 452-454 (see pp. lvi-lvii below for a detailed analysis), the analysis of reflection for the first case is complete. The remaining analysis of concave spherical mirrors centers on the case in which A and B lie on separate normals. Accordingly, in proposition 37, pp. 454-458, Alhacen shows, first, that if A and B lie outside the mirror on different normals, only one reflection can occur within the given plane. On the other hand, he continues, if those points lie inside the mirror, and if they are equidistant from the mirror's centerpoint, there can be as few as two or as many as four (but not three) reflections, depending on whether the circle passing through the center of sight, the object-point, and the center of the mirror intersects arc KL.

Thus, as illustrated in figure 12, p. 529, the circle passing through A, G, and B intersects arc KL at points D_1 and D_2, and these constitute legitimate points of reflection for A and B (see pp. lvii-lix below for a detailed analysis). So in this case there will be four points of reflection: D_1, D, and D_2 on arc KL, and D′ on arc HF. As Alhacen points out, circle AGB will intersect arc KL if and only if lines BD and AD drawn from the object-point and the center of sight to the midpoint of arc KL form acute angles with their respective normals on the side of G. Thus, when the circle is either tangent to arc KL, as represented by circle A′GB′ passing through viewpoint A′ and object-point B′, or when it falls short of that arc, reflection can occur from only two points, one of which is D on arc KL, the other D′ on arc HF. And in that case, as is clear from the diagram, lines A′D and B′D form obtuse angles (i.e., DA′G and DB′G) with their respective normals. Since, therefore, circle AGB necessarily touches or intersects arc KL at one or at most two points (or none if it falls short), and since in the one case (i.e., when circle AGB is either tangent to or falls short of arc KL) there will be one reflection only on arc KL, whereas in the other case there will be three on that arc, reflection cannot possibly occur from three points (including D′ on arc HF) in the entire circle. Implicit in this analysis, of course, is the method for finding the appropriate points of reflection according to the bisection of arcs KL and HF by line DGD′ and the intersection of arc KL by circle AGB (see pp. lviii-lix below for a detailed analysis).

Starting with proposition 38, pp. 458-459, the most complicated situation of all—that in which the center of sight and the object-point lie on different normals at different distances from centerpoint G of the mirror—becomes the focus of analysis. After showing in that proposition how to determine point of reflection D′ on arc HF when A and B lie at different distances from G, as illustrated in figure 13, p. 529 (see pp. lx-lxii below for the detailed analysis), Alhacen demonstrates in proposition 39, pp. 459-460, that any such point D′ on arc HF, except for point Z on line XZ bisecting angle KGL, can serve as a reflection-point for an infinite number of point-pairs A and B on normals KH and LF—provided, of course, that those points are not equidistant from G. He then establishes in proposition 40, p. 460, that, if D′ is the point of reflection for specific points A and B on those normals, there can be no reflection from B to A at any other point on arc HF.

So much for arc HF, now to arc KL. In proposition 41, p. 461, Alhacen shows that, if A in figure 14, p. 530, is given on normal KH, and if normal LF is also given, there will be some point of reflection such as D or D′ on arc KL for some object-point such as B or B′ on normal LF. Or, to put it another way, from every point on normal LF reflection will occur to point A at some point on arc KL. Alhacen goes on in proposition 42, pp. 461-462, to establish that, when both BG and B′G in figure 14 are unequal to AG and when reflection occurs from both B and B′ to A at points D and D′, respectively, angles BDA and B′D′A may be larger or smaller than angle LGH adjacent to angle KGL subtended by arc KL. Then, in proposition 43, pp. 462-463, he demonstrates that, while angles BDA and B′D′A can be either larger or smaller than angle LGH, they can never be equal to it under the specified conditions—i.e., with BG and B′G ≠ AG. On the basis, finally, of the points established in the preceding two propositions, Alhacen demonstrates in proposition 44, pp. 463-465, that, if reflection occurs from B to A at two points D and D′ on arc KL, as illustrated in figure 15, p. 530, angles BDA and B′D′A cannot both be smaller than angle LGH. As will become clear shortly, this does not mean that either of the angles need be smaller than LGH; they may both be larger.

The next two propositions, which are absolutely critical to the analysis of reflection from arc KL, are somewhat confusing, because their true intent is not clear from the way they are presented. In proposition 45, pp. 465-467, for instance, the ostensible point to be demonstrated is that, for any two points A and B lying unequal distances from G on normals KH and LF, reflection can occur from one to the other at two points on arc K. Implicit in the construction on which the proof is based, however, is that A and B in figure 16, p. 531, must be disposed on their normals in such a way that circle AGB passing through them will intersect arc KL. Implicit as well is that the lines from A and B to midpoint X of arc KL must form acute angles with their respective normals on the side of G. Accordingly, the points of intersection C and E must lie on opposite sides of X, where GX bisects angle KGL and arc KL along with it. If, however, circle AGB fails to intersect arc KL, or if it is merely tangent to it, as represented in figure 16a, only one reflection is possible, in which case angle BDA formed at the point of reflection will be smaller than angle LGH. This follows from the fact (demonstrated in Euclid, III.22) that angle BTA at tangent point T on circle AGBT and angle AGB at the opposite vertex of quadrilateral ATBG in the same circle sum up to two right angles, as do KGL and adjacent angle LGH. But we know from proposition 43 that reflection cannot occur from an angle, such as BTA, that is equal to LGH, so T cannot be a legitimate point of reflection. We also know that all angles, such as BDA, that intersect on arc KL outside tangent circle BGA will be more acute than BTA and thus more acute than LGH. Therefore, angle BDA must be smaller than LGH.

Likewise, if circle AGB in figure 16b, p. 532, intersects arc KL at bisection-point X, no reflection can occur within the segment of arc KL between X and point C of intersection to the left of it. This fact is not immediately apparent, because any angle, such as BD′A, within that segment of arc KL will be greater than angle LGH and will therefore fulfill the specification of proposition 43 that both angles BDA and BD′A not be smaller than LGH. So let us suppose that D′ is a legitimate point of reflection. The test of whether it is follows from certain points established earlier in proposition 44 and can be easily understood by recourse to point D in figure 16b. Let D in figure 16c, p. 532, represent that same point on the same circle KLHF with A and B posed on their respective normals as before. For a start, we know by previous conclusions that angle BDA < angle LGH, and we know by construction that angle of incidence BDG = angle of reflection ADG. Draw line AB connecting A and B, bisect it at M, and construct circle ADB passing through B, A, and point of reflection D. Draw PMN orthogonal to AB through midpoint M to form a diameter in circle ADB. PMN will therefore bisect arc AB at point P, leaving arc AP = arc BP. From Euclid, III.27 we know that equal arcs subtend equal angles, so we know that angle BDP subtended by arc BP is equal to angle ADP subtended by arc AP. Therefore, DP bisects angle BDA. But since normal DG also bisects that angle, DP must coincide with that normal. Therefore, point P will lie at the intersection of normal DG and circle ADB.

Applying the same test to D′ in figure 16b, p. 532, we locate that same point D′, in figure 16d, p. 533, on the same circle KLHF with A and B posed on their respective normals as before. We then construct circle AD′B, connect A and B with line AB, bisect it at point M, and produce diameter PMN through point M. Being orthogonal to AB at M, PMN bisects arc AB at P, leaving arc AP = arc BP. Therefore, D′P bisects angle BD′A. But D′P does not coincide with normal D′G, so D′G does not, bisect angle BD′A, leaving angle of incidence BD′G unequal to angle of reflection AD′G. D′ has therefore failed the test. Moreover, under the specified conditions, every other point D′ chosen to the left of X on arc KL will fail this test.[29] Unless circle AGB intersects arc KL on both sides of point X, then, reflection can occur from only one point D on arc KL, and the resulting angle BDA < angle LGH. However, as mentioned just above, having circle AGB intersect arc KL on both sides of X is not enough. Lines BX and AX dropped from the object-point and the center of sight to the midpoint of arc KL must also form acute angles with their respective normals. So the ulterior intent of proposition 45 is to prove that, as long as circle AGB does intersect arc KL on both sides of X, and as long as both BX and AX form acute angles with their normals, there can be *at least* two reflections from that arc.

Against this background, both the purport and significance of proposition 46, pp. 467-470, become crystal clear. For in that proposition Alhacen proves that, when there is one point of reflection D on arc KL such that angle BDA > angle LGH, there will necessarily be another point D′ on that arc such that angle BD′A > angle LGH. The proof itself turns on the fact, demonstrated in proposition 20, lemma 2, that two lines, such as AD and AD′ in figure 5, p. 525, can be drawn from a given point A such that segments ED and E′D′ will be equal.[30] As was shown in the previous theorem, any such point of reflection must lie on the segment of arc KL between midpoint X and the point to the right of X where circle AGB intersects it (i.e., between X and E in figure 16, p. 531). Consequently, Alhacen actually proves two things in this proposition, one explicitly, the other implicitly. Explicitly, he demonstrates that, if there is one point D of reflection such that angle BDA > angle LGH, there must be another point D′ such that angle BD′A > angle LGH. Implicitly, he demonstrates that, if circle AGB intersects arc KL to the right of X at some point E, there will necessarily be two points of reflection D and D′ within segment EX of that arc such that both angles BDA and BD′A are greater than angle LGH.

As he sums them up in proposition 49, pp. 472-475, the points established in propositions 44-46 are as follows. First, if circle AGB fails to intersect arc KL to the right of midpoint X, there can only be one point of reflection D. It will lie on segment CK of arc KL in figure 16b, p. 532, and it will yield angle BDA < angle LGH. Second, if circle AGB does intersect arc KL to the right of midpoint X, there will be two points of reflection D and D′ on segment EX of arc AK in figure 16, p. 531, and they will yield angles BDA and BD′A > angle LGH. Third, when circle AGB cuts arc KL to the left of midpoint X, as in figure 16, p. 531, there will be a third reflection from some point on segment KC of arc KL such that it yields an angle < FGD. If, however, circle AGB intersects arc KL at point K (which would thus coincide with A) or fails to cut arc KL to the left of midpoint X, then, as long as arc XE in figure 16 is exposed to point A, there will be two reflections from segment XE but none from any point on remainder KX or LE of arc KL. And, finally, if there is a third reflection, it must occur within the segment lying on the opposite side of X from the segment within which the other two reflections occur. Thus, as represented in figure 16, since points D and D′ lie on segment EX to the right of X, the third reflection must occur from segment CK to the left of X, not from segment EL to the right. That this must be the case follows from the proof that, although angle BD′A in figure 16b, p. 532, is larger than angle LGH, no point D′ within segment XC of arc KL can be a legitimate point of reflection if point D on arc CK is a legitimate point of reflection (see pp. xli-xlii above).[31]

Consequently, by the end of proposition 46, Alhacen has established definitively both that and under what conditions there can be one, two, or three reflections from arc KL when AG ≠ BG. If there is only one, it will yield an angle BDA < angle LGH, if there are two, they will yield angles BDA and BD′A > angle LGH, and if there are three they will yield one angle BDA < LGH and two angles > angle LGH. All that is left to do is show how to determine those points of reflection for any properly disposed O and E. This Alhacen does in proposition 47, pp. 470-471, for the two reflection-points that yield angles > LGH (see pp. lxiii-lxiv below for a detailed analysis), but in the text as it stands, no such determination is given for the point of reflection yielding an angle less than LGH. That determination can be easily reconstructed on the basis of the method provided in proposition 38 for determining the point of reflection on arc GD opposite arc KL (see note 138, pp. 513-514, for a detailed analysis)

Compared to the cluster of five propositions culminating in 47, the last six theorems of book 5—including propositions 52 and 54, where the points of reflection are determined for concave cylindrical and conical mirrors (see pp. lxiv-lxvi below for a detailed analysis)—are at best anticlimactic and therefore call for no discussion. What does remain to be discussed before this overview concludes is a set of three issues that crop up earlier in book 5. The first has to do with the location of images seen along the normal. As Alhacen points out in several places, these particular images can only be of the spot on the cornea through which the orthogonal line of sight passes to the mirror (see, e.g., propositions 2 and 8, pp. 399-401 and 404-405). According to the cathetus-rule, such images should lie where the cathetus dropped from that spot on the cornea intersects the line of reflection. But in this case the two lines coincide, so there is no point of intersection. How, then, do we judge the location of images along that line? Alhacen's response is that we do so within the context of proximate points, whose images are defined by appropriate intersections. Thus, when we look straight down into a mirror, we locate the image of the cornea's centerpoint perceptually by referring it to the images of surrounding points whose locations are definite. That way we perceive the image of the cornea's centerpoint to lie the same distance from the center of sight as the images of the rest of the points on the cornea's surface (see, 5.2.38, pp. 396-397). The ulterior point here is that, even when the image has no definite location according to the cathetus-rule, it will be perceptually located as if it did.

The second issue is why we see a single image with both eyes in curved mirrors when each center of sight receives the image of any given object-point from a different reflection-point on the mirror's surface and therefore perceives it along a different line of reflection. As a result, the object's image-location is different for each eye. Why, then, is it not perceived double? The answer, according to Alhacen, is that, under normal conditions, the separate image-locations are so close that the visual faculty naturally melds both images together to make a single composite from them (see, e.g., 5.2.221, p. 432). Therefore, as far as the cathetus-rule is concerned, what obtains in single-viewpoint vision essentially obtains in binocular vision.

The third and final issue involves image-formation in concave mirrors. The problem here is that, although there should be an image for every point of reflection, this does not always seem to be the case in concave mirrors, where some images are clearly apparent while others are not. As Alhacen shows in proposition 32, pp. 446-449, this problem can be resolved geometrically. Suppose that O_5R in the lower diagram of figure 5.5, p. 218, represents a line of incidence, RE a line of reflection, E a center of sight, and C the center of the mirror. When the form of object-point O_1 reflects from R to E, normal CO_1 will intersect line of reflection RE at I_1 behind the mirror. When, however, the form of O_2 reflects to E, normal CO_2 will be parallel to line of reflection RE, so there will be no intersection and thus no definite image for O_2. According to cathetus O_3C, meanwhile, the image of O_3 will lie at I_3 beyond the eye, whereas the image of O_4 will lie at the center of sight itself, because its cathetus O_4C passes through point E. The image of O_5 finally, will lie at I_5 between the reflecting surface and the center of sight. Consequently, of all five images, only two—I_1 and I_5—will be clearly perceived. The rest will either go unperceived, or they will be seen on the reflecting surface itself, in which case all that is seen on that surface is a blur the same color as the object whose image it is. Consequently, in concave mirrors, the apparent lack of images for certain reflection-points is only apparent because of the physical conditions under which they can or cannot be seen.

2. *Alhacen's Solutions to "Alhazen's Problem"*

In 1669, on the basis of his reaction to book 5 of the *De aspectibus*, Christiaan Huygens formulated the following problem, which quickly gained currency as "Alhazen's Problem": *Given a spherical convex or concave mirror, and given a point of sight and a point on a visible object, to find the point of reflection on the surface of the mirror.*[32] This problem exercised not only Christiaan Huygens but also several other seventeenth-century mathematicians, their interest piqued by impatience with Alhacen's solution of it. Or, rather, I should say "solutions," because Alhacen was forced to approach the problem from several different directions when it came to concave spherical mirrors. The root of his difficulty lay in the fact (already discussed at length in the previous section) that, depending on how the object-point and the center of sight are disposed with respect to the center of curvature in such mirrors, there can be as many as four points of reflection. Furthermore, both the number and determination of these points depend on how the object-point and the center of sight are disposed with respect to the reflecting surface itself.

Faced with an extraordinarily complex phenomenon, then, Alhacen felt compelled—and understandably so—to address it on a case-by-case basis. Moreover, he was concerned not only with convex and concave spherical mirrors, but with convex and concave cylindrical and conical mirrors as well. Alhacen's problem, in short, was considerably broader in scope and more demanding than "Alhazen's Problem." In this section, we will examine Alhacen's approach to this problem in its full scope, dealing in order with convex spherical, convex cylindrical, convex conical, concave spherical, concave cylindrical, and concave conical mirrors. Insofar as possible, we will follow Alhacen's actual train of analysis, although at times it will be necessary to disrupt that train for the sake of clarity.

Convex Spherical Mirrors: In proposition 18, p. 415, Alhacen addresses the problem of how to find the point of reflection on the surface of a convex spherical mirror when the point-source of radiation and the center of sight are equidistant from the mirror's center. The solution to this problem is so simple as to be trivial. Let B in figure 17, p. 534, be the object-point, A the center of sight, and G the center of the mirror. Pass a plane through all three points to form a great circle on the mirror. Draw normals AG and BG, and bisect angle AGB with line GDE. That D is the point of reflection is clear from the fact that triangles BDG and ADG are equal, as are their corresponding angles BDG and ADG. Hence, the adjacent angles BDE (the angle of incidence) and ADE (the angle of reflection) are equal.

Having set the stage with the series of six lemmas (propositions 19-24, pp. 415-427) discussed earlier on pp. xxxvi-xxxviii above, Alhacen addresses the same problem in proposition 25, pp. 427-432, but this time with points A and B lying different distances from the mirror's center. Thus, as illustrated in figure 17a, p. 534, normals AG and BG are not equal. Unlike the previous case, this one is far from trivial, and Alhacen's method for solving it is commensurately complex.

Take random line D′M′, and divide it at point Q′ so that M′Q′:Q′D′ = BG:AG. Bisect line D′M′ at point N′, and draw perpendicular B′N′T′ through it. From endpoint D′ of line D′M′ draw D′T′ to B′N′T′ to form angle D′T′N′ equal to half of angle BGA. As given in the figure, GE′ represents the actual bisector of angle BGA, so angle D′T′N′ = angle BGE′. Through point Q′ on D′M′ draw a line meeting B′T′ at B′ and D′T′ at G′ such that B′G′:G′D′ = BG:GD (i.e., the radius of the great circle on the mirror). This Alhacen has shown us how to do in the preceding theorem (proposition 24, lemma 6, pp. 426-427). Line B′G′ will form angle B′G′D′ with line D′T′. At the center of the circle form angle BGE equal to B′G′D′. Point D, where leg GE of that angle intersects the circle, will be the sought-after point of reflection.

Alhacen's proof that D is in fact the point of reflection is based on figure 17b, p. 535. The gist of the proof—and here I will take some liberties with his actual procedure—is as follows. For a start, it is clear by construction that triangles BGD and B′G′D′ are similar. Draw line BT to form angle BTD equal to half of angle AGB. Drop perpendicular DN to line BT, and continue DN to M so that DN = MN. DM will intersect BG at point Q. Accordingly, the entire figure consisting of triangle BDGT and line DQNM highlighted in bold will be similar to the entire figure consisting of triangle B′D′G′T′ and line D′Q′N′M′ to the left of the circle. Thus, angle BTD = angle B′T′D′ = half of angle BGA = angle E′GB (or E′GA) from the previous figure. From this it follows that angle GBT = angle E′GE from the previous figure—i.e., the difference between angle BGD and half of angle AGB.

Draw line BM to form triangle BMN. Since MN = DN, by construction, and since angles BNM and BND are right, and thus equal, by construction, triangle BMN = triangle BND. Since, moreover, angle DBN = angle DBG + angle E′GE from the previous figure (i.e., the difference between angle BGD and half of angle AGB), and since angle E′GE = angle GBT, angle DBM =

2DBG + 2GBT. From point D draw line DS parallel to BM, and extend BG to meet it at point S. Since vertical angle BQM = vertical angle DQS, and since alternate angles BMD and MDS are equal, triangles BQM and DQS are similar, leaving their respective sides proportional. Hence, BQ:QS = MQ:QD. But angle BMQ = angle BDQ, since triangle BMD is isosceles by construction. Therefore, given the equality of alternate angles BMQ and QDS, it follows that angle BDQ = angle QDS, so DQM bisects angle BDS. Therefore, by Euclid, VI.3, BD:DS = BQ:QS. But we have already established that BQ:QS = MQ:QD, and MQ:QD = BG:GA by construction, so BQ:QS = BG:GA.

Draw line DP such that angle PDS = angle BGA, and draw XDY tangent to the circle at point D. The remainder of the proof depends on showing that vertical angle XDP, which equals vertical angle ADY, also equals angle BDX. This follows from the fact that angle XDQ = angle NTD, because XDQ + QDG (i.e., NDT) sums up to a right angle, and so does NTD + QDG (i.e., NDT). Thus, since angle NTD = one-half angle BGA by construction, angle XDQ = one-half angle BGA. Now, angle PDS = angle BGA by construction, so angle QDS = angle BGA – angle PDQ. But angle QDS = angle BDQ, since QD bisects angle BDS, so angle BDQ + angle PDQ = angle BGA. Since angle BDQ = angle BDX + angle XDP + angle PDQ, then angle BDX + angle XDP + 2 angle PDQ = angle BGA. But we have already established that angle XDQ (which = angle XDP + angle PDQ) = one-half angle BGA. Thus, angle XDB + angle PDQ + one-half angle BGA (i.e., angle XDP + angle PDQ) = angle BGA. It follows, then, that angle XDB + angle PDQ = one half angle BGA, so angle XDB = angle XDP.

The next step is to demonstrate that, if PD is extended toward point A, it will intersect GA at that very point. That it does rests on showing that the resulting triangle PAG will be similar to triangle PDS, which follows from the fact that angle PDS = angle BGA by construction, while angle APG is common to both triangles. Therefore, PDA is a straight line, so vertical angles XDP and ADY are equal. But angle XDB = angle XDP, so angle XDB = angle ADY. Accordingly, given that angles XDE and YDE are both right by construction, and given that their parts XDB and ADY are equal, it follows that remainders BDE (the angle of incidence) and ADE (the angle of reflection) will be equal.

The key to Alhacen's method for finding D in this case is the construction of triangle B′G′D′ with constituent angle B′G′D′, and the construction of similar triangle BDG with constituent angle BGD at the center of the mirror. And the key to this construction lies in dividing line D′M′ into segments M′Q′ and Q′D′ that are proportional to normals BG and AG. Both that and how this expedient works is clear enough. Far less obvious is how and why Alhacen battened on to it in the first place. Did some unfathomable flash of inspiration lead him to realize that forming triangle BGD according to the conditions just laid out would yield the appropriate angle B′G′D′? Perhaps so, but there is a more mundane explanation based on the particular construction he used for the proof and some fairly obvious patterns that emerge from that construction. In order to uncover these patterns, we need to backtrack from the construction and proof as completed to the initial grounds upon which they are laid.

Let us therefore take point D as given, and let us begin with the limiting case in which A and B are equidistant from the mirror's center, as represented in figure 17c, p. 535. Since normal GDE bisects angle AGB, there is no difference between angle BGD and half of angle AGB. Therefore, line BNT from the previous figure will coincide with line BQG, so N will coincide with Q, and G with T. Carry out the construction as before, drawing line DQ perpendicular to BT (which coincides with BG) and extending perpendicular DN (which coincides with DQ) to point M so that MN (i.e., MQ) = DN (i.e., DQ). On base MN form triangle BMN equal to triangle BND. Draw DS parallel to BM and extend BG to meet it at S. Triangles BMN and DNS will thus be similar (through equality) leaving their corresponding sides proportional. Hence, BN:NS = BD:DS, which is to say that BQ:QS = BD:DS, from which it follows that DQ bisects angle BDS (by Euclid, VI.3). Likewise, since GE bisects angle AGB, BE:EA = BG:GA. Thus, if we connect line AS and draw a line parallel to it from E, that second line will pass through point Q, cutting both lines BA and BS equiproportionally. In other words, BE:EA (which = BG:GA) = BQ:QS.

Triangles BGA and BDS therefore correspond in two crucial ways: the lines bisecting the angles at their respective vertices G and D cut equiproportional segments from them (i.e., BE:EA = BG:GA, and BQ:QS = BD:DS), and the segments cut off by those bisectors on respective bases BA and BS are equiproportional (i.e., BE:EA = BQ:QS). Therefore, BQ:QS = BG:GA. But triangles BMQ and DSQ are similar (through equality), so their corresponding sides are proportional, leaving BQ:QS = MQ:QD. Hence, MQ:QD = BG:GA.

If, therefore, we apply Alhacen's method for finding D in this case, we proceed as follows. Take random line D′M′, and divide it at point Q′ such that M′Q′:Q′D′ = BG:AG. Since BG = AG, M′Q′ = Q′D′, so line D′M′ is already bisected (i.e., points N′ and Q′ from figure 17b coincide). Through point Q′ pass a line perpendicular to M′D′, and from point D′ draw line D′T′ to that perpendicular so as to form angle B′T′D′ = angle BGD = one-half angle AGB. From point B′ on the perpendicular draw line B′G′ through point Q′, intersecting D′T′ at point G′ so that B′G′:G′D′ = BG:GD (i.e., the radius of the great circle on the mirror). When angle BGD is formed at the center of the mirror equal to the resulting angle B′G′D′, point D where leg GD of that angle intersects the circle will be the point of reflection.

Now, using this simple limiting case as a guide, let us return to the original case in which AG and BG are unequal, and let us compare this case to the one just discussed. These two cases are illustrated in figure 17d, p. 536, with the limiting case backgrounded in light gray. Let A_1 be the new center of sight on line of reflection AD, and let B remain the object-point. Since A_1 is on the same line of reflection as in the previous case, point D is given. Bisect angle A_1GB with line GE_1, connect BA_1, and let bisector GE_1 intersect it at E_1. Again, from Euclid, VI.3 we know that bisector GE_1 cuts BA_1 in such a way that $BE_1:E_1A_1 = BG:GA_1$. At point B form angle GBN_1 equal to angle E_1GE (i.e., the difference between half of angle A_1GB and angle BGD), and extend BN_1 to meet DG at T_1. Hence, angle BT_1D = angle BGE_1 = half of angle BGA_1. From point D draw DN_1 perpendicular to BT_1, and extend it to M_1 so that $M_1N_1 = N_1D$.

Carry out the construction as before, drawing DS_1 parallel to BM_1 and extending BG to meet it at point S_1. Extend BT_1 to meet the extension of DS_1 at V_1. Accordingly, because of the equality of corresponding angles and sides, triangle DN_1V_1 = triangle BN_1M_1, which equals triangle BDN_1 by construction, so triangle BDN_1 = triangle V_1DN_1. Consequently, DN_1 bisects angle BDV_1. Earlier we established that $BE_1:E_1A_1 = BG:GA_1$, so, if we draw a line from point E_1 parallel to A_1S_1, it will pass through point Q_1, cutting lines A_1B and S_1B equiproportionally. Hence, $BE_1:E_1A_1 = BQ_1:Q_1S_1$, from which it follows that $BQ_1:Q_1S_1 = BG:GA_1$. But Q_1 is where DM_1, which bisects angle BDS_1, cuts line BS_1. Therefore, given the similarity of triangles BM_1Q_1 and Q_1DS_1, it follows that corresponding sides are proportional. Hence., $BQ_1:Q_1S_1 = M_1Q_1:Q_1D$. But $BQ_1:Q_1S_1 = BE_1:E_1A_1 = BG:GA_1$, so $M_1Q_1:Q_1D = BG:GA_1$. Triangles BDS_1 and BGA_1 thus correspond in the two crucial ways mentioned before: the lines bisecting the angles at respective vertices G and D cut equiproportional segments from them (i.e., $BE_1:E_1A_1 = BG:GA_1$, and $BQ_1:Q_1S_1 = BD:DS_1$), and the segments cut off by those bisectors on respective bases BA_1 and BS_1 are equiproportional (i.e., $BE_1:E_1A_1 = BQ_1:Q_1S_1 = BG:GA_1$). Finally, since $BQ_1:Q_1S_1 = M_1Q_1:Q_1D$, it follows that $M_1Q_1:Q_1D = BG:GA_1$. And the same pattern holds for any other center of sight chosen on line of reflection AD. No matter where the new point A lies on that line, it will always be the case that BE:EA = BG:GA = BQ:QS = MQ:QD. The ratio of normals BG and GA, in short, is absolutely fundamental and systemic to the construction.

So far we have taken point D of reflection, object-point B, center of curvature G, and line of reflection AD as given and, using D as the anchor-point, constructed triangle $BDGT_1$ and line $DQ_1N_1M_1$ highlighted in bold lines in the lower diagram of figure 17d. To find point D when only A, B, and G are given is simply a matter of recapitulating that construction outside the mirror and then importing it back into the mirror. This we do by creating an exact replica in triangle $B'D'G'T'$ and line $D'Q_1'N_1'M_1'$ in the upper diagram of figure 17d. To that end we need to know, first, the ratio of BG to AG, and second, the measure of angle BGA. These, of course, are given in the problem as it is set up, so points Q_1' and N_1' on $M_1'D'$ are given, as is angle $D'T_1'N_1'$ (= half of angle BGA) on perpendicular $T_1'N_1'$ passing through the midpoint of $M_1'D_1'$. The sticking-point lies in knowing how to generate line $B'G'$ through point Q_1' to $D'T_1'$ such that $B'G':G'D'$ = BG:GD. The real test of Alhacen's ingenuity in determining the location of D was therefore to figure out how to generate this line, and he passed it with flying colors in proposition 24, lemma 6. At bottom, then, that lemma forms the crux of Alhacen's solution to the reflection-problem for convex spherical mirrors.

It is worth noting, finally, that the ratio of normals is also fundamental to reflection from plane mirrors. In this case, of course, there is no center of curvature, so the normals will never intersect. Thus, in figure 17e, p. 537, if D is the point of reflection, DE the normal to that point, and AD the line of reflection, and if A, A_1, and A_2 represent various viewpoints on that line, then normals AG, A_1G_1, and A_2G_2 dropped from them to the mirror will be parallel to one another as well as to normals DE and BG_3. Since the resulting triangles AGD, A_1G_1D, A_2G_2D, and BG_3D are all similar, their corresponding sides will be proportional. Thus, for example, sides AD, AG, and GD in triangle AGD will be proportional to corresponding sides A_1D, A_1G_1, and G_1D in triangle A_1G_1D, as well as to corresponding sides BD, BG_3, and G_3D in triangle BG_3D.

Now, according to the conditions specified, angle BDE = angle ADE, since DE bisects angle BDA. If we connect line BA, then, according to Euclid, VI.3, DE will intersect that line at point E such that BE:EA = BD:AD. The same holds for the remaining points A_1 and A_2: $BE_1:E_1A_1 = BD:A_1D$, and $BE_2:E_2A_2 = BD:A_2D$. So the location of D is ultimately determined by the ratio of normals BG_3 and AG, BG_3 and A_1G_1, or BG_3 and A_2G_2. Accordingly, given points B and A, A_1, or A_2 facing reflecting surface G_1G_3, we can find point D by dropping the normals from B and A, or from B and A_1, or from B and A_2, and cutting line G_3G, G_3G_1, or G_3G_2 according to the ratio of those normals: i.e., $G_3D:DG = BG_3:AG$, $G_3D:DG_1 = BG_3:A_1G_1$, or $G_3D:DG_2 = BG_3:A_2G_2$. And the same holds for any other point A chosen on line of reflection AD. There is thus a close analogy between this analysis of reflection from plane mirrors and Alhacen's analysis of reflection from convex spherical mirrors. The key thing in both cases is the ratio of normals and the recapitulation of that ratio in lines BA, BA_1, and BA_2 through bisection, although in the case of reflection from plane mirrors it is the angle of reflection rather than the angle formed by the intersection of normals that is bisected.

Whether Alhacen saw the patterns and connections just discussed is an open question, because he never mentions them explicitly. But they are clearly implicit in the construction upon which he based his proof, and that construction is clearly implicit in the construction for the limiting case. Granted, this latter construction is not self-evident, but it is fairly obvious—obvious enough, I suggest, that Alhacen could have seen in it the model for his solution to the problem of finding D when AG and BG are unequal. To suppose that he did takes some of the mystery out of that solution and its derivation, but it takes nothing whatever from its ingenuity or originality.

Convex Cylindrical Mirrors: Alhacen implicitly addresses the problem of finding the point of reflection in convex cylindrical mirrors in proposition 28, pp. 435-437, where he subdivides the problem according to three cases, depending upon how the plane of reflection cuts the mirror and its axis. On the one hand, if that plane cuts the cylinder along a line of longitude, it will include the entire axis. Thus, as illustrated in figure 18, p. 538, plane of reflection AXYB passing through object-point B and center of sight A contains line of longitude ST and axis XY. Point D of reflection must therefore lie on line ST. Furthermore, since ST is the common section of the plane of reflection and the plane tangent to the mirror along line ST, finding point D reduces to finding the point of reflection on a plane mirror with ST as the common section of the reflecting surface and the plane of reflection. Alhacen has already provided the solution to this problem in proposition 1, p. 399. Accordingly, once point of reflection D is found and normal FDE is erected at that point, angle BDE of incidence will equal angle ADE of reflection.

If, on the other hand, the plane of reflection intersects the axis orthogonally, it will cut a circle on the cylinder's surface. This case is represented in figure 18a, p. 538, where plane ABG of reflection intersects axis XY orthogonally at point G and cuts the circle in light gray on the cylinder's surface. Point D will therefore lie on that circle, and to find it we need only apply the method in proposition 25 based on the proportionality of normals BG and AG. Consequently, with normal GDE produced, angle BDE of incidence will equal angle ADE of reflection.

In the third and most complex case, which is represented in figure 18b, p. 538, the plane of reflection ABHK intersects the cylinder's axis obliquely at point F and therefore cuts the cylinder's surface along an ellipse. Here the solution requires passing a plane through point A so as to form the circle with centerpoint G represented in light gray on the cylinder's surface. Then line BB′ is dropped from point B to that plane parallel to the edge of the cylinder and thus perpendicular to its base-planes. Points A and B′ will therefore face the circle centered on G within the same plane. To find point D′ on that circle where the form of point B′ would reflect to point A is, as before, a matter of applying the method outlined in proposition 25 according to the proportionality of normals B′G and AG. Having determined that point, we then pass line of longitude D′D through it. Point D, where the plane of reflection intersects this line of longitude, will be the point of reflection. Hence, if we extend normal FDE through that point, angle BDE of incidence will equal angle ADE of reflection.

Convex Conical Mirrors: As with convex cylindrical mirrors, so with convex conical mirrors, points A and B can be disposed to the mirror in various ways. The problem of finding the point of reflection can thus be subdivided into a variety of cases. The first and most trivial of these has the plane of reflection cutting a line of longitude on the cone's surface. As with the first case in cylindrical mirrors, so with this one, the problem of finding point D reduces to finding the point of reflection on a plane mirror. The remaining cases, which number six according to Alhacen's analysis, depend upon where A and B are situated with respect to the plane passing perpendicular to the cone's axis through its vertex. Let us deal with these cases in order as they occur in proposition 31, pp. 441-446.

In the first case (pp. 441-442) A and B are assumed to lie below that plane, as represented in figure 19, p. 539, where lines SXT and UXV define the plane passing perpendicular to axis XY through vertex X of the cone. Let B be the object-point and A the center of sight, and let ABHK be the plane of reflection. Through point A pass a plane orthogonal to axis XY of the cone, and let it meet the axis at point Y. That plane will thus cut a circle with centerpoint Y on the cone's surface. Drop a line from vertex X of the cone to point B, extend it to the plane of the circle, and let it intersect that plane at point B′. B′ and A will therefore face the circle centered on Y within the same plane, so, by applying the method in proposition 25 according to the ratio of normals B′Y and AY, we can find point D′ where the form of B′ would reflect to A. In this case, of course, the circle must be treated abstractly, as if it lay on the surface of a spherical or cylindrical mirror, because in its actual situation normal YE′ strikes the cone's surface obliquely, which means that plane B′AY cannot be a true plane of reflection. Now, draw line of longitude D′X. Point D, where plane of reflection ABHK intersects line of longitude D′X, will be the point of reflection. Hence, if we erect normal FDE at that point, angle of incidence BDE will equal angle of reflection ADE. Needless to say (which is undoubtedly why Alhacen does not), the construction just outlined holds if A and B lie in a plane, such as AYB′, that cuts the axis orthogonally rather than obliquely and, therefore, that already forms a circle on the cone.

In the second case (pp. 442-443), points A and B in figure 19a, p. 539, are both assumed to lie in the plane that passes perpendicular to axis XY through vertex X of the cone. In this case, X can be treated abstractly as an already-defined point of reflection, so the angles of incidence and reflection are determined by bisecting angle AXB with line XE. Pass a plane through line XE along axis XY to produce line of longitude XK on the cone's surface. From point E drop a normal to line of longitude XK, and extend it to point F on the axis. BEAF will thus be the plane of reflection, D the point of reflection, and FDE the normal to that point. Accordingly, angle of incidence BDE will equal angle of reflecton ADE.

In the third case (pp. 443-444), points A and B are both situated above the plane passing perpendicular to axis XY through vertex X, as represented in figure 19b, p. 540. Produce the opposite section of the original cone. If line AB is disposed orthogonally with respect to axis XY, then pass a plane through it parallel to the base of the lower cone to form the circle centered on T in the opposite, upper cone. If line AB is disposed obliquely with respect to axis XY, then let A lie lower than B. From X draw a line through A and extend it until it meets line AB at new point A such that line AB is disposed orthogonally with respect to axis XY. Either way, A and B will lie in the plane of the circle. Find point D′ on the concave section of that circle where the form of B would reflect to A, that circle being taken abstractly as if it were in a concave spherical or cylindrical mirror. Alhacen's method for determining this point will come later, when he analyzes reflection in concave mirrors, so for now we must take D′ as properly determined.[33] Accordingly, with normal D′E produced, angle BD′E of incidence will equal angle AD′E of reflection. From point D′, as appropriately determined, draw line of longitude D′X in the upper, opposite cone, and continue it to point L in the lower cone. From point E on line BA drop normal ED to line of longitude XL, and extend it to point F on axis XY. D will therefore be the point of reflection so that, with normal FDE produced, angle BDE of incidence will equal angle ADE of reflection.

In the fourth case (pp. 444-445), one of the points is assumed to lie in the plane passing perpendicular to axis XY through vertex-point X, the other below that plane. Let B, in figure 19c, p. 540, lie in that plane and A below it. Pass a plane through point A to intersect axis XY orthogonally, thus forming the circle centered on Y on the cone's surface. Draw line BX, and drop BL perpendicular to the plane of the circle just formed. Connect LY, which is thus parallel to BX, and from A drop a line to LY, cutting it at Q such that D′Q = QY. To do this we follow the procedure Alhacen has outlined previously in proposition 23, lemma 5. Through point D′, where line AQ intersects the circle, extend line YE′. From point L draw line LB′ parallel and equal to YD′, and then draw lines BB′ and B′D′. LB′D′Y will therefore be a parallelogram, so angle LB′D′ = opposite angle LYD′, and angle B′LY = opposite angle B′D′Y. Likewise, angle B′D′E′ = alternate angle LYD′. But angle LYD′ = angle QD′Y, since, by construction, triangle D′QY is isosceles on base D′Y. Therefore, angle B′D′E′ = angle QD′Y. But angle AD′E′ is also equal to angle QD′Y, since they are vertical angles. Hence, angle B′D′E′ = angle AD′E′, so D′ is the point from which the form of B′ would reflect to A. Draw line of longitude XD′. Point D, where plane of reflection ABHK intersects it will be the point of reflection for B and A. Hence, with normal FDE produced, angle BDE of incidence will equal angle ADE of reflection.

The fifth case (p. 445) is illustrated in figure 19d, p. 541, where B lies in the plane passing perpendicular to axis XY through vertex-point X, and A lies above it. Form the complementary cone above the base cone, and pass a plane through A perpendicular to the axis so as to produce the circle centered on T. Drop perpendicular BB′ to the plane of that circle. Hence, A and B′ face the concave surface of the circle. Find point D′ on that surface where the form of B would reflect to A. As before, Alhacen's method for determining this point has yet to be revealed, so we must take D′ as given. With normal D′E′ drawn, angle of incidence B′D′E′ = angle of reflection AD′E′. From point D′, as appropriately determined, draw line of longitude D′X on the upper cone and extend it to L on the lower one. Point of reflection D will therefore lie where plane of reflection ABHK intersects line of longitude XL. Thus, with normal FDE produced, angle of incidence BDE will equal angle of reflection ADE.

In the sixth and final case (pp. 445-446), the two points lie on either side of the plane passing perpendicular to axis XY through vertex X. Hence, as represented in figure 19e, p. 541, A lies below that plane, B above it. Form the opposite cone above the base cone, and pass a plane through B to intersect the axis orthogonally at point T. Point T will therefore be the center of the circle cut by that plane on the cone's surface. From point A draw line AK through vertex X and continue it until it meets the plane of that circle at point K. Then, on the basis of proposition 22, lemma 4, find point D′ through which tangent SD′C passes in such a way as to form angle SD′K equal to angle SD′B. Through centerpoint T of the circle draw diameter D′TE′, which is therefore perpendicular to tangent SD′C. From point D′ draw line of longitude D′X on the upper cone and extend it to point L on the lower cone. Produce line AA′ parallel to that line of longitude, and let it meet the plane of the upper circle at point A′. Since line AK and line of longitude D′L intersect at X, they are in the same plane, and since line AA′ is parallel to line of longitude D′L, all three lines AK, D′L, and AA′ lie in the same plane, so points K, D′, and A′ will lie in the same plane. Connect them with line KD′A′.

Now, angle KD′S = angle BD′S by construction, and angle KD′S = vertical angle A′D′C, so angle BD′S = angle A′D′C. But, since D′TE′ is perpendicular to SD′C by construction, angle SD′E′ = angle CD′E′, both being right. Hence, because their corresponding parts BD′S and A′D′C are equal, remainders BD′E′ and A′D′E′ will be equal. Accordingly, D′ is the point from which the form of B would reflect to A′. The point of reflection D for points A and B will therefore lie at the intersection of the plane of reflection and line of longitude D′XL. Hence, with normal FDE produced, angle of incidence BDE will equal angle of reflection ADE.

In all, therefore, Alhacen's method for finding the point of reflection in convex cylindrical and conical mirrors is based on defining the appropriate line of longitude and determining the point at which the plane of reflection intersects it. To define the appropriate line of longitude, in its turn, requires passing a plane through either A or B to cut the cone or its complement along a circle and then projecting the other point onto that plane.[34] With both points thus facing the circle, the final step is to find the point on that circle at which the form of the one would reflect to the other and produce the line of longitude from that point. Granted, this method is sometimes complex at the operational level, but it is fairly simple at the conceptual level.

Concave Spherical Mirrors: As mentioned earlier, the analysis of concave spherical mirrors is considerably more complex than that of convex spherical mirrors because of the multiplicity of possible reflection-points, depending on how the source-point of radiation and the center of sight are disposed with respect to one another as well as to the center of curvature and the reflecting surface itself. As will become clear in the course of our analysis, however, there are fundamental linkages between Alhacen's approach to convex mirrors and his approach to concave ones.

Let us start with the case in which both A and B are equidistant from the center of curvature. The most trivial form of this case, which Alhacen addresses in proposition 36, case 1, p. 452, occurs when A and B lie on the same normal at equal distances from centerpoint G of the mirror, as in figure 20, p. 542. Given that AG = BG, finding the normal to the point of reflection is a matter simply of dropping DG perpendicular to AB at centerpoint G of the mirror. From the equality of triangles AGD and BGD it is obvious that angle of incidence BDG = angle of reflection ADG. So D will be a point of reflection for A and B, and, by symmetry, so will D_1. In fact, if line DD_1 is rotated on axis AGB, the form of B will reflect to A from every point on the resulting circle formed by D on the mirror's surface. That it will reflect from no other point D_2 on the mirror is obvious enough from the diagram to require no proof. Furthermore, it makes no difference whether A and B lie inside or outside the circle of the mirror as long as point D on the reflecting surface is exposed to them (i.e., as long as BD can reach the mirror and DA can reach A).

But what if A and B do not lie on the same normal? Alhacen addresses this case in proposition 37, pp. 454-458. The first and most trivial instance is represented in figure 20a, p. 542, where A and B lie far enough outside the great circle of the mirror that the line passing through them does not intersect that circle. To find the point of reflection under these conditions, we need only bisect angle AGB with line GE and extend the bisector to point D. Like its corollary in case one for convex mirrors, the problem in this case is so simple in its solution and proof as to need no elaboration. From the equality of triangles AGD and BGD it is obvious that angle BDE of incidence is equal to angle ADE of reflection. That the form of B cannot reflect to A from any point, such as D_1, on arc FD follows from the fact that, no matter where D_1 lies on that arc, normal GD_1 dropped to it will not form equal angles with AD_1 and BD_1. Nor can the form of B reflect to A from any point D_2 on arc FK, because normal GD_2 will lie outside angle AD_2B and will thus not bisect it. The same holds by symmetry for arcs DH and HL. Consequently, since reflection cannot occur from any point other than D on the entire circle, D is the only possible point of reflection in this instance. It should be noted that in this case, as well as in every other case in which A and B lie on different normals, reflection will occur only in the great circle formed on the mirror by the plane of reflection. In other words, there is no circle of reflection as in the previous case.

Now, if we shift A and B toward G along the same normals so that the line passing through them intersects the circle, and if A and B remain equidistant from G, as represented in figure 20b, p. 543, it is obvious that the previous analysis holds for arc FH: i.e., that the form of B will still reflect to A from point D but from no other point on arc FH. However, as long as A and B lie below XY, which is tangent to the circle at point D_1, where bisector GE intersects the circle opposite D, points A and B also face some segment of arc KL bounded by the extensions of normals AG and BG. Accordingly, if we assume that the reflecting surface extends through arc KL, then it is obvious from the diagram that the form of B will reflect to A from point D_1 and that the same holds for any other points—such as A′ and B′—that lie below tangent XY. It is also evident that the form of B will not reflect to A from any point D_2 on arcs FK or HL, for in that case normal D_2G would lie outside angle AD_2B. In this instance, therefore, the form of B will reflect to A from at least two points, D and D_1, the former on arc FH and the latter on arc KL.

Whether the form of B can reflect to A from any point other than D_1 on arc KL depends upon whether the circle passing through points A, B and G intersects the circle of the mirror within that arc, as discussed in the previous section (pp. xl-xli). Moreover, as we pointed out in that discussion, circle ABG will intersect arc KL if and only if the lines dropped to points A and B from the midpoint of arc KL form acute angles with their respective normals. Thus, as illustrated in figure 20c, p. 543, when A and B are disposed as they are, circle ABG never touches the circle of the mirror. Assume that D_2 is a legitimate point of reflection. Draw normal GD_2, extend it to P on the opposite side of the mirror, and let it pass through the inner circle at point T. Connect AT and TB. Angle ATG is subtended by arc AG, and angle BTG is subtended by arc BG, and both arcs are equal by construction (i.e., chord AG = chord BG). Thus, by Euclid, III.21, angle ATG = angle BTG.

If D_2 is a legitimate point of reflection, then angle of incidence AD_2G should equal angle BD_2G. Extend AD_2 to M and BD_2 to N. Within the circle of the mirror, then, arc MP subtended by angle AD_2G should equal arc NP subtended by angle BD_2G. Extend line MO perpendicular to normal D_2P, and let R be where the two lines intersect. Accordingly, since angle MRP = angle ORP (both being right by construction), arcs PM and PO subtending those angles are equal. But arc PM subtends angle PD_2M, whereas arc PN subtends angle PD_2N, and arc PN > arc PO. It therefore follows that angle PD_2N, which coincides with angle of incidence BD_2G, is greater than angle PD_2M, which coincides with angle of reflection AD_2G, so the form of B cannot reflect to A from point D_2 or, by extension, from any other point on arc KL except D_1.

If, however, circle ABG does intersect the arc, it will necessarily do so at two points. Those two points will be points of reflection, as claimed in the previous section. Accordingly, as represented in figure 20d, p. 544, points D_2 and D_3, where circle ABG intersects arc KL, will be points of reflection, because the arcs on circle ABG subtending angles AD_2G and BD_2G within circle ABG—i.e., arcs AG and BG—are equal, by construction, and those same arcs subtend angles AD_3G and BD_3G within that same circle. In the previous case we excluded reflection from arcs KD_2 and LD_3, where circle ABG does not touch the circle of the mirror. We can also exclude reflection from any point D_4 on arc D_2D_1 according to the inequality of arcs MP and PN, since arcs MP and MO are equal by construction, and we can extend that exclusion to arc D_3D_1 by symmetry. In this instance, then, the form of B can reflect to A from points D_1, D_2, and D_3 on arc KL and from point D on arc FH. Four is therefore the maximum number of reflections when A and B lie inside the circle at equal distances from G.

From this sketchy analysis it should be evident that, even in the relatively trivial case of equality between AG and BG, the determination of D is complicated by the need to take into account not only where A and B lie with respect to G, but also where they lie with respect to the surface of the mirror. It should be clear as well that the problem of finding D is intimately connected with the problem of determining precisely how many legitimate points of reflection there can be under the various conditions specified. Yet, despite these complicating factors, once the conditions are properly specified, the two methods for finding the appropriate point or points of reflection—i.e., through the bisection of angle AGB and the intersection of circle ABG—are almost intuitively obvious in their simplicity.

Things become more complicated when we turn to the case in which normals AB and AG are unequal, as represented in figure 20e, p. 544. There are, however, certain parallels between this case and the preceding one. For one thing, as we showed in the previous section (p. xli), the form of B can reflect to A from one, and only one, point on arc FH, and it cannot reflect from any point on arcs FK or HL. For another, if the line passing through A and B does not touch the circle of the mirror, the only reflection that can occur will be from the given point on arc FH. For yet another, when A and B lie inside the circle of the mirror, reflection can occur from as many as, but no more than, four points: one from arc FH and the rest from arc KL. And for yet another, the circle passing through points A, B, and G is instrumental in determining how and where reflection can occur from arc KL.

Despite these parallels, there are certain limitations that are specific to the case at hand. For instance, if the form of B reflects to A from some point D on arc KL in figure 20e, then angle ADB (to which I will henceforth refer as the "reflected angle") cannot be equal to angle LGH adjacent to KGL. Alhacen proves this point in proposition 43, pp. 462-463, which was discussed earlier on p. xli. Furthermore, as was shown at the same point in the previous section, if the form of B reflects to A from two points on arc KL, it is impossible for both reflected angles to be less than angle LGH. In addition, there can be two reflections from arc KL such that both reflected angles are greater than adjacent angle LGH. Thus, when the circle passing through points A, G, and B intersects arc KL, two reflections may occur in arc CE cut by that circle (but only on segment XE), and in both of them the reflected angle will be greater than adjacent angle LGH. An additional reflection will occur from arc CK, and the reflected angle in that case will be less than adjacent angle LGH. Altogether, then, there can be three reflections from arc KL and one from arc FH. Four, in short, is the maximum number of reflections that can occur when AG $\neq$ BG. As was also pointed out earlier (pp. xl-xli), if circle AGB intersects arc KL at midpoint X, as represented in figure 16b, p. 532, or if it is tangent to or falls short of that arc, as represented in figure 16a, p. 531, the form of B will reflect to A from only one point on arc KL, and the resulting reflected angle BDA will be less than adjacent angle LGH.

With these specifications in mind, let us turn to the actual problem of determining the point or points of reflection in concave sperical mirrors when points A and B lie inside the circle of the mirror and normals AG and BG are unequal. The first and simplest case, which Alhacen addresses in proposition 36, case 2, pp. 453-454, has A and B lying on the same normal, as represented in figure 20f, p. 544. Extend line AG through and beyond point F on the circle of the mirror. Find point E on that extension such that BE:GE = BG:AG. In other words, BE is to GE in the same ratio as the normals dropped to the center of the mirror from the object-point and the center of sight. From point E draw a circle passing through points D, G, and D_1 within the mirror. The form of A will reflect to B from the two points of intersection D and D_1. The proof rests on showing that within the circle of the mirror, normal DG bisects angle BDA. This point, in turn, rests on the fact that DB:DA = BG:AG.[35]

Furthermore, as happens when A and B lie equidistant from G on the same normal, reflection in this case occurs not just from D and D_1 but from the entire circle produced by D as line DAD_1 rotates about axis AGB. That the form of A cannot reflect to B from any other point D_2 on the circle of the mirror follows from the fact that, when normal D_2G is drawn, $BD_2:AD_2 \neq$ BG:AG. At bottom then, the determination of D and D_1 comes to ground in the proportionality between normals AG and BG, just as it does for convex spherical mirrors.

Now, let A and B lie on different normals at unequal distances from G, as represented in figure 20g, p. 545, where B lies on normal KGH and A on normal LGF. Let line XGY bisect angle FGH. As Alhacen formulates it in proposition 38, pp. 458-459, the method for finding point D of reflection on arc FH is as follows. Take some line D′M′, and divide it at point Q′ such that M′Q′:Q′D′ = BG:AG. Bisect D′M′ at point N′, and through that point draw line N′B′ perpendicular to D′M′. At point D′ form angle N′D′T′ equal to half of angle LGH adjacent to FGH.[36] Then, following the procedure Alhacen has already outlined in proposition 21, lemma 3, draw a line from point Q′ that meets perpendicular N′B′ at point B′ and passes through D′T′ at point G′ such that B′G′:D′G′ = BG:DG (i.e., the radius of the mirror). At centerpoint G of the mirror form angle BGD equal to angle B′G′D′. Point D, where leg DG of that angle intersects the mirror, will be the point of reflection.

The construction according to which Alhacen proves that D is in fact the point of reflection begins with the formation of angle GDQ equal to angle G′D′Q′, as represented in figure 20h, p. 545. Thus, as is obvious from the diagram, triangles QDG and Q′D′G′ are similar. Likewise, if we draw BN perpendicular to the extension of DQ at point N, it is obvious from the diagram that triangles QBN and Q′B′N′ are similar. Equally obvious from the diagram is that triangles BGT and B′G′T′ are similar. Extend DN to M so that DN = NM. Accordingly, the entire figure constructed on line D′M′ to the left is recapitulated in the figure highlighted by bold lines in the circle of the mirror to the right.

Connect MB, draw line DS parallel to it, and continue BH until it meets DS at S. Since vertical angles SQD and MQB are equal, and since alternate angles DSQ and MBQ are equal, triangles DSQ and MQB are similar, so their corresponding sides will be proportional. Hence, SQ:QB = DQ:QM. Moreover, since DM is bisected by perpendicular BN, triangles BDN and BMN are equal, so their corresponding angles BDN and BMN are equal. But angle BMD = alternate angle MDS, so angle BDQ = angle SDQ, from which it follows that DQ bisects angle SDB. The rest of the proof is based on constructing angle GDP equal to angle GDB and then showing that line AP is a rectilinear continuation of line DP, from which it follows that angle of incidence BDG = angle of reflection ADG.

The similarity between Alhacen's method for determining D in this case and his method for determining D in convex spherical mirrors becomes clear if we extend line BN in figure 20k, p. 546, to meet the extension of line DS at point V and repeat the process with lines B′N′ and D′S′. Then, with line D′B′ produced, the entire figure B′D′S′V′ formed on line D′M′ to the left will be recapitulated in figure BDSV formed on line DM in the circle of the mirror.

Now, if we join AB, it is clear that XY, which bisects angle KGL—and thus angle FGH—will intersect AB at point E. But, by Euclid, VI.3 we know that AE:EB = AG:BG. We also know that MQ:QD = BG:AG by construction. But we have established earlier that MQ:QD = BQ:QS, and, since DQ bisects angle SDB, DB:DS = MQ:QD = BQ:QS. Hence, BQ:QS = BD:DS = BG:AG. Triangles BDS and BGA thus correspond in the two crucial ways discussed earlier in the context of convex spherical mirrors: the lines bisecting the angles at respective vertices G and D cut equiproportional segments from them (i.e., BE:EA = BG:GA, and BQ:QS = BD:DS), and the segments cut off by those bisectors on respective bases BA and BS are equiproportional (i.e., BE:EA = BQ:QS = BG:GA). Furthermore, it should be obvious that dropping line Q′G′B′ from point Q′ on D′M′ through point G′ on D′T′ so that D′G′:G′B′ = DG:BG is tantamount to producing line B′Q′S′ so that B′S′:S′D′ = BS:SD. In other words, implicit in Alhacen's construction for determining D in this case is his construction for determining D in convex spherical mirrors.[37]

So much for arc FH, in which there is only one possible point of reflection. Arc KL is another matter altogether, for, as we have already observed, the form of B may reflect to A from as many as three points on that arc if the circle passing through A, G, and B intersects the arc. Let A and B in figure 20n, p. 547, lie on normals FGL and KGH at unequal distances from G, and let them be disposed in such a way that circle ABG does intersect arc KL. Let us start with the case in which the point of reflection lies outside circle ABG so that, according to previous discussion, the resulting reflected angle ADB is less than angle LGH. Oddly enough, Alhacen never addresses this case explicitly, but the method to be applied to it is essentially the obverse of the one applied to reflection from arc KH. Accordingly, take some line D′M′, and cut it at point Q′ so that M′Q′:Q′D′ = AG:BG. Bisect D′M′ at N′, draw perpendicular N′T′ through that point, and form angle N′D′T′ equal to half of angle LGH. From point Q′ drop line Q′A′G′ through N′T′ so that it meets D′T′ at point G′ in such a way that Q′G′:D′G′ = BG:GD (i.e., the radius of the circle of the mirror). Alhacen has already demonstrated how to do this in proposition 24, lemma 6. At centerpoint G of the mirror, form angle AGD equal to angle A′G′D′ (i.e., Q′G′D′). D will be the point of reflection.

The proof that D actually is the point of reflection is essentially the same as that in the previous case of reflection from arc FH. The construction is illustrated in figure 20p, p. 547. From that diagram it is obvious that the entire figure DGTANQM highlighted in bold in the mirror of the circle is similar to figure D′G′T′A′N′Q′M′, so all the corresponding triangles are similar. Draw DA and AM. Since AN is perpendicular to DN and bisects it according to construction, triangles DAN and MAN are equal. Draw DS parallel to AM, and continue GAQ to meet it at point S. Therefore, angle SDM = alternate angle DMA, which equals angle MDA by construction, so DQM bisects angle SDA. Furthermore, since angle SDM = alternate angle DMA, while angle DSQ = alternate angle QAM, triangles DSQ and QAM are similar, so their corresponding sides will be proportional. It follows, then, that SQ:QA = DQ:QM. But DQ:QM = BG:AG by construction, so SQ:QA = BG:AG. The rest of the proof is based on constructing angle GDP equal to angle GDA and then showing that line DB is a rectilinear continuation of BP, from which it follows that angle of incidence BDG equals angle of reflection ADG.

The patterns that underly the two previous methods underly this one as well, the two associated triangles being ADS and BAG in figure 20q, p. 548. In the former, bisector DQ cuts base AS at point Q such that DS:DA = QS:AQ. But, since it has been established that triangles DQS and AQM are similar, then QS:AQ = DQ:QM = BG:AG. Likewise, XY bisects angle BGA to intersect line BA at point E so that BE:EA = BG:GA. Therefore, corresponding bases AS and BE of the two triangles are cut according to the proportionality of normals BG and AG. That proportionality, in short, is systemic throughout the construction.[38]

We are now left with the problem of finding the two remaining points of reflection on the arc inside the intersecting circle. However, as mentioned earlier, just because circle ABG intersects arc KL does not necessarily mean that reflection will occur from the arc within the intersecting circle. There is one additional restriction. If reflection is to occur from more than one point on arc KL (in which case it will necessarily occur from three) the lines drawn from the point of bisection on arc KL to A and B must form acute angles with their respective normals on the side of G. In other words, angles XBG and XAG in figure 20t, p. 550, must be acute.[39] Let A and B be properly disposed according to this final restriction. As given in proposition 47, pp. 470-471, the method for finding the two points of reflection in this case is as follows.

Take some line D′M′, in figure 20t, p. 550, and divide it at Q′ such that M′Q′:Q′D′ = BG:AG. Bisect D′M′ at N′ and pass perpendicular line B′N′T′ through that bisection-point. Form angle M′D′T′ equal to half of angle LGH. Draw line B′Q′G′ through point Q′ so that it intersects B′T′ at point B′ and meets D′T′ at G′ in such a way that B′G′:G′D′ = BG:GD. This we can do by following the procedure laid out by Alhacen in proposition 24, lemma 6. Form angle BGD in the circle equal to angle B′G′D′. D will be a point of reflection.

Now, as we established earlier in the discussion of proposition 24, lemma 6, it is possible to project two lines through point Q′ to meet the specified conditions. Accordingly, in the figure below and to the right of the circle, line D1′M″ is cut proportionate to line D′M′ in the figure above and to the left of the circle, so that M″Q″:Q″D1′ = M′Q′:Q′D′ = BG:AG, and M″D1′ is bisected at point N″, with line B″N″T″ passing through it perpendicular to M″D1′. Under these conditions, it is possible to pass a line B″Q″G″ through point Q″ such that B″G″:G″D1′ = BG:GD in the circle. Thus, B″G″:G″D1′ = B′G′:G′D′. Nevertheless, the two resulting angles B″G″D1′ and B′G′D′ will be unequal.[40] Accordingly, form angles BGD and BGD1 in the circle equal to angles B′G′D′ and B″G″D1′, and D and D1 will both be points of reflection.

Not only does Alhacen not mention the construction of this second angle in proposition 47, but he does not demonstrate that D is in fact a legitimate point of reflection. He does suggest the direction the reader might take in order to prove that fact, but the approach he suggests, which is based on conclusions drawn in proposition 46, is different from the one we have been taking to this point in the analysis. The construction for the proof according to our approach is given in figure 20x, p. 551. As is clear from the diagram, the entire figure D'G'T'B'Q" formed on line D'M' is recapitulated in figure DGTBQ formed on line DM in the circle of the mirror. Accordingly, DS is parallel to BM by construction, so angle BDQ = angle QDS, from which it follows that DQM bisects angle BDS. Hence BQ:QS = BD:DS = MQ:QD = BG:AG. By the same token, BE:EA = BG:GA, so triangles BDS and BAG correspond in the two ways discussed earlier: i.e., the lines bisecting the angles at respective vertices G and D cut equiproportional segments from them (i.e., BE:EA = BG:GA, and BQ:QS = BD:DS), and the segments cut off by those bisectors on respective bases BA and BS are equiproportional (i.e., BE:EA = BQ:QS = BG:GA). The same construction applies *mutatis mutandis* to reflection-point D_1, and it also applies to the limiting case in which normals BG and AG are equal and the reflection-point lies at X, where arc KL is bisected.[41]

Concave Cylindrical Mirrors: Alhacen's method for finding the point, or points, of reflection in concave cylindrical mirrors is analogous to his method for finding the point of reflection in convex cylindrical mirrors. Accordingly, the simplest case involves A and B lying in a plane of reflection that cuts the mirror along a line of longitude and includes the axis. In this case, of course, the problem reduces to finding the point of reflection on a plane mirror in which the line of longitude represents the common section of the reflecting surface and the plane of reflection. In the second case, the plane of reflection containing A and B is orthogonal to the axis so that it cuts the cylinder's surface along a circle. Since that circle lies in the plane of reflection, finding the point or points of reflection on it is a matter of applying the methods for determining such points in concave spherical mirrors. Thus, the form of B can reflect to A from as few as one point and as many as four, depending on where A and B lie in relation to the center of the circle and the mirror's surface.

The third and final case, which Alhacen addresses in proposition 52, pp. 478-482, involves A and B lying in a plane of reflection that intersects the axis obliquely and therefore cuts the cylinder's surface along an ellipse. Let the cylinder with axis XG in figure 21, p. 553, represent the mirror, and let B lie above A inside the cylinder. Through A pass a plane orthogonal to axis XG so as to cut the cylinder's surface along circle FKLH centered on G. From point B drop a perpendicular to the plane of the circle, and let it meet that plane at B′. Connect lines AB′ and AB.

To find the point or points on circle FKLH from which the form of A would reflect to B′, we need only apply the methods for finding such points on concave spherical mirrors. Let D′ be one of those points. Accordingly, normal E′D′G will bisect angle AD′B′ to render the angles of incidence and reflection equal. From point D′ produce line of longitude D′T on the cylinder. Then, within plane D′GXT formed by the axis and the line of longitude draw line YDE through AB parallel to E′D′G. Since line E′D′G is normal to D′T, line YDE will also be normal to D′T, and it will lie in the same plane as AB, which it intersects. Point D, will therefore be the point of reflection within plane of reflection ADBY, from which it follows that angle BDY of incidence will equal angle ADY of reflection. Quadrangle ADBY, with its associated diagonals AB and YD, is thus the projection of quadrangle AD′B′G, with its associated diagonals AB′ and GD′, onto the plane of reflection.

The same procedure can be applied to every other point from which the form of A would reflect to B′ within the base-circle centered on G. For each such point, there will be an appropriate point D on the cylinder, and for each such point D the plane of reflection will cut an ellipse on the cylinder's surface, each ellipse being particular to the given plane of reflection. Suffice it to say, the number of actual reflections will depend upon how much of the reflecting surface is exposed to A and B.

Concave Conical Mirrors: As is the case with all the other curved mirrors, if A and B lie in a plane that includes the axis, that plane will cut a line of longitude on the surface of the cone. Finding the point of reflection thus reduces to finding the point of reflection on a plane mirror in which the line of longitude forms the common section of the reflecting surface and the plane of reflection.

On the other hand, if A and B lie on a line through which a plane can be passed orthogonal to the cone's axis so as to cut a circle on the cone's surface, then the first step is to find the point or points of reflection within that circle. Let A and B in figure 22, p. 554, be two such points, and let them lie in the plane of circle FKLH centered on point G. That plane, of course, will be orthogonal to axis XGY of the cone. Applying the methods for determining the point, or points, of reflection in concave spherical mirrors, find the point, or points, within circle FKLH from which the form of B would reflect to A. Let D′ be one of them. From that point produce line of longitude D′X on the mirror's surface, and join line AB. Then, within the plane formed by line of longitude D′X and axis XG of the cone, draw line YDE through AB normal to line of longitude D′X. That line will intersect axis XG at point Y, and point D, where it intersects D′T, will be the point of reflection within the plane of reflection ADBY. Accordingly, angle of incidence ADY will equal angle of reflection BDY. Quadrangle ADBY, with its associated diagonals AB and DY, is therefore just a projection of quadrangle AD′BG, with its associated diagonals AB and GD′, onto the plane of reflection, and AB is the common section of the planes containing the two quadrangles. The same procedure can be applied to every other point from which the form of A would reflect to B within the base-circle centered on G.

In the third and final case, A and B are disposed so that the plane passing through them cuts the axis obliquely. For instance, let B in figure 22a, p. 555, lie above A so that the plane passing through line AB strikes axis XG at a slant. Pass a plane through A orthogonal to axis XG to form circle FKLH with centerpoint G on the cone's surface. From vertex X draw XB and extend it until it meets the plane of circle FKLH at point B′. Applying the methods for finding the point, or points, of reflection in concave spherical mirrors, determine the point, or points, on circle FKLH from which the form of point B′ would reflect to A. Let D′ be one of them. From point D′ produce line of longitude D′X on the cone's surface. Within the plane formed by D′X and axis GX of the cone, produce line YD through AB to intersect D′X orthogonally. Point D, where that line meets D′X, will be the point of reflection within the plane formed by ADBY, so angle of incidence BDY will equal angle of reflection ADY. Thus, quadrangle ADBY, with its associated diagonals AB and YD, is simply a projection of quadrangle AD′B′G, with its associated diagonals AB′ and GD′, onto the plane of reflection, and line AM passing through the intersection of DY and D′G is the common section of those two planes. The same procedure can be followed for any other point of reflection on circle FKLH, so there can be as few as one and as many as four such points on the cone, and each point will lie on a particular conic section.

3. *The Sources of Alhacen's Reflection-Analysis*

The clearly identifiable sources for Alhacen's analysis of reflection in books 4 and 5 of the *De aspectibus* are extremely limited in both type and number, consisting of Ptolemy's *Optics*, Euclid's *Elements*, Apollonius' *Conics*, and Serenus' *On the Cone and Cylinder*. Since these sources fall into two categories by type—optical and mathematical—the following examination of how Alhacen used them will be organized according to that division.

Optical Sources: As we noted in the discussion of sources in the previous edition of books 1-3, Alhacen drew on a variety of Greek and Arabic optical sources in formulating his overall theory of visual perception. We also noted that, rather than cite those sources specifically, Alhacen referred to them according to such generalities as "mathematicians" or "natural philosophers."[42] Still, it is possible to isolate probable sources on the basis of internal and external evidence. Such evidence, for instance, points to Ptolemy as the crucial authority among the "mathematicans" drawn upon by Alhacen in constructing the theory of visual perception and illusions laid out in books 1-3.[43]

The same problem applies to identifying the optical sources for Alhacen's analysis of reflection in books 4 and 5 of the *De aspectibus*. The possibilities are legion, but internal evidence suggests strongly that Ptolemy's account of reflection in books 3 and 4 of the *Optics* was the major, if not the sole, optical source for Alhacen. In fact, as will be clear in fairly short order, Ptolemy's account of reflection served as nothing less than a model for Alhacen's, determining not only the general structure, but also specific elements of analysis. This is not to say that Alhacen followed Ptolemy slavishly in every particular, but it would, I think, be fair to say that Alhacen's account of reflection represents a critical elaboration on Ptolemy's in terms of both rigor and comprehensiveness.

A case in point is Alhacen's experimental confirmation of the equal-angles law of reflection in book 4 of the *De aspectibus*. Alhacen sets the stage in chapter 2 with a series of simple experiments to show that light and illuminated color reflect in a consistent way, no matter the source or type of illumination (i.e., primary or secondary light). While these experiments have no counterpart in Ptolemy's *Optics*, the actual set of confirmatory experiments in chapter 3 of book 4 does. Described in *Optics* III, 8-11 (Smith, *Ptolemy's Theory*, 134-135), Ptolemy's version of these experiments is based on the following apparatus.

Let bronze disk BEDG in figure 23, p. 556, be divided into quadrants by lines BD and GE, each quadrant being subdivided into 90 degrees. From three thin strips of iron form three mirrors, one plane, one convex cylindrical, and one concave cylindrical.[44] Place these mirrors upright on the disk at point A such that line BAD is perpendicular to their bottom edges. Line GE will thus coincide with the reflecting surface of the plane mirror and will be tangent to the reflecting surfaces of the two curved mirrors TK and ZH at point A. Let line BA be drawn in white, and in arc BG draw some line LA in another color. Fix a sighting device at point L, and establish a line of sight along LA to point A, which is marked with a pin. Finally, while sighting along line LA, move a tiny colored marker along arc BE until its image I appears to lie perfectly in line with LA, as represented for all three mirrors in figures 23a-23c, p. 556, where cathetus MI intersects line of reflection AL or its extension in the case of the plane and convex mirrors. Let point M be where the colored marker lies on arc BE. Under those conditions, arc MB = arc LB, which is to say that angle of reflection MAB = angle of incidence LAB. No matter what point L is chosen on arc BG, point M, where the colored marker ends up, will be such that the equality of angles is preserved for all three mirrors.

There are, of course, significant differences in detail between this experiment and the one described by Alhacen in the third chapter of book 4, the most obvious being that Ptolemy's experiment measures the reflection of visual rays, while Alhacen's measures the reflection of light-rays. By choosing different things to measure, however, Ptolemy and Alhacen had to measure them in fundamentally different ways. Since visual rays are impossible to detect by direct observation, Ptolemy necessarily fell back upon an indirect method for confirming the equal-angles law based on image-location. In measuring the reflection of light-rays, on the other hand, Alhacen could approach the phenomenon directly by restructuring the experiment as he did so as to highlight the passage of light-beams through the wooden ring to the mirror and thence to the ring's inner wall. Even so, in the case of secondary color-radiation, Alhacen had no choice but to confirm the equal-angles law by indirect means because of the weakness of that radiation.[45]

A second major difference between the two experiments is that Ptolemy's involves only plane, convex cylindrical, and concave cylindrical mirrors, while Alhacen added convex and concave spherical and conical mirrors to bring the total of tested mirrors from three to seven. In this case, of course, the addition of spherical and conical mirrors represents an effort to make the experiment more comprehensive and general. To some extent, moreover, the need to accommodate the four new mirrors—particularly the convex and concave spherical ones—to the experiment explains the greater sophistication and complexity of Alhacen's apparatus over that of Ptolemy's.[46] By the same token, the fact that Alhacen, unlike Ptolemy, was measuring the passage of light-rays through his apparatus helps explain the extensiveness of his tests for both primary and secondary light- and color-radiation. It also helps explain his inclusion of the preliminary experiments in chapter 2. These experiments, of course, would have made no sense within the context of Ptolemy's analysis of reflection because that analysis was not based on light-rays.

By no means exhaustive, this brief discussion should nonetheless be adequate to show not only that Alhacen's core experiment for validating the equal-angles law of reflection was based on Ptolemy's, but that, where the two experiments differ, they do so in degree rather than in kind. In other words, those differences stem from differences both in what the two experiments were intended to measure and in the universality of the resulting conclusions. It is therefore at the level of implementation rather than conception that Alhacen actually broke from Ptolemy in his approach to optical experiment.

This point is clearly exemplified in the series of experiments Alhacen proposes at the beginning of book 5, paragraphs 2.1-2.20, to confirm that, no matter which of the seven mirrors is tested, all point-images seen in those mirrors will appear on the normal, or cathetus, dropped from the object-point to the reflecting surface. These experiments are suggested by Ptolemy in *Optics*, III, 4, where he observes that, "if we stand long, straight objects at right angles to the surface of mirrors, and if the distance is moderate, the images will appear [to lie] perfectly in line with those objects [as they] are properly viewed outside the mirror."[47] Hence, in establishing this point empirically for each of the seven types of mirror, Alhacen was simply following up on Ptolemy's suggestion and, in the process, validating it as comprehensively and rigorously as possible.

Even clearer than the link between Alhacen's and Ptolemy's empirical justification of the principles of reflection are the links between their respective applications of those principles to the mathematical analysis of reflection. These links are especially clear in the way Ptolemy and Alhacen approach the analysis of concave spherical mirrors in order to determine how many points of reflection there can be for a given center of sight and object-point.

As we saw on pp. lvi-lvii above, Alhacen begins this analysis in proposition 36, pp. 452-454, with the center of sight and the object-point lying on the same diameter within the mirror and flanking the center of curvature. Accordingly, he turns first in proposition 36, case 1, to the situation in which the two points are equidistant from the center of curvature, as illustrated in figure 5.2.36, p. 255, where E is the center of sight, H the object-point, and D the center of curvature, and where ED = DH. On that basis, he demonstrates that within the circle containing diameter ZEDHA, points G and B will be points of reflection for E and H and, furthermore, that if circle GABZ containing those points is rotated about diameter ZEDHA as axis, G and H will form a circle of reflection within the sphere of the mirror. Alhacen then concludes that no other point, such as C, within circle AZBG can be a point of reflection.

Ptolemy demonstrates the first two points in essentially the same way as Alhacen by showing in theorem IV.2 (Smith, *Ptolemy's Theory*, 176-177) that, if points H, Z, and E in figure 24, p. 557, represent the center of sight, the object-point, and the center of curvature on diameter BHEZD, and if HE = EZ, point A will be a point of reflection. Like Alhacen, he goes on to explain that, if triangle HAZ is rotated about axis BD, point A will form a circle of reflection centered on E. To prove the third point—i.e., that no point outside of the circle of reflection formed by A can serve as a point of reflection for Z and H—Ptolemy takes a slightly different tack from Alhacen's. Thus, in theorem IV.3 (Smith, *Ptolemy's Theory*, 177), he shows that, if ZE > HE, as represented in figure 24a, p. 557, A cannot be a point of reflection. He then shows in theorem IV.4 (Smith, *Ptolemy's Theory*, 177) that, if ZE > EH, as represented in figure 24b, p. 557, there will be no point of reflection, such as K, in arc AD of circle ADGB. On that basis, he concludes in theorem IV.9 (Smith, *Ptolemy's Theory*, 181-182) that, if there is some point of reflection, such as L, for H and Z in semicircle BAD in figure 24c, p. 557, there can be no other point of reflection, such as S, within that semicircle—which means by extension that, if A is the point of reflection for H and Z when HE = HZ, as in figure 24, p. 557, there can be no other point of reflection in semicircle BAD. He closes that theorem by observing that, if triangle LHZ in figure 24c, p. 557, is rotated about axis BZ, point L will produce a circle of reflection for H and Z.

Having demonstrated the previous three points for the case in which the center of sight and the object-point lie equidistant from the center of curvature on the same diameter within the spherical mirror, Alhacen then turns in proposition 36, case 2, to the situation in which the two points lie different distances from the center of curvature on the same diameter. Given these conditions, Alhacen wishes to show, first, how the point of reflection within the circle containing the diameter can be determined, and second, that there can be only one such point within the circle. Accordingly, if E and H in figure 5.2.36a, p. 256, represent the center of sight and object-point on diameter ADZ of circle AGZB within a concave spherical mirror, and if ED > HD, then we first locate point Q on the extension of AZ such that EQ:DQ = ED:DH. With point Q as center, we draw a circle of radius QD. Points G and B, where that circle intersects circle AGZB of the mirror, will be points of reflection. Those points, moreover, are the points at which the line drawn from point Q′ on diameter Q′D of the cutting circle are tangent to circle AGZB of the mirror. Furthermore, no points other than G and B (e.g., C) can serve as points of reflection for E and H within circle AGZB. It therefore follows, according to Alhacen, that, if triangle EHG is rotated about axis

QE, point G will describe a circle of reflection for points E and H, which is the same point established by Ptolemy in theorem IV.9 discussed above.

At first glance, Ptolemy's version of this determination appears markedly different from Alhacen's, but under closer inspection it turns out to be essentially the same. For instance, in theorem IV.5 (Smith, *Ptolemy's Theory,* 177-179), Ptolemy locates the point of reflection as follows, according to figure 25, p. 558. Let Z and H be the center of sight and object-point on diameter BED, let E be the center of the mirror, let ZE > HE, and let point T be taken on ZE such that ET = EH. Find point K on the extension of diameter BD according to which ZT:EH = ZH:KH. It therefore follows that ZT:ZH = EH:KH, from which it follows that KZ (i.e., KH + ZH):KH = ZE (i.e., ZT + ET):ET = ZE:HE, since ET = HE. When line KL is then drawn tangent to circle ABGD, point L, where it touches arc AB of the circle, will be a point of reflection, and by implication there will be a corresponding point of reflection where the tangent from point K touches arc BG of the circle.

Now, if we reletter the diagram for Alhacen's demonstration to correspond with the diagram for Ptolemy's, as in figure 25a, p. 558, Alhacen's original proportion for determining the point of reflection becomes QZ:QE = ZE:HE. But QZ:QE = KZ:KH, so we end up with KZ:KH = ZE:HE, which is the proportion upon which Ptolemy's determination in theorem IV.5 is based. Accordingly, in the next theorem (IV.6, in Smith, *Ptolemy's Theory,* 179), Ptolemy shows that, if KZ:KH ≠ ZE:EH, there will be no reflection within circle ABGD, which is tantamount to saying that, if some point other than L is chosen on arc AB of the circle, and if the tangent is dropped from that point to the extension of diameter BD, point K, where it intersects that diameter, will not yield the appropriate ratio.

Having fully analyzed the case in which the center of sight and object-point lie on the same diameter at equal or different distances from the center of sight, Alhacen turns to the case in which they lie on different diameters at equal or unequal distances from the center of curvature. As in the previous account, the simpler version of this case, which Alhacen addresses in proposition 37, cases 2 and 3, pp. 455-458. has the center of sight and object-point equidistant from the center of curvature. Under these circumstances, Alhacen shows first that, if H and T in figure 5.2.37b, p. 258, are the two points, and if they are situated in such a way that the circle passing through them and center of curvature D does not cut facing arc GB, there can be only one point of reflection—i.e., E. The proof centers on showing that if some other point O is chosen on arc GB, angles TOD and HOD will be unequal. On the other hand, if H and T are situated such that the circle passing through them and center of curvature D does cut the facing arc, as represented in figure 5.2.37c, p. 259, there will be three points of reflection on arc BG—i.e., E and points M and L where circle HDT intersects arc BG.

The Ptolemaic counterparts of these two propositions are to be found (in reverse order) in theorems IV.10 and IV.17 (Smith, *Ptolemy's Theory*, 182-183 and 187-188). In theorem IV.17, Ptolemy demonstrates that, if Z and H in figure 26, p. 559, represent the center of sight and object-point, if ZE = EH, and if circle ZDH does not cut facing arc LM of the mirror, there can be only one point of reflection, B, from that arc.[48] Like Alhacen, Ptolemy proves this by showing that, if some other point T is chosen on LM subtending angle LDM, the resulting angles HTD and ZTD will be unequal. When circle ZDH does cut arc LM, Ptolemy demonstrates in theorem IV.10, there will be three points of reflection for Z and H, those points being B, T, and K in figure 26a, p. 559. Ptolemy then goes on in theorems IV.11 and 12 (Smith, *Ptolemy's Theory*, 183-184) to show that no more than three such reflections are possible on arc LM.

When we move to the more complex case in which the center of sight and object-point lie on different diameters at unequal distances from the center of curvature, the links between Alhacen and Ptolemy become more tenuous, in great part because Alhacen's ultimate goal in analyzing this case is to set up the conditions for determining precisely where the points of reflection are located and thus to solve part of his eponymous problem. As a result, Alhacen's analysis is far richer and more sophisticated than Ptolemy's because, unlike Ptolemy, Alhacen is concerned not only with how many points of reflection there can be for a particular configuration of center of sight and object-point on their respective diameters, but also with the size of the reflected angles vis-à-vis the angle adjacent to the angle subtended by the arc facing the two points.[49] Nevertheless, there are some obvious points of convergence between the two analyses, and the most effective way to illustrate those convergences is by recourse to Alhacen's summation of the results of his analysis in proposition 49.

For instance, in proposition 49, case 1, p. 472, Alhacen establishes that, if A and B on diameters CG and HG in figure 5.2.49, p. 277, represent the center of sight and object-point facing arc CH of the mirror, if BG $\neq$ AG, and if circle AGB passing through the mirror's center of curvature does not cut that arc, there can be only one point of reflection. Moreover, in proposition 49, case 2, pp. 472-473, where circle AGB in figure 5.2.49a, p. 277, touches arc CH at point T (and thus represents a special case of the previous one), and given that AG > BG, Alhacen establishes that the single point of reflection—i.e., T′—will fall on arc CT.

Ptolemy establishes these two points in theorem IV.18 (Smith, *Ptolemy's Theory*, 188-189), as follows. Let Z and H in figure 27, p. 560, represent the center of sight and object-point facing arc ML of the mirror, let ZE > EH, which means that ZD > HD, and let circle ZDH passing through those two points and center of curvature D not cut arc ML. In addition, let KTD bisect line ZH at point T. Under those conditions, Ptolemy goes on to demonstrate, there can be only one reflection, and it will necessarily take place at some point N within arc KL.

In proposition 49, subcase 3b, pp. 473-474, Alhacen addresses the situation in which circle AGB cuts arc CH at points E and F in figure 5.2.49c, p. 277. According to this configuration, he concludes, it is possible for there to be as many as three reflections from arc CH, two of them within arc EF and one within arc EC or arc FH. However, as we noted in our discussion of propositions 45 and 46 on pp. xli-lxiv above, the fact that circle AGB cuts arc CH in two places is not enough by itself to ensure that there will be three reflection-points. Point F, where it intersects the arc on the right-hand side, must lie to the right of the midpoint of that arc. Thus, as represented in figures 16a and 16b, pp. 531-532, if that intersection-point falls on or to the left of midpoint X, there can be one, and only one reflection-point, and it will lie on the arc between K and the point where circle AGB intersects arc CH on the left-hand side (note that in figure 16a the intersection-point T represents the tangent).

Ptolemy establishes this latter point in theorem IV.16 (Smith, *Ptolemy's Theory*, 186-187) according to figure 27a, p. 560, where Z and H represent the center of sight and object-point, B the midpoint of arc LM, and ZDH the circle passing through Z, H, and center of curvature D. In that case, Ptolemy goes on to demonstrate, there can be no reflection from either arc BM or TB, so the only possible reflection will occur from arc TL. On the other hand, as Ptolemy proves in theorems IV.12 and IV.13 (Smith, *Ptolemy's Theory*, 183-185), if circle ZDH in figure 27b, p. 560, cuts arc LM at points T and K on either side of midpoint B, there will be three reflections altogether, two from arc BK and one from arc LT. This, of course, is the same conclusion that would follow from an Alhacenian analysis of the same configuration for center of sight Z and object-point H.

This relatively brief comparative account of Alhacen's and Ptolemy's analyses of reflection from concave spherical mirrors should suffice to show how deeply indebted Alhacen was to Ptolemy not only for fundamental points of analysis (e.g., the number of possible reflections for a particular configuration of the center of sight and object-point), but also for analytic techniques (e.g., the cutting circle that passes through the center of sight, the object-point, and the center of curvature). Alhacen's genius and originality thus lay not in reconstructing the science of optics from the ground up but in building upon the foundations already laid by Ptolemy and restructuring the subsequent analysis to tie up various loose ends left by Ptolemy. Foremost among these, of course, was determining exactly where the point or points of reflection will fall in the six curved mirrors Alhacen chose for analysis when the center of sight and object-point face those mirrors at random locations.

Mathematical Sources: Alhacen mentions only two mathematical source-authorities by name in books 4 and 5 of the *De aspectibus*, and then only sparingly. The first and more frequently cited of these, Euclid, is named only seven times, and only once is a locus in the *Elements* provided.[50] The second, Apollonius (or "Ablonius"), is named three times, and in only one instance is a locus in the *Conics* given this time by both book and proposition.[51]

Among unmentioned sources, there are numerous possibilities. To explain various details of Alhacen's mathematical reasoning in books 4 and 5, for example, Friedrich Risner, the late-sixteenth-century editor of the *De aspectibus*, refers to Theon's recension of Euclid's *Elements* and commentary on the *Almagest*, Proclus' commentary on the *Elements*, Theodosius' *Sphaerics*, and Serenus' *On the Section of a Cylinder*. In citing these authorities, however, Risner was simply appealing to sources that would have helped contemporary readers follow his explanation of Alhacen's logic. In no way was he attempting to isolate Alhacen's actual sources—a point borne out by his citation of Campanus of Novara's thirteenth-century edition of the *Elements*, a work to which Alhacen could not possibly have had access.[52]

It is, of course, highly likely that Alhacen at least knew of, and was even familiar with, the ancient sources cited by Risner. It is equally likely that these and many other ancient Greek and contemporary Arabic sources contributed in various ways to the mathematical knowledge Alhacen brought to bear on his analysis of reflection. The problem is whether and how he might have drawn on specific points within these sources in the course of that analysis. Was he actually thinking of Proclus' commentary on *Elements*, I.29 when he concluded in proposition 25 (paragraph 2.212, p. 430) that, if line BI in figure 5.2.25, p. 239, is drawn parallel to line DL, then line DQ will intersect BI? Perhaps so. But if we ask whether he needed to refer to Proclus' commentary in order to reach this conclusion, the answer is clearly no. The most rudimentary understanding of Euclidean geometry is sufficient warrant for that conclusion.

If, therefore, we restrict our examination to mathematical sources that seem directly and necessarily pertinent to Alhacen's analysis, the list reduces to three: Euclid's *Elements*, Apollonius' *Conics*, and Serenus' *On the Section of a Cylinder*. The choice of the first two is of course dictated by

Alhacen's own citation of them. The choice of the third follows from Alhacen's awareness that a plane cutting a cylinder obliquely will form a true ellipse, not simply an ellipse-like section, on the cylinder's surface. Since this fact is by no means self-evident, it stands to reason that Alhacen knew it either directly or indirectly from Serenus' demonstration in proposition 20 of *On the Section of a Cylinder.*[53]

Alhacen's explicit use of Apollonius in the fifth book of *De aspectibus* is limited to a handful of propositions in book 2 of the *Conics,* where Apollonius deals with hyperbolic sections. Twice, in proposition 19, lemma 1, and proposition 20, lemma 2, Alhacen cites proposition II.4 of the *Conics.* He also cites proposition II.8 twice, both times in proposition 19, lemma 1, and once he has recourse to proposition II.16, in proposition 20, lemma 2. Other than these particular references, there are no other instances in books 4 and 5 of the *De aspectibus* where specific loci in the *Conics* can be pinpointed with certainty.[54] There is no question, however, that Alhacen's narrative account of conical and cylindrical mirrors in book 4, chapter 5 of the *De aspectibus* is based on a thorough understanding of conic sections and their focal properties. This is hardly surprising, given that Alhacen had adequate mastery of the *Conics* to essay a reconstruction of the missing eighth book of that treatise.[55]

Euclid's *Elements* is by far the most widely and consistently applied source for Alhacen's analysis of reflection in book 4 and, more to the point, book 5. This claim is supported by the fact that, despite Alhacen's almost unvarying failure to cite specific loci, it is easy to tie particular steps in his constructions and proofs to particular theorems in the *Elements,* a process made even easier by Risner's explanatory interpolations.[56] Practically speaking, however, it would be pointless to tie every step in Alhacen's constructions and proofs to its relevant propositional source in the *Elements.* It makes virtually no sense, for instance, to assume that every time Alhacen mandated erecting a line perpendicular to another at a given point on it he was consciously drawing on *Elements,* I.11. Nor is it likely that he expected his readers to have explicit recourse to that proposition in order to grasp the instruction and its full meaning. After all, the required construction is conceptual, not practical.

Barring such relatively insignificant links, Alhacen's mathematical reasoning at various points throughout book 5 of the *De aspectibus* is based on—or at least implicitly supported by—some 42 Euclidean propositions, virtually all of them distributed among the first six books of the *Elements.* Or, to put it in slightly different terms, without tying various logical steps in Alhacen's reflection-analysis to these particular propositions, it is extremely difficult, if not impossible, to fully comprehend the mathematical reasoning upon which that analysis is based. Take as an example Alhacen's construction for proposition 25, as illustrated by figure 5.2.25, p. 239. In the course of this construction, Alhacen concludes that, since angle BDG of triangle BDG is bisected by DQ, it follows that BQ:QG = BD:DG. Far from evident in its own right, this conclusion become evident only in the light of *Elements*, VI.3, where Euclid demonstrates that, if an angle of a triangle is bisected by a line that intersects its base, the ratio of the segments cut from the base by that bisector is the same as the ratio of the other two sides of the triangle. Like this particular proposition, each of the remaining 41 just mentioned clarifies specific steps taken by Alhacen when the logic underlying those steps is not immediately apparent.

Although the 42 Euclidean source-propositions isolated according to this criterion all fall within the first six books of the *Elements*, their distribution among those books is highly irregular. Only one proposition in book IV (IV.5) figures in Alhacen's analysis, and it occurs only once. Likewise, propositions from book II of the *Elements* rarely crop up in the course of book 5, and the ones that do are limited to II.1-II.3. Book I of the *Elements* is represented by nine propositions, four of which (I.4, I.19, I.26, and I.32) recur with moderate frequency in Alhacen's analysis—I.26 and I.32 six times, I.4 five times, and I.19 four times. The lion's share of propositions—29 in all—is thus distributed among books III (12), V (8), and VI (9). Within book III, the most frequently recurring propositions are III.22 (nine times) and III.31 (six times), and within book VI, propositions VI.3 and VI.4 recur most frequently by far, the first showing up sixteen times and the second twenty-one times. Within book V, proposition V.22 recurs six times, V.16 and V.18 four times, and IV.17 three times. None of the rest crops up more than twice.[57]

As far as total use is concerned, the propositions in book V, which deal generally with ratios and proportionality, and book VI, which deal with the proportionality of areas (as applied to triangles in particular), figure somewhat more prominently than those from the preceding four books. The reason for this prominence is the extensive use to which Alhacen puts proportionality theory throughout his analysis. Indeed, proportionality theory lies at the very heart of Alhacen's most intricate and technically demanding constructions and demonstrations. From a purely Euclidean standpoint, Alhacen's reliance on compound ratios at certain points in his analysis is somewhat unusual, if not idiosyncratic, but there is ample precedent for that in Apollonius' *Conics*.[58] Otherwise, Alhacen's treatment of proportions and ratios is Euclidean to the core.

All told, then, it is possible to follow Alhacen's analysis in book 5 to the letter on the basis of a remarkably small fund of mathematical sources,

Euclid's *Elements* foremost among them. And the same holds for his optical sources, which reduce essentially to Ptolemy's *Optics*. One might therefore expect that, given the paucity of sources upon which he relied more or less explicitly, Alhacen's application of those few sources to his analysis of reflection would be commensurately limited. But, as should be eminently clear from previous discussion, the very opposite is true. Alhacen's account of reflection is extraordinarily complex and sophisticated in terms not only of structure and content but also of conception. In those regards, in fact, it far outstrips the sources upon which it was based. Rather than detracting from Alhacen's achievement in book 5, the fact that he drew on so few sources and yet managed to make so much of them is a testament to his logical acuity and rigor as well as to his powers of invention.

4. *The Reception of Alhacen's Reflection-Analysis in the Latin West*

The account of reflection in the fifth book of the *De aspectibus* represents a critical turning point for Alhacenian optics. Everything, or virtually everything, up to proposition 1 of that book (with the possible exception of chapter 5 of book 4) is readily understandable to an intelligent but mathematically untutored reader. Such is clearly not the case with the theorematic portion of book 5. Many of the conclusions drawn in that portion are difficult to grasp because they are so deeply buried in dense thickets of mathematical reasoning. And the same holds for the ulterior intent of several of the theorems, propositions 45 and 46 being prime examples.[59] In order to understand the optical content of book 5 in its full array of implications and ramifications, therefore, it is necessary to understand the underlying analysis as fully as possible. This simply cannot be done without a fairly solid grasp of the first six books of Euclid's *Elements* and a reasonably good understanding of the basic geometry of conic sections.

More to the point, without a fairly sure grounding in the Euclidean theory of proportions laid out in book 5 (and to some extent in book 6) of the *Elements*, even the most assiduous reader is likely to flounder quite early in the course of Alhacen's reflection-analysis. This point has a crucial bearing on the reception and use of the *De aspectibus* during the thirteenth and succeeding centuries. Because of its conceptual difficulty, the fifth book of the *Elements* was regarded by medieval students as the *pons asinorum* of Euclidean geometry. Its crossing was therefore considered to be an achievement. But to cross it was not necessarily to command it. Those who had truly mastered Euclidean proportionality theory thus represented a limited segment of the university-educated elite of medieval Europe. In addition, there is no firm evidence that Apollonius' *Conics* was available in Latin translation during the Middle Ages.[60] If it was, it had almost no impact on mathematical learning in the Latin West before the Renaissance. Likewise, there is little or nothing to suggest that a Latin version of Serenus' *On the Section of a Cylinder* was in circulation during the Middle Ages.

By the early thirteenth century, then, lack of appropriate texts and a somewhat superficial training in geometry left most scholastic readers unprepared to make real sense of Alhacen's mathematical analysis in book 5 of the *De aspectibus*. There was at least one exception, though. Thanks to Marshall Clagett we know that, in proposition IV.20 of his *De triangulis*, Jordanus Nemorarius appealed to proposition 20, lemma 2, of book 5 of the *De aspectibus* to support a step in his method for trisecting an angle.[61] Although the dating of this work is problematic, it was certainly written before 1260 and perhaps as early as the 1230s.[62] Whichever the case, Jordanus' use of the Alhacenian lemma indicates that the *De aspectibus* was actually being read by the mid-thirteenth century, not only for its optical content, but also for its mathematical content. It also suggests that by this time there was an adequate fund of mathematical knowledge—and therefore, presumably, of appropriate texts—to make that reading meaningful.[63]

A truly gifted mathematician, Jordanus was exceptional for his time, so his ability to mine the *De aspectibus* for its mathematical treasures tells us little about the general level of mathematical learning in thirteenth-century Europe. A better indicator is Witelo, whose *Perspectiva* of c. 1275 was so closely modeled on Alhacen's *De aspectibus* that the late-sixteenth-century scientific impressario Giambattista della Porta dubbed him "Alhazen's ape."[64] Born in Silesia, probably in Wroclaw, in the early-to-mid-1230s, Witelo apparently received his formative education in Poland before entering the University of Paris as an arts student in 1253. Master's degree in hand, Witelo seems to have returned home for a short while before enrolling for higher study, perhaps in canon law, at the University of Padua. There he remained from the very early to the very late 1260's, at which time he moved to the papal court at Viterbo. It was during his sojourn there that Witelo composed the *Perspectiva* under the mentorship of William of Moerbeke, who provided him both encouragement and practical help in the form of textual translations.[65]

Porta's flippant dismissal of Witelo as "Alhazen's ape" was both unfair and inaccurate. In fact, Witelo wrote the *Perspectiva* not simply to recapitulate Alhacen's optical analysis, but to make it readily accessible to contemporary readers. To that end he subdivided Alhacen's analysis into relatively short, digestible theorematic chunks, restructured it accordingly, and added elaboration or explanation whenever he saw fit. As part of this effort to render the *De aspectibus* accessible to contemporary scholars, Witelo devoted the first book of the *Perspectiva* to a set of 137 theorems (plus sixteen definitions and five postulates) that were clearly intended to provide the requisite mathematical foundations for the subsequent account of reflection and image-formation in books 5-9 of the *Perspectiva.*[66]

Some of the theorems among the set provided by Witelo in the opening book were undoubtedly included for the sake of rigor rather than for actual instructional purposes. The fact that two spheres can touch at only one point is virtually self-evident to anyone with a rudimentary understanding of spheres and their properties. But self-evidence is not proof, so Witelo offers a demonstration in I.76 (Unguru, 102). The same holds for the fact that two planes parallel to the same plane are parallel to each other, which Witelo demonstrates in I.24 (Unguru, 61).[67] Such demonstrations of "self-evident" points are in the minority though. Most of the theorems in book 1 show things that are not immediately obvious in order to lay the ground for their later application in the body of the *Perspectiva*. For instance, in propositions I.3-13 (Unguru, 49-55), Witelo demonstrates a number of points about proportions and the manipulation of their terms in order to explain such operations as alternation, composition, separation, and compounding of ratios, all of which play a crucial role in the subsequent analysis of reflection.[68] But propositions I.3-13 add little or nothing to what Euclid had already established in the fifth book of the *Elements.*[69] Why, then, did Witelo bother to include them? High among the reasons, I suggest, was a concern on Witelo's part that his readers might not be adequately schooled in proportionality theory to grasp the point of these propositions without help.

It is surely with this consideration in mind that Witelo undertook the analysis of cones and cylinders presented in propositions I.89-118 (Unguru, 110-136). The fundamental purpose of that analysis is twofold: first, to provide a framework for understanding how planes of reflection passing at various angles through cones and cylinders create particular sections within which reflection occurs in determinate ways, and second, to introduce the appropriate conditions for solving the six lemmas provided in propositions 19-24 of the *De aspectibus*. Forming the capstone of book 1 of the *Perspectiva,* those six lemmas are recapitulated in the final ten propositions: I.128 and I.130 (Unguru, 145-146 and 147-152) corresponding to proposition 19, lemma 1; I.133 (Unguru, 153-155) corresponding to proposition 20, lemma 2; I.134 (Unguru, 155-158) corresponding to proposition 21, lemma 3; I.135 (Unguru, 158-162) corresponding to proposition 22, lemma 4; I.136 (Unguru, 163-165) corresponding to proposition 23, lemma 5; and I.137 (Unguru, 165-167) corresponding to proposition 24, lemma 6.

Witelo's analysis of cones and cylinders in the second half of book 1 is especially interesting for the insights it provides into the level of Witelo's mathematical expertise and thus the state of mathematical development in his day. After all, the study of conic sections represented a major advance in post-Euclidean geometry and, as such, was regarded as a higher-order subdiscipline within mathematics well up into the Renaissance and beyond. The fact that Witelo was able to deal with cones (and their extension to cylinders) would seem to indicate a surprisingly high level of mathematical sophistication on his part. It also suggests that he was at least aware of, if not familiar with, Apollonius' *Conics* and Serenus' *On the Section of a Cylinder*. But an examination of the actual theorems included by Witelo in his analysis of cones and cylinders belies the first of these suppositions and therefore brings the second into question. Almost all of the proofs offered in the course of that analysis can be formulated on the basis of Euclid's *Elements* and a fairly rudimentary understanding of the generation of right cones from the rotation of isosceles triangles about the axis formed by the line bisecting the angle at the vertex and the generation of cylinders from the rotation of rectangles about one of their sides as axis.[70]

There is, however, at least one intriguing exception. In I.98 (Unguru, 115-116), Witelo explains the generation of the three principal conic sections descriptively according to the obliquity of the plane cutting the cone. He then denominates each of the resulting sections in three ways: according to whether they are generated by a right-angled, obtuse-angled, or acute-angled planar cut; according to whether that cut produces a parabola, hyperbola, or ellipse on the cone's surface; or according to whether the generated section conforms to "what the Arabs call" *mukefi*, increased *mukefi*, or diminished *mukefi*.[71] Three things are worth noting about these denominations. First, the characterization of conic sections according to the type of angle is pre-Apollonian, so its use by Witelo suggests a pre-Apollonian source (perhaps Archimedes) for at least some of his knowledge of conic sections. Second, because the terms parabola, hyperbola, and ellipse were coined by Apollonius in the *Conics*, Witelo's use of them suggests some familiarity either with the *Conics* itself or with a post-Apollonian source. And third, Witelo's adverting to the Arabic terminology (which reflects the Apollonian terminology) indicates that at least some of what he knew about conic sections came through Arab channels other than Alhacen, who refers to all conic sections generically as "sectiones piramidis."[72]

All of this suggests that Witelo actually knew Apollonius' *Conics*, or at least the section of book 1 dealing with the generation of the three principal conic sections (i.e., I.11-13 at a minimum) and the relevant analysis of hyperbolic sections in book 2 (i.e., I.4, I.8, and I.16 at a minimum). The problem is that, in terms of its actual application in book 1 of the *Perspectiva*, most of this knowledge, particularly as regards book 2 of the *Conics*, could have come from Alhacen rather than from the *Conics*.[73] In fact, the propositions in book 1 of the *Perspectiva* where the traces of that knowledge are especially prominent (i.e., I.128-133—Unguru, 145-155) are based entirely upon propositions 19-20 (lemmas 1 and 2) of the *De aspectibus*, and it is precisely here that Alhacen brings book 2 of the *Conics* to bear in his analysis.

That this may in fact have been the route, or at least the main route, by which Witelo became acquainted with Apollonius' analysis of hyperbolic sections finds support in certain peculiarities of Witelo's analysis. For one thing, *Conics*, II.8 is absolutely central to the construction of the figure (5.2.19c, p. 231) that accompanies proposition 19, lemma 2 (see esp. case 2, pp. 416-417). The key point in this construction is to establish that MO = LC, which follows from Apollonius' proof in II.8 that, if a line cuts a hyperbolic section at two points, and if it is extended on both sides to the asymptotes, the segments cut off on each side by the hyperbola and the respective asymptote will be equal. This proof is relatively straightforward, yet Witelo fails to include it among the propositions given in book 1 of the *Perspectiva*. Instead, in proposition I.129 he offers an extraordinarily maladroit description of what it demonstrates, concluding that the proof itself is not worth giving, "since it depends on many preliminary principles of [the *Conics*]" (Unguru, 147).

Witelo is correct. Proposition II.8 does depend on several preliminary principles, but for all practical purposes these can be reduced to the four points demonstrated in *Conics*, I.32, II.3, II.4, and II.7.[74] Hence, given his apparent familiarity with the *Conics*, Witelo's failure even to take a stab at recapitulating the proof in II.8 is puzzling. On the other hand, if proposition 19, lemma 2, of the *De aspectibus* was the sole, or even the main, conduit through which Witelo had access to *Conics*, II.8, and if, therefore, he learned *what* it demonstrated without learning *how* it was demonstrated, then that failure becomes far less puzzling.[75] None of this precludes the possibility that Witelo had a version of the *Conics* at hand in some form or another, but it does, I think, raise serious doubts about how well he understood the contents of that version.[76]

That Witelo was not particularly adept in the geometry of conic sections should not blind us to the fact that he knew enough to make basic sense of the two Alhacenian lemmas in which it figures prominently. Nor should it blind us to the likelihood that Witelo availed himself of textual resources beyond Alhacen's *De aspectibus* to acquaint himself at least to some extent with that geometry. Indeed, Witelo seems to have drawn on a surprising variety of mathematical sources in the course of book 1. Included among these, according to Unguru's count, may be Pappus' *Mathematical Collection*, Campanus' recension of the *Elements*, Theon's commentary on the *Elements*, Eutocius' commentary on Archimedes' *On the Sphere and the Cylinder*, Jordanus Nemorarius' *Geometry*, and Theodosius' *Sphaerics*.[77]

As Unguru rightly observes, Witelo was a mathematician of no great originality or genius. Many of his interpolations and elaborations (e.g., the "proof" of the parallel postulate) are at best misconceived and at worst logically insupportable.[78] But what Witelo lacked in innate talent he made up for in perseverance, and it was this trait combined with a more-or-less adequate fund of mathematical resources that enabled him to cope with Alhacen's reflection-analysis as well as he did. That, in a nutshell, is what makes Witelo so significant as a gauge of the intellectual temper of his times. The very fact that a mathematician of such limited talent could nonetheless make good sense of Alhacen's reflection-analysis indicates that, by the last quarter of the thirteenth century, mathematics had reached a high enough level of development in the Latin West to permit mathematicians of far lower caliber than Jordanus a generation or so earlier to read the *De aspectibus* from cover to cover with almost perfect comprehension. Such readers were doubtless thin on the ground, but there were enough to ensure the entrance of Alhacen's treatise into the scholastic mainstream by the close of the thirteenth century.

Like Newton's *Principia*, however, Alhacen's *De aspectibus* had more than one audience. On the one hand, there were those, like Witelo, who absorbed the work in all its technical, mathematical detail. The traces of this group can be seen in the copious annotations and corrections to be found in the surviving manuscripts of the *De aspectibus*. On the other hand, there were those (undoubtedly a significantly larger group) who were more interested in the gist than in the minute details of Alhacen's analysis. The two foremost examples of this latter group are Witelo's contemporaries, Roger Bacon and John Pecham.[79] Of the roughly 168 pages comprising the modern Latin edition of Bacon's *Perspectiva*, for example, only about seventeen are devoted to reflection. This amounts to around ten percent of the entire treatise, as opposed to something over fory-five percent of Alhacen's and Witelo's counterparts. Furthermore, in the course of his brief account of reflection, Bacon barely mentions the key points in book 5 of the *De aspectibus* concerning the number of possible reflections in the six curved mirrors, and he makes no mention whatever of the method for locating them.[80] But Bacon was less interested in providing a comprehensive analysis of reflection than in explaining image-distortion as clearly and briefly as possible on the basis of the cathetus-rule. Moreover, his interest in optics was spurred less by scientific than by theological concerns.[81] So it was not necessarily out of mathematical incompetence that Bacon avoided the technical issues of Alhacen's reflection-analysis. It was out of a desire to follow particular analytic imperatives that included those issues only peripherally.[82]

Pecham's case is somewhat different from Bacon's. For one thing, he devoted far more of his optical compendium (roughly thirty percent) to reflection than did Bacon. For another, as he put it himself, Pecham's primary goal in writing the *Perspectiva communis* was "to compress into concise summaries the teachings of perspective, which [in existing treatises] are presented with great obscurity. . . ."[83] Unlike Bacon, in short, Pecham focused on the scientific rather than the theological implications of optics. As a result, he confronted some of the technical issues that Bacon had sidestepped.

For instance, in his discussion of concave spherical mirrors in *Perspectiva communis*, II.45, II.46, and II.48 (Lindberg, *Pecham*, 197-205), Pecham establishes the conditions under which one, two, or four images can be seen in such mirrors when the center of sight and object-point lie on different diameters at equal distances from the center of curvature. These three theorems correspond to the three cases presented in proposition 5.37 of the *De aspectibus*. Accordingly, in proposition II.45 (cf. 5.37, case 1, pp. 454-455), Pecham shows that, when the center of sight and object-point lie outside the sphere of the mirror, only one reflection is possible. Not only is his proof essentially the same as Alhacen's, but the figure on which he bases it is virtually identical in form and lettering to the one used by Alhacen (cf. figure 5.2.37, p. 396). Pecham then goes on in proposition II.46 (cf. 5.37, case 2, pp. 455-456) to show that, if the center of sight and object-point both lie inside the sphere of the mirror, there can be as many as two reflections. In this case, Pecham's demonstration is such a rough distillation of Alhacen's as to be no demonstration at all. Finally, in proposition II.48 (cf. 5.37, case 3, pp. 456-458), Pecham shows that there can be as many as four reflections, the proof depending on a figure that is clearly abstracted from the one used by Alhacen (cf. figure 5.2.37c, p. 259).

All three of these propositions are predicated on having the center of sight and object-point equidistant from the center of curvature. But what about the more complex situation in which the two points lie different distances from that center? Pecham addresses this case obliquely in proposition II.47, where the ostensible purpose, as stated in the enunciation, is to show that three images are possible in concave spherical mirrors. Entirely descriptive, the body of this brief proposition appears to be based somewhat loosely on proposition 5.49 of the *De aspectibus*, where Alhacen offers a summary of results from propositions 43-48. Pecham's exposition is worth quoting in full:

> Take two points on different diameters, one inside the circle and the other on the circumference of the circle or outside. And if a [second] circle is drawn including these two points along with the center of the mirror, then if the [second] circle intersects the circle of the mirror in one place, reflection will take place from one arc only; if intersection occurs in two places, reflection can take place from one point of the arc interposed between the diameters or from two or from three or sometimes from four.[84]

What Pecham seems to have in mind for the first situation is illustrated in figure 5.2.49d, p. 278. In this case, if A and B′ represent the relevant points, with B′ lying on the mirror's circumference and A inside the mirror so that at point E circle AGB intersects arc CB′ (i.e., "the circle of the mirror"), then reflection can only occur from segment EB′ of arc CB′. Thus, to use Pecham's own words, "reflection will take place from one arc only" (i.e., EB′) and not from both CE and EB′. In addition, as we established earlier on pp. xlii-xliv, there will be two reflections in this case, both from arc EB′. One reflection will also occur from arc KD, bringing the total to three.

On the other hand, if "intersection occurs in two places" on arc CH, then both A and B must lie inside the circle of the mirror, as represented in figure 5.2.49c, p. 277.[85] Under these conditions, as we established on p. xliv above, reflection can occur from segment FE as well as from segment CE or HF of arc CH, depending on whether A or B is closer to centerpoint G. Moreover, within segment FE there can be two reflections, so if we add the single reflection from segment CE or HF, there can be as many as three reflections from arc CH. Add to those the reflection from arc KD, and the number rises to four.

However, as we noted earlier on pp. xli-xlii above, unless point F, where circle AGB cuts arc CH, lies to the right of the midpoint of CH, there can be no reflection from segment FE, in which case reflection is only possible from segment CE or FH. With the single reflection from KD added, there will thus be a total of two reflections. Altogether, then, counting the case in which both A and B lie outside the mirror (as in II.46), "reflection can take place from one point . . . or from two or from three or sometimes from four."

Suffice to say, this interpretation of Pecham's intent in proposition II.47 is forced to the point of torture, and its forcing depends upon considerable knowledge of what Alhacen establishes in propositions 43-46 of the *De aspectibus*. But, as far as the actual mathematical demonstrations are concerned, those propositions exemplify "the great obscurity" of optical analysis that Pecham was at such pains to avoid, which is doubtless why he ignored them.[86] Hence, in his effort to streamline the account of reflection in concave spherical mirrors, Pecham chose as his paradigm the simpler case, in which A and B are equidistant from center of curvature G. In the interest of comprehensiveness, however, he was compelled to give at least lip service to the more complex case, in which A and B lie different distances from G. Both simplistic and confused, the resulting account misses the real import of that case, and, as a result, Pecham failed to grasp the central point of Alhacen's reflection-analysis, which was to formulate a general method for locating the point or points of reflection in the six curved mirrors chosen for analysis in books 4-6 of the *De aspectibus*.

Hence, while Pecham's account of reflection is somewhat less superficial than Bacon's, it is only marginally so in comparison to the one provided in Alhacen's *De aspectibus* and Witelo's *Perspectiva*. Yet it was precisely because of their superficiality (which passed for simplicity) that Bacon's and Pecham's optical treatises enjoyed considerably wider diffusion in scholastic circles than those of Alhacen or Witelo. Indeed, as remarked earlier, Pecham's *Perspectiva communis* became a standard text for the teaching of optics in the Middle Age precisely because of its apparent simplicity and comprehensiveness. Accordingly, many medieval scholars were introduced to Alhacenian optics through what amounted to popularizations in the works of Pecham and Bacon, who focused on the basic conclusions of Alhacen's reflection-analysis with virtually no regard for the deep mathematical structure of analysis. There were thus two distinct avenues for the dissemination of Alhacenian, or Perspectivist, optics in the Middle Ages. The one mapped out by Alhacen and Witelo takes the reader through the science of optics in all its analytic detail and complexity. The other mapped out by Bacon and Pecham offers a short cut that intentionally bypasses such analytic detail and complexity.

David Lindberg has done yeoman service in showing how these four treatises shaped the Perspectivist optical tradition that lasted well into the seventeenth century.[87] It is largely through his efforts, for instance, that we have a fairly complete census of the manuscripts and subsequent printed editions of all four treatises.[88] This census, of course, tells us a great deal about the reception and circulation of those works throughout the Middle Ages and Renaissance. We also know a fair amount about their use during that period, either from direct citations or from the context within which they crop up. Hence, we have a number of ways to trace the influence of these works, some of them direct, some indirect.

The problem is that, although we have ample evidence *that* these works were read and used, and are even able to pinpoint their actual use, we know very little about *how* they were read and used. This problem is especially acute in the case of Alhacen's *De aspectibus* and its proxy, Witelo's *Perspectiva*. Both were cited copiously during the Middle Ages and Renaissance, and both were drawn upon in a variety of ways without explicit reference.[89] But citing a work or drawing on it selectively does not require mastery of it. Nor for that matter does it require more than passing familiarity based on a knowledge of certain key points in the text. It may well be the case, then, that, as they were picked over during the decades after their initial appearance, Alhacen's *De aspectibus* and Witelo's *Perspectiva* dwindled in scope to a relatively small set of canonical loci in the same way as happened to so many other authoritative texts over the period.[90]

This being so, we would expect to find few if any traces of the technical details of Alhacen's reflection-analysis in later works inspired by the *De aspectibus* or Witelo's *Perspectiva*. This expectation is, in fact, met. There is no evidence, apart from the manuscripts themselves of the two treatises, that the hard, mathematical core of Alhacen's analysis was studied in earnest before the early fifteenth century. As far as we know, the earliest witness to such study is an Oxford statute of 1431 that allowed Alhacen's or Witelo's treatises to be substituted for Euclid's *Elements* in the teaching of geometry at the undergraduate level.[91] Otherwise, although there are vague indications that the two works were used for university instruction during the fourteenth and fifteenth centuries, we have no clear idea of how they were employed for that purpose. In view of the optical commentaries and epitomes that survive from this period, though, it seems clear that the two works were overwhelmingly favored for their optical rather than their mathematical content.[92]

That this latter content is all but missing in derivative studies of the period suggests that very few scholastic thinkers appreciated the complexity of Alhacen's reflection-analysis, particularly as it pertains to the most original and ingenious part of that analysis: the method for locating points of reflection. This suggestion is reinforced by the fact that, despite their relative lack of theoretical and analytic sophistication, Euclid's *Catoptrics* and Ptolemy's *Optics* were even more widely studied than Alhacen's *De aspectibus* and Witelo's *Perspectiva* during the Middle Ages and Renaissance. Indeed, interest in these two works, Ptolemy's *Optics* in particular, appears to have burgeoned during the Renaissance.[93] Why? For one thing, the visual-ray theory of Euclid and Ptolemy is mathematically equivalent to the light-ray theory of Perspectivist optics, so, as far as the basic geometry of reflection is concerned, there is nothing to choose between the two. For another, at a practical level, Euclid's and Ptolemy's works provide a perfectly adequate account of reflection. To understand the essential rules of reflection and image-location does not require the sort of intricate mathematical analysis provided in the Alhacenian account.[94]

That the most mathematically sophisticated and interesting portion of Alhacen's reflection-analysis failed on the whole to capture the interest of Renaissance scholars was not due to a corresponding lack of interest in mathematics during that period. On the contrary, from the mid-fifteenth century on, the study of mathematics underwent explosive development through the recovery of Greek texts and their eventual translation and publication. For instance, the Latin version of Euclid's *Elements,* based on Campanus' translation from the Arabic, appeared in print for the first time in 1482 and was followed in 1505 by a translation from the Greek. The first Greek edition was published not long after, in 1533.[95] Four years later, in 1537, books 1-4 of Apollonius' *Conics* appeared in print for the first time in Latin translation, and in 1566 Federico Commandino's magisterial Latin version was published along with Eutocius' commentary and Pappus' *lemmata.* Serenus' *On the Section of a Cylinder* was also included in that edition.[96] Along with these key mathematical sources, moreover, Witelo's *Perspectiva* saw the first light of publication in 1535, the resulting edition having been popular enough to warrant reprinting sixteen years later.[97]

Within the context of these newly published sources, Friedrich Risner's tandem edition of Alhacen's *De aspectibus* and Witelo's *Perspectiva,* which appeared in 1572 under the title *Opticae Thesaurus,* is doubly significant. Friend and protegé of Petrus Ramus, who was one of the foremost mathematicians of the sixteenth century, Risner was inspired by his mentor to undertake the edition in the first place. Ramus, in fact, provided him with the necessary manuscripts upon which to base the edition.[98] A notable mathematician in his own right, Risner was considered eminent enough in that discipline to have been earmarked by Ramus for a chair in mathematics that he had endowed at the University of Paris.

Risner's quality as a mathematician and his familiarity with the sources just discussed come clear throughout the tandem edition of Alhacen's and Witelo's treatises. In short, the *Opticae Thesaurus* represents the cutting edge in mathematics for its day. On the other hand, Risner was no less an optician than a mathematician and, as such, collaborated with Ramus in the production of an optical tract that was eventually published well after both men had died.[99] The *Opticae Thesaurus* thus represents the cutting edge in optics for its day. In its double capacity as a mathematical and an optical text, therefore, Risner's edition of Alhacen and Witelo offered the best of both disciplines for several generations to come.

The list of those who were introduced to Alhacen and Witelo through Risner's edition includes a remarkable array of scientific luminaries of the later sixteenth and seventeenth centuries. Among these, of course, Johan

Kepler, René Descartes, and Christiaan Huygens loom large, but many equal or lesser lights, such as Willibrord Snel and Thomas Hariot, carry that list to the very end of the seventeenth century.[100] But Alhacen's model of visual imaging had been pretty thoroughly discredited by the mid-seventeenth century on the basis of Kepler's searching critique in the *Ad Vitellionem paralipomena* of 1604. Popularized by the likes of René Descartes, Kepler's alternative model of retinal imaging, which struck at the very heart of Alhacen's account of visual imaging, become more or less canonical by mid-century.[101] Moreover, as part of his critique, Kepler raised serious doubts about the applicability of the cathetus-rule to the determination of image-location in convex and concave mirrors.[102] These doubts, which called into question Alhacen's entire account of specular and refractive image-formation, were taken seriously by subsequent opticians as they attempted to make better sense of mirrors and lenses. Consequently, the analysis of reflection and refraction was approached in new, more sophisticated (and complex) ways, as witness the series of optical lectures Isaac Barrow gave as Lucasian Professor of mathematics at Cambridge in 1667 and published two years later under the title *Lectiones Opticae*.[103]

The last effective traces of Alhacen's reflection-analysis are to be found in the discussion of "Alhazen's Problem," which centered on a general solution conveyed by Christiaan Huygens to Henry Oldenbourg, secretary of the Royal Society, in June of 1669.[104] Huygens' solution involved the generation of opposite branches of a hyperbola within particular asymptotes constructed according to the placement of the center of sight and object-point with respect to the given spherical mirror. The four points where the two branches cut the circle of the mirror constitute the points of reflection.[105] A little over a year later, in July, 1670, René François de Sluse of Liège wrote a letter to Oldenbourg in which he claimed to have been inspired to solve the same problem in the course of reading Isaac Barrow's *Lectiones Opticae* of 1669.[106] Indefatigable correspondent that he was, Oldenbourg shared Hugyens' solution with a number of scholars, including Sluse, to whom he sent a copy in September, 1670.[107] Sluse then responded in November of that year, informing Oldenbourg that Huygens had "followed just the same line of analysis" as he had, although, "as two means of effecting [the solution] present themselves as a result, each by a hyperbola upon asymptotes, he [i.e., Huygens] chose one as the easier to use and I the other."[108] Thus began a spate of correspondence between the two through Oldenbourg's mediation that lasted until 1673, when Oldenbourg saw fit to publish excerpts from it in the *Philosophical Transactions*.[109] The resulting controversy, a mild one at that, focused on which of the two solutions was the easier and more "natural," but this very controversy was predicated on the understanding that, being difficult and unnatural because of its inelegance, Alhacen's method was fatally flawed.[110] In the end, therefore, Alhacen's last significant bequest to the Latin West was a problem to which his solution was rendered, if not wrong, at least superfluous. Henceforth, Alhacen would be little more than a footnote in the history of optics and mathematics, a history in which he had nonetheless played a crucial formative role. After that (to rephrase Sarton on Apollonius' fate at the end of the seventeenth century), the Alhacenian tradition was lost in the new optics of the time, like a river in the ocean.[111]

5. *Conclusion*

Viewed from a comfortable post-Keplerian perch, books 4 and 5 of Alhacen's *De aspectibus* appear to represent little more than wasted effort. Most of the basic principles established in these two books are so obvious now as to be axiomatic. That light reflects at equal angles within a plane normal to the surface of reflection needs no proof today, nor does the fact that the equal-angles law applies to all light-radiation, regardless of source or type. Light, after all, is light—or, rather, a composite of discrete colors. Worse, as Kepler showed in his critique of Alhacen's reflection-analysis, Alhacen generalized the cathetus-rule of image-location beyond its legitimate scope to include convex and concave mirrors and, by extension, refracting media. Even Alhacen's solution to his eponymous problem, though mathematically correct, is so inelegant by today's standards as to be all but worthless except as a quaint historical artifact.

Within the context of his time, however, Alhacen's analysis of reflection in books 4 and 5 was a *tour de force* in terms not only of empirical and mathematical rigor, but also of originality. This point becomes eminently clear when we compare Alhacen's analysis to that of its primary source in Ptolemy's *Optics*. Having substituted light-rays for visual rays as the basis for analysis, Alhacen was faced with certain issues at the empirical level that Ptolemy never had to confront. Consequently, Alhacen was forced to take into account all forms of visible radiation, from direct sunlight to the shining of brightly illuminated color. He was also forced to re-establish the principles of reflection for such radiation. Much of what today strikes us as wasted effort in his account thus derives from the need to demonstrate those principles for all possible forms of radiation from reflecting surfaces of all possible shapes.

The substitution of light-rays for visual rays also forced Alhacen to recast Ptolemy's reflection-experiment. Part of that recasting involved an effort at generalizing the experiment to include convex spherical and conical mirrors, as well as concave spherical and conical mirrors. But the brunt of that recasting involved the creation of an experimental apparatus within which the actual passage of light to and from the selected mirrors could be rendered visible—hence the hollow ring with its inserted register and selection of holes through which the light could be directed to the mirrors. That this apparatus was probably never constructed as described (and, therefore, that the various tests based on that apparatus were probably never conducted as described) does not detract from Alhacen's ingenuity in devising the experiment as he did, knowing full well that such an instrumental setup *would* have yielded the appropriate results.

As with his attempts to validate the principles of reflection empirically, so with his application of those principles to the mathematical analysis of reflection, Alhacen far outstripped Ptolemy in terms of rigor, scope, and ingenuity. As we have seen, Alhacen bent Ptolemy's analysis to the specific end of locating the point or points of reflection in the seven mirrors chosen for analysis. To that end, particularly as it applies to concave spherical mirrors, Alhacen went much further than Ptolemy in specifying the conditions under which the number of possible points varies according to the placement of eye and object within the mirror. He also went further in restricting those possibilities according to the reflected angle and, on that basis, demonstrated conclusively that there can be no more than four reflections from a concave spherical mirror for any given center of sight and object-point. This restriction is implicit in Ptolemy's analysis, but it is never rigorously demonstrated.

The most salient difference between Alhacen's and Ptolemy's reflection-analysis, of course, lies in Alhacen's having tied a major loose end left dangling by Ptolemy: defining with absolute mathematical precision the relevant point or points of reflection on convex and concave spherical mirrors. As we have seen, this is a hideously complex problem, requiring four separate solutions, although in fact those solutions break down into two closely related pairs. But it is crucial to understand that, having inherited the problem from Ptolemy, Alhacen also inherited the analytic *structure* of that problem from him. Within the confines of that structure, Alhacen's method for defining the points of reflection is truly remarkable for its ingenuity, originality, and rigor. It is grossly anachronistic, therefore, to tax him with not having resolved the problem according to some other analytic structure, such as that within which Huygens and Sluse operated toward the end of the seventeenth century. It is also grossly anachronistic to tax Alhacen with failing to fully grasp the implications of his solution, which eventually led to its subsequent streamlining. Judged, therefore, according to the conceptual and analytic tools that were effectively available to Alhacen, his reflection-analysis stands as a landmark not only in the history of mathematical optics but also in the history of science overall.

NOTES

[1]For a brief discussion of this cone and its implications for visual perception, see A. Mark Smith, *Alhacen's Theory of Visual Perception*, Transactions of the American Philosophical Society, 91.4 and 5 (Philadelphia: American Philosophical Society, 2001).

[2]As Alhacen observes later on in book 4, pp. 320-321 below, points have no dimensions and are therefore invisible. In order to be seen, light and color must radiate from physical spots on the surfaces of luminous or illuminated bodies. Those surfaces, in turn, are seen according to the light and color radiated from all such spots on them, and it is through the perception of their surfaces that we perceive the bodies themselves. Nonetheless, for the sake of analytic convenience, we can treat those spots as if they were points.

[3]I. L. Heiberg, ed., *Euclidis Opera Omnia*, vol. 7 (Leipzig: Teubner, 1985), 286-289; for an English translation, see A. Mark Smith, *Ptolemy and the Foundations of Ancient Mathematical Optics*, Transactions of the American Philosophical Society, 89.3 (Philadelphia: American Philosophical Society, 2001), 80. This "proof," which is based on visual rays rather than light-rays, depends on presupposing that normal EA dropped from center of sight A to the mirror (as represented in the top diagram of figure 2, p. 522) is to ED as normal BG dropped from object-point B to the mirror is to GD (i.e., AE:ED = BG:GD).

[4]For this demonstration, see Smith, *Ptolemy and the Foundations*, 80-81. Far more ingenious than Euclid's, Hero's proof, which is also based on visual rays, assumes that the distance the visual ray travels from the center of sight to the object-point via the point of reflection will be the shortest possible. Accordingly, R_3 will be the point at which the lines of incidence and reflection will add up to the least amount among all possible combinations according to other points on the mirror. From this it necessarily follows that the angles of incidence and reflection will be equal. A major shortcoming of this proof is that it does not work for concave mirrors.

[5]See *Optics* III, 3-5, in A. Mark Smith, *Ptolemy's Theory of Visual Perception*, Transactions of the American Philosophical Society, 86.2 (Philadelphia: American Philosophical Society, 1996), 131-132. Ptolemy also applies the point-analysis described above to image-formation for object-surfaces in plane and curved mirrors.

[6]See ibid., 134-136; see also the account on pp. lxvii-lxviii above. Although Ptolemy does not specify that the two curved mirrors be cylindrical, it is clear from the structure of the experiment that the curved mirrors used were in fact cylindrical rather than spherical or conical.

[7]For some discussion of Alhacen's reliance on Ptolemy's *Optics* as a source, see A. Mark Smith, "Alhazen's Debt to Ptolemy's *Optics*," in T. H. Levere and W. R. Shea, eds., *Nature, Experiment, and the Sciences* (Dordrecht: Kluwer, 1990), 147-164. See also the discussion on pp. lxvii-lxxiv above.

[8]According to Alhacen's analysis, primary light is the light inherent in luminous sources, whereas secondary, or accidental, light is the light cast on opaque objects from those sources. However, the light in objects illuminated by luminous sources becomes primary when that light shines on other objects to create secondary light in them. See Smith, *Alhacen's Theory*, liii-liv.

[9]A digit is approximately 1.9 cm. or 3/4 in. This measure is confirmed by implication in book 3 of the *De aspectibus*, where the distance between the pupils of the eyes is assumed to be four digits; see Smith, *Alhacen's Theory*, 573-574.

[10]As a measure, a full grain of barley is around .42 cm. or 1/6 in, so half a grain of barley is about .21 cm. Accordingly, the circle formed here on the ring's inner wall is somewhat less than 4 cm. above the ring's base.

[11]Alhacen gives no specification for the thickness of this block, but later construction makes it clear that it must be thicker than one digit.

[12]As will become clear later on, the virtue of these particular shapes is that planar cuts through them will yield points (if through the vertex), straight lines, circles, or the three conic sections: i.e., ellipses, parabolas, or hyperbolas. These sections can be mathematically analyzed on the basis of Euclid's *Elements* or Apollonius' *Conics*, both of which Alhacen was thoroughly familiar with (see pp. lxxiv-lxxvii above).

[13]Again, Alhacen specifies no particular thickness, but it makes sense that the mirror be as thin as possible for convenience's sake.

[14]Yet again, Alhacen specifies no particular thickness, but it is clear from subsequent use that these panels must be less than two digits thick yet thick enough to stand perfectly upright without wobbling.

[15]This, of course, follows from the fact that the top of the register has been inserted into its notch so that its face lies two digits minus half a grain of barley above the base of the ring. Since the panel has been stood on the floor of the square cavity in the block at the ring's base, and since that cavity is precisely one digit deep, the midpoint of each panel will lie precisely three digits above the panel's base, where the midpoints of the mirrors lie. Thus, the centerpoint of each mirror will lie half a grain of barley higher than the face of the register.

[16]See notes 49 and 50, pp. 354-355, for an explanation of this point.

[17]See note 50, pp. 354-355, for the derivation of this figure.

[18]See notes 60 and 61, pp. 356-357, for a diagrammatic account of the points just made about the parallels and perpendiculars contained by the experimental apparatus when it is properly set up.

[19]*Ibn al-Haytham's Optics: A Study of the Origins of Experimental Science* (Minneapolis: Bibliotheca Islamica, 1977).

[20]See, however, notes 28 and 29, pp. 351-352.

[21]Although there is evidence for crucible smelting of iron in the Arab world of Alhacen's day, this process is a far cry from the sort of sophisticated casting that would have been needed to produce the mirror-forms Alhacen describes; see T. Rehren and O. Papakhristu, "Cutting-Edge Technology—The Ferghana Process of medieval crucible steel smelting," *Metalla* 7 (2000): 55-69.

[22]Bronze technology was advanced enough, even in Ptolemy's day, that he was able to produce satisfactory plane and concave mirrors for his reflection-experiments.

Until the Renaissance the best mirrors were formed of bronze or silver, both being malleable and relatively easy to polish. Nevertheless, the few extant mirrors we have from the period between Antiquity and the Renaissance are of fairly poor reflective quality, and even during the Renaissance, when glass mirrors with metal backing become increasingly common, the results were far from perfect because of imperfections in the formation of the glass. Problems with the quality of glass extended to the formation of lenses as well, and they became especially acute in the late sixteenth and seventeenth centuries when efforts to develop and perfect telescopy and microscopy became a central concern of opticians. For an account of pre-modern mirrors and their low quality, see Sara J. Schechner, "Between Knowing and Doing: Mirrors and their Imperfections in the Renaissance," *Early Science and Medicine* 10 (2005): 137-162. For an account of the development and use of lenses during the Middle Ages and Renaissance, see Vincent Ilardi, *Renaissance Vision from Spectacles to Telescopes*, forthcoming in *Memoirs of the American Philosophical Society*, 2006.

[23]That Alhacen was fully aware of this phenomenon is clear from his having written treatises on concave spherical and parabolic burning mirrors; see items 1 and 2 on p. xvii in Smith, *Alhacen's Theory*. Alhacen's treatise on parabolic burning mirrors was in fact translated into Latin and disseminated fairly widely in that form; for a critical edition of this version, along with a German translation, see I. L. Heiberg and E. Wiedemann, "Ibn al-Haitams Schrift über parabolische Hohlspiegel," *Bibliotheca Mathematica*, ser. 3, vol. 10 (1910): 201-237.

[24]Presumably, this scattered and absorbed light constitutes secondary light, but the model of dispersion and absorption here seems inconsistent with Alhacen's earlier account of how secondary light mixes with the color on the surfaces of things; see Smith, *Alhacen's Theory*, liv-lv.

[25]That the reflectivity of mirrors is due to something like physical hardness, but not to physical hardness itself, is clear from the fact that light reflects intensely from water (4.3.99, p. 321).

[26]If, however, the line of sight does not coincide with the the axis but lies above the cone and enters it through its vertex, the plane containing that line of sight and the axis will form two corresponding lines of longitude on opposite sides of the mirror's surface (4.5.34, p. 338).

[27]In this case, of course, if the line of sight coincides with the axis of the cone, all planes passing through the viewpoint along that axis will form lines of longitude on the exposed portion of the reflecting surface. From any other viewpoint, though, only one plane of reflection will form a line of longitude on that surface, as is the case with convex conical mirrors.

[28]See p. xliii.

[29]The same test can be applied to any point on arc KL between X and where circle AGB intersects KL to the right of X. Take point D in figure 16, p. 531, as an example, and let it be represented in figure 16e, p. 533, with A and B posed on their respective normals as before within the same circle KLHF. Draw circle ADB through the relevant points, connect AB, and bisect it at point M with diameter PMN so that arc AP = arc BP. When normal DG is extended, it intersects PN at point P on circle ABD, leaving

angle of incidence BDGP and angle of reflection ADGP subtended, respectively, by equal arcs BP and AP. Those angles are therefore equal, so D passes the test as a legitimate point of reflection.

[30]See esp. 5, 2.457-2.461, p. 470.

[31]As should be clear from this analysis, the two points of reflection that yield angles greater than LGH lie on the same side of arc KL as the point, whether object-point B or center of sight A, that is closer to center of curvature G. By the same token, the point of reflection that yields an angle less than LGH lies on the same side of arc KL as the other point on the normal that lies farther away from the center of curvature. Thus, if A and B were to switch places in figure 16, p. 531, the two points of reflection D and D′ that yield angles ADB and AD′B greater than LGH will lie on the same side as the center of sight and thus on the side opposite the object-point.

[32]"Dato speculo sphaerico convexo aut cavo, datisque punto visus et puncto rei vise, invenire in superficie speculi punctum reflexionis," *Oeuvres complètes de Christiaan Huygens*, vol. 20 (La Haye: M. Nijhoff, 1940), 265.

[33]This determination is made in proposition 38, pp. 458-459, which is analyzed on pp. lx-lxii.

[34]Case two of this determination, as analyzed on p. liv, is an apparent exception in that points A and B lie in the plane that passes perpendicular to axis XY through vertex-point X, which serves as a hypothetical point from which the form of A would reflect to B. But this exception is less exceptional than it may appear at first glance, because circles and points are both degenerate conic sections.

[35]For the actual demonstration of this point, see proposition 36, case 2, paragraph 2.344, p. 453.

[36]This, of course, is tantamount to forming angle D′T′N′ equal to half of angle BGA, as was the case in the construction for proposition 25, pp. 427-432.

[37]This same method can be applied to finding D when normals AG and BG are equal. Thus, as illustrated in figure 20m, p. 546, we start by cutting line D′M′ at point N′, which coincides with Q′, such that M′N′ (i.e., M′Q′):N′D′ (Q′D′) = BG:AG. At point D′ we form angle N′D′T′ (i.e., Q′D′G′) equal to half of angle LGH (leaving angle D′T′N, which = D′G′Q′, equal to half of angle BGA), and then from point B′ we pass a line through D′G′ to point Q′ so that B′G′:D′G′ = BG:DG (i.e., the radius of the circle). We then form angle BGD in the circle of the mirror equal to angle B′G′D′, and D will be the point of reflection. Accordingly, it is clear that the case in which normals BG and AG are equal is simply a special, limiting case that can be subsumed under the more general case in which normals AG and BG are to one another in any ratio we please.

[38]This same method applies in the case where normals AG and BG are equal, as illustrated in figure 20r, p. 548. Take some line D_2′M′, and cut it at point A′ such that A′M′:A′D_2′ = BG:AG. Since BG = AG, then A′M′ = A′D_2′, so Q′ and N′ in the previous figure will both coincide with A′. At D_2′ form angle M′D_2′T′ equal to half of angle LGH, and from A′ (i.e., Q′ in the previous figure) drop line A′G′ to line D_2′T′ such that A′G′:D_2′G′ = AG:D_2G (i.e., the radius of the circle). At point G in the circle form angle D_2GA equal to angle D_2′G′A′. D_2 will therefore be a point of reflection. The second point of reflection between D_1 and K can be found by reversing the process and form-

ing the relevant figure on point B rather than A. Thus, as in the previous analysis, this is simply a special, limiting case that can be subsumed under the more general case in which normals AG and BG are to one another in any ratio we please.

[39]This is equivalent to the stipulation that, when AG = BG, as in figure 20d, p. 544, angles D_1AG and D_1BG must both be acute if reflection is to occur from three points on arc KL. Otherwise, A and B will be too close to the center of curvature for circle ABG to intersect that arc, and, as we have seen, the two points where circle ABG cuts arc KL are reflection-points. The case in which AG ≠ BG is a bit more complicated, because A and B can be posed in such a way that, although circle ABG does intersect arc KL on both sides of midpoint X, there can only be one reflection from arc KL, and it will occur from some point D according to which reflected angle BDA < angle LGH. Take the case represented in figure 20n, p. 547, where circle ABG intersects arc KL on both sides of X, and yet A is close enough to G that angle XAG is obtuse rather than acute. According to our stipulation, there can be no reflection other than that from point D. If such reflection were possible, then, as we have already established in the previous section (pp. xli-xliii), the reflected angle would have to be greater than LGH, and the reflection would have to occur to the right of point X. That such reflection is impossible in this case can be established on the basis of the test applied in the previous section. Let the conditions in figure 20n be recapitulated in figure 20s, p. 549, so that A and B are equivalently disposed within circle KLHF, circle ABG passes through the same points on arc KL, and angle XAG is obtuse rather than acute. The issue in this case is whether there can be any reflection from segment EX of arc KL. Assume that point D in the top diagram of figure 20s is such a point. Applying the test, we pass circle ABD through D, connect AB, bisect it at M, and draw diameter PMN through it so that arc AB is bisected at P. Since the extension of normal DG does not intersect circle ABD at point P, it is clear that angle of incidence BDG and angle of reflection ADG are subtended by unequal arcs, so they are unequal. Likewise, if we apply the test to point E in the middle diagram, or to point X in the bottom diagram, it is clear that those two points cannot be points of reflection for the same reason: i.e., that normals EG and XG do not intersect their respective circles ABE and ABX at point P, so they cut unequal arcs from APB, leaving the angles of incidence and reflection unequal. There is thus no point between X and E at which the normal will intersect the circle at P and therefore no point at which the angles of incidence and reflection will be equal.

[40]Thus, as illustrated in figure 20v, p. 550, lines B′G′ and B″G″ can be projected through Q′ to form equiproportional segments: i.e., B′G′:G′D′ = B″G″:G″D′. Clearly, however, the resulting angles B′G′D′ and B″G″D′ are not equal. It should be noted that only two such lines can be projected through Q′ so as to fulfill the requisite conditions, so only two angles can be appropriately formed, from which it follows that only two points of reflection are possible.

[41]The construction in this case is given in figure 20y, p. 552, where points Q and N coincide, because MQ and QD are equal, given the equality of normals BG and AG. Thus, line BQG is perpendicular to DM, so the angle it forms with DG is simply half of angle BGA formed by the normals. This, of course, represents the limiting case for the more general situation in which AG and BG can be in any ratio we please.

[42]See Smith, *Alhacen's Theory*, xxv.

[43]Ibid.

[44]It is interesting that, like Alhacen, Ptolemy specifies iron as the material from which to form the test-mirrors. As pointed out in n. 22 above, bronze was the normal material from which metal mirrors were made in Ptolemy's day because, unlike iron, it is reasonably easy to work. However, forming plane, convex cylindrical, and concave cylindrical mirrors from strips of iron, as Ptolemy describes, would have been immeasurably less demanding at the practical level than forming the spherical and conical mirrors Alhacen describes.

[45]See the experiment detailed in book 4, paragraph 3.108, pp. 323-324, where the equal-angles law is confirmed for the reflection of secondarily illuminated color by viewing the color's image through the appropriate hole in the wooden ring.

[46]Note, however, that Alhacen's addition of mirrors to test was limited by his ability to analyze them mathematically; see pp. xx-xxi above.

[47]Smith, *Ptolemy's Theory*, 132.

[48]Ptolemy's construction in this case has H and Z lie on a line, AG, that falls between the mirror's surface and its center of curvature D. Radius BED of the mirror is normal to that line, and H and Z are situated at equal distances from intersection-point E. Alhacen, on the other hand, places the two points on diameters DL and DM so that their distances ZD and HD from the center of curvature are equal. Suffice to say, both methods of construction are equivalent insofar as Z and H are equidistant from both E and D.

[49]See the discussion on p. xli above.

[50]See paragraphs 5, 2.151 and 2.152 (pp. 417-418), 2.155 (p. 418), 2.208 (p. 429), 2.343 (p. 453), and 2.363 (pp. 455-456). The reference to the *Elements*, which is found in paragraph 5, 2.151, cites the book (3) but not the proposition.

[51]See paragraphs 5, 2.147 (p. 417), 2.160 (p. 419), and 2.165 (p. 420). The reference to the *Conics* (book 2, proposition 4) is found in paragraph 5, 2.47.

[52]For the two references to Campanus cited by Risner, see *Opticae thesaurus*, proposition 33, p. 144, and proposition 71, p. 168. That Risner was not attempting to pinpoint Alhacen's actual sources is reinforced by Sabetai Unguru's claim that Risner's citation of sources for Witelo's *Perspectiva*— which was published in tandem with the *De aspectibus* in the *Opticae thesaurus*—"should not . . . be taken to mean that Risner thought all of them were sources actually employed by Witelo" (*Witelonis* Perspectivae *liber primus*, Studia Copernicana, XV [Warsaw: Ossolineum, 1977], 28).

[53]See I. L. Heiberg, ed. and trans., *Sereni Antinoensis opuscula* (Leipzig: Teubner, 1896), 58-65.

[54]It bears noting that the Arabic version of the *De aspectibus* also contains explicit references to *Conics*, V.34 and V.61 that are missing in the Latin version; see A. I. Sabra, "Ibn al-Haytham's Lemmas for Solving 'Alhazen's Problem'," *Archive for History of Exact Sciences* 26 (1982): 310, 318.

[55]See Jan P. Hogendijk, *Ibn Al-Haytham's* Completion of the Conics (New York: Springer, 1984). In section 7.7 of his introduction to this edition, Hogendijk ties Alhacen's approach to the solution of "Alhazen's Problem" to certain of the problems he addresses in his attempt to reconstruct book 8 of the *Conics*; see pp. 105-113.

[56]See p. cix below.

[57]Aside from the few explicit references given by Alhacen, the figures given in this paragraph are based on the citations I interpolated into the English translation where I felt they were needed to clarify Alhacen's mathematical reasoning.

[58]See, e.g., *Conics*, I.11-I.13.

[59]See pp. xli-xlii above for a discussion of the ostensible vs. the ulterior intent of these two propositions.

[60]There is, however, a fragmentary Latin version of the *Conics* attributed to Gerard of Cremona and consisting of the definitions and some of the enunciations from book 1; see I. L. Heiberg, *Apollonii Pergaei Quae Graece Existant Cum Commentariis Antiquis*, vol. 2 (Leipzig: Teubner, 1893), lxxv-lxxx. For an English translation of this version, see Marshall Clagett, *Archimedes in the Middle Ages*, vol. 4 (Philadelphia: American Philosophical Society, 1980), 3-13. In addition, there are a couple of medieval Latin texts containing snippets on conic sections that are based on Apollonius, but there is nothing to indicate that they were circulated in any significant way during the Middle Ages; see, e.g., Marshall Clagett, "A Medieval Latin Translation of a Short Arabic Tract on the Hyperbola," *Osiris* 11 (1954): 359-366.

[61]See Clagett, *Archimedes in the Middle Ages*, vol. 1 (Madison: University of Wisconsin, 1964), 673-675, esp. 675. In fact, Jordanus appeals to "*figuram 19 quinti perspective*," which I have renumbered to proposition 20. Presumably, then, Jordanus based his citation on a manuscript within which the figure for the proposition was labeled "19"; see p. cviii below.

[62]For a brief discussion of the problematic dating of Jordanus' life and works, see the article by Edward Grant in *Dictionary of Scientific Biography*, ed. Charles Gillispie, vol. 7 (New York: Scribner's, 1973), 171-172.

[63]The Latin translation itself of the *De aspectibus* raises some intriguing questions about the level of mathematical development in the Latin West at the beginning of the thirteenth century. The most obvious is whether the translator actually understood the mathematical content of what he was translating or whether he was mindlessly following the Arabic lead. The rendering "Ablonius," which is the Latin transliteration of the Arabic transliteration of "Apollonius," suggests (weakly) that he may have been unfamiliar with the *Conics*. On the other hand, the very decision to undertake the translation must have been motivated by a recognition on either the translator's part or that of his commissioner that the work was worth translating. That recognition must have been based on a reasonably informed evaluation of its content.

[64]*De refractione optice partes libri novem* (Naples, 1593), 76.

[65]For more detailed accounts of Witelo's life, see D. C. Lindberg's introduction to the reprint edition of Risner's *Opticae thesaurus* (New York: Johnson Reprint, 1972), vii-xiii, esp. vii-ix, Sabetai Unguru, *Witelonis* Perspectivae *liber primus*, Studia Copernicana 15 (Warsaw: Ossolineum, 1977), 12-19; and Sabetai Unguru, "Witelo," in Thomas F. Glick, Steven J. Livesey, and Faith Wallis, eds, *Medieval Science, Technology, and Medicine: An Encyclopedia* (New York: Routledge, 2005), 520-522. Among the works that Moerbeke translated for Witelo is Hero of Alexandria's *Catoptrics*, which was finished in late 1269 and which Witelo cites specifically in book 5 of the *Perspectiva*; see A. Mark Smith, ed. and trans., *Witelonis* Perspectivae *liber quintus*, Studia

Copernicana 23 (Warsaw: Ossolineum, 1983), 17. Moerbeke translated other Greek mathematical works, many of them Archimedean, that may have been of use to Witelo; for details, see Marshall Clagett, *Archimedes in the Middle Ages*, vol. 2 (Philadelphia: American Philosophical Society, 1976).

[66]The following discussion of the mathematical content of book 1 of the *Perspectiva* is based on Sabetai Unguru's edition in *Witelonis liber primus*, henceforth cited as "Unguru."

[67]Surely the most blatant example of Witelo's zest for rigor is to be found in his attempt to prove the fifth postulate (i.e., the "parallel" postulate) of book 1 of the *Elements*; see *Perspectiva* I.14 (Unguru, 55).

[68]Propositions I.119-I.128 also deal with proportionality theory as applied to the cutting of lines according to specific proportions and the formation of proportional figures from proportional line-segments.

[69]The relevant propositions in Euclid are V.8, V.16, V.19, and V.27-29 (through Campanus' recension). *Elements*, VI.3 also enters in, as does VI, definition 5 (through Eutocius' *Commentary on the Sphere and Cylinder of Archimedes*—see Clagett, *Archimedes in the Middle Ages*, vol. 2, 15-21).

[70]The generation of a right cone from the rotation of an isosceles triangle can easily be inferred from various constructions in book 5 of the *De aspectibus*; see, e.g., paragraph 5, 2.527, p. 482. Alhacen explicitly describes the method for generating a cylinder from the rotation of a rectangle in paragraph 4, 5.21, p. 333.

[71]Unguru, 116. Witelo's description of the conic sections according to obliquity of angle is reminiscent of the pre-Apollonian description of those sections according to a perpendicular planar cut through a cone whose angle at the vertex is right (parabola), obtuse (hyperbola), or acute (ellipse); for a discussion of the pre- and post-Apollonian definitions of the conic sections, see Michael Fried and Sabetai Unguru, *Apollonius of Perga's* Conica: *Text, Context, Subtext* (Leiden: Brill, 2001), 74-90. It should be noted, however, that Witelo stresses the obliquity of the planar cut itself on the basis of a right cone whose vertex-angle is right.

[72]As Unguru observes in "A Very Early Acquaintance with Apollonius of Perga's Treatise on Conic Sections in the Latin West," *Centaurus* 20 (1976): 121-122, the Arabic terms based on *mukefi* are provided in the fragmentary Latin version alluded to in note 60 above. Clagett is convinced that Witelo in fact "depended significantly" on this version for his knowledge of conic sections; see Clagett, *Archimedes in the Middle Ages*, vol. 4, 64.

[73]It is worth noting (as in fact Unguru does in note 3 to proposition 91, p. 190) that Witelo's analysis of conic sections is based on right cones only and is thus far more limited than Apollonius' analysis, which is generalized to all cones, including oblique and scalene ones. This limitation on Witelo's part may have been due to his failure to understand the full implications of Apollonius' analysis, but it may also have been due (and I suggest this as the likelier alternative) to his having relied too heavily on Alhacen, whose concern with cones and conic sections in the *De aspectibus* was centered on the analysis of planar cuts through right conical mirrors, according to his description of the formation of such mirrors on p. xxi above.

[74]The main principles upon which the proof of II.8 rests are: that the line drawn tangent to a conic section at its vertex is parallel to the ordinates of that section (I.32);

that when such a tangent intersects the asymptotes of a hyperbolic section, it is bisected at the point of tangency (II.3); that a hyperbola can be erected at any point lying between two intersecting lines, which will form its asymptotes (II.4); and that the ordinate of any hyperbolic section is bisected by the diameter of that section, from which it follows that the diameter of a hyperbolic section will pass through the midpoint of the ordinates (II.7). The resulting proof would not be perfectly rigorous, to be sure, but it would get the point across with adequate clarity. Furthermore, if Witelo had restricted his proof of II.8 to hyperbolas generated in right cones, it would have been especially easy, albeit lacking in generality.

75Another instance that reveals some acquaintance on Witelo's part with the *Conics* is to be found in proposition I.131 of the *Perspectiva*, where the point Witelo makes is clearly based on *Conics*, II.16. In this case too, I suggest, Witelo's knowledge of that proposition and its point could have come from his reading of Alhacen.

76In arguing for Alhacen as the primary source for Witelo's knowledge of the *Conics*, I am parting ways somewhat with Unguru, who believes that Witelo gained that knowledge through an as-yet-undiscovered Latin translation of the *Conics* produced, perhaps, by William of Moerbeke; see Unguru, "Very Early Acquaintance," 122. Although Unguru's theory is not implausible, I find its suppositional basis needlessly complex. If, however, Witelo did have some version of the *Conics* at hand—other than the one alluded to in note 60 above—then it may have consisted only of the enunciations without the accompanying proofs. Whatever the case, Witelo's grasp of conic sections seems less sure than it should have been under the assumption that he had access to a complete version of books 1 and 2 of the *Conics*.

77In addition to these sources, Unguru includes Serenus' *On the Section of a Cylinder*, but I see nothing concrete in Witelo's analysis of cylinders to indicate that he actually used it. Not only does he not attempt to prove that an oblique section through a cylinder produces a true ellipse, but in proposition I.103 (Unguru, 120), where it would have been natural for him to mention this equivalency explicitly, he fails to do so. Instead, he concludes by saying that "we shall, therefore, call that section *conic* [*pyramidalem*] in cones and *cylindrical* [*columpnarem*] in cylinders. Even so, that section in cones has been called before . . . an acute-angled section, or ellipse." As we have seen, Alhacen draws the same distinction between *sectiones piramidales* and *sectiones columpnares* in his discussion of the two kinds of ellipse.

78See Unguru, 32-35.

79Comprising part 5 of his *Opus majus*, Bacon's *Perspectiva* was completed by no later than 1267, but it evidently drew on ideas developed much earlier, perhaps even from the 1240's. In tandem with the *De multiplicatione specierum* (c. 1262), the *Perspectiva* was an influential source for optical lore in succeeding centuries. John Pecham's primary contribution to optics, the *Perspectiva communis*, was completed sometime around 1280 and enjoyed widespread circulation as a standard text for teaching in succeeding centuries. For critical editions of the two Baconian works, see David C. Lindberg, ed. and trans., *Roger Bacon and the Origins of* Perspectiva *in the Middle Ages* (Oxford: Clarendon, 1996) and *Roger Bacon's Philosophy of Nature* (Oxford: Clarendon, 1983). For a critical edition of Pecham's *Perspectiva communis*, see Lindberg, *John Pecham and the Science of Optics* (Madison: University of Wisconsin, 1970).

[80]See, for instance, Bacon's brief, general account of reflection in concave spherical mirrors in *Perspectiva*, III.i.4 (Lindberg, *Bacon and the Origins*, 269-275, esp. 271). Note that Bacon's mathematical explanation of image-locations in concave spherical mirrors is based on virtually the same diagram (figure 42, p. 273), even down to the lettering, as Alhacen's explanation of the same thing in proposition 32 of the *De aspectibus* (cf. figure 5.2.32, p. 250).

[81]See Lindberg, *Bacon and the Origins*, xx-xxiii.

[82]That Bacon was apparently able to follow the intricacies of Alhacen's mathematical reasoning is evidenced by his assertion in *Perspectiva*, III.i.3 that, although images in convex spherical mirrors generally look smaller than their objects, they sometimes appear equal or larger (Lindberg, *Bacon and the Origins*, 267). This assertion is based on *De aspectibus*, 6, prop. 6 (Risner, *Opticae thesaurus*,190-197), where Alhacen offers an extraordinarily intricate mathematical justification of it. Bacon may, of course, have simply taken Alhacen's word in the "enunciation" of the proposition (*Quod autem forma in his speculis* [i.e., convex spherical] *aliquando videatur maior re visa*), although there is no mention at this point in the theorem of the possibility of equality. Or he could have skipped to the very end of the proposition, where Alhacen does raise that possibility (*Igitur in his speculis imaginem aliquando equalem rei vise, aliquando maiorem esse*). Bacon was convinced that mathematics held the key to a full understanding of nature and the traces of God's creative impulse in it, so it is difficult to believe that he could not, much less would not, have read the entire proposition with critical care.

[83]Lindberg, *Pecham*, 61.

[84]Ibid., 203.

[85]Note, however, that placing both points inside the mirror to satisfy the condition set here (i.e., having circle AGB cut arc CH in two points) flouts the condition set at the beginning of the proposition, where one of the points is to lie either at or beyond point H.

[86]In fact, without a firm grasp of the points established in 5.43-48, particularly 5.44-46, one can easily be confused or misled by Alhacen's summary in 5.49. It seems likely, therefore, that Pecham based the garbled account in II.47 on that summary with little or no understanding of the points established in the preceding theorems.

[87]Along with numerous articles on particular aspects of Perspectivist optics and its reception in the Latin West, David Lindberg's *Theories of Vision from Al-Kindi to Kepler* (Chicago: University of Chicago, 1976) still shapes our understanding of the Perspectivist tradition and its development from the late thirteenth to the early seventeenth century.

[88]See Lindberg, *A Catalogue of Medieval and Renaissance Optical Manuscripts* (Toronto: University of Toronto, 1975).

[89]For elaboration, see Lindberg's introduction to the reprint edition of Risner's *Opticae Thesaurus*, xxi-xxiii.

[90]This distillation process is most clearly exemplified in the development of *Sentences* commentaries according to a fairly rigid canon of *quaestiones* based on certain key points in Lombard's text. Much the same thing happened with commentaries on Aristotle's works, which increasingly focused on specific *quaestiones*.

91As Lindberg points out in his introduction to the reprint edition of Risner's *Opticae Thesaurus*, there is evidence for the continued use of Witelo's *Perspectiva* as a text in mathematics even to the late sixteenth century in Cambridge; see p. xxiii. Nevertheless there is no way of determining precisely how either Alhacen's *De aspectibus* or Witelo's *Perspectiva* would actually have been used for the teaching of Euclidean geometry. Were they closely analyzed from cover to cover, or were they dipped into at specific points? If the latter, then at what specific points? To these questions we have no definitive answer, nor does it seem likely that we ever will.

92It should be noted, however, that many of these commentaries are slanted toward issues in Aristotelian natural philosophy, so the optical questions raised in them are rather narrowly defined by that context. See Lindberg, *Theories*, 122-139. See also Smith, *Alhacen's Theory*, xciv-c.

93The conclusion that Euclid's *Catoptrics* and Ptolemy's *Optics* were more widely read than Alhacen's *De aspectibus* and Witelo's *Perspectiva* is based on the fact that the surviving manuscripts (both complete and incomplete) of the first two works significantly outnumber the surviving manuscripts of the second two; see Lindberg, *Catalogue*, 47-50, 74 vs. 17-18, 77-79. The increasing popularity of these works during the Renaissance is especially clear in the case of Ptolemy's *Optics*, the overwhelming majority of whose surviving manuscripts date from the sixteenth and seventeenth centuries.

94It bears noting that, despite its relative mathematical complexity, the analysis of parabolic burning mirrors did capture the imagination of medieval and Renaissance scholars, presumably because that analysis was seen to have practical applications, whereas the exact determination of reflection-points has little or none. The primary source for the study of burning mirrors in the Latin West was Alhacen's *De speculis comburentibus*, although Bacon also wrote a tract that bears the same title; see the list of respective manuscripts in Lindberg, *Catalogue*, 20-21, 39-40. Witelo also offers an analysis of paraboloidal burning mirrors at the end of book 9 of the *Perspectiva*; see Risner, *Opticae Thesaurus*, 392-403. In addition to the mathematical analysis of such mirrors, Alhacen's and Witelo's accounts include "practical" advice on how to manufacture them. Clagett finds in Witelo's discussion of paraboloidal burning mirrors a strong indication that Witelo had access to at least some of the *Conics*, although his understanding was confused; see *Archimedes in the Middle Ages*, vol. 4, 91-98; for a detailed account of the tradition of burning mirrors and the associated study of parabolas in the later Middle Ages and Renaissance see the entire volume.

95For publication-details of these early editions of the *Elements*, see George Sarton, *A History of Science*, vol. 2 (New York: Norton, 1970), 43-45.

96For publication-details of these editions of the *Conics*, see ibid., 96-97. Not until Edmund Halley's edition of 1710 did a Greek edition of the *Conics* appear in print.

97For publication-details, see Unguru, *Witelonis liber primus*, 41-42.

98See Risner's preface to the *Opticae Thesaurus*, folio 2r, l. 31-folio 2v, l. 37.

99*Opticae libri quatuor ex voto Petri Rami novissimo per Fridericum Risnerum* . . . . (Kassel, 1606).

[100]See Lindberg, introduction to the reprint edition of *Opticae Thesaurus*, xxiv-xxv.

[101]See Smith, *Alhacen's Theory*, lxxxii-civ. See also Smith, "What Is the History of Medieval Optics Really About?" *Proceedings of the American Philosophical Society* 148 (2004): 180-194.

[102]*Paralipomena*, chap. 3; for an English translation, see William Donahue, *Johannes Kepler, Optics:* Paralipomena to Witelo and Optical Part of Astronomy (Santa Fe: Green Lion, 2000), 75-91.

[103]For an English translation of the *Lectiones Opticae*, see H. C. Fay, trans., and A. G. Bennett and D. F. Edgar, eds., *Isaac Barrow's Optical Lectures, 1667* (London: The Worshipful Company of Spectacle Makers, 1987).

[104]See letter 1213 in A. R. Hall and M. B. Hall, ed. and trans., *The Correspondence of Henry Oldenbourg*, vol. 6, (Madison: University of Wisconsin, 1969), 42-46.

[105]For the Latin version of Huygens' solution, see *Philosophical Transactions (1665-1678)*, 8 (1673), 6119; for an English translation, see Hall and Hall, *Correspondence of Oldenbourg*, 7 (1970), 191-192. Based on figure 28, p. 555, which is taken directly from the one provided in the *Philosophical Transactions*, Huygens' method for solving the problem is as follows. Let the top circle DdDd centered on A represent a great circle within the sphere of the mirror. Let B and C represent the center of sight and object-point, and let circle ABC be drawn through them and centerpoint A of the mirror. Let point z be the center of that circle. Draw AER normal to BC. Cut AR at point N such that AR:AO = AO:AN, and through point N draw line MN parallel to BC. Cut AR at point I such that AI:AO = AO:4AE. Find point Y above A such that IY = IN, and through point Y draw line MY parallel to line AZ connecting the centers of the two circles. Finally, find points S and X on line AR such that IS = IX, and the square formed on either of them = one-half $AO^2 + AI^2$. Lines MY and MN will therefore form the asymptotes of a hyperbola whose opposite branches pass through X and S. Points D and d where each of those branches cuts the circle of the mirror will be potential points of reflection for B and C. According to Hugyens' placement of B and C in the figure, then, if the mirror is convex, reflection can only occur from point D on the convex surface facing B and C, so if a normal is dropped from A through that point D, it will bisect angle BDC. Suffice it to say that neither of the two points d can be a reflection-point because either ray Cd or ray Bd would have to pass through the surface to reach it. By the same token, if reflection occurs from the concave portion of the mirror facing B and C, reflection can only occur from point D, since B and C lie entirely outside the circle of the mirror. Thus, if a normal is dropped from A to that point D, it will bisect angle BDC. The same will not hold, of course, for either point d. On the other hand, if B and C are located inside the great circle of the mirror, as illustrated in figure 28a, p. 556, the same procedure can be followed to define the two asymptotes, MN and MY, as well as points S and X on them. The two branches of the hyperbola passing through those points will each cut the circle on the mirror at points D and d, and if a normal is dropped from A to the two points D and the two points d, it will bisect the respective angles BDC and BdC. Hence, the two pairs of points D and d will yield four reflections altogether. Since circle ABC in Hugyens' solution is the same as Alhacen's cutting circle, the number of possible reflections from the arc on the mirror that subtends angle BAC will depend on whether circle ABC cuts that arc to the left or

right of its midpoint, so it is possible for B and C to be placed in such a way that only one reflection will occur; see the discussion on pp. xli-xliv above.

[106]See letter 1489 in *Correspondence of Oldenbourg,* vol. 7., 73-81.

[107]See letter 1528 in ibid., 177-193. Oldenbourg evidently sent a copy of Hugyens' solution to John Wallis, who wrote back in July of 1669, admitting that "Mr. Hugen's optical Probleme I have not had time yet to consider of" but adding rather snidely that "it doth not seem, at first view, to be a matter of very great difficulty"; letter 1260 in ibid., 159-161.

[108]See letter 1548 in ibid., 246-256.

[109]These excerpts were published in two parts in *Philosophical Transactions (1665-1678),* 8 (1673), 6119-6126, 6140-6146.

[110]The notion that Alhacen's method for finding reflection-points is needlessly unwieldy is reflected in lecture IX of Barrow's *Lectiones Opticae,* where he deals with reflection from spherical convex mirrors. After outlining his own method for finding the point of reflection in such mirrors, he provides a faithful but abbreviated version of Alhacen's method in proposition 5.25 of the *De aspectibus* in order to provide something "acceptable to [the] taste" of geometers while stripping it "of that horrible combination of prolixity and obscurity, and of . . . the uncouth barbarity of speech" so characteristic of the Alhacenian original; See Fay et al., *Barrow's Optical Lectures,* 118-121. For a survey of various approaches to Alhazen's Problem since the mid-seventeenth century, see J. A. Lohne, "Alhazens Spiegelproblem," *Nordisk Matematisk Tidskrift* 18 (1970): 5-35.

[111]Sarton's original sentence, which I found irresistably pithy, reads: "After that, the Apollonian tradition was lost in the new geometry of the time, like a river in the ocean" (*An Introduction to the History of Science,* vol. 2 [Baltimore: Williams & Wilkins, 1927], 28).

MANUSCRIPTS AND EDITING

In the previous edition of books 1-3 of Alhacen's *De aspectibus* I provided a detailed account of the available manuscripts and outlined my procedures for selecting particular representatives from among them for collation in the critical text. I concluded that the seventeen complete or virtually complete manuscripts could be broken into three family groups, the first consisting of six members (*F*, *P1*, *Va*, *V2*, *L2*, and *S*), the second of four (*Er*, *C1*, *O*, and *M*), and the last of seven (*E*, *P3*, *P2*, *L3*, *C2*, *L1*, and *V1*).[1] I also concluded that the first family lies closest to the *Urtext* and, furthermore, that among its members *F* is closest to the family progenitor. *F* was therefore the logical candidate to represent this family in the critical text, but since it lacks a considerable portion of books 1 and 2, I was forced to fall back on its nearest relative, *P1*. For the second family the choice was less clear, but I eventually decided on *Er*, thus bringing the total for collation to two. The third family was even more problematic, but I was finally drawn to *E*, *P3*, and *L3* as the most suitable choices for collation. To the resulting list of five manuscripts I added *S* and *C1*, because both seemed to form inter-family links, *S* between the first and second families, *C1* between the second and third. Altogether, then, I based my critical text on seven manuscripts—*P1*, *S*, *E*, *P3*, *L3* , *Er*, and *C1*—using *O* for occasional cross checking when necessary.

Over the course of editing the text on the basis of these seven manuscripts, I was led to modify my initial conclusions somewhat.[2] For one thing, it became clear to me that *E* and *P3* are so close as to be virtually identical, the latter having most likely been copied directly from the former. It was therefore obvious that *P3* added nothing of substance to the critical text. I also discovered that *P1*, which I initially took to be the arch-representative of the first family, was less reliable as a textual witness than its distant relative *S* and, furthermore, that *O*, which I had marginalized somewhat in my initial evaluation, would have been preferable to *Er* as a representative of the second family. On the basis of these insights, I decided in retrospect that, were I to do it all over again, I would substitute *O* for *Er* and ignore *P3*. And that is precisely what I have done here in the critical edition of books 4 and 5, exchanging *O* for *Er* and dropping *P3* from consideration. I have also added *F* to the mix, since it includes the entirety of books 4 and 5, thus bringing the revised list of manuscripts to be collated back to seven: *F*, *P1*,

S, E, L3 , O, and *C1*. Sample pages from these seven manuscripts are reproduced on pp. cxxi-cxxvii below, each page containing the incipit of proposition 32, pp. 141 (Latin) and 446 (English) along with the relevant diagram, which is not included in the Saint-Omer manuscript (see p. lxxi).

As before, so now, the process of establishing the critical text has led me to reconsider my already-reconsidered assumptions about the selected family-representatives and their place in the manuscript-tradition. Central to this reconsideration is that the text shifts quite early in book 5 from narrative explanation to mathematical demonstration. In narrative explanation, of course, two different, sometimes even contradictory, readings can make perfect sense in a given context. Choosing the "right" reading is thus dictated more often than not by consensus of manuscripts (appropriately weighted for authority) rather than by the logic of the passage. In mathematical discourse, however, there is little or no ambiguity, so the right reading is dictated more often than not by logic, not consensus. In many cases, in fact, consensus is simply wrong. Accordingly, as the text of book 5 unfolded, it became increasingly clear to me that manuscripts *F* and *P1* were even less authoritative and reliable than I had expected and, conversely, that *O* was commensurately more so.

This re-evaluation of *F, P1,* and *O* in light of the critical text of book 5 raises questions about the authenticity of *F* and *P1* as witnesses to the *Urtext*. It could be, of course, that the two manuscripts reflect flaws in the *Urtext* itself, flaws that were corrected as the text passed from hand to hand in its subsequent transmission. With their heavy burden of redactions, *O* and *E* in particular seem to bear this possibility out. It could also be that, although still closer to the *Urtext* than the rest of the manuscripts, *F* and *P1* suffered from the maladroit efforts of the original scribe to copy mathematical theorems that made little or no sense to him. Or it could be that *F* and *P1* represent a particular textual compilation, parts of which were more or less authentically tied to the *Urtext* and parts of which were not, a possibility already raised in the previous edition of books 1-3.[3] Whatever the case, it is by now evident that the textual tradition of Alhacen's *De aspectibus* is complicated enough to resist a simple, definitive reconstruction, which is hardly surprising, given the size and complexity of the text in question. Yet, despite such qualms about the textual tradition and its accurate reconstruction, I have no misgivings about my organization of the manuscripts according to families or my selection of representatives from those families. I am, in short, confident that the critical text is appropriately critical.

The Critical Text: The gross format for books 4 and 5 is clear and clearly stated at the beginning of each book, where the number of chapters (or parts), and a brief description of their content is given explicitly in most of the manuscripts. Accordingly, book 4 is properly divided into five chapters, book 5 into two. As far as book 4 is concerned, all but two of the manuscripts used in the critical text agree on both the number and placement of the chapter-breaks, and—as can be seen in Table 4, Appendix 2, in *Alhacen's Theory*, p. 658—the two exceptions, *L3* and *C1*, diverge only by splitting the fifth chapter into two, the new chapter opening with the phrase "in speculis autem columpnaribus."

Book 5 is a different matter altogether. For a start, it consists of only two chapters, the first of which occupies a single paragraph. In addition, despite the claim at the beginning of the book that it "is divided into two parts," a claim repeated in all but two of the manuscripts (*F* and *P1*), chapter 2 is subdivided in several different ways in the relevant manuscripts. A look at Table 5 of the appendix just cited shows that *C1* leads the way by opening a new chapter (designated as 2b) with the phrase "restat iam ut loca ymaginum." This is followed in *O* by chapter 2c, whose incipit is "in speculis exterioribus pyramidalibus," after which *S*, *P1*, *L3*, and *F* interpolate chapter 2d, beginning with "in speculis spericis concavis." Then comes chapter 2e in *P1*, *L3*, *O*, and *F*, its opening phrase being "in speculis columpnaribus concavis." In *C1*, finally, a weak break that signals chapter 2f occurs at the phrase "in speculis pyramidalibus concavis."

Barring the stray chapter-breaks, which are evidently spurious, the Latin text of these two books presents a relatively blank face broken into two main segments, i.e., books 4 and 5, and seven lesser segments consisting of the five chapters in book 4 and the two in book 5. What remains is a succession of long, unrelieved swaths of text whose internal punctuation is as sporadic as it is haphazard. Some of these swaths, moreover, are dauntingly long, the most egregious example being chapter 2 of book 5, which occupies well over half the combined text of books 4 and 5. Left in this rather sterile format, the critical text would have been as faithful as possible to the original, to be sure, but it would have been essentially inaccessible to contemporary readers, who are accustomed to a far more punctuated format than their medieval forebears. The trick is to impose such punctuation without traducing the intent of the original. Fortunately, this task is made easy by Alhacen's rigorously systematic approach, according to which books 4 and 5 are further broken, albeit implicitly, into two clear sub-levels of organization.

The first of these is topical and is determined by an analytic passage through the seven mirrors discussed earlier in the introduction. Thus, in both books, Alhacen analyzes reflection and its various aspects according to a specific order, starting with plane mirrors, passing to convex spherical, convex cylindrical, and convex conical mirrors, and ending with concave spherical, concave cylindrical, and concave conical mirrors. Ultimately, I decided not to punctuate book 4 according to this topical format because it is short enough and the organizational principles are clear enough to need no reinforcement. Not so for book 5, however. Not only is it more than half again as long as book 4, but virtually all of it occupies a single chapter, i.e., chapter 2, whose inordinate length is matched by its analytic complexity. In order, therefore, to break this textual segment into manageable chunks, I imposed strong breaks between topical sections. These breaks, in fact, occur precisely where the various interpolated chapters do in the manuscripts, so in a sense I was merely following their lead.

The second sub-level of organization pertains to book 5 alone and is based on the fact that the vast majority of it consists of geometrical constructions and proofs. On the face of it, dividing the text into its constituent propositions should be simple enough, given Alhacen's penchant for ending proofs with phrases such as "quod est propositum" or "et ita propositum." Moreover, several of the manuscripts key the diagrams by number to their appropriate theorems. The problem is that not all of the manuscripts do number the diagrams and, worse, that among those that do, the numbering is inconsistent. Another problem is that, in several instances, what could be construed as individual propositions in succession can also be construed as specific cases falling under the head of a single, more general proposition.

An obvious solution to these problems would be to follow the format of the Arabic text, but unfortunately there is as yet no critical edition of that text available, although one has been in the works for some time now.[4] Nor do the Latin manuscripts offer much hope, since they are so obviously inconsistent. I was therefore forced to decide on my own how to parse the text by propositional elements. After some trial and error, I eventually decided that the most appropriate breakdown results in a set of 54 propositions, a few of which are so long and involved that they need further subdivision into cases and even subcases. Accordingly, between individual propositions I have inserted fairly strong spacing-breaks, whereas between cases and subcases within a given proposition I have inserted weaker spacing-breaks. Each proposition is further demarcated by a numerical designation (e.g., [PROPOSITIO 1]). Altogether, then, I have organized the text according to an order of spacing-breaks from strong to weak: strongest between chapters, less strong between topical segments, weaker yet between propositions, and weakest between intra-propositional elements—all with the hope of making the complex analytic structure of the text as transparent as possible.

At the lowest level of external punctuation, division into paragraphs, I had to fall back on my own devices for lack of an appropriate guide either in the manuscripts themselves, where such punctuation is random at best, or in a critical Arabic edition. Hence, unlike the previous text and translation of books 1-3, this one is not keyed to the Arabic version in its paragraph-structure. Nevertheless, I have followed the convention established in my previous edition of numbering the paragraphs for easy internal reference. Regrettably, the lack of coordination between Latin and Arabic texts at this level of punctuation will make future comparison of the two versions more difficult for books 4 and 5 than for the preceding three.

As to internal punctuation, finally, I have tried insofar as feasible to break the text up according to both the sense and syntactic structure of the Latin. At times, of course, the syntax of the Latin is so convoluted that I have had no real choice but to break sentences up into more digestible chunks. I have also taken liberties with the punctuation itself, following modern conventions by capitalizing words at the beginning of sentences, inserting commas, and so forth. In addition, I have taken the liberty of capitalizing letter-designations in the text, so that readings such as "linea ab" or "angulus gnd" in the actual manuscripts have been transformed to "linea AB" or "angulus GND" in the critical text.

Both honesty and admiration compel me to close my discussion of the critical text with a few remarks about Friedrich Risner's 1572 edition of the *De aspectibus*. As I pointed out earlier, Risner's aim in creating this edition was not to make it critical by modern historical standards. It was, rather, to upgrade the work to contemporary, late-Renaissance standards by revising the grammar and vocabulary (albeit fairly lightly), breaking the text into propositional elements, and inserting commentary when he deemed it necessary.[5] Nowhere is this latter modification clearer than in book 5, where Risner explains virtually every propositional conclusion either by providing specific citations to the appropriate mathematical source, mostly Euclid's *Elements*, or by explaining in detail the logical steps leading to that conclusion. So deep was Risner's understanding of the work and its analytic structure, in fact, that at one point he was able not only to recognize that a brief passage had been improperly transposed in the manuscripts but also to restore the passage to its rightful place.[6] It is therefore without shame that I acknowledge my debt to Risner. Without his clear and virtually inerrant guidance, I would have been far harder pressed than I was to make sense not just of certain propositional elements, but of the overall analytic structure of the second chapter of book 5.

Diagrams: Perhaps the knottiest problem I faced as an editor was how to handle the diagrams accompanying the text. For one thing, the number of diagrams varies widely among manuscripts. In the fourth book, for example, *F* has no diagrams whatever, *C1* three, *L3* four, *E* fourteen, *S* fifteen, *O* eighteen, and *P1* twenty-five. In the fifth book the number of diagrams is considerably greater, although the variation in number among manuscripts is commensurately less. Hence, the first and most obvious issue I had to address was which, if any, of the diagrams in the two books should be included as an integral part of the text and which should be regarded as mere ancillary illustrations and, therefore, treated as marginal glosses. In fact, I had already faced this problem on a much smaller scale in my previous edition of books 1-3, and the criterion I followed there is the one I followed here: if the letter-designations in the diagram reflect equivalents in the text, then that diagram is to be considered integral.[7] On that basis, I effectively eliminated all the diagrams in book 4, since they are clearly meant to illustrate technical descriptions not couched in specific geometrical format—i.e., where, instead of saying "if line EA is extended from center of sight E through vertex A of the cone," the text simply says "if a line is extended from the center of sight through the vertex of the cone." Book 5 is entirely different, in that most of it consists of geometrical propositions with specific letter-designations for points in the construction described. In this case, then, the obvious choice was to include all diagrams that reflect the letter-designations given in the propositions. According to that standard, I was able to pare the number of relevant diagrams to 83, still sizeable but significantly smaller than it would have been had I taken an all-inclusive approach.

Having resolved the problem of quantity, I was faced with the more vexing issue of quality. In a few cases, the requisite diagrams are straightforward enough that, when produced with ruler and compass, they adequately reflect the conditions specified in the proposition (e.g., parallelism or equality of lines, equality of angles, etc.). These diagrams actually "look" like what they are meant to represent. In most cases, though, the structure of the theorems and the constructions on which they are based are too complex for the simple expedients of compass and ruler, especially when those constructions are three-dimensional. The interrelationships among angles, lengths, and areas within the diagram are so intricate and exacting as to defy such easy or straightforward reproduction. Consequently, most of the figures provided in the manuscripts misrepresent, often grossly, what they are intended to illustrate.

Take, for example, FIGURE 5.2.25, p. 581, which is adapted directly from the drawing provided on folio 64v of manuscript *O*. This diagram is intended to illustrate "Alhazen's Problem" as applied to convex spherical mirrors: i.e., given a convex spherical mirror with centerpoint G, and given point-source B of radiation and center of sight A, to find point D of reflection on the mirror's surface. According to the construction given in the theorem, which has already been discussed on pp. xclvii-xlviii of the introduction, the resulting figure ought to reflect a variety of key conditions. For instance, the length of MF compared to that of FK should be equivalent to the length of BG compared to that of AG. Angle FKC should be precisely half of angle BGA. Line BZ should be perpendicular to line DI. Line DZ should be the same length as line ZI, and, as a result, triangle BDZ should be identical in size and shape to triangle BZI. Most important of all, angle of incidence BDE should equal angle of reflection EDA. Clearly, none of these conditions has been met in the figure. The closest the diagram comes to representative accuracy is a gross approximation of equality between DZ and ZI, and even in that case the degree of accuracy falls short of 75%.

One more example should suffice. In FIGURE 5.2.31, p. 587, which is adapted from a more complicated diagram on folio 67r of manuscript *O*, point G represents the vertex of a convex conical mirror with circle DEZ as its base and GZ and DZ as its outer edges. Point T represents the center of the cone's base-circle, TG the axis of the cone, TR the normal to point E on the base-circle, and GE a line of longitude on the cone's surface. MGN represents a plane parallel to that of the base circle, leaving source-point A of radiation and center of sight B positioned below it. KCF, finally, represents a line normal to line of longitude GE.

The intent of this construction is to illustrate that, under the conditions just specified, C is the appropriate point of reflection for A and B. But, as actually represented, point T is nowhere near the center of base-circle DEZ and, consequently, axis GT of the cone is badly misplaced. This, in turn, renders the diagram virtually useless as an aid to understanding why in this particular case C must be the point of reflection for A and B, because the proof for this claim depends on TER's being normal to circle DEZ—which it is obviously not in the diagram—from which it follows that angles HER and AER should be equal—which they are obviously not in the diagram.

Whether and how such misrepresentations might have affected a medieval reader's ability to make sense, or at least immediate sense, of the propositions they purport to illustrate are open questions. In ancient and medieval times diagrams were not always provided in mathematical texts, but we cannot be certain whether or to what extent such omission might have been intentional. If intentional, then the reader would have been expected to construct the diagrams for himself, either mentally or physically, and he would have been expected to do so unerringly from the verbal description. This interpretation is borne out to some extent by book 4 of the *De aspectibus*, where the geometrical descriptions provided by Alhacen are so detailed and punctilious as to render illustrative diagrams almost superfluous. Nevertheless, the very fact that such diagrams are supplied in some of the manuscripts indicates that more than a few medieval scholars regarded them as useful, if not necessary adjuncts to the descriptions they illustrate.

Unlike those of book 4, the mathematical constructions in book 5 are often so intricate and involved that it is diffcult to believe Alhacen expected anyone to make sense of them without the aid of diagrams. The issue here, however, is not whether Alhacen actually provided diagrams, much less whether he provided the diagrams (or their equivalents) that appear in the Latin manuscripts. The definitive resolution of this issue awaits the appearance of A. I. Sabra's long-expected edition of the Arabic text of books 4 and 5.[8] The relevant question is whether the diagrams in the Latin manuscripts constitute an integral part of the *Urtext*. Were they, in other words, explicitly produced, either through copying from the Arabic exemplar or through original construction, to accompany the initial Latin translation of the *Kitab al-Manazir*?

The remarkable consistency among figures across the spectrum of manuscripts suggests strongly that they were. Not only is there general agreement among the manuscripts about the point-by-point letter-designations in individual figures, but there is also general agreement about how those figures should be configured and presented.[9] Such consistency indicates (not surprisingly) that, like the text itself, the figures were simply copied from manuscript to manuscript rather than made to order for each manuscript. Furthermore—and again, not surprisingly—the figures seem in most cases to have been produced apart from the text, scribe and illustrator working independently rather than in collusion.[10]

With these points in mind, and having determined which figures to include with the text, I was left with one final issue. What form should those figures take? I had three clear options. One was to include a set of generic figures appropriately keyed to the text. These I had at hand in the form of diagrams I had laboriously reconstructed as I worked my way through the text, proposition by proposition. The problem with this choice is that I had taken great pains, with the aid of a fairly versatile drawing program, to make the reconstructed diagrams as accurate and true-to-description as possible—which means that they misrepresent the representations provided in the manuscripts. The second option was to reproduce the figures directly, by scanning, from one or more of the manuscripts. This option has the signal advantage of presenting the figures warts and all, with minimal editorial intervention on my part. But there are disadvantages as well. One concerns clarity. The scans would have to come from microfilm copies of the manuscripts, and the best examples available in that form are too light and nebulous to be useful. Another disadvantage is that, as they appear in the microfilms, many of the figures are so small or so awkwardly oriented that it is difficult to reconcile them with the text. Such figures could, of course, be magnified or reoriented but at the sacrifice of clarity and ease of reading.[11]

The option I finally lit on was to trace the figures directly from the scanned versions, re-orienting the abstracted diagrams as needed, and relettering them suitably, the result being crisp, clear, and eminently readable. I chose manuscript *O* as the representative basis for this process both because it includes almost all of the requisite figures and because those figures have not been editorially adjusted, as is sometimes the case in other manuscripts, *E* and *P3* in particular. On rare occasions I had to look beyond *O* to *P1* and *P3* for what I needed, but the lion's share of the adapted diagrams come directly from *O*. As far as basic configuration and lettering are concerned, therefore, the figures accompanying the Latin text are absolutely faithful to the originals, although they do not necessarily reflect the orientation of those originals. I should add that not all of the figures included with the Latin text are abstracted from originals. Only those that are too complex to have been accurately, or at least adequately, produced by compass and ruler have been treated this way. The rest come from my own stock of reconstructed diagrams. I should also add that, in the case of FIGURES 5.2.31, 5.2.31a, and 5.2.31b on pp. 587 and 588, I have actually abstracted each diagram from the composite figure provided on folio 67r of ms. *O*.[12] Generally speaking, though, I have treated the adapted figures with a relatively light editorial hand.

The Critical Apparatus: Since the conventions I used for the critical apparatus in this edition are precisely the same as those I used in the edition of books 1-3, I will not repeat them here. They can be found in *Alhacen's Theory*, pp. clxii-clxiv.

The Translation and Commentary: As far as my basic approach to translation is concerned, I have nothing to add to my discussion in the previous edition of books 1-3.[13] But a few words about mathematical notation are in order. Much of Alhacen's mathematical reasoning in book 5 is based on the analysis of ratios and proportions given in the fifth book of Euclid's *Elements* and subsequently applied to triangles and parallelograms in the sixth. Ratios and proportions are, of course, readily convertible to fractions and equations, so that, for instance, the expression $a:b :: c:d$ translates almost automatically to $a/b = c/d$. But consider the implications of such conversion. The first expression calls for comparison (i.e., magnitude *a* compares in size to magnitude *b* in the same way that magnitude *c* compares in size to magnitude *d*). The second calls for division (i.e., magnitude *a* divided by magnitude *b* leaves the same quotient as magnitude *c* divided by magnitude *d*). Not only do the two expressions invoke different operations (comparison vs. division), but the operative terms are different. In the first expression it is the magnitudes themselves that are being manipulated; in the second it is the numerical quantity of the magnitudes, abstracted from the magnitudes themselves, that is being manipulated.

To see how fundamentally incompatible the two expressions are at both the operational and conceptual level, we need look no further than *Elements* V.16, where Euclid demonstrates that, if *a:b* :: *c:d*, then, by alternation, *a:c* :: *b:d*. In fractional form, this means that, if $a/b = c/d$, then $a/c = b/d$, which is arrived at by multiplying both sides of the equation by b/c. So far so good, but suppose that *a* and *b* represent areas, while *c* and *d* represent lengths. In fractional form this poses no problem whatever, since areas can be divided by lengths (and vice-versa), leaving the quotients equal. But it poses an insuperable problem as a statement of proportionality, because there is no meaningful comparison between lines and areas. The absurdity of such a comparison becomes even clearer if *a* and *b* are taken to represent angles. For, while it makes sense to say that angle *a* compares in size to angle *b* in the same way that line *c* compares in size to line *d*, it makes no sense whatever to say that angle *a* compares in size to line *c* in the same way that angle *b* compares in size to line *d*. At bottom, then, the language of ratios and proportions is as different from that of fractions and equations as, say, Attic Greek is from modern English. Each has its distinct grammar, syntax, and vocabulary.

Not only did Alhacen think in the language of ratios and proportions; he "spoke" it fluently. I have therefore resisted the urge to recast his discourse in modern algebraic form, not just because such translation would be inauthentic and misleading, but because it would stand in the way of a proper appreciation of the ingenuity and elegance with which Alhacen manipulated ratios and proportions in his analysis of reflection. I have, however, streamlined the expression of proportions in the text, rendering such verbal statements as "erit proportio BN ad NO sicut proportio BM ad MO" in the symbolic form BN:NO = BM:MO, the equal sign substituting for the double colon so that the expression should be taken as "BN is to NO as BM is to MO" rather than "BN to NO is equal to BM to MO." Likewise, in denominating the square created from a given length, I have used the exponent for the sake of convenience (i.e., "quadratum AG" is rendered "AG^2"), but the result is meant to be understood as "the square composed from side AG" rather than "AG squared" in the algebraic sense. As to rectangular areas, finally, I have chosen to denominate them by placing an unspaced comma between the constituent sides, so that "ductus BD in DG" is rendered BD,DG and is to be construed as "the rectangle formed by sides BD and DG" rather than "BD times DG."

Aside from streamlining mathematical expressions, I have taken a few other liberties with the English translation. For one thing, I have reinforced the organizational breaks discussed earlier by interpolating parenthetical headings, such as [CASE 1] or [SUBCASE 1a], to clarify the analytic structure of the text. In addition—and here I must again acknowledge my debt to Risner—I have provided external and internal citations at appropriate points in the translation, inserting them parenthetically and setting them in brackets. In a few instances I have done no more than fill in the blanks. For example, where the text reads "angle GAN = angle GBA, as Euclid demonstrates in the third," I have appended the specific reference parenthetically (i.e., " . . . third [book of the *Elements*, prop. 32]"). Likewise, when the text adverts to earlier internal conclusions (for instance, toward the beginning of paragraph 2.201, p. 273, when it reads "according to what we established earlier," I have inserted the appropriate reference parenthetically (i.e., " . . . established [in proposition 24, lemma 6]"); and at various points in the theorems I have inserted "by previous conclusions" after certain claims to indicate that they have already been justified and, therefore, do not come out of thin air. But most of the interpolated citations amount to commentary and, as such, are intended to steer the reader to the specific theorem in Euclid (or, far more rarely, in Apollonius or Serenus) that justifies the conclusion at that point in the text. Thus, for example, at the end of paragraph 2.72, p. 238, I have given the particular theorem upon which the conclusion rests, so the resulting passage reads, "Therefore AG:DG = AE:ED [by Euclid, VI.3]." In some cases the conclusion and its basis struck me as so obvious, even to a reader unschooled in Euclidean geometry, that it required no annotation at all, but such cases are fairly limited

In addition to these source-citations, I have also inserted brief explanations at spots in the text where I considered such parenthetical interruptions short enough not to interfere unduly with the logical flow. The remainder of the commentary I have remanded to endnotes. At a few places in the translation, where the same letter is used to designate different points in the Latin version, I have distinguished the letters in the English translation by adding prime or double-prime signs and adjusting the designations in the accompanying diagrams to match (see, e.g., proposition 38, pp. 458-459 and its accompanying figure on p. 260). Moreover, I have done my best to make the diagrams that go with the English translation reflect the actual conditions specified in the construction and proof. They are, in short, meant to look as much like what they represent as possible. Accordingly, I have felt free to add letter-designations when necessary and even to add lines to the originals, all with the aim of making the analysis illustrated by the figure as easy to follow as possible. As a result, the diagrams included with the English text are interpretations of the originals in much the same way that the English text itself is an interpretation of the Latin original. To distinguish the figures adapted from the Latin text from those reconstructed for the translation, I have designated the former in capital letters (i.e., FIGURE 5.2.25), the latter taking the form figure 5.2.25. In book 4, the numbering of figures is determined by book and chapter, so that figure 4.2.1 refers to the first figure pertaining to chapter 2 of book 4. In book 5, most of the figures are determined by book, chapter, and proposition, so that figure 5.2.25 is keyed to proposition 25, which occurs in the second chapter of book 5. Because the first six figures in book 5 illustrate points that are not keyed to propositions, I have designated them neither by chapter nor proposition. I have simply listed them in order from 5.1. to 5.6.

So numerous are the diagrams used in this edition (well over 300) and, in some cases, so involved that inserting them directly into the text in appropriately reduced form struck me as too problematic to be worthwhile. I therefore decided to include them separately, the result occupying a total of 191 pages. In order to ease the burden of referring to them, though, I have put all the diagrams pertaining to the English translation and commentary at the end of the volume that contains the introduction and the Latin text. Conversely, I have placed virtually all the diagrams pertaining to the Introduction and Latin text at the end of the volume that contains the English translation and commentary. In a few cases, however, diagrams at the end of the volume containing the introduction are referred to in the introduction. Enabled thus to consult the diagrams continually in the one volume while progressing through the theorems in the other, the reader will be spared the inconvenience and irritation of flipping back and forth from page to page in an effort to reconcile text and figure.

The reference-aids provided in this edition are the same as those provided in the previous one. Accordingly, I have included a detailed topical synopsis at the beginning of each book in the English translation. In addition to the bibliography and general index, I have provided a Latin-English index keyed to technical terms in both Latin text and English translation. I have also provided an English-Latin glossary for cross referencing. In all, I have done my best throughout this edition to strike a balance between the needs and demands of experts in the field of medieval optics and mathematics and those of the larger community of interested scholars who require guidance to navigate that field effectively. Perhaps I have leaned too far in this latter direction, offering more guidance than necessary, but I took that risk to ensure that even a relatively inexpert reader could follow the intricacies of Alhacen's analysis and, in the process, gain a true appreciation of its ingenuity, rigor, and elegance.

NOTES

[1]For a complete description of the manuscripts and my method for grouping them according to families, see Smith, *Alhacen's Theory*, clv-clxxi. The actual manuscripts designated by the listed sigla are as follows: ***F*** Florence, Biblioteca Nazionale Centrale, ms II.III.324; ***P1*** Paris, Bibliothèque Nationale, ms Lat. 7247; ***Va*** Vatican, Biblioteca Apostolica, ms Palat Lat. 1355; ***V2*** Brugge, Stedelijke Openbare Biblioteek, ms 512; ***L2*** London, British Library, ms Sloane 306; ***S*** Saint-Omer, Bibliothèque Municipale, ms 605; ***Er*** Erfurt, Wissenschaftliche Bibliothek, ms Ampl F.392; ***C1*** Cambridge, Trinity College, ms 0.5.30; ***O*** Oxford, Corpus Christi College, ms 150; ***M*** Munich, Bayerishe Staatsbibliothek, ms CLM 10269; ***E*** Edinburgh, Royal Observatory, Crawford Library, ms Cr3.3; ***P3*** Paris, Bibliothèque Nationale, ms Lat. 7319; ***P2*** Paris, Bibliothèque Nationale, ms Lat. 16199; ***L3*** London, Royal College of Physicians, ms 383; ***C2*** Cambridge, University Library, ms Peterhouse 209; ***L1*** London, British Library, ms Royal 12.G.7; and ***V1*** Vienna, Österreichische Nationalbibliothek, ms 5322.

[2]See ibid., clxvii-clxix.

[3]See ibid., clxv.

[4]A. I. Sabra has in fact been working on this edition for well over 20 years, as indicated by the analysis of six propositions from book 5 that he published, along with an English translation, in "Ibn al-Haytham's Lemmas," 299-324, esp. 315-324. These six propositions correspond to the ones numbered 19-24 in the present Latin edition.

[5]See Smith, *Alhacen's Theory*, clx-clxi.

[6]The passage in question occurs at the beginning of paragraph 2.451 but is found at the end of paragraph 2.445 in all the manuscripts.

[7]See Smith, *Alhacen's Theory*, clxxvi.

[8]It is worth noting that the diagrams Sabra adapted from the Arabic originals in "Lemmas for Solving 'Alhazen's Problem'" correspond in all essential respects to their counterparts in the Latin manuscripts.

[9]The figures in Risner's edition are somewhat misleading in this regard, because, although most of them are based on the manuscripts he used (within the family represented by *E*), he felt free on occasion to reletter them; see, for example, p. 132 of Risner's edition for his version of figure 5.2.1, p. 220, where all the letters have been unaccountably changed. Risner also included many of the figures provided in the manuscripts for book 4. On the other hand, he made significant strides toward improving the accuracy of the more involved figures in the manuscripts; see, e.g., p. 150 of Risner's edition for his version of figure 5.2.25, p. 239. In several instances I found his improved figures to be indispensable as I worked my way through Alhacen's mathematical analysis.

[10]The production of illustrated and / or rubricated manuscripts tended to be a small-scale assembly-line operation, each phase of which, from writing to illuminating and

illustrating, was carried out by a separate specialist. The resulting lack of communication between scribe and illustrator certainly helps explain the divergences in letter-designations between text and diagrams in so many of the manuscripts. It also helps explain why, in such manuscripts as *O* and *E*, extensive textual correction was necessary.

[11]On the one hand, when such digitally scanned diagrams are magnified, the lines tend to fragment and blur. On the other hand, when those diagrams are reoriented, the accompanying letters are reoriented as well, so, if the diagram is inverted, the letters will also be inverted. These problems can, of course, be resolved in various ways by enhancement and cut-and-paste options, but, given the number and complexity of the diagrams involved, I rejected that option as too time consuming to be worthwhile.

[12]See figure 5.6, p. 219, for this composite figure, the portion highlighted by heavier lines forming the basis for FIGURE 5.2.31, p. 587. Note that, in addition to abstracting the figure from its context in the composite diagram, I inverted it to stand the cone upright on its base and adjusted the lettering to conform with this inversion.

[13]See Smith, *Alhacen's Theory*, clxxiv-clxxv.

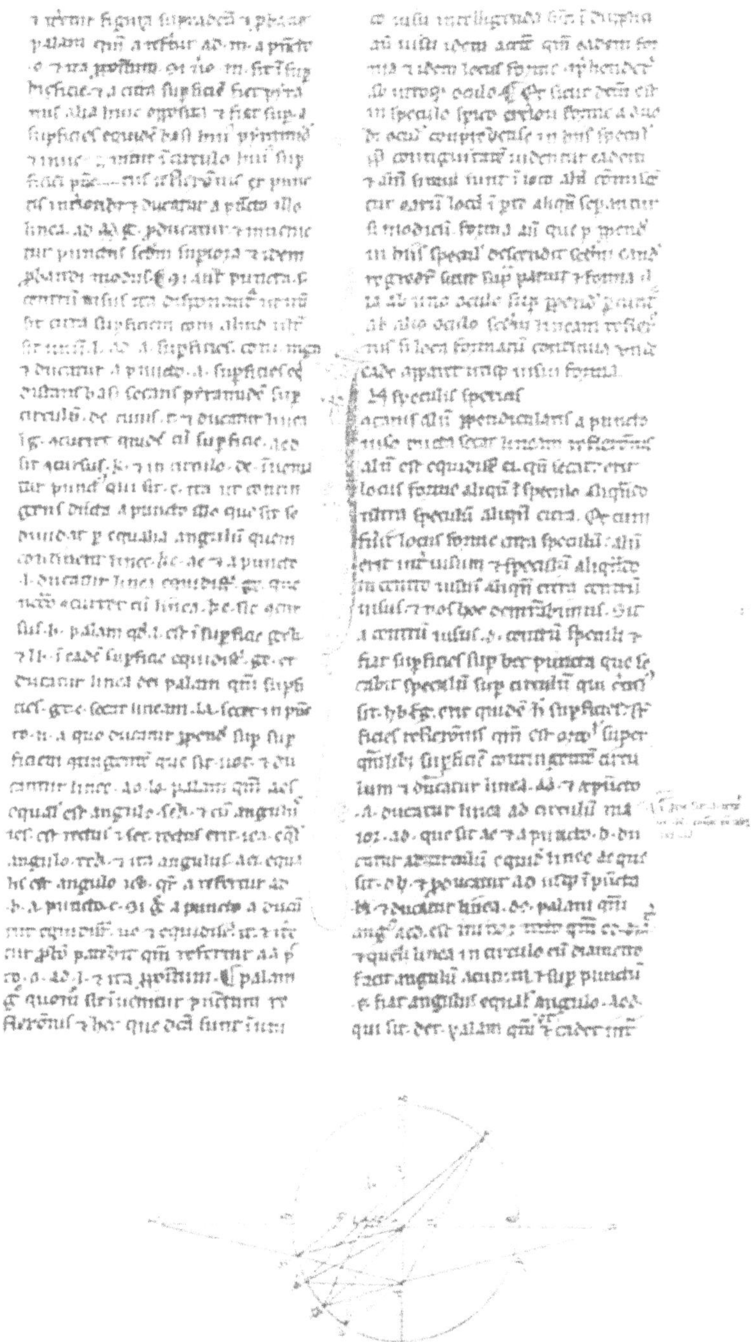

F Florence, Bibl. Naz. Cent., MS II.III.324, f 72 (71)v

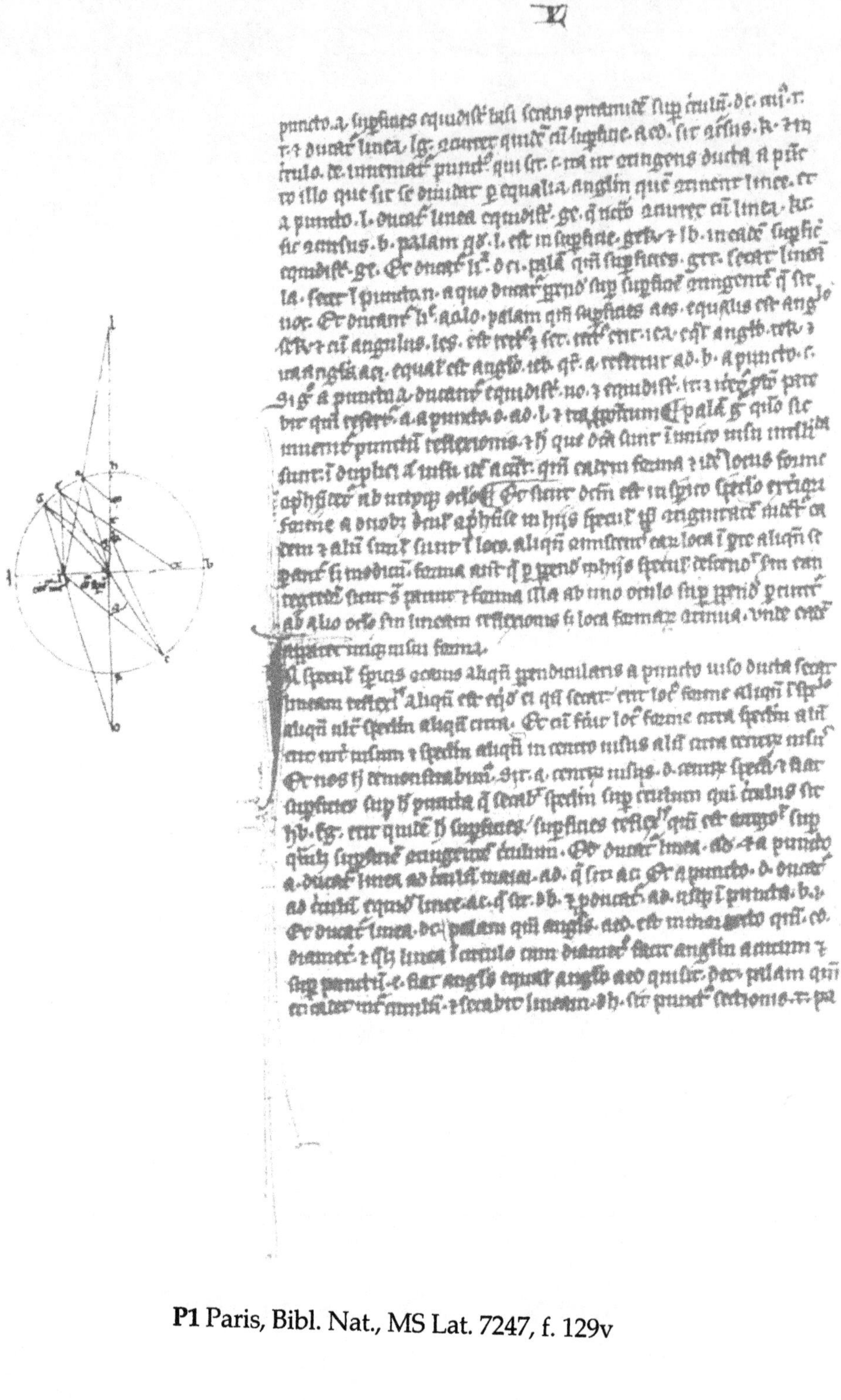

P1 Paris, Bibl. Nat., MS Lat. 7247, f. 129v

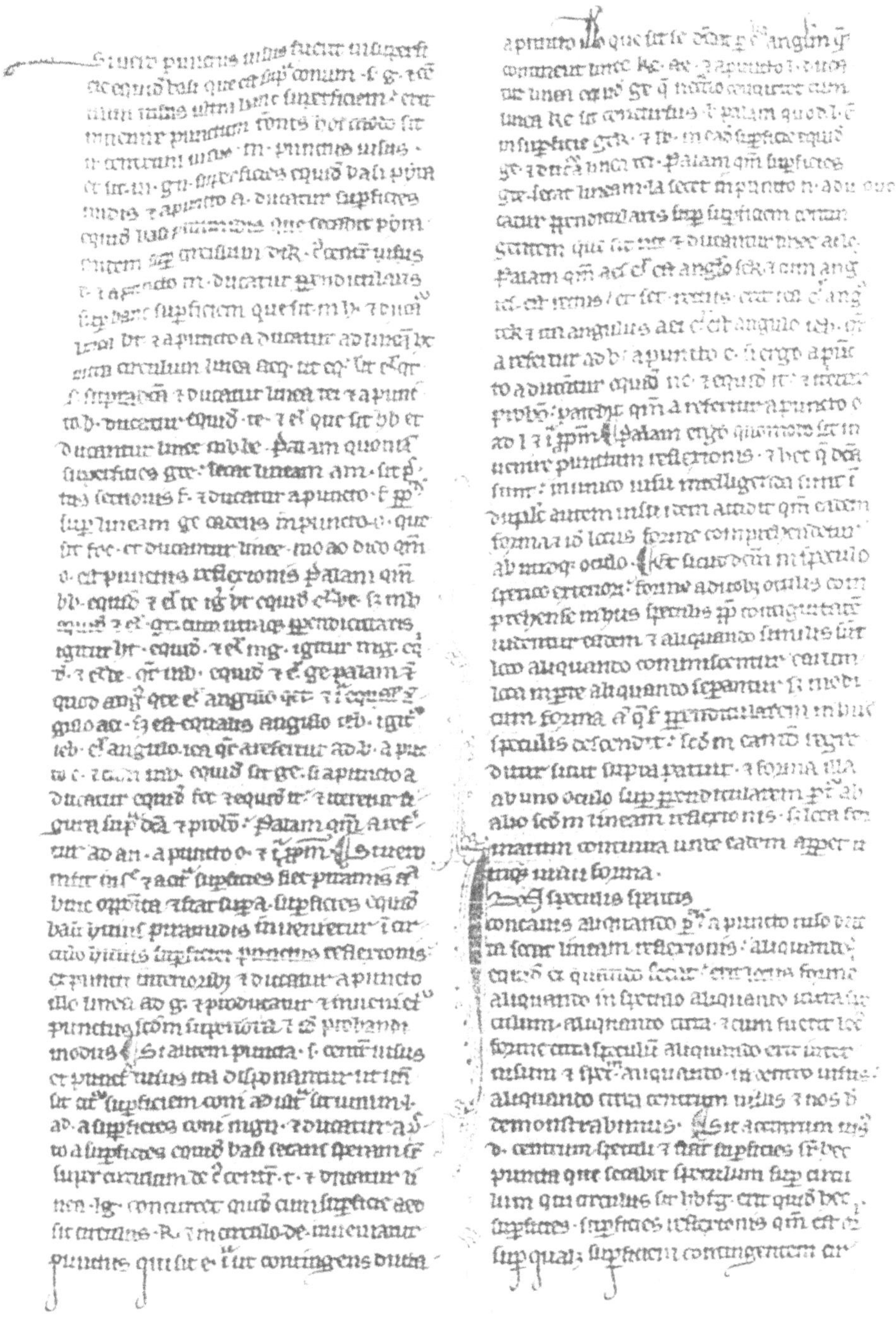

S Saint-Omer, Bibl. Munic., MS 605, f 96r

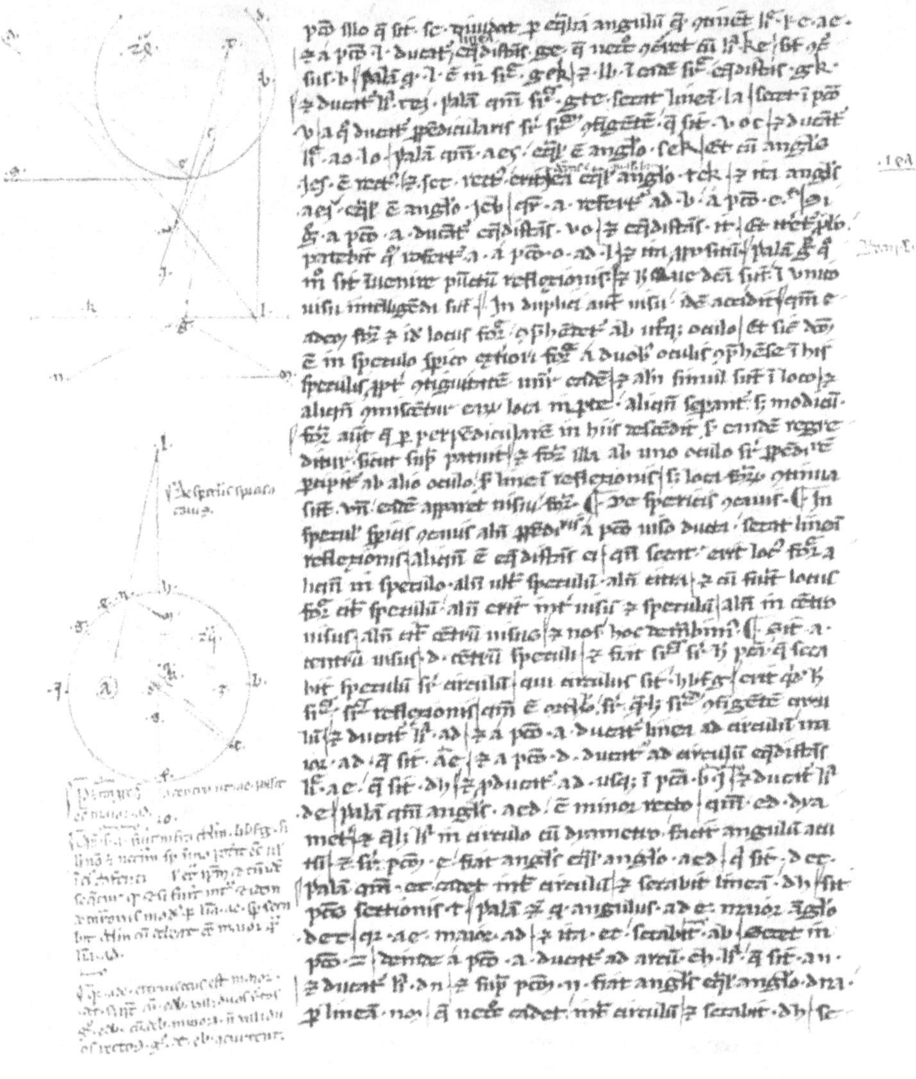

E Edinburgh, Crawford Library, Royal Observatory, MS Cr3.3, f. 122v

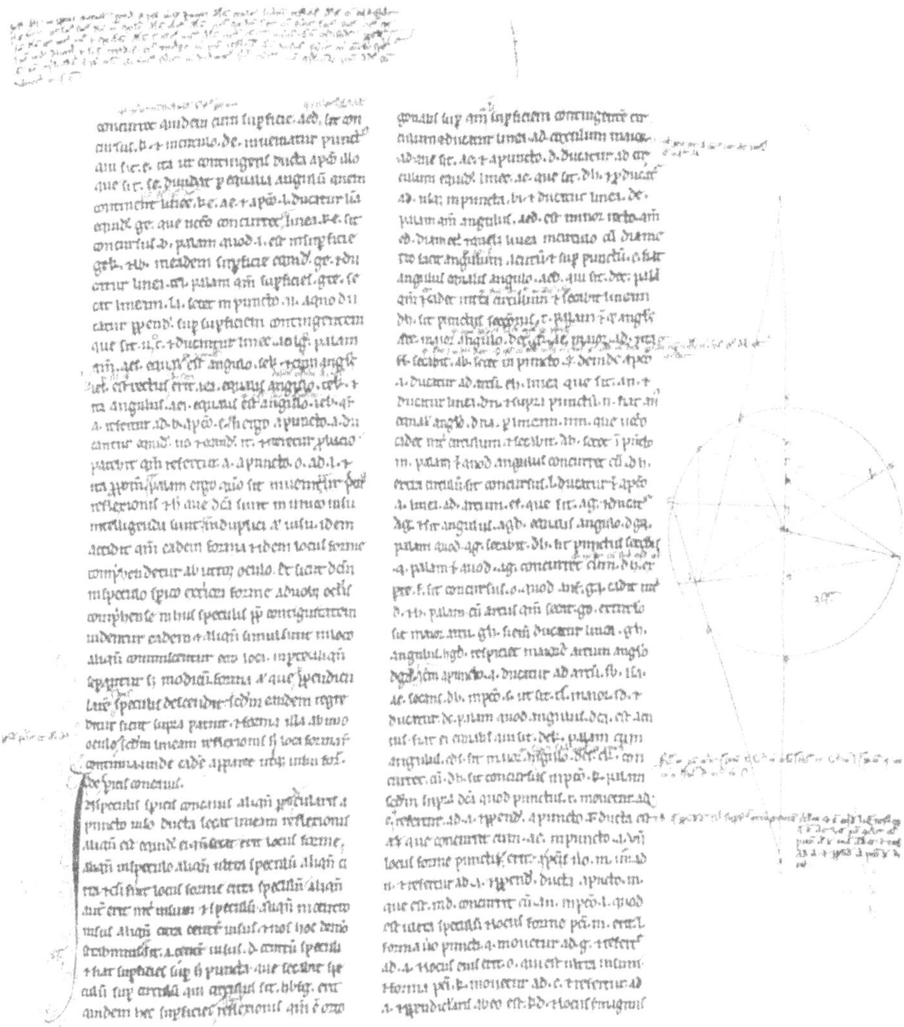

L3 London, Royal College of Physicians, MS 383, f. 77r

O Oxford, Corpus Christi College, MS 150, f. 68v

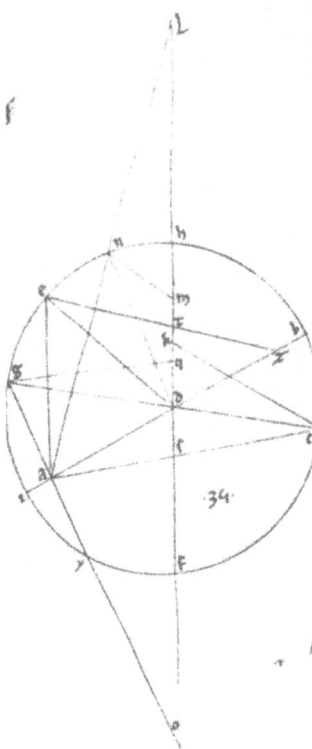

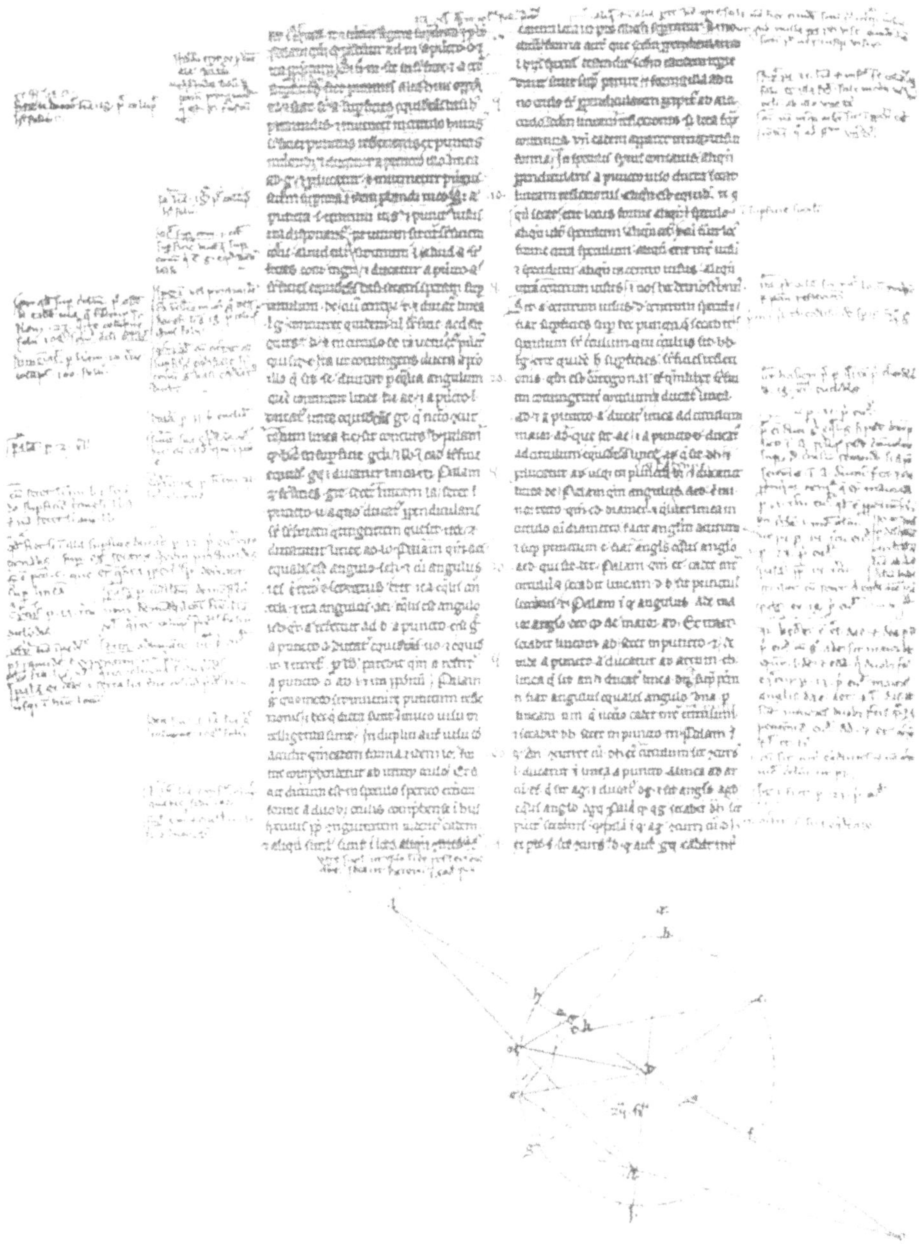

C Cambridge, Trinity College, MS 0.5.30, f. 105v

ALHACEN'S
DE ASPECTIBUS
LATIN TEXT

[QUARTUS TRACTATUS]

Liber iste dividitur in quinque partes. Pars prima proemium libri; secunda in declaratione quod lucis accidit reflexio ex politis corporibus; tertia in modo reflexionis forme; pars quarta in ostensione quod comprehensio forme ex corporibus politis non est nisi ex reflexione; pars quinta in modo comprehensionis formarum per reflexionem.

[CAPITULUM 1]

[1.1] Iam explanavimus in libris tribus modum comprehensionis formarum in visu cum fuerit directus, et enumeravimus singula que in rebus visis comprehendit visus. Sed diversificatur adquisitio visus tripliciter: aut enim directe, sicut diximus; aut per reflexionem in politis corporibus; aut per penetrationem, ut in raris, quorum non est raritas sicut raritas aeris. Nec potest diversificari visus nisi hiis modis tribus.

[1.2] Et hiis duobus modis posterioribus comprehendit visus in rebus visis ea que supra exposuimus et quorum adquisitionem in visu directo patefecimus. Et forsitan visus in hiis incurret errorem aut consequitur veritatem. Et nos assignabimus in hoc libro quomodo per reflexionem fiat formarum adquisitio, et quomodo erit reflexio, et quis linearum reflexarum situs. Et preponemus quedam antecedentia preponenda.

1 *post* prima *add.* est *C1R; inter. O* 2 *ante* in *add.* est *R*/lucis: luci *C1R*; lucibus *O*/accidit: accidet *R*/ex: a *R*/politis *corr. ex* polititis *S* 3 *post* tertia *add.*est *R*/pars *om. R*/quod *mg. a. m. F; om. P1* 5 pars *om. R*/*post* quinta *add.* est *R*/reflexionem *corr. ex* flexionem *O; a. m. S* 6 libris tribus *transp. SOL3C1E*/modum: modos *C1* 7 directus *corr. ex* ductus *O* 10 *post* raris *add.* corporibus *C1* 11 sicut raritas *mg. L3*/*post* sicut *add.* est *C1*/raritas *om. O*/nec *corr. ex* non *E*; et non *R*/*post* nisi *add.* in *ER*/tribus . . . modis (12) *om. P1* 13 ea *om. R*/exposuimus *corr. ex* exposuerimus *C1* 14 forsitan: forte *FP1*/incurret *corr. ex* curret *a. m. S; alter. ex* intret *in* incurrit *a. m. E*; incurrit *R*/*ante* errorem *add.* in *R* 15 consequitur *alter. in* consequetur *C1*/quomodo *corr. ex* quando *a. m. E*; quando *R* 17 reflexarum *mg. F*/reflexarum situs *transp. P1*/antecedentia *corr. ex* accedentia *S*; accedentia *E*; accidentia *OL3R*

[CAPTITULUM 2]

[2.1] Planum ex libro primo quoniam lux a corpore lucido luce ei propria vel accidentali dirigitur in omne corpus ei oppositum, et eodem modo color cum in eo lux fuerit mittitur. Igitur corpore polito opposito corpori lucido, mittitur ad ipsum lux lucidi mixtim cum colore, et reflectitur lux cum colore, sive fuerit fortis sive debilis, sive prima sive secondaria.

[2.2] Et quod fiat in luce forti reflexio patere potest opposito luci forti speculo ferreo, et etiam oppositus sit paries speculo; et descendat super ipsum lux declinata non recta. Videbitur in pariete lux fortis reflexa, que quidem non videbitur super eundem locum si speculum auferatur, nec videbitur super eundem locum si speculum moveatur; immo secundum motum speculi mutabitur locus lucis reflexe in pariete. Quare palam reflexionem fieri in luce forti.

[2.3] In luce debili patere potest defacili. Intra domum aliquam per foramen unicam a terra elongatum, sed non multum, descendat lux diei, non solis, super aliquod corpus. Et circa corpus statuatur speculum ferreum, et circa speculum corpus aliquod album. Apparebit in secundo corpore albo lux maior quam sine speculo, et augmentum illius lucis non est nisi ex speculi reflexione, quoniam ablato speculo, sola lux secundaria debilis apparebit in corpore albo.

[2.4] Amplius, si diligens figatur intuitus in lineis per quas a corpore primo lux in speculum mittitur, perpenditur quidem linearum illarum declinatio super speculum et super idem linearum reflexionis declinatio eadem. Et est proprium reflexioni ut eadem sit declinatio et idem angulus linearum venientium et reflexarum. Quod si moveatur

19 *post* planum *add.* est *R; inter. a. m. S*/*post* ex *add.* hac *E*/libro primo *transp. FP1*/primo *inter. O*/ quoniam: quod *R*/ei *om. ER* 20 dirigitur: dirigatur *R*/in . . . mittitur (21) *mg. a. m. C1* (color *om. C1*) /*post* omne *add.* ad quod reflectantur luces vel colores ad (ad[1], ad[2] *inter.*) *O*/ei *corr. ex* enim *O* 21 lux fuerit *transp. FP1*/igitur: itaque *R*/polito *corr. ex* posito *OL3*/corpori *corr. ex* corporis *C1* 22 lucidi *om. R*/mixtim: mixtum *FP1; corr. ex* mixtum *L3C1* 23 sive[2]: si *E* 24 secondaria: secundaria *R* 25 in *om. FP1*/patere potest *transp. ER* 26 et etiam: si *R*; et cum *inter. a. m. E*/sit *om. FP1*; fuerit *R*/descendat: descenderit *R* 28 *post* si *add.* moveatur *O*/*post* speculum *inter.* vel *O* 29 auferatur *alter. in* movetur *a. m. E*/nec . . . speculum *om. R*/speculum moveatur *transp. O*/*ante* moveatur *add.* vel *R*/moveatur: removeatur *C1* 30 secundum *inter. a. m. E*/*post* speculi *add.* et *L3C1*/*post* locus *scr. et del.* speculi *O* 31 *post* palam *inter.* est *O; mg. a. m. SC1*/luce forti *transp. S* 32 defacili: facile *R*/*ante* intra *add.* si *R*/unicam: unicum *R* 34 *post* diei *add.* ut *O*/et: sed *C1*/circa *alter. ex* citra *in* contra *a. m. C1*/speculum ferreum (35) *transp. L3C1* 35 circa: citra *C1*/ferreum . . . speculum *rep. FP1* 36 secundo *inter. L3*/illius lucis (37) *transp. P1* 37 lucis *mg. L3; om. ER* 39 *post* figatur *add.* ?? *C1* 40 perpenditur: perpendetur *R* 41 illarum: aliarum *O*/*post* linearum *add.* punctum *R* 42 reflexioni: reflexionis *R*/eadem sit *transp. R*/sit *om. C1* 43 quod *corr. ex* et *a. m. E*

corpus album a loco reflexionis in alium locum, tamen circa speculum, non videbitur in eo lucis augmentum, nec videri poterit nisi in illo situ tantum. Quare planum proprium esse reflexioni hunc situm.

[2.5] Hoc idem poteris videre secundaria luce si predictum speculum sit argenteum et corpus tertium album sit ex alia parte speculi. Apparebit quidem super corpus tertium lux secundaria, et super corpus secundum lux maior illa, et palam huius maioritatis causam solam esse reflexionem. Patebit autem lucis reflexio in omni loco ubi super corpus descendat per foramen aliquod lux fortis, adhibito luci speculo et ei corpore albo opposito modo supraposito.

[2.6] Verum locum reflexionis proprium et linearum situm explanabimus. Iam patuit in libro primo quod lux reflexa sequitur rectitudinem linearum, quare ex corporis politis fit reflexio secundum processum rectitudinis in situ proprio.

[2.7] Amplius, planum ex superioribus quod lux secunda a corpore illuminato accidentali luce procedens secum fert colorem corporis. Ab omni igitur corpore illuminato seu lucido color mixtus cum luce ad corpora opposita polita mittitur, et mixtim in partem debitam reflectitur.

[2.8] Et huius rei fides poterit fieri si intra domum unius foraminis tantum descendit lux solis super corpus forti et specioso colore. Et statuatur circa ipsum speculum ferreum, et circa speculum corpus concavum ad ciphi modum intra quod sit corpus album, et aptetur hoc vas in loco reflexionis ut lux reflexa incidat in corpus album. Apparebit quidem super faciem albi corporis color illius in quo fit descensus lucis, quod quidem non accideret si extra proprium reflexionis situm statuatur corpus album. Et secundum diversas colorum species hec probatum invenies, velut colori celesti, rubore, viriditate, et huiusmodi. Quare planum colorem mixtum cum luce remitti, et certior est coloris reflexi apparentia si speculum fuerit argenteum.

44 circa: citra *C1* 45 nisi *mg. a. m. F* / illo situ *transp. SOL3C1E* 46 *post* planum *add.* est *R* / reflexioni: reflexionis *R* / *post* situm *scr. et del.* h *P1* 47 poteris: poterit *FP1SR* / videre: videri *R* / *post* videre *inter.* in *a. m. L3E; add. C1R* 50 illa *corr. ex* illarum *E* / et *om. E* / causam: causa est *O* 51 autem *inter. a. m. L3* / lucis reflexio in omni loco: in omni loco lucis reflexio *R* / ubi *corr. ex* visi *L3* 52 descendat *alter. in* descendit *C1E*; descendit *R* / *post* aliquod *scr. et del.* super *C1* / luci *corr. ex* lucis *E* 53 opposito *om. O; mg. L3; inter. E* / supraposito: superposito *L3*; supradicto *OR* 54 proprium *om. R* 55 *post* lux *scr. et del.* et *C1* 56 corporis: corporibus *R* 58 *post* planum *add.* est *R* 59 ab omni igitur (60): igitur ab omni *C1* 60 igitur: ergo *O* / mixtus: mixtim *SOL3R*; mixtum *P1*; *corr. ex* mixtum *F*; *alter. ex* mixtum *in* mixtim *C1E* / ad *corr. ex* et *S* 61 *post* corpora *scr. et del.* ad *S* / mittitur *corr. ex* mittere *O* / mixtim: mixtum *L3*; *corr. ex* mixtum *C1* / partem: parietem *P1* 62 huius: huic *R* 63 descendit: descendat *FP1SR* / solis *om. R* 66 ut *corr. ex* et *a. m. C1* 67 *post* illius *scr. et del.* corporis *S* / quo: quod *OR* / *post* fit *add.* primo *R* 68 accideret: accidet *R* / reflexionis situm *transp. R* 69 hec: hoc *L3C1R* 70 velut: ita in *R* / colori: colore *OER* 71 mixtum: mixtim *O* / cum *om. L3* / certior: certiorem *ER* / est: esse *SL3ER*; *corr. ex* esse *FO*; *alter. ex* esse *in* esset *C1* 72 apparentia: apparentiam *ER* / argenteum *corr. ex* argentumteum *F*

[2.9] Quare autem non appareat hec probatio—scilicet, quod comprehendatur color reflexus cuicumque corpori opponatur speculum et ei adhibeatur album—hec est ratio. Sicut supradictum est, colores debiles, licet simul cum luce mittantur, non sentiuntur, quia forme que reflectuntur debiliores sunt formis a quibus reflexio oritur. Et hoc in luce potest patere, quoniam luce forti in speculum cadente et reflexa in pariete, debilior videbitur lux parietis quam speculi, et notabilis est inter eas proportio.

[2.10] Idem patebit in luce debili pari modo. In domo prima dispositione prima, si corpus tertium album ponatur loco speculi ferrei vel circa ipsum, maior apparebit lux super hoc corpus quam super secundum, quod non accideret nisi reflexio lucem debilitaret.

[2.11] Sed dicet aliquis causam huius rei esse nigredinem speculi ferrei, que admixta luci in speculum cadenti ipsam obumbrat, et inde, reflexa in corpus secundum, debilis et fusca apparet. Sed in corpus tertium loco speculi vel circa positum, non descendit lux nisi a corpore primo nulli admixta nigredini. Verum quod hoc non sit in causa palam eo quod, loco speculi ferrei argenteo posito, eadem accidet probatio.

[2.12] Pari modo reflexus color debilior erit colore a quo fit reflexio, quod in domo reflexionis coloris patere poterit, si corpus album loco speculi ponatur vel circa. Fortior apparebit in ipso color quam in corpore albo intra vas posito. Et idem patebit si in loco ferrei argenteum ponatur speculum. Igitur reflexio debilitat et luces et colores, sed colores amplius quam luces secundum utrumque speculum. Et est quoniam colores accedunt debiliores quam luces, unde facile efficiuntur in reflexione debiliores.

[2.13] Amplius, color debilis, cum ad speculum pervenit, miscetur colori eius, quare reflexus apparet debilis et tenebrosus; et forme

73 appareat hec: apparet hoc *L3C1E* 74 opponatur *corr. ex* opponati *L3* 75 et: in *FP1*; sed *R* 76 debiles *inter. a. m. E* / non sentiuntur *om. FP1* / quia: in *O*; *inter. L3*; *om. ER* / *post* forme *inter.* enim *a. m. E*; *add. R* / que *alter. in* et *O* 77 debiliores: debilioris *F*; delioris *P1*; deliores *S* / hoc *inter. a. m. E* / hoc in *transp. L3* / hoc in luce (78): in luce hoc *S* / in *inter. a. m. C1* / *post* in *add.* hoc *ER* 78 in[2] *inter. O* 81 *post* modo *add.* ut *R* / *post* domo *add.* in *R* / prima dispositione (82) *inter. L3*; *inter. a. m. O* 82 prima *om. R* / *post* corpus *add.* tersum *R* / tertium *corr. ex* tersum *a. m. E*; *alter. ex* tirsum *in* tersum *a. m. C1* / ponatur: ponamus *ER* 84 accideret: accidet *P1* / reflexio *corr. ex* reflexionis *C1* 85 huius: huiusmodi *C1E* 86 inde *corr. ex* in *a. m. E*; *om. R* 87 sed *corr. ex* si *a. m. E* 88 tertium *alter. in* tersum *a. m. C1* / vel circa *om. R* / circa: citra *C1* / *post* positum *scr. et del.* d *F* / non *corr. ex* duo *F* 89 primo: proprio *O* / *post* palam *add.* ex *OR* 90 *post* eo *add.* est *R* / argenteo: argento *FP1C1* / *ante* posito *scr. et del.* posito *F* / accidet *om. FP1*; accidit *ER* 91 fit *inter. a. m. E* / reflexio: reflexionis *E* 92 reflexionis coloris: et vase ut antea *R* / poterit: potest *R* 93 ponatur vel circa: vel circa ponatur *O* / *post* in[1] *add.* ea *FP1* / quam: quem *FP1* 94 si *om. FS*; *inter. P1O* / in *om. R* / *post* ferrei *add.* speculi *R* / argenteum: argentum *O* 96 *post* luces *scr. et del.* unde facile efficiuntur in reflexione *O* / quoniam: quia *O* 97 accedunt debiliores *transp. OC1* / debiliores: debiles *S*; *inter. a. m. L3* / unde . . . reflexione (98) *mg. O* 99 cum . . . (eius: speculi) . . . aliquis (103) *mg. S* / ad speculum pervenit: pervenit ad speculum *R* 100 apparet: apparebit *ER* / *post* forme *scr. et del.* enim *O*

debiliores sunt reflexe quam in loco reflexionis, et reflexio causa est debilitatis.

[2.14] Poterit aliquis dicere non esse debilitatem formarum in reflexione nisi ex elongatione earum a sua origine. Sed explanabitur quod, licet ab ortu equaliter elongentur lux directa et lux reflexa, tamen debilior erit reflexa.

[2.15] Intret radius solis domum aliquam per foramen, et opponatur foramini in aere speculum ferreum minus foramine. Et lux foraminis residua cadat in terram super corpus album, et lux a speculo reflexa cadat in corpus album elevatum. Hoc observato ut eadem sit elevati et iacentis a foramine longitudo, videbitur quidem super elevatum lux minor quam super iacens. Et huius minoritatis non potest assignari causa nisi reflexio sola. Idem accidet si speculum fuerit argenteum.

[2.16] Idem in colore potest patere, luce solis in domum aliquam per foramen descendente super corpus coloris fortis cui circa adhibeatur speculum, et aliud corpus concavum intra quod sit corpus album in quod cadit reflexio. Et statuatur in domo aliud corpus album eiusdem modi cum eo quod est in concavo, et sit elongatio huius albi a corpore colorato in quod cadit lux foraminis eadem cum elongatione albi quod est in concavo ab eodem, et elongatione speculi ab eodem. Perpendi quidem poterit color debilior in albo quod est intra concavum quam in eo quod est extra, licet equidistant ab actu suo—id est a corpore colorato. Et in causa est reflexio colorem debilitans.

[2.17] Amplius, lux reflexa fortior est luce secundaria, licet eiusdem sit elongationis ab origine sua. Luce etenim reflexa cadente in corpus aliquod, si aliud eiusmodi corpus ponatur extra locum reflexionis, et

101 sunt *inter. O* / *post* reflexe *scr. et del.* ?? *C1* / quam *corr. ex* quasi *F* / reflexio *corr. ex* reflexionem *O* / causa est *transp. O* / causa *corr. ex* causam *L3* / est *om. L3* 103 formarum *om. P1; alter. ex* foraminis *in* forme *O* 105 quod *om. FP1; inter. a. m. E* / *post* quod *add.* solam *mg. L3* / *post* ortu *add.* suo *O*; sui *mg. a. m. C1* / elongentur: elongetur *O* / tamen *om. FP1O* 107 *post* solis *add.* in *R* / et *inter. O* 108 ferreum *corr. ex* ferenum *E* 109 cadat[1]: cadit *SL3; corr. ex* cadit *a. m. E* / a speculo reflexa: reflexa a speculo *FP1* / cadat[2]: cadit *L3* 110 album *alter. in* aliud *inter. a. m. C1* / *post* album *add.* a terra *O* / *post* et *add.* latentis *FP1E* 111 iacentis *mg. a. m. F; om. P1E* / a *inter. a. m. C1* / *post* foramine *scr. et del.* et *L3C1* / super elevatum *corr. ex* elevatum super *E* 113 accidet *corr. ex* accidit *P1*; accidit *ER* / fuerit *om. E*; sit *R* / argenteum *corr. ex* arteum *mg.F* 114 idem: item *L3C1* / luce: lux *R* / *ante* solis *add.* enim *R* / domum: domui *O* 115 descendente: descendat *R* / circa: contra *C1; om. R* / adhibeatur: adhibeantur *L3; corr. ex* adhibeantur *C1* 116 aliud: quod *O* / *post* aliud *inter.* sit *O* / corpus[1] *om. E* / sit *corr. ex* si *S* 117 cadit: cadat *O* / statuatur: statuatuatur *S* / aliud *corr. ex* aliquid *O* / eiusdem: eius *E* 118 cum eo *inter. O* / huius: huiusmodi *C1* / a *mg. a. m. E* / *post* corpore *scr. et del.* s *O* 120 ab[1] *corr. ex* cum *L3E* (*a. m. E*) / eodem[1]: eadem *E* / *post* et *add.* cum *R* / elongatione: elongatio *OL3C1E* / eodem[2]: eadem *L3E* / perpendi *alter. ex* perpendicularis *in* erit *O* / perpendi quidem poterit (121): tunc comprehendetur *R* 121 quidem poterit *transp. E* 122 equidistant: equidistent *OL3C1ER* / actu *alter. in* arcu *L3*; ortu *C1ER* / a: in *FP1* 124 fortior est *transp. O* / est *inter. SOC1* (*a. m. C1*); *om. L3E* 125 sit: sint *R* / etenim: enim *R* 126 eiusmodi: huiusmodi *O*; eiusdemmodi *L3C1*

sit cum eo eiusdem elongationis a speculo, videbitur super ipsum lux minor quam in illo.

[2.18] Idem etiam planum erit in domo, si deprimatur in terra in directo foraminis speculum quod accipit totam foraminis lucem. Erit lux fortior super corpus in loco reflexionis positum quam super aliud eiusdem modi extra hunc locum tantumdem a speculo elongatum.

[2.19] Eodem modo, si excedat lux foraminis quantitatem speculi, et cadat circa speculum lux in terram aut corpus album a quo aliud corpus tantum elongatur quantum corpus reflexionis a speculo, debilior apparebit in eo lux quam super reflexionis corpus.

[2.20] Similiter accidit in colore. Si corpus aliquod tantum distet a speculo extra situm reflexionis quantum aliud ei simile quod est in situ, apparebit quidem super corpus quod est in situ reflexionis color reflexus; super aliud forsitan nullus. Si enim ferreum fuerit speculum aut fere nullus videbitur, aut omnino nullus; si vero argenteum fuerit speculum, apparebit super ipsum color aliquis, sed valde debilis, et longe debilior quam in corpore quod est in situ reflexionis.

[2.21] Et iam igitur planum quod forme lucium et colorum reflectuntur ex corporibus politis et in reflexione debilitantur. Et erit forma directa fortior reflexa cum eadem earum origo et equalis ab ea elongatio. Et reflexa fortior secundaria cum idem vel equalis earum ortus et par elongatio.

[CAPITULUM 3]

Pars tertia: in modum reflexionis formarum in corporibus politis

[3.1] Politum est lene multum in superficie, et lenitas est quod sint partes superficiei continue sine pororum multitudine. Lenitas intensa

129 etiam *corr. ex* est *O*; et *L3* / deprimatur: deponatur *R* / terra: terram *R*   130 quod *alter. in* et *inter. O* / accipit *alter. in* accipat *O*; accipiat *R*   132 tantumdem *corr. ex* tandtumdem *F* / a speculo *mg. L3E* (*a. m. E*) / a speculo elongatum: elongatum a speculo *ER*   133 foraminis *om. O*   134 cadat: cadit *SL3*; *corr. ex* vadat *O* / in terram *corr. ex* interior *a. m. E* / *post* aut *inter.* in *a. m. C1*   135 quantum: quam *E*   139 *post* situ[1] *add.* reflexionis *R* / apparebit . . . situ[2] *om. P1* / color reflexus (140) *om. O*   140 forsitan: forte *FP1* / nullus: vel *O*   141 aut . . . speculum (142) *om. O*; *mg. a. m. L3* / fere nullus *om. R* / *ante* videbitur *add.* modicus *R*   142 sed: et *O*   143 corpore *corr. ex* corde *F*   144 et[1] *om. OC1* / *post* forme *scr. et del.* lucis *P1* / reflectuntur (145): inflectuntur *FP1*; *om. ER*   145 *post* politis *add.* reflectuntur *ER*   146 *post* cum *inter.* sit *O* / *post* eadem *add.* fuerit *R* / *post* ea *add.* origine *R*   147 *post* cum *add.* est *R* / vel *inter. a. m. E* / equalis earum *transp. C1* / earum: eorum *O*; *inter. a. m. L3*; *om. ER* / *post* ortus *add.* est *C1* / *post* et[2] *scr. et del.* p *S*   149 pars . . . politis (150) *om. FP1S*   151 lene: leve *R* / lenitas: levitas *R* / quod: ut *R*   152 pororum *inter. a. m. E* / lenitas: levitas *R*

est ubi multa partium superficiei continuitas et pororum parvitas et paucitas, et finis lenitatis est privatio pororum et privatio divisionis partium. Igitur politas est politiva continuitas partium superficiei cum poris raris et exiguis, et finis politive est vera continuitas partium et privatio pororum.

[3.2] In omnibus politis superficiebus, licet diversis subiaceant figuris, accidet reflexio, et idem reflexionis modus et eadem proprietas: et est quod in omni polita superficie ab omni puncto fit reflexio; et sumpto quocumque puncto in superficie a quo fiat reflexio, linea accessus forme alicuius ad illum punctum et linea reflexionis in eadem superficie erunt cum linea perpendiculari super illud punctum erecta; et tenebunt hee linee eundem situm respectu perpendicularis et equalitatem angulorum. Et volo dicere perpendicularem que sit perpendicularis super superficiem tangentem corpus politum in illo puncto, et due linee cum perpendiculari sunt in eadem superficie ortogonaliter cadente super superficiem corpus politum in puncto a quo fit reflexio tangentem.

[3.3] Si autem linea per quam accidit ad speculum forma cadat perpendiculariter super illud, fiet reflexio forme per ipsam, non per aliam, et hoc est proprium in omni reflexione in omni polito corpore. Si ergo corpus politum fuerit planum, superficies tangens punctum reflexionis erit una et eadem cum superficie corporis. Si vero fuerit columpnare speculum interius aut extra politum, erit contactus superficiei speculi et superficiei contingentis linea tantum secundum longitudinem speculi intellecta. Idem in speculo piramidali intus vel extra polito. In sperico, sive concavo interius sive exterius polito, contingens superficies tangit in solo puncto.

153 est *om. SL3; alter. in* et *inter. O* / ubi *corr. ex* ut *L3* / *post* ubi *add.* est *R* 154 lenitatis: levitatis *R* 155 igitur: itaque *R* / politas: politio *R* / est: vel *O* / politiva *alter. in* politura *a. m. C1* / *post* politiva *scr. et del.* igitur politas *F; inter.* est *O* 156 politive *corr. ex* polite *O; corr. ex* politionis *a. m. E;* politionis *R* 157 *post* pororum *add.* omnium *C1* 158 *post* politis *scr. et del.* licet *O* 159 reflexio et idem *om. FP1* / et[1] *inter. a. m. E* / *post* proprietas *add.* est *R* 160 in: ab *R* / polita: politi *ER* / *post* superficie *add.* et quolibet eius *R* / ab omni *om. R* / ab . . . superficie (161) *mg. a. m. S* 161 *post* superficie *scr. et del.* a quo in superficie *P1* / quo: qua *R* / fiat: fit *ER* 162 alicuius *om. ER* / illum: illud *R* / *post* eadem *scr. et del.* lo *F* 163 superficie *inter. O* 164 respectu: respecti *F* 165 angulorum *corr. ex* angelorum *F* / sit *inter. a. m. E* 167 cum *inter. C1* 168 *post* superficiem *add.* visi *O* / in *inter. OL3E (a. m. L3)* 170 per quam *om. O* / accidit: accedit *C1R* / *post* speculum *inter.* ad *O* / cadat: cadet *E* 172 ergo: igitur *R* 174 et *om. OL3E* / superficie *corr. ex* super forme *O* / *post* vero *scr. et del.* corporis *S* / *post* fuerit *add.* columna *F* 175 extra: exterius *R* / erit: erunt *FP1SL3E; corr. ex* erunt *O* / *post* superficie *scr. et del.* et *F* / speculi: speculum *O* 176 et superficiei *om. O* / contingentis *alter. in* tangentis *a. m. C1* / linea *corr. ex* lineam *SL3;* lineam *C1* 177 intellecta: intellectam *OE* / speculo piramidali *transp. O* 178 concavo *om. R* / *post* concavo *add.* sive *C1* / interius *om. O; inter. L3; inter. a. m. E* / tangit *corr. ex* tangi *O* 179 puncto *corr. ex* speculo *O*

[3.4] Quomodo autem ad oculum pateat hic modus reflexionis in speculis omnibus explanabimus. Accipe tabulam eneam spissam ut firmior sit, eius longitudo non minus quam 12 digitorum, et sit latitudo sex digitorum. Et fiat linea equidistans extremitati longitudinis et circa illam extremitatem. Et super punctum huius linee medium ponatur pes circini, et fiat semicirculus cuius semidyameter sit latitudo tabule.

[3.5] Et extrahatur a puncto quod est centrum linea ortogonaliter super dyametrum iam factum. Et erit linea illa semidyameter dividens semicirculum per equalia. Et in hoc semidyametro sumatur mensura unius digiti et, posito pede circini super centrum, fiat semicirculus secundum quantitatem partis residue semidyametri, residue scilicet secundum semidyametrum quinque digitorum.

[3.6] Et dividantur semicirculi primi medietates in quot libuerit partes ita quod sibi respondeant in qualitate prima—scilicet prima prime, secunda secunde, et sic de aliis—et protrahantur linee a centro ad puncta divisionum.

[3.7] Deinceps in semidyametro mensura digiti signetur, et ex parte centri, et super punctum signatum protrahatur linea equidistans dyametro semicirculi, sive tabule extremitati, quod idem est. Et secetur ex tabula quod interiacet hanc lineam et semidyametrum usque ad centrum et lineas primas ad divisiones semicirculi protractas—id est ad lineas tales semidyametro propinquiores.

[3.8] Post secetur tabula circa semicirculum maiorem ut solum remaneat semicirculus. Et secetur tabula sub centro ubi centri locus acuatur quasi punctus hoc tamen modo ut in eadem superficie plana remaneat cum semicirculo et aliis lineis.

[3.9] Post sumatur tabula lignea plana excedens eneam in longitudinem duobus digitis, et sit quadrata, et eius altitudo, sive spissitudo,

180 quomodo: quoniam *FP1S* / autem: et *E* / *post* autem *add.* etiam *R* 181 explanabimus: declanabimus *P1* / *post* spissam *add.* paululum *O* 182 *post* longitudo *inter.* sit *O*; *add.* sit *R* / *post* non *inter.* sit *a. m. C1* / minus: minor *R* / quam *om. R* / 12: 22 *P1* / digitorum: digitis *R* / et *om. OL3ER* / *post* sit² *inter.* que *a. m. E*; *add.* que *R* 185 circini: circuli *FP1*; *inter. E* / semicirculus *corr. ex* semicircularis *O* / semidyameter: semidyametri *E* 187 factum: factam *R* 188 semicirculum *corr. ex* circulum *L3C1* (*a. m. C1*) / hoc: hac *R* / semidyametro: dyametro *O* / sumatur *corr. ex* sumsit *F* 189 circini: circuli *FP1*; *om. C1* 190 *post* semidyametri *inter.* huius *a. m. S* / residue *inter. a. m. S* / residue scilicet *om. R* / scilicet *mg. a. m. C1*; *inter. a. m. E* 191 *ante* semidyametrum *scr. et del.* d *C1* / semidyametrum *corr. ex* semi *L3E* (*a. m. L3*) / quinque: 2 *O* 192 dividantur: dividentur *E*; dividiantur *R* / semicirculi *corr. ex* semidantur *F* / medietates: medietatis *FP1SOL3* 193 quod: ut *R* / respondeant: remaneant *FP1SL3* / qualitate: quantitate *O*; equalitate *R* / prima *scr. et del. O*; *om. C1R* 196 digiti *corr. ex* digitis *OL3* 198 quod item est *inter. L3*; *om. OC1ER* 200 semicirculi *corr. ex* semicirculis *S* / protractas: promotas *O*; pertractas *C1* 201 *post* tales *scr. et del.* dyametri *L3C1* / propinquiores *om. P1* 202 *post* post *add.* illud *C1* 203 et *inter. O* / ubi: ut *C1ER* 204 quasi: qua *L3* / punctus: punctum *R* / plana *om. R* 205 *post* remaneat *inter.* centrum *a. m. C1* 206 longitudinem (207): longitudine *R* 207 et sit quadrata *inter. O* / eius: ei *O*

septem digitorum. Signetur ergo in hac tabula punctum medium, et super ipsum fiat circulus excedens maiorem circulum tabule enee super quantitatem digiti magni. Et fiat super idem centrum circulus equalis circulo minori tabule enee.

[3.10] Et dividatur circulus maior in partes in equalitate respondentes partibus semicirculi tabule enee, ut scilicet prima respondeat prime, secunda secunde, et sic de aliis. Et secetur circumquaque tabula lignea ut solum remaneat maior circulus, et erit hec sectio usitato secandi modo. Secetur etiam pars tabule circulo minori contenta, et modus sectionis erit ut huic tabule associetur alia tabula ita ut linea a centro huius ad centrum illius transiens sit perpendicularis super illam. Et adhibito tornatili instrumento centris earum, fiat sectio partis circularis iam dicte. Est autem alterius tabule associatio, ut fixa stet in sectione.

[3.11] Igitur restabit tabula quasi anulus circularis cuius latitudo duorum digitorum, longtitudo 14, altitudo septem, et sit hec altitudo optime circulata ad modum columpne. Remanent autem in latitudine huius anuli linee dividentes circulum eius secundum divisionem semicirculi tabule enee.

[3.12] A capitibus harum linearum producantur linee in superficie altitudinis exterioris perpendicularis super superficiem latitudinis, et poterit hoc modo fieri. Queratur regula bene acuta cuius capiti linee adhibeantur, et regula moveatur donec tangat superficiem altitudinis in qualibet parte acuminis. Signa eius capita, et fac lineam, quoniam illa erit perpendicularis quam queris. Et eadem sit operatio secundum quamlibet dividentem lineam.

[3.13] Aliter poterit hoc idem fieri. Ponatur pes circini super terminum linee dividentis, et fiat semicirculus secundum altitudinem anuli, qui dividatur per equa. Et protrahatur a puncto ad punctum

208 signetur: significetur *L3* 209 super (210): secundum *O*; *alter. in* secundum *L3*; *om. R* 210 quantitatem: quantitate *R*/et *om. O* 212 et . . . enee (213) *mg. a. m. S* 213 ut *om. O*/respondeat (214) *corr. ex* respondet *C1* 214 secetur circumquaque *transp. ER* 215 erit: fiet *R*/usitato: usitati *P1* 216 secetur: seceat *O*/etiam: et *O*/circulo minori: minore circulo *R*/contenta: contempta *O*; *corr. ex* contempta *L3* 217 alia *corr. ex* talia *O* 218 huius *om. P1*/super *corr. ex* per *O*/illam: ipsam *L3C1* 220 dicte: dicto *O*/*ante* est *scr. et del.* autem *O*/autem *inter. O* 222 quasi: sicut *FP1*; ?? *O*/*post* latitudo *add.* erit *R* 224 optime *om. P1*/columpne: calumpne *F*/remanent: remanet *FC1* 225 anuli: anguli *E* 227 *post* capitibus *add.* autem *R*/harum linearum *transp. R*/linee: linea *R* 228 perpendicularis *alter. in* perpendiculares *O*; perpendiculares *R* 229 regula *om. FP1*/bene: lene *S*/*post* cuius *scr. et del.* s *F* 230 adhibeantur: adhibeatur *FP1O*/regula: reliqua *FP1*; *alter. in* ita *a. m. E*/*post* regula *scr. et del.* in *S*/tangat: transeat *ER* 231 signa *inter. O* 232 illa *inter. a. m. E*/queris *corr. ex* quamvis *O*/et . . . lineam (233) *om. R* 234 circini: circuli *FP1* 235 *post* dividentis *add.* circulum *R*/*post* altitudinem *scr. et del.* anguli *P1* 236 equa: equalia *SR*/ad: in *R*

linea, et ita in singulis. Pari modo, a terminis illarum dividentium protrahantur perpendiculares ex parte interioris altitudinis.

[3.14] Amplius, sumatur in altitudine interiori ex parte faciei non divisa altitudo duorum digitorum, et in perpendicularibus fiat signum. Et in signis illis fiat circulus equidistans faciei anuli hoc modo. Tabula aliqua plana fiat circularis equalis circulo minori tabule enee, et secetur ex ea pars aliqua usque ad centrum quasi triangulus ex duobus semidyametris et arcu circuli secundum quod libuerit ut possis tabulam cum manu imponere et locis assignatis aptare. Apta ergo locis illis ut sit equidistans faciei anuli, et fac circulum secundum ipsam.

[3.15] Sumatur etiam in hunc circulum altitudo medietatis grani ordei, et fiant signa, et in punctis signatis fiat circulus per aptationem tabule. Et in hoc circulo postremo fiat circularis concavitas, et sit unius digiti eius profunditas et altitudo tamquam altitudo tabule enee. Et sit altitudo hec intra altitudinem duorum digitorum ut eadem sit postremi circuli et concavitatis superficies.

[3.16] Aptetur autem huic concavitati tabula enea, que quidem intrabit concavitatem usque ad circulum minorem, cum distantia minoris a maiori sit unius digiti, et concavitas similiter. Igitur circulo postremo et tabule enee communis erit superficies, et linee perpendiculares in altitudine anuli tangunt lineas divisionis tabule enee, et cadent perpendiculariter super tabulam eneam. Sit autem facies tabule enee divisa ex parte faciei anuli divise.

[3.17] Amplius, in exteriori altitudine anuli signetur punctus a longitudine duorum digitorum, et posito pede circini super punctum signatum, fiat circulus secundum quantitatem unius grani ordei. Et instrumento ferreo cuius similiter latitudo sit quantitas unius grani ordei perforetur foramine columpnari. Et baculus ligneus foramini aptetur,

237 et *inter. E* / in: de *R* / *post* singulis *inter.* sectionibus *L3; add.* sectionibus *C1* / pari *alter. ex* ponatur *inter.* eodem *O* / *post* dividentium *add.* circuli *C1* 238 perpendiculares: particulares *E* 240 divisa: diverse *O*; divise *R* 242 plana: plena *L3* / equalis circulo *transp. OL3C1R* / *post* minori *add. circulo O* 243 ea: eo *FP1SO* / triangulus: triangulis *FP1L3*; triangus *S*; triangulum *R* / duobus: duabus *R* 244 quod *inter. a. m. E* 245 *post* assignatis *add.* imponere et *E* / ergo: igitur *O* / locis *inter. a. m. S* 246 ipsam: ipsum *C1* 247 etiam: ergo *L3C1E* / in: infra *R* / circulum: modum *E* / medietatis grani *transp. C1* 248 punctis: ?? *O* / signatis: assignatis *OL3C1ER* 249 circulo postremo *transp. R* / *post* circulo *add.* possumus *P1* / *post* fiat *add.* circulus *P1* / concavitas *corr. ex* cavitatis *O* 250 *post* digiti *inter.* et sit *a. m. E* / eius *inter. a. m. E* / et altitudo *inter. O* 251 altitudo hec *transp. R* 252 circuli: speculi *E* / concavitatis *corr. ex* continuitatis *L3* / superficies: species *R* 253 intrabit *alter. ex* intrat *in* intret *E*; intret *R* 254 *post* minorem *add.* et *R* / cum *inter. a. m. C1* 257 anuli: anguli *FP1* / tangunt: tangent *R* 258 facies: superficies *R* 259 *post* divisa *scr. et del.* ex parte *O* 260 altitudine *corr. ex* planitudine *a. m. E*; planitudine *R* / signetur: signatur *SOE*; significatur *L3*; *corr. ex* signatur *C1* / punctus *om. O* / a *om. C1* 261 longitudine *corr. ex* altitudine *a. m. C1* / circini: circuli *FP1* 262 secundum *corr. ex* secundus *S* / et . . . ordei (263) *mg. a. m. E* 263 *post* cuius *scr. et del.* ci *F* / unius: cuius *FP1* 264 perforetur: perfororetur *F* / *post* foramine *scr. et del.* circulari *O*

qui quidem, cum transierit ad interiorem concavitatem, tanget tabule enee superficiem. Pari modo, super singulas exterioris altitudinis perpendiculares similia et equalia efficiantur foramina in quantitate et altitudine.

[3.18] Deinde sumatur tabula lignea quadrata cuius latus est equale dyametro anuli, et protrahatur in eius superficie linea dividens quadratum per medium equidistans lateribus. Et ab una parte sumatur longitudo duorum digitorum, et fiat signum. Post sumatur longitudo semidyametri minoris circuli tabule enee, et posito pede circini, fiat circulus transiens per signum, qui circulus erit equalis minori circulo tabule enee et concavitati anuli.

[3.19] Deinde supra centrum huius circuli sumatur longitudo duorum digitorum, et infra centrum similiter, et signentur puncta. Ab utroque in utramque partem protrahatur linea equidistans lateribus quadrati, et in utraque harum linearum signetur longitudo duorum digitorum ex utraque parte puncti signati. Et a punctis unius linee signatis protrahantur linee equidistantes ad puncta alterius linee signata, et fiet quadratum quatuor digitorum. Fodiatur hoc quadratum secundum altitudinem unius digiti, et concavationis latera efficiantur plana et ortogonalia, et fundus similiter planus.

[3.20] Deinde aptetur hec tabula faciei anuli ita ut circulus minor applicetur foramini anuli, et extremitas eius extremitati. Et firmetur hec applicatio cum clavis ut immota maneat tabula. Nota quod in omnibus predictis duorum digitorum mensura certa debet esse et determinata, et ob hoc in linea aliqua fiat immutabili ne ex mutatione mensure error accidat.

[3.21] Amplius, fiat columpna ferrea concava plana aliquantulum spissa ut nec statim intret nec immutari queat, et sit quantitas dyametri circuli eius unius grani ordei. Et ponatur columpna in foraminibus, que quidem cum ad interiora anuli pervenerit, continget lineas in tabula

265 cum . . . singulas (266) *mg. O* / tanget: tangit *L3C1E* 266 *post* modo *add.* si *FP1SC1; inter.* si *L3* / super: per *C1* 267 et equalia *om. O* 269 *post* tabula *scr. et del.* ignea *C1* / est: fit *O;* sit *L3C1ER* 271 quadratum *om. O* / quadratum per medium: per medium quadratum *ER* / per medium *inter. L3E; a. m. E* / sumatur *corr. ex* sumuatur *F* 272 *post* signum *add.* et *R* 273 circini: circuli *FP1* 274 transiens . . . circulus[2] *om. P1* / *post* qui *add.* quidem *R* 276 supra: super *E* 277 duorum *inter. E* / infra: infixa *L3* / signentur: assignentur *L3* 278 *post* partem *add.* et *R* / protrahatur *corr. ex* pertrahatur *S* 279 *post* utraque *scr. et del.* istarum *P1* / harum *corr. ex* secundarum *L3* / duorum: et *S* 280 parte puncti signati: puncti signati parte *O* 281 protrahantur linee equidistantes: protrahatur linea equidistans *O* / equidistantes *corr. ex* equidistan *F* 282 fiet: fiat *L3ER* / fodiatur: cavetur postea *R* / *post* fodiatur *scr. et del.* in *E* 283 *post* altitudinem *scr. et del.* illius *P1* / latera *corr. ex* litera *O* 286 anuli: eius *R* 287 nota: notandum *R* / *ante* quod *add.* vero *R* 288 predictis: punctis *O* / duorum: dictorum *R* / certa *inter. a. m. E* / et *inter. C1E; a. m. C1* 289 et ob hoc *mg. a. m. E* / *post* ob *scr. et del.* bet *O* / hoc *om. FP1* / linea: lignea *SL3E* / fiat: fit *L3E* / immutabili: immutabuli *F; corr. ex* immutabuli *P1S* / ne: nec *P1* 292 nec[1] *om. ER* / et *om. E* 294 quidem *corr. ex* idem *L3*

enea factas. Et erit operis eius complementum si linea tabule enee sit contingens circulo columpne in puncto linee altitudinis anuli perpendicularis super tabulam eneam et transeuntis per centrum circuli columpne.

[3.22] Fiat autem in capite columpne anulus aut repagulum quod non permittat columpnam intrare nisi ad locum determinatum. Fit autem huiusmodi longitudinis columpna ut procedens supra tabulam eneam attingat lineam equidistantem dyametro tabule intra quas facta est sectio. Et est linea illa equidistans basi trianguli tabule enee.

[3.23] Amplius, fabricentur septem specula ferrea quorum unum planum; duo sperica, unum concavum intra politum, aliud extra; duo columpnaria, unum concavum, aliud in superficie politum; duo piramidalia, unum politum in facie, aliud in concavitate. Speculum autem planum fit circulare, et sit eius dyameter longitudinis trium digitorum.

[3.24] Speculum columpnare politum in superficie sit lucidum et perfecte politum, et sit dyameter circuli longitudinis sex digitorum, qui circulus est basis eius. Longitudo autem columpne sit trium digitorum. In base columpne sumatur corda longitudinis trium digitorum. Similiter in base eiusdem columpne opposita sumatur equalis huic corda et ei opposita ut linee a capitibus unius corde ad capita alterius producte sunt recte. Et secetur hec columpna secundum harum linearum processum ut restet nobis pars columpne cuius capita portiones cordarum earum, aut altitudo axis portionis remanentis minus quam dimidii digiti. Axem autem dico lineam a medio puncto arcus ad medium corde punctum productum.

[3.25] Columpne concave longitudo sit trium digitorum, et dyameter basis eius sex digitorum, et in ea sumatur corda trium digitorum, et

295 factas: stans *FP1O* / eius *om. L3; mg. a. m. C1* / complementum *corr. ex* completum *S* / sit *om. R* 296 contingens circulo: contingat circulum *R* 297 transeuntis *corr. ex* transeuntes *E* 300 permittat *corr. ex* permittit *E* / fit: sit *R* 1 autem *om. FP1* / huiusmodi: huius *SL3C1ER* / supra: super *FP1ER* 2 *post* tabule *add.* linee *P1* / intra: inter *OE* / quas: quam *R* 3 est[1] *inter. F* / *post* et *add.* hec *R* / *post* illa *scr. et del.* et est *P1* / basi *mg. a. m. C1* / *post* basi *scr. et del.* linearum *C1* / trianguli *corr. ex* triangulum *O* / *post* trianguli *scr. et del.* linea tria *P1* / tabule enee *transp. C1* / enee *inter. a. m. S* 4 *post* unum *scr. et del.* speculum *O* 5 intra . . . extra *inter. a. m. E* / *post* extra *add.* duo piramidalia unum politum in facie aliud in concavitate *R* / duo . . . politum (6) *om. E* 6 superficie: specie *FP1S; corr. ex* specie *O* / duo . . . concavitate (7) *om. R* 8 fit: sit *R* / sit: fit *FP1O* / dyameter *corr. ex* diametre *S; corr. ex* diametri *C1* / longitudinis *om. R* 9 *post* digitorum *add.* in base columpne sumatur corda longitudinis trium digitorum *mg. a. m. E* 10 superficie: facie *O* 12 *post* columpne *scr. et del.* en *O* 13 in . . . digitorum (14) *scr. et del. E; ante* in *add.* similiter *E; post* columpne *add.* eiusdem opposita *E;* corda *om. E* / base: basi *R* 14 base: basi *R* 15 equalis huic *transp. R* / equalis huic corda: corda equalis huic *L3C1* / opposita: opposite *O* 16 sunt: sint *SOER; alter. in* sint *C1* / secetur: sececetur *F* 17 restet: restat *L3; corr. ex* restat *OC1E* 18 *post* capita *add.* sint *R* / earum aut *om. R* / aut: autem *O* / *post* altitudo *scr. et del.* basis *L3; add.* autem *R* / portionis remanentis *transp. R* 19 minus: minor *R* / *post* quam *add.* altitudo *R* / autem *om. O* 20 productum: productam *R* 21 *post* columpne *add.* autem *FP1* / sit *inter. a. m. E* 22 et[1] . . . digitorum *om. O* / ea: eo *FP1*

fiat sectio sicut in prima. Et erit altitudo axis partis remanentis minus quam dimidii digiti. Sit autem in hiis omnibus politura exquisita et equalitas omnimoda.

[3.26] Speculum piramidale queratur dyameter basis cuius quantitas sit sex digitorum, et corda trium, et longitudo piramidalis quatuor digitorum et dimidii. Et fiat sectio secundum lineas rectas, et axis portionis altitudo minor quam dimidii digiti, et hoc in unoquoque piramidali intellige.

[3.27] Speculum spericum sit portio sperica cuius dyameter sit sex digitorum, et dyameter basis huius speculi trium digitorum, et erit axis minus quam dimidii digiti. Item operare in speculo sperico concavo.

[3.28] Deinde facias septem regulas ligneas planas quarum latera equidistantia et ortogonalia super capita equidistantia in fine possibilitatis, et sit longitudo regularum sex digitorum, latitudo quatuor. Postea quadrato concavo adaptetur alia regularum ita ut ortogonaliter cadat super inferiorem concavi quadrati superficiem, et vide ut facile intret quadratum ne comprimens immutetur.

[3.29] Cadat igitur super faciem lateris regule acumen tabule enee, et ubi continuabitur ei fiat signum, et a puncto assignato producatur in extremitates regule linea equidistans lateribus regule ut sit linea illa linea longitudinis regule. Deinceps in longiori parte illius linee circa punctum sumptum sumatur altitudo medii grani ordei, et fiat punctum. Dico quod ille est punctus medius regule, qui etiam centro foraminum opponitur recte.

[3.30] Probatio: quoniam centra foraminum elongantur super superficiem tabule enee in medii grani quantitate, et distant a superficie anuli per duos digitos, igitur punctus ille distat ab eadem per duos

23 partis *inter. L3C1E* (*mg. a. m. C1; a. m. E*)/minus: minor *R* 24 *post* quam *add.* altitudo *R*/ politura: politia *O*/*post* omnimoda *add.* in *R* 26 speculum piramidale: speculo piramidali *R* 27 piramidalis *om. R*/*post* quatuor *add.* sit *O* 29 *post* altitudo *scr. et del.* timor *F; add.* sit *R*/ *post* quam *add.* altitudo *R*/hoc *inter. L3E* (*a. m. E*); hec *R*/in *om. E*/unoquoque: unoque *S*; utraque *L3C1ER* 30 intellige *corr. ex* intelligere *L3* 31 superius *O* 36 quatuor:7 *O* 37 concavo *om. R*/alia: aliqua *L3R*/*post* alia *scr. et del.* rum ita *F*/regularum *corr. ex* regularium *L3* 38 concavi quadrati *transp. O*/superficiem *corr. ex* super faciem *O* 39 comprimens: compressa *R*/immutetur: immittetur *E* 40 cadat *alter. in* cadet *O*/super faciem: superficiem *FP1*; superficie *O*/faciem *inter. a. m. E*/*post* faciem *scr. et del.* in *F* 41 ubi *inter. O*/a *om. C1*/assignato: signato *O* 42 *post* extremitates *scr. et del.* enee *P1*/regule[1] *alter. in* linee *a. m. E*/ut: et *FP1O*/*post* linea[2] *scr. et del.* longitudinis regu *F* 43 linea *om. O*/linea longitudinis: linealis *FP1; inter. L3*/longiori: longiore *R* 44 sumptum *om. C1* 45 ille *om. O*; illud *R*/est *om. FP1SO; inter. L3E* (*mg. L3; a. m. E*)/punctus medius: punctum medium *R*/*post* regule *inter.* puncto *O*/qui: quod *R*/etiam: in *O*/centro: centris *R* 46 opponitur *corr. ex* opposita *O*/recte: directe *O* 47 probatio *om. OL3C1ER*/quoniam: quam *FP1*/*post* quoniam *add.* enim *R* 48 distant: distans *O*/*post* superficie *add.* tabulee enee superficies (superficies *inter.*) *O* 49 punctus ille: punctum illud *R*

digitos. Et regula in quadrato concavo, per digitum unum. Quare ab extremitatibus regule ad punctum sunt tres digiti, quare punctus ille est medius. Super hunc medium punctum producatur in utramque partem linea secundum latitudinem equidistans extremitatibus. Et medietates linee longitudinis super quam hec est perpendicularis dividantur per equalia per lineas latitudinis perpendiculares extremitatibus equidistantes. Et ita divisa erit regula in quatuor equales partes. Similis fiat in aliis regulis operatio.

[3.31] Hiis completis, adaptetur speculum planum uni regularum. Et est ut sit regula cavata secundum altitudinem speculi ita ut superficies speculi sit in eadem superficie cum superficie regule, et ita ut medium superficiei speculi punctum directe supponatur medio superficiei regule puncto, et ita ut linea dividens superficiem regule in duo equalia dividat etiam superficiem speculi per equalia, et ut continuentur partes speculi cum linea dividente. Et observetur in possibilitatis fine.

[3.32] Deinde speculum columpnare politum in facie applicetur alicui regule ita ut medius eius punctus cadat super medium regule punctum, et ita ut linea in longitudine speculi sumpta dividens ipsum per equalia continuetur cum partibus linee longitudinis superficiei regule eque dividenti, et ut media longitudinis speculi linea sit in superficie regule. Et hoc sic fieri poterit utriusque basis speculi arcus per equalia dividantur et a puncto divisionis signato ad oppositum signatum linea producatur, et linee medie longitudinis regule aptetur et continuetur.

[3.33] Speculum columpnare concavum aptetur regule ut media longitudinis eius linea secundum equalem arcuum basium divisionem sumpta equidistans sit medie linee longitudinis regule, et etiam ut

50 et *inter. L3*/regula: reliqua *FP1*; ita *E*; *om. R*/*post* regula *inter.* est *L3*; *add.* est *C1*/ab: pro *E* 51 regule: tabule *O*/punctus . . . medius (52): punctum illud erit medium *R* 52 est: erit *E*/hunc: hoc *R*/medium punctum *transp. FP1* 53 secundum latitudinem *mg. a. m. E*; secundum altitudinem vel latitudinem *inter. L3*/latitudinem: altitudinem *O* 54 quam: quas *C1*/hec est *transp. C1ER*/perpendicularis *inter. a. m. O*; *corr. ex* perpendiculariter *L3* 55 perpendiculares *corr. ex* propter *a. m. L3* 56 *post* ita *scr. et del.* et ita *S*/in *inter. O* 58 completis: expletis *O*/planum *inter. a. m. E*/uni *inter. L3* 59 regula *om. O*/ita *corr. ex* prima *S*/ita . . . speculi (60) *inter. L3*; ita . . . superficie[1] (60) *mg. a. m. C1*/*post* ita *scr. et del.* s *F* 61 *post* speculi *add.* sit in eadem superficie speculi *C1* 62 regule[1] *mg. a. m. C1*/ut *om. E*/regule[2] *inter. a. m. E* 63 etiam: et *OL3E*/et *om. E*/continuentur: continuetur *P1*; *corr. ex* continentur *OL3* 64 *post* et *add.* hoc *R* 66 ita ut *corr. ex* ut ita *E*/medius: medium *R*/eius punctus *transp. L3C1R*/punctus: punctum *R*/*post* punctus *add.* eius *E* 67 punctum *om. C1*/ut *inter. L3*; *om. E* 68 continuetur *corr. ex* continetur *L3*/superficiei: superficiem *O* 69 regule *inter. a. m. E*/*post* regule *add.* per *O*/dividenti: dividentis *O* 70 sic *om. OL3*; sicut *E*/*post* poterit *add.* ut *O*; *inter.* sic *a. m. L3*; *inter.* ut *a. m. C1* 71 a: ex *O*/signato: signata *FP1* 72 *post* signatum *add.* punctum *R*/regule *om. R* 74 columpnare: columpnarem *S* 75 longitudinis *om. O*/arcuum *corr. ex* arcu *C1*/arcuum basium *transp. R* 76 *post* equidistans *add.* linee *E*/medie linee *transp. R*/linee *om. E*/*post* longitudinis *scr. et del.* linea *C1*

utriusque arcus corda cum lineis longitudinis extremis sint in superficie regule.

[3.34] Piramidale speculum extra politum applicetur regule ut acumen eius sit in termino medie longitudinis regule linee, et linea dividens portionem piramidis per equa—que scilicet a cono ad medium arcus basis punctum producitur—sit in superficie continuata cum parte restante medie linee longitudinis regule.

[3.35] Speculum piramidale concavum applicetur regule ita ut acumen eius sit in directo medie linee longitudinis regule; corda vero arcus basis sit in superficie, scilicet regule. Linea a cono ad medium arcus basis punctum ducta sit equidistans medio linee longitudinis regule. Cum autem longitudo piramidum sit quatuor digitorum et dimidius, restabunt ex longitudine regule digitus et medius.

[3.36] Adaptandum regule speculum spericum extra politum, fiat in regula circulus secundum quantitatem trium digitorum. Eius sit centrum medium regule punctum. Et cava et apta speculum ut medium superficiei eius punctum sit in superficie regule et in medio puncti medie linee longitudinis regule, quod quidem sciri poterit per applicationem alterius regule acute equalis huic in longitudine et divise per equalitatem et applicate medie linee longitudinis regule ita ut medius huius regule acute punctus tangat medium speculi sperici punctum.

[3.37] Spericum concavum, facto in regula circulo secundum quantitatem trium digitorum cuius centrum medius regule punctus, cavato circulo, imponatur ita ut circulus basis speculi sit in superficie regule, et punctum medium concavitatis speculi directe oppositum medio regule puncto. Et dyameter basis speculi continuetur medie linee regule, quod ita perpendetur. In regula acuta punctus signetur, et ab illo puncto

77 corda *inter. a. m. E* / lineis: linee *R* / sint: sunt *P1* 79 piramidale speculum *transp. L3C1* / *post* piramidale *scr. et del.* s *F* 80 *post* termino *add.* linee *R* / *post* longitudinis *scr. et del.* linee *C1* / regule: eius *O* / linee *corr. ex* lignee *a. m. C1*; *om. R* 81 piramidis: piramidalis *OL3ER* / cono: vertice *R* 82 sit: fit *FS* 83 medie linee *transp. R* 84 piramidale *inter. L3* / acumen (85): arcuum *L3* 86 sit: fit *FP1S* / scilicet *om. L3C1R* / scilicet regule *om. O*; *scr. et del. E* / regule *inter. L3* / *post* regule *add.* et *R* / *post* linea *scr. et del.* sit *C1* / cono: vertice *R* / ad *inter. a. m. E* 87 punctum *corr. ex* puncta *O* / medio: medie *R* 88 autem longitudo *corr. ex* a longitudine *a. m. E* / piramidum: pyramidis *R* / dimidius: dimidii *OL3C1ER* 89 medius: dimidius *R* 90 fiat *corr. ex* fifiat *C1* 91 regula: respectu *FP1*; *inter. a. m. E* / secundum: secundus *F* / trium *om. O* / sit: fit *FP1S* / sit centrum (92) *transp. ER* 92 et cava *om. R* / apta: aptetur *R* / ut *mg. a. m. E* 93 puncti: puncto *OC1R* 94 *post* regule *add.* per *L3C1*; *scr. et del.* per *E* / quidem *om. O* / *post* quidem *add.* est *L3C1* / sciri poterit: scire poteris *O* / poterit *om. FP1* 95 *post* applicationem *scr. et del.* alterius *O* 96 medius (97): medium *ER* 97 acute *inter. a. m. E* / punctus *corr. ex* pres *O*; punctum *ER* 99 *post* concavum *add.* aptatur *R* / in regula: integro *O* 100 medius: medium *R* / punctus: punctum *R* 101 ut *om. S* / basis *om. L3C1*; *inter. a. m. E* 102 speculi: speculum *FP1* / *post* speculi *add.* sit *R* / *post* directe *add.* sit *S* 103 continuetur *corr. ex* continetur *L3I* / *post* regule *scri. et del.* o *F* 104 quod: que *R* / perpendetur *corr. ex* perpendeatur *O* / punctus: punctum *R* / signetur: significetur *L3C1*

longitudo semidyametri basis speculi notetur ex utraque parte. Et ita hec acuta regula medie linee regule applicetur ut punctum signatum in ea directe opponatur medio concavitatis speculi puncto et dyameter in ea factus similis sit cum basis dyametro.

[3.38] Hiis peractis, in semidyametro tabule enee triangulum per equalia dividente signetur ab acumine eius longitudo equalis axi huius speculi concavi, et fiat punctum. Axis autem sic dinoscitur. Regula acuta superficiei applicetur ut acuitas directe sit super mediam longitudinis lineam, puncto eius super medium concavi punctum directe statuto. Deinde acus recta et subtilis secundum illud regule acute punctum perpendiculariter cadat in speculum. Descendet quidem super medium concavi punctum. Signetur autem in acu punctum quod post eius descensum tangit acuitas regule sive punctum signatum, et sit modicum declinata regula ut certius possit fieri in acu signum. Postea secundum longitudinem acus a puncto signato in ea metire ab acumine tabule enee in linea triangulum dividente, et fac punctum.

[3.39] Deinceps hanc regulam facias intrare quadratum concavum ita ut acumen tabule enee descendat supra speculum; et adhibeatur regula acuta ut signetur punctum in linea dividente triangulum, quem tetigerit ex ea regula acuta, cum acumen trianguli descenderit usque ad superficiem speculi concavi. Signa igitur punctum.

[3.40] Erit autem hoc secundum punctum minus distans ab acumine quam primum, superficies enim tabule enee distat a superficie anuli sive tabule in qua est quadratum concavum per duos digitos minus medietate grani ordei. Punctus autem medius regule directe est oppositus medio speculi sperici concavi puncto, qui quidem distat ab

105 longitudo *om. O* / semidyametri *corr. ex* semedyametri *O* 106 *post* applicetur *add.* et *L3*; *scr. et del.* et *C1* / ut: et *O* 107 directe *corr. ex* directa *F* 108 factus: facta *R* / similis *corr. ex* simul *F*; *alter. in* simul *O*; simul *R* / sit *om. FP1S*; *inter. a. m. E* 109 tabule:tunc *P1*; *om. S*; *corr. ex* regule *OL3* 110 signetur: signatur *FP1*; significetur *L3* / eius: est *O* 111 sic *inter. a. m. E* / dinoscitur: dignesserit *O*; dignoscitur *R* 112 *post* acuta *inter.* speculi *a. m. C1* / *post* superficiei *add.* speculi *R* / sit *om. FP1* 113 lineam: ferream *O* / super: similiter *FP1S*; supra *OL3C1* / *post* concavi *add.* speculi *R* / *post* punctum *scr. et del.* medium *F* 114 secundum *alter. in* super *C1*; *alter. ex* contra *in* super *E*; super *R* / illud: aliud *O* 115 in *mg. F* 116 concavi: concavum *FP1* / acu: actu *E* 117 post *corr. ex* potest *S* / eius: suum *R* / descensum: decessum *OL3*; recessum *E* / tangit: tangat *FP1SR* / acuitas: concavitas *ER* / *post* regule *add.* sue *C1* / sive *corr. ex* sue *L3E* / sive . . . signatum *om. R* 118 sit: sic *L3* / *post* ut *add.* cuius *FP1* 119 *post* signato *add.* et *O* 120 tabule: regule *OL3E* / enee *corr. ex* ?? *E* 122 *post* tabule *scr. et del.* tabule *F* / supra: super *L3C1* 123 linea: lineas *L3* / triangulum *om. O* / quem: *corr. ex* quoniam *S*; quam *O*; quod *R* 124 descenderit: descendet *C1* 125 superficiem *corr. ex* semidyametrum *L3* / speculi *inter. a. m. S*; *om. OL3C1ER* / *post* igitur *add.* erit hoc autem secundum *OL3C1* (hoc autem *transp. O*); *add.* hoc vero secundum *R* 126 erit . . . hoc *om. C1* / erit . . . punctum *om. FP1OL3R* / punctum *om. E* 127 *ante* quam *scr. et del.* minus *E* / enee *corr. ex* eene *F* 129 punctus: punctum *R* / medius: medium *R* / regule directe *corr. ex* recte regule *L3* 130 oppositus: oppositum *R* / *post* medio *add.* vel *FP1* / speculi *om. OL3C1E* / *post* speculi *scr. et del.* or *P1* / sperici *om. P1R* / qui: quod *R*

eadem superficie tabule per duos digitos. Cum ergo acumen tabule enee ortogonaliter descendat, non cadet super medium concavi, qui est terminus axis, sed in puncto altiori, quare propositum.

[3.41] Signetur vero in speculo concavo punctum in quod accidit acumen tabule enee, et extracto in puncto illo foramine ortogonaliter descendente et modico ad hanc quidem mensuram ut in eo descendat acutum donec acuitas regule adhibite contingat punctum linee dividentis triangulum primo signatum. Quod cum fuerit, erit quidem acumen tabule enee in eadem superficie cum termino axis speculi, que superficies sit equidistans superficiei regule. Et erit linea a termino axis ad acumen ducta perpendicularis super superficiem tabule enee. Axis autem speculi in eadem superficie cum centris foraminum, quoniam distantia eorum a superficie anuli duorum est digitorum, et medius terminus axis similiter.

[3.42] Hiis cum diligentia preparatis, poterit videri quod promisimus. Immitatur anulo regula super quam est speculum planum donec acumen tabule enee cadat super speculum, et infigatur regula quadrato concavo, et in eo subtus regulam aliquid opponatur quod ei firmitatem conserat ne vacillet. Deinde opponatur pargamenum foraminibus, et cum digito fiat impressio ut obturentur et impressionem percipere possis. Et signum foraminis fiat in pargameno cum incausto, vel aliquo alio. Unum autem foraminum relinquatur apertum declinatum non super regulam mediam, et adhibeatur radio solis foramen apertum. Certioratum autem erit huius rei comprehensio si adhibeatur radio solis per foramen domus intranti.

131 tabule[2] *corr. ex* regule *O* 132 enee *om. R* / *post* descendat *scr. et del.* descendat *S* / cadet: cadat *L3; alter. ex* cadat *in* cadit *E* / *post* concavi *add.* punctum *R* / qui: quod *R* 133 *post* est *scr. et del.* punctus *P1* / terminus: terminis *S* / puncto altiori: punctum altius *R* / *post* quare *add.* patet *R* / propositum: propo *F*; propter *P1*; proprium *O; corr. ex* proprium *L3*; liquet *C1* 134 signetur *corr. ex* servetur *L3* / vero: ergo *C1* / concavo *corr. ex* concavum *S* / quod *corr. ex* quo *C1* / accidit: accidat *OL3; corr. ex* accidat *C1*; incidit *R* 135 extracto: extra *FP1*; extracta *S; corr. ex* extracta *C1E* / *post* extracto *scr. et del.* regula *L3* / *post* illo *scr. et del.* cavent *L3* 137 acutum: acumen *C1ER* / regule: tabule *L3; corr. ex* tabule *a. m. C1; alter. ex* tabule *in* linee *a. m. E*; linee *R* 138 cum *inter. L3; inter. a. m. E* / fuerit: fuerint *E* 139 enee *inter. C1; inter. a. m. L3* / superficie: sic *P1* 140 sit: est *R* / equidistans: distans *FP1* 141 ducta perpendicularis: dicta perpendiculariter *P1* 142 *post* eadem *add.* est *O; add.* erit *R* 143 distantia: distantie *FP1* / a *inter. a. m. S* 144 medius *om. OC1R* / terminus *om. L3; inter. a. m. C1* 147 enee cadat *transp. S* / *post* et *add.* sit *R* / infigatur: infixa *FP1SR; corr. ex* infixa *a. m. E* 148 aliquid: aliquo *L3* / *post* aliquid *add.* deinde *C1* / opponatur: apponatur *C1ER* / ei *inter. O* / *post* ei *add.* conferat *R* / firmitatem: confirmitatem *FP1*; firmitudinem *O* / firmitatem conserat (149) *transp. E* 149 conserat *om. R* / opponatur: apponatur *C1* / pargamenum: pergamenum *R* 151 pargameno: parchameno *FP1*; pergameno *R* / *post* incausto *scr. et del.* longitudo *L3* / vel *inter. L3* / aliquo alio (152) *transp. C1* 152 alio; albo *O; inter. S; om. L3E* / relinquatur: relinquetur *FP1* 153 non *om. OL3; corr. ex* nec *E* / *ante* super *add.* mediam *E* / regulam mediam *transp. R* / *post* regulam *add.* non *OL3; inter.* non *a. m. C1* / mediam *om. E* / et *inter. O* 154 certioratum: certior *OR*; certiorata *E* / autem . . . intranti (155) *inter. a. m. E*; huius rei: huiusmodi *E; post* si *add.* hoc *E* 155 intranti: intrare *O*

[3.43] Cum igitur radius foramen intrans ad speculum pervenerit, videbis ipsum reflecti ad foramen illud respiciens super lineam tabule enee equalem angulum continentem cum linea triangulum per equa dividente et angulo quam tenet linea a foramine discooperto cum illo tabule semidyametro. Si vero foramen in quod fit reflexio discoopertum opponas radio priore cooperto, videbis reflecti radium in coopertum.

[3.44] Si vero foramini imponatur columpna ferrea concava, quam ad quantitatem foraminum fieri precipimus (ut firmius stet modicum cere circa eam apponatur), descendet lux per columpne concavitatem sicut descendit per foramen. Et reflectetur in foramen respiciens, et super lineas tabule enee erit descensus et reflexio pari modo, ut prius. Et si ad secundum foramen columpnam transtulerimus, in primum lucem reflexam videbimus. Erit autem debilior lux per columpnam descendens quam sine columpna per foramen. Erit autem videre eundem reflectendi modum in debiliori luce.

[3.45] Obturetur foramen cum cera ut modicum circa centrum eius restet vacuum, et videbitur lucis reflexio in foramine, sive circa centrum. Pari modo, si concavitatem columpne cum cera obturaveris ut remaneat quasi terminus solius axis, descendet lux super axem columpne et reflectetur ad centrum foraminis similis. Eodem modo, alterata columpna imposita, cum descenderit lux super axem unius foraminis, reflectetur super axem similis, centrum enim foraminis directe axi opponitur. Et cum lucis reflexio cadat in centrum nec moveatur nisi per lineam rectam, oportet ut procedat secundum axem.

[3.46] Obturatis autem foraminibus singulis preter medium quod directe super tabulam eneam incidit, fiat baculus columpnaris ad quantitatem foraminis, et extremitas eius acuatur ut remaneat solus terminus axis eius. Et descendat per foramen, et signa punctum speculi in quod ceciderit. Deinde descendat radius solis per foramen illud. Cadet quidem super punctum signatum, et circa ipsum efficiet circulum.

157 videbis: videbit *FP1SOL3*; ?? *C1* 158 continentem: continente *O* / *post* cum *scr. et del.* pe *C1* 159 et: ei *OL3C1ER* / angulo: triangulo *FP1* / quam: quem *R* / illo: eadem *R* 160 quod: quo *O* 161 opponas *corr. ex* opponis *a. m. C1* / videbis: videbit *OL3* / reflecti radium *transp. R* 162 concava *om. O* 163 foraminum *om. FP1* / precipimus: precepimus *SO* / *post* precipimus *add.* que *R* / firmius *corr. ex* firmus *C1* 164 cere *corr. ex* scere *L3*; cera *scr. et del. O* / *post* eam *inter.* cera *O* / apponatur: opponatur *FP1* / descendet: descendat *E* 165 respiciens: inspiciens *E*; sibi respondens *R* 166 erit *om. O* 170 debiliori: debiliore *R* 171 eius: ei *R* 172 videbitur *corr. ex* videtur *a. m. E* / sive: simili *L3C1R*; simili *inter. a. m. E* 173 *ante* pari *inter.* eius *a. m. E* / cum *om. O* 174 solius *corr. ex* solus *L3* 176 alterata: altera *R* / columpna imposita: columpne imposite *O* / unius . . . axem (177) *inter. L3* 177 reflectetur *corr. ex* reflectitur *E* / reflectetur . . . foraminis *mg. a. m. S* 180 quod: g *E* 181 incidit: incedit *C1* 183 descendat *corr. ex* descendet *E* / *post* foramen *add.* ad speculum *R* / signa: signetur *R* / *post* signa *add.* tabule *OE*; *scr. et del.* tabule enee *L3* / *post* punctum *add.* tabule enee *FP1S*; *add.* enee *OE* / speculum *om. R* 184 quod *corr. ex* quo *C1* / ceciderit: cecideret *L3* 185 super: superi *P1* / *post* super *add.* quod *L3* / efficiet *corr. ex* efficit *S*

[3.47] Signetur autem in fine huius lucis circularis punctum, et secundum quantitatem linee interiacentis puncta signata fiat circulus. Erit quidem circulus iste maior circulo foraminis, quoniam processus lucis per foramen ingredientis est in modum piramidis. Verum in nullo foraminum videbitur lucis reflexio, unde palam quod lux descendens per axem reflectitur super eundem. Verumptamen apparebit lux circularis circa basem interioris foraminis maioris quidem capacitatis radio, maioris etiam lucis interioris circulo.

[3.48] Et palam hanc lucem apparentem esse per reflexionem, verum non per reflexionem lucis super axem descendentis, quod ex hoc poterit patere. Obturata utraque foraminis base ut quasi sola remaneat axis via, et radio solis per viam axis descendente, non apparebit lux illa circularis circa inferiorem basem foraminis, quare non procedebat ex reflexa luce axis.

[3.49] Amplius, supra quamdam regulam supposuimus ut ortogonaliter caderet in quadratum concavum. Si aliquantulum ex eis auferatur ut regula declinetur ita ut extremitas a quadrato remotior sit dimissior radio descendente super foramen medium, non cadet perpendiculariter supra speculum, et apparebit lux reflexa a foramine medio remota. Et quanto maior erit declinatio maior erit lucis reflexe a foramine remotio. Si vero ad rectitudinem regula reducatur, lux reflexa circa inferiorem foraminis basem, ut prius, videbitur.

[3.50] Palam igitur quod luce super speculum perpendiculariter cadente, regreditur ad foramen per quod ingressa est. Cum vero lux axis declinata ceciderit, reflectitur non ad foramen, sed apparebit super lineam superficiei anuli perpendicularem super tabulam eneam et descendentem per centrum foraminis medii.

186 autem: igitur *R* / huius: eius *L3* / *post* huius *scr. et del.* s *C1* 187 signata: signa *L3* / fiat circulus *om. FP1S* / circulus: circulum *O* / erit . . . circulus (188) *inter. L3* 188 quidem: que *S* / *post* circulus *scr. et del.* punctum *S* / lucis *om. P1* 189 in[1]: per *ER* / verum . . . reflexio (190) *om. R* 191 *post* eundem *add.* axem *C1* 192 basem interioris: basim interiorem *R* / *post* capacitas *add.* luce incidente vel *R* 193 *post* radio *add.* et *R* / lucis *om. O* / *post* interioris *add.* lucis *R* 194 *post* palam *add.* est *R* / apparentem *om. R* / *post* per *scr. et del.* in *S* / reflexionem *corr. ex* inflexionem *a. m. E* 196 *ante* patere *scr. et del.* com *C1* / base: basi *R* 197 non *om. FP1; inter. L3; inter. a. m. E* 198 inferiorem basem: interiorem basim *R* / procedebat: precedebat *FP1S*; procedat *C1* 199 luce axis: lucis axe *R* 200 *post* amplius *add.* ut *ER* / quamdam: quidem *R* / regulam *om. ER* / *post* ut *add.* regula *R* 201 ex eis: inde *R* 203 dimissior: demissior *R* / descendente: descente *S* / cadet: cadat *OL3*; *corr. ex* cadat *E* 204 supra: super *ER* 205 *post* quanto *add.* erit *C1* / maior: maiorem *S* / declinatio: declaratio *FP1S* / *post* declinatio *add.* tanto *ER* 206 regula *om. S*; linea *OL3E* / *post* regula *add.* cum speculo *C1* 207 inferiorem: interiorem *R* / basem: basim *R* 208 *post* igitur *scr. et del.* luce *P1* / *post* luce *add.* quod *C1* / super: supra *E* 209 per . . . foramen (210) *rep. E* (axis[1] *inter. a. m.*; reflectitur non[1,2] *transp.*) 210 reflectitur non: non reflectetur *R* / *post* foramen *scr. et del.* per *P1*; *add.* per quod ingressa est *R* / *post* sed *add.* tamen *R* / *post* apparebit *add.* centrum lucis semper *R* 211 superficiei: superficie *O* / *post* superficiei *add.* concave *R* / perpendicularem: perpendiculariter *E*

[3.51] Quecumque autem dicta sunt in duobus foraminibus primis declinatis intellige in singulis. Et quod dictum est in speculo plano, luce per foramen declinatum seu medium descendente, regula recta seu declinata, in aliis speculis intellige.

[3.52] Si autem regula in qua fuerit speculum columpnare extra politum declinetur in quadrato ita ut non ortogonaliter cadat super quadratum sed declinetur super partem dextram vel sinistram, apparebit tamen lux reflecti super foramen simile eius descensui, et medium lucis super medium foraminis, sicut visum est regula non declinata.

[3.53] Regulam in quam situm est columpnare concavum impones, et descendat acumen tabule enee donec tangat superficiem speculi, et declinabis hoc speculum secundum latus suum sicut declinasti extra politum.

[3.54] Idem in speculis piramidalibus concavis operaberis.

[3.55] Spericum concavum aptetur donec descendat acumen tabule enee in foramen speculi factum secundum acuminis descensum.

[3.56] Spericum extra politum sic imponatur ut acumen tabule enee sit in superficie regule et in eadem superficie cum medio speculi puncto, quod sic fieri poterit. Adhibeatur regula acuta regule et puncto speculi medio, et descendat acumen tabule enee quousque sit in directo acuitatis regule. Et tunc cogatur sistere.

[3.57] In speculis columpnaribus videbis reflexionem hoc modo. Aptetur speculum, sicut dictum est, et per foramen medium descendat baculus columpnaris, sicut factum est in speculis planis. Cadet quidem baculus super mediam longitudinis speculi lineam, et erit eius terminus in superficie regule. Super mediam lineam signetur punctum in quod cadit, et ab hoc puncto in superficie regule sumatur longitudo semidyametri circuli facti in regula ad discernendum circularem lucis casum. Et ex alia parte puncti sumatur longitudo eadem, et habebitur linea equalis dyametro predicti circuli. Videbitur autem lux cadens

213 dicta sunt *transp. FP1S* 214 in[1]: de *C1* / et *om. S* / dictum est *L3C1* / *post* plano *add.* de *R* 215 *post* foramen *add.* seu *R* / seu: sive *FP1* / regula . . . declinata (216) *inter. a. m. E* / *post* regula *add.* seu *R* 216 seu *om. L3* / declinata: declanata *R* 218 politum: positum *O*; *corr. ex* punctum *a. m. C1* / non *om. O* / cadat super: cadet supra *FP1* 219 super: secundum *O* / dextram *corr. ex* destram *S*; *corr. ex* dexteram *O* 220 reflecti: reflexa *O* 221 *post* est *inter.* in *E*; *add.* in *R* 223 regulam: regula *C1* / quam: qua *L3C1R* 224 et: ut *R* / descendat *alter. in* descendet *E* 225 declinabis: declinabit *FP1OL3E*; *corr. ex* declinabit *C1* / declinasti: declinati *O* 227 idem: item *L3* / operaberis *om. P1* 230 enee *om. FP1* 231 regule *inter. a. m. E* / *post* regule *add.* enee *E* / et: est *O* 233 acuitatis *corr. ex* acuitas *O* 234 regule *corr. ex* regula *E* 235 videbis: videbit *FP1OL3* 236 *post* aptetur *add.* hoc *P1* 237 factum *inter. E* 238 longitudinis *corr. ex* lineis *P1* / speculi *inter. a. m. E* / et . . . lineam (239) *mg. a. m. E* 239 *post* mediam *scr. et del.* longitu *P1*; *add.* igitur *R* 241 semidyametri *corr. ex* diametri *a. m. C1* / circuli *corr. ex* circulis *L3* 242 parte *inter. O* / et[2] *inter. a. m. E* 243 equalis *inter. OL3*; *inter. a. m. E* / dyametro: semidyametro *O* / predicti: predicto *FP1SL3C1*

extendi supra predictam lineam tantum, et reflectitur ad foramen medium. Et circa eius basem inferiorem videbitur lux circularis maior circulo inferiori, sicut in speculis planis visum est.

[3.58] Idem in speculis piramidalibus videre poteris.

[3.59] Pari modo in speculis spericis, luce per foramen medium descendente, fiat circulus in superficie regule ad quantitatem circuli iam dicti. Et videbitur lux extendi super hunc circulum et reflecti ad foramen medium modo iam dicto. Et apparebit in hiis omnibus rectis reflexionibus linea perpendicularis in interiori superficie anuli secare lucem circularem reflexam et dividere circulum eius per medium.

[3.60] Quod dictum est de luce naturali videri poterit in luce accidentali. Domus unici foraminis opponatur parieti in quam descendit solis radius, et applicetur instrumentum foramini cum intraverit lux accidentalis per foramen non medium. Videbitur reflecti per eius oppositum, et si aptetur instrumentum ut intret per duo foramina, reflectetur per duo similia.

[3.61] Verum ut possis perpendere lucem cum intraverit directe et ad ipsam transierit, appone superius pargamenum album, et inclina instrumentum donec videas lucem cadentem super pargamenum. In speculis etenim non plene comprehenditur lucis accidentalis casus propter debilitatem eius. Idem autem in hac luce patebit quod in naturali patuit, non enim est diversitas in earum natura nisi quod una fortis et alia debilis.

[3.62] Palam ergo quod luces propter diversas lineas ad specula accidentes per diversas reflectuntur lineas. Et si eadem parte ad speculum venerit, in eandem gradiuntur partem, et declinatio linearum reflexionis equalis declinationi erit linearum accessus. Et planum quod linee lucis reflexe et advenientis sunt in eadem superficie ortogonaliter

244 supra: super *R* / reflectitur: reflectetur *OR* 245 basem inferiorem: basim interiorem *R* 246 inferiori: interiori *R* / *post* in *add.* speris *L3* / speculis *corr. ex* speris *a. m. O*; *inter. L3*; speris *E* 247 speculis: speris *FSL3*; *corr. ex* speris *a. m. O*; spericis *P1* / *ante* piramidalibus *inter.* speculis *L3* 248 *post* in *add.* speris *L3* / speculis *corr. ex* speris *O*; *inter. L3* 252 in *inter. S* / interiori: interiore *R* 253 et *om. SOL3C1E* 254 *post* quod *add.* autem *R* / de: in *O* / videri: videre *L3C1* / poterit: poteris *C1* 255 parieti *om. C1* / quam: qua *L3*; quem *R* / descendit: descendat *OL3C1* 256 solis radius *transp. R* / *post* cum *add.* ergo *R* 258 oppositum *corr. ex* oppositis *P1*; *corr. ex* oppositionem *a. m. E* 260 verum *om. S* / et . . . transierit (261) *om. R* 261 pargamenum: pergamenum *R* / inclina: declina *E* 262 cadentem: carentem *FP1SC1* / pargamenum: pergamenum *R* 263 speculis *corr. ex* peris *L3*; *corr. ex* speris *SE* (*a. m. S*) / etenim: enim *R* / plene *corr. ex* bene *a. m. E* 264 autem: aut *P1* / *post* quod *scr. et del.* qi *P1* 265 *post* quod *scr. et del.* in *E* 266 *post* fortis *add.* est *R* 267 propter: per *C1R* / ad *om. C1* / specula: speras *FOL3E*; *alter. ex* speram *in* speras *P1* / *corr. ex* speras *a. m. S*; speculi *C1* 268 accidentes: accedentes *C1R* / per: in *inter. L3* / et si *inter. a. m. E* / si: ab *C1*; *corr. ex* sub *OL3* (*a. m. O*) / si . . . partem (269) *om. R* / *post* si *add.* sub *E* 269 *ante* et *add.* quod secundum rectam perpendicularem incidentes secundum eandem regrediuntur *R* / et . . . equalis (270) *inter. a. m. E* / *post* et *add.* quod *R* 270 *post* reflexionis *add.* est *R* / *post* equalis *add.* est *C1E* / erit *om. OL3C1ER* 271 ortogonaliter: ortogonali *R*

super superficiem politi et contingenti punctum a quo fit reflexio. Et si super perpendicularem venerit, reflectetur super perpendicularem, et in quemcumque punctum cadit reflectitur in superficie perpendiculari super superficiem tangentem illud punctum.

[3.63] Et semper linea reflexa cum perpendiculari super illud punctum equalem tenet angulum angulo quem includit linea veniens cum eadem perpendiculari. Et huius rei probatio est quia palam quod, si descendat lux quecumque per foramen aliquod, reflectitur per ipsum respiciens. Et si constringatur foramen ut restet quasi solus axis, reflectitur per axem respicientis foraminis. Et si fiat alteratio descensus lucis, reflectitur lux per lineas per quas prius descenderat. Et palam quod foramina se respicientia eundem habent situm respectu medii, et cum non procedat lux nisi per lineas rectas, planum quod reflectitur per lineas eiusdem situs respectu medii cum lineis descensus.

[3.64] Unde cum accidit per ortogonale, per eam reflectitur solam, quare semper linee reflexionis eundem servant situm cum lineis descensus respectu superficiei contingentis punctum reflexionis. Et hoc substantiale sive in substantiali sive in accidentali luce, sive forti sive debili, et generaliter in omni.

[3.65] Et nos ostendemus ydemptitatem situs. Iam scimus quod superficies regule cadit super tabulam in qua quadratum fecimus ortogonaliter. Igitur linea media tabule ortogonaliter est super lineam communem ei et regule, et est super lineam latitudinis regule. Et tabula equidistans enee tabule, et linea eius media equidistans linee medie tabule enee, et est linee a centro tabule enee producte et dividentis arcum per equalia.

272 super *corr. ex* et *L3* / *post* et[1] *add.* aut superficiem contingentem *R* / *post* punctum *add.* politi *R* / *post* si *add.* lux *R* 273 venerit . . . perpendicularem *om. O* / reflectetur: reflectitur *ER* 274 quemcumque: quodcumque *OC1R* / cadit *corr. ex* cadat *O*; ceciderit *R* / reflectitur: reflectuntur *FP1SL3C1* 276 et *inter. O* / semper . . . punctum (277) *om. O* 278 eadem perpendiculari *transp. L3C1* / rei *om. FP1* / est *om. FP1* / quia: quoniam *O* / si *inter. O* 279 *post* per[2] *add.* aliud *R* 280 si *om. FP1SOE; inter. L3* / solus: solis *O*; *corr. ex* solis *L3* 281 respicientis: respiciens *O*; *corr. ex* respiciens *S* / foraminis *om. OL3C1ER* / si *inter. a. m. E* 282 lux *om. ER* / *post* per *scr. et del.* per *F* / descenderat: descendebat *S* 283 se *om. FP1SL3; inter. a. m. E* / *post* respicienia *inter.* se *L3* / habent *inter. a. m. E* / *post* medii *add.* cum lineis *C1* 283 et . . . medii (285) *mg. a. m. L3; inter. a. m. E* / lineas rectas *transp. OER* / planum: palam *ER* / *post* quod *add.* non *E* / reflectitur: reflecti *S* 285 eiusdem *inter. E* / *post* medii *add.* foraminis *ER* / *post* lineis *scr. et del.* respectu *C1* 286 *post* unde *scr. et del.* cum *C1* / accidit: accedit *SC1R* / ortogonale: ortogonalem *SC1R* / per eam reflectitur: reflectitur per eam *O* / reflectitur *inter. L3; inter. a. m. E* 287 *post* quare *add.* linee *C1* 288 *post* hoc *add.* est *C1* 289 substaniale: verum est *R* / in[1] *om. FP1* 291 ostendemus: ostendimus *O* 293 *post* tabule *inter.* enee *L3; add.* quadrati ortogonalis *R* 294 communem *corr. ex* que *O* / communem . . . lineam *inter. a. m. E* / *post* communem *add.* sectioni *E; add.* sectionis ipsius *R* / ei *om. R* / *post* et[2] *add.* illa res *C1* / est *om. R* / *post* regule[2] *scr. et del.* et est linea acutus tabule enee *C1* / tabula *corr. ex* regula *L3E* (*a. m. E*) / *post* tabula *inter.* est *a. m. C1* 295 equidistans: quadrati equidistat *R* / enee tabule *transp. L3C1* / et . . . linee (296) *inter. a. m. E* / *post* eius *add.* id est tabule quadrate *E; add.* id est tabule quadrate concave *R* / *post* media *add.* est *C1* / equidistans: equidistat *R* 296 et[1]: que *R* / et . . . enee[2] *om. C1* / linee: linea *R* / *post* linee *add.* enee medie *E* / producte: producta *R* / dividentis: dividens *R* / arcum: semicirculum *R*

[3.66] Linea autem communis tabule enee et regule, que est linea latitudinis regule, est equidistans linee communi tabule et regule, quare linea media tabule enee cadit perpendiculariter super lineam communem regule et tabule enee. Et regula perpendicularis est super superficiem quadrati, et superficies quadrati equidistans superficiei tabule, quare superficies tabule ortogonaliter super superficiem regule.

[3.67] Et similiter superficies tabule enee ortogonaliter super eandem, et linea media longitudinis regule est perpendicularis super latitudinem eius, quare linea media tabule erit perpendicularis super mediam longitudinis regule lineam, cum cadit super eam; et similiter linea media tabule enee erit perpendicularis super eandem. Et ita media linea tabule enee est perpendicularis super superficiem regule et super mediam longitudinis eius lineam, et ita est perpendicularis super superficiem speculi plani et super mediam longitudinis eius lineam.

[3.68] Amplius, superficies tabule enee est equidistans superficiei descendenti per centra foraminum, quoniam longitudo centrorum a superficie tabule enee eadem—id est medietatis unius grani ordei—et dyameter foraminis est unius grani ordei. Similiter latitudo superficiei columpne est unius grani, que superficies descendens per centra foraminum secat columpnam per medium. Et ita axis columpne est in superficie illa, et columpna descensu suo tangit lineam in tabula enea cui quidem equidistat axis, quoniam axis est equidistans cuilibet linee superficiei columpne.

[3.69] Et axis columpne cadit in punctum superficiei regule, a quo puncto linea ducta ad centrum tabule enee est perpendicularis super

298 *post* communis *add.* superficiei *R* / *post* enee *add.* superficiei *R* / *post* regule *add.* in *R* / que: qua *R* / est *inter. a. m. E* 299 regule[1] *om. C1ER* / *post* linee *scr. et del.* com *S* / *post* communi *add.* concave *R* / regule[2] *corr. ex* linee *L3* 300 cadit: cadet *ER* / perpendiculariter *corr. ex* perpendicularis *O* 1 et regula *mg. L3* 2 et superficies quadrati *mg. L3* / *ante* equidistans *add.* est *C1* 3 *post* tabule[1] *inter.* enee *a. m. S; add.* enee *R* / *post* tabule[2] *inter.* est *L3; scr. et del.* enee *C1; add.* enee *R* / ortogonaliter: ortogonalis *R* / *post* ortogonaliter *inter.* est *O; add.* est *C1R* / regule: tabule *L3* / *post* regule *inter.* ortogonaliter *a. m. E* 4 et . . . eandem (5) *om. R* / similiter *corr. ex* super *C1* / *post* enee *inter.* est *a. m. C1* 5 longitudinis *inter. a. m. E;* latitudinis *R* / *post* longitudinis *add.* tabule *O; scr. et del.* tabule *L3* / est: erit *E* / perpendicularis *alter. in* perpendiculariter *a. m. E* 6 latitudinem . . . super *inter. a. m. E; om. R* 7 *post* longitudinis *add.* tabule vel *L3; scr. et del.* eius *C1* / cum . . . eam *om. R* 8 erit: est *L3C1R* / perpendicularis *alter. in* perpendiculariter *a. m. E* / super . . . perpendicularis (9) *mg. SL3* (*a. m. S*) / perpendicularis *alter. in* perpendiculariter *a. m. E* 10 longitudinis *alter. in* eius *a. m. E* / eius *om. E* / et *om. R* / ita *om. E* / ita est *transp. FP1;* est ergo *R* 12 est *om. S* 13 descendenti *mg. F* / quoniam: nam *R* / *post* a *add.* et *FS* 14 *post* enee *inter.* est *a. m. C1; add.* est *R* / *post* eadem *inter.* est *a. m. L3* / medietatis: medietatem *FP1S* / ordei *corr. ex* ei *a. m. C1* / et . . . ordei (15) *inter. a. m. E* 15 ordei *om. O* 16 est *corr. ex* sibi *L3* / *post* est *add.* igitur *L3* / que: et *R* 17 medium: mediam *L3* 18 descensu *corr. ex* descensuo *P1* / *post* descensu *add.* illo *E* 19 est *om. O* / linee *corr. ex* ille *a. m. C1* 21 et axis columpne *om. O* / *post* axis *scr. et del.* et *P1* / punctum superficiei *transp. L3C1* 22 perpendicularis *alter. in* perpendiculariter *a. m. E*

tabulam eneam, quoniam, per quodcumque foramen descendat columpna, axis eius cadit super mediam longitudinis regule lineam, et omnes ille perpendiculares sunt equales.

[3.70] Et linea protracta a puncto regule in quem cadit axis per centrum foraminum est equidistans linee protracte a centro tabule enee ad terminum dyametri foraminis. Quoniam linea inter punctum illud et centrum est ortogonaliter super superficiem tabule enee, cum sit pars linee medie longitudinis regule, et etiam super axem. Et huic linee interiacenti centrum tabule enee et punctum est equidistans linea anuli transiens per centra foraminum et perpendiculariter cadens in superficiem tabule enee, quare equidistantes erunt linee cadentes in terminos linee anuli et longitudinis regule equalium et equidistantium.

[3.71] Pari modo in singulis foraminibus, quare linee a puncto regule in quem cadit axis producte ad centra duorum foraminum se respicientium sunt equidistantes duabus lineis a centro tabule enee ad extremitates dyametrorum eorumdem foraminum protractis, quare hee due linee equalem tenent angulum cum illis lineis.

[3.72] Et si a termino axis erigatur linea ad centrum foraminis, erit in superficie per centra descendente, et erit equidistans medie linee tabule enee. Quoniam linea inferior interiacens capita eorum est perpendicularis super tabulam eneam et equalis superiori eadem capita interiacenti et super tabulam eneam perpendiculari. Et est equidistans ei, quare linea a centro foraminis medii ad terminum axis columpne est equidistans medie linee tabule enee, et illa est perpendicularis super regulam, quare et ista. Igitur hec linea et latera alterum angulum continentia sunt equidistantes medie linee tabule enee et alteri linearum

23 quodcumque: quocumque *O* 24 columpna *corr. ex* columpnat *E*/cadit: cadat *C1*/*post* super *add.* lineam *C1*/regule *corr. ex* eius *L3C1* (*a. m. C1*)/lineam *om. C1*/et . . . equales (25) *om. R* 26 *post* regule *add.* enee *O*/quem: quam *FP1O*; quod *R*/per . . . foraminum (27) *om. O*/centrum (27): centra *R* 28 quoniam linea *mg. F*/et *inter. P1* 29 ortogonaliter: ortogonalis *R*/super: in *E*/superficiem *corr. ex* superficie *E* 30 *post* medie *scr. et del.* foraminis quoniam linea *O*/et[1]: est *E*/etiam . . . et[2] *om. R* 31 *post* centrum *inter.* et *O*/et *om. O*/punctum *corr. ex* ipsum *a. m. C1* 32 centra: centrum *E*/in: super *R* 33 superficiem: superficie *E*/*post* quare *scr. et del.* d *F*/in . . . equidistantium (34) *om. R* 34 anuli: anguli *FP1SOL3E*/anuli . . . regule *mg. a. m. C1*/longitudinis *corr. ex* longitudinibus *P1*/*post* regule *scr. et del.* et anguli *C1*/*post* equidistantium *add.* sunt equales *C1*; *add.* a puncto regule ad centra foraminum lineis a tabule enee centro ad terminos diametrorum eorundem foraminum in superficie tabule ductis *R* 35 *post* puncto *scr. et del.* in *F*; *scr. et del.* regula *P1* 36 quem: quod *R*/ad: et *O*/centra *corr. ex* centrum *E*; centrum *R*/respicientium (37) *corr. ex* respicientes *L3* 37 *ante* sunt *add.* tantum *OL3*/sunt *om. FP1SR*/duabus: duobus *S* 38 quare . . . linee (39) *om. R* 39 tenent . . . lineis: cum his lineis tenent angulum *R* 40 si *inter. a. m. E*/a *inter. C1* 41 centra: centrum *R*/descendente: descendentia *O* 42 linea *om. P1*/eorum: earum *R* 43 *post* super *scr. et del.* capi *F*/tabulam *corr. ex* capita *F*/equalis: equali *FS*/*post* superior *add.* et *O* 44 perpendiculari *corr. ex* perpendiculariter *P1* 45 *post* ei *add.* et *R*/quare *inter. a. m. E*; similiter *R*/*post* foraminis *scr. et del.* et *C1* 46 illa est *transp. ER* 47 *ante* regulam *scr. et del.* lineam *E*/latera: laterum *E*; altera *R*/latera alterum *transp. OL3E*/alterum *om. R* 48 continentia sunt equidistantes: continentes equidistant *R*/alteri: alterus *O*; alterum *C1*/linearum: linee *R*

in tabula enea angulum continentium, quare quasi partiales sibi oppositi sunt equales.

[3.73] Igitur linea media tabule enee dividit angulum suum per equalia, quare linea a centro foraminis medii dividit angulum suum per equalia. Et cum certum sit quod lux foramen declinatum intrans per illas lineas angulum continentes moveatur, planum quod lux omnis reflectitur per lineas que cum lineis descensus sunt in eadem superficie ortogonali super superficiem reflexionis et angulum equalem facientibus cum perpendiculari cum lineis descensus.

[3.74] Et lux perpendiculariter descendens reflectitur per perpendicularem. Et hoc generale in omni luce.

[3.75] Si autem declinetur regula non in latus suum sed in caput ut axis foraminis medii non sit perpendicularis super regulam, reflectitur lux, et videbitur super lineam altitudinis anuli perpendicularem et per centrum foraminis transeuntem. Et quanto maior fuerit declinatio, maior erit lucis reflexe a foramine vel axe elongatio. Et si diminuatur declinatio, diminuetur elongatio, et ita donec situs regule ad rectitudinem regrediatur, super perpendicularem illam reflectitur lux.

[3.76] Quod autem in hac declinatione axis foraminis medii et linea reflexionis sunt in eadem superficie ortogonali super superficiem reflexionis planum per hoc quoniam axis foraminis medii est perpendicularis super latitudinem regule—id est super lineam communem superficiei regule et superficiei per centra foraminum descendentis—et media linea tabule anuli est equidistans huic axi et equidistans medie linee tabule enee.

[3.77] Et media linea tabule enee est perpendicularis super latitudinem regule, et est super lineam communem superficiei regule

49 *post* enea *add.* reliquum *R* / *post* angulum *inter.* reliquum *L3* / continentium: continenti *R* / quasi: anguli *R* / partiales: percipiales *FP1* / oppositi *corr. ex* opposi *E* 51 linea *om. F* / linea media *transp. P1OC1* / media tabule enee: tabule enee media *R* / *post* dividit *scr. et del.* angulum *F* 52 quare . . . equalia (53) *inter. a. m. L3* / angulum suum *transp. L3C1* 53 per: super *FP1* / *post* lux *add.* per *P1* 54 illas lineas *transp. O* / lux omnis *transp. L3C1* 55 eadem *om. R* 56 super *om. O*; *inter. L3* 57 facientibus: facientes *R* / *post* cum[1] *add.* linea *R* / *post* perpendiculari *add.* angulo quem continet perpendicularis *R* / cum: tamen *F*; *alter. in* et *O*; *corr. ex* et *L3* 58 *post* et *add.* quod *R* / descendens: descensus *FP1* / per *om. FP1*; *inter. OL3* 59 generale: generaliter *E* / *post* generale *add.* est *FP1R* 60 si autem *corr. ex* suam *O* / *post* regula *add.* cum *L3* / caput: capite *E* 61 medii: medium *FP1S* / reflectitur: reflectatur *S*; reflectetur *OR*; *alter. ex* reflectatur *in* reflectetur *C1* 63 transeuntem: transeuntes *O* 64 *ante* maior *add.* tanto *ER* / *post* diminuatur *scr. et del.* elo *O* 65 diminuetur elongatio *mg. L3* / *post* elongatio *add.* lucis a foramine *C1* 66 regrediatur: egrediatur *O*; aggrediatur *L3*; *corr. ex* egrediatur *a. m. E* / *post* regrediatur *add.* et *FP1R* / super *corr. ex* per *a. m. S*; *inter. L3E* (*a. m. E*) / reflectitur: reflectatur *R* 67 *post* in *add.* declinatio *E* / declinatione: declaratione *FP1*; *corr. ex* declaratione *L3* 68 *post* reflexionis *scr. et del.* in *F* / sunt: sit *P1*; sint *R* 69 *post* planum *add.* est *C1* / *post* quoniam *add.* enim *R* 70 super[2] *inter. O* 71 superficiei[2] *inter. a. m. E* 72 *post* tabule *inter.* scilicet *a. m. E*; *add.* scilicet *R* / anuli *inter. a. m. E* 73 *ante* equidistans *add.* est *C1* / equidistans *corr. ex* quidem *L3* 75 est *om. R*

et superficiei tabule enee, quare superficies in qua sunt media linea tabule enee et axis foraminis medii ortogonalis est super superficiem regule. Et in hac superficie est linea perpendicularis in altitudine anuli, quoniam transit per terminos equidistantium—scilicet medie tabule enee et axis foraminis medii.

[3.78] Palam igitur quod lux reflexa que apparet in perpendiculari altitudinis anuli reflectitur per lineam que cum axe per quem fit descensus est in superficie ortogonali super superficiem regule. Luce ergo descendente in speculum planum, fit reflexio secundum lineas quarum eadem declinatio super superficiem speculi, et ipse cum perpendiculari in superficie ortogonali super superficiem speculi.

[3.79] In speculo columpnari exteriori eadem penitus probatio que est in plano—scilicet quod acumen tabule enee cadit super lineam longitudinis speculi ortogonalis, et similiter columpna descendens super eandem. Et pars illius linee super hos casus est ortogonaliter super tabulam eneam. Et semper, sive per foramen medium sive per declinatum descendet lux, reflexio eius cum descensu erit in eadem superficie ortogonali super superficiem contingentem lineam longitudinis speculi.

[3.80] In piramidali vero exteriori, cum superficies regule sit in eadem superficie cum linea media longitudinis piramidalis, sicut in columpnari, erit idem situs linearum superficierum et idem reflexionis modus, sicut in speculo plano, et eadem penitus probatio.

[3.81] In speculo autem columpnari concavo descendit acumen tabule enee usque ad lineam longitudinis eius mediam, et super eandem cadit axis cuiusque foraminis. Et linea pars illius inter hos casus est ortogonalis super superficiem tabule enee, et axis foraminis et media linee tabule enee sunt ortogonales super superficiem tangentem speculum illud in linea longitudinis, que est locus reflexionis, et equidistantes superficiei regule.

76 enee *om. FP1* / *post* linea *scr. et del.* regule et *C1* 77 enee *inter. a. m. E* / *post* medii *add.* et *E*; *add.* etiam *R* / est *om. SOE*; *inter. L3* 78 et *om. S* / *post* hac *scr. et del.* in *E* / est *om. FP1S* 81 apparet *corr. ex* apparent *S* 82 *post* anuli *scr. et del.* e *C1* / reflectitur: reflectetur *S* / fit: sit *L3* / descensus (83): de census *R* 83 est *inter. a. m. C1* / ergo: igitur *FP1* 84 descendente: descente *O* / fit: sit *SL3* 85 *post* ipse *add.* sunt *C1ER* / cum *inter. a. m. L3* 86 superficiem speculi *transp. ER* 88 quod *inter. a. m. E* / cadit: cadat *R* 89 ortogonalis: ortogonali *O*; ortogonaliter *C1ER* 90 eandem: eadem *O* / super[1] *corr. ex* per *L3C1* (*a. m. C1*); inter *R* / ortogonaliter: ortogonalis *R* 91 sive[1] *corr. ex* fuit *L3*; *mg. a. m. C1* / medium: mediam *FP1S* 92 descendet *corr. ex* descendit *a. m. C1*; descenderit *R* / *post* descensu *scr. et del.* erit *E* / superficie . . . super (93) *mg. O* 93 *ante* superficiem *scr. et del.* superficiebus *O* 94 vero *om. O* / superficies: superficie *FP1SL3* / *post* regule *scr. et del.* regule *O* 95 media *om. FP1SR* / piramidalis: pyramidis *R* / sicut: sit *FP1* 96 *post* idem[1] *scr. et del.* ?? *E* / superficierum: superficiei *L3ER* 97 speculo plano *transp. R* 98 autem *om. OL3C1ER* / descendit *corr. ex* descendet *O*; *alter. in* descendet *a. m. C1* 100 linea pars illius: pars illius linea *R* 102 linee: linea *L3C1ER* / super *om. FP1* 103 et *om. C1* / equidistantes: equidistantem *OER*; *alter. in* equidistantem *C1*

[3.82] Et ita idem modus probandi qui prius—quod scilicet reflexio et descensus sunt in eadem superficie ortogonali super superficiem loci reflexionis, et eiusdem sunt declinationis, et quod descensus per medium efficit reflexionem per ipsum. Et declinato capite regule, erit reflexio super perpendicularem anuli, sicut dictum est in plano.

[3.83] In speculo piramidali concavo eadem in omnibus probatio.

[3.84] In speculo sperico exteriori palam quod medius eius punctus est in superficie regule, et axis cadit in punctum illud, et erit in eo idem situs linearum. Et aliorum penitus quod in plano, et eadem demonstratio.

[3.85] In speculo sperico concavo iam determinatum est quod axis foraminis descendit ad punctum eius medium, et acumen tabule enee transit per foramen in speculo iam factum usque dum sit in eadem superficie cum puncto illo medio. Et linea a puncto illo ad acumen protracta est equidistans medie linee longitudinis regule, et ita descensus et reflexio sunt in superficie ortogonali super superficiem contingentem speculum in illo puncto medio et equidistantem superficiei regule. Et eadem probatio penitus ut in aliis.

[3.86] Palam ergo quod omnis lux in quodcumque speculum eorum cadit reflexio et descensus sunt in eadem superficie ortogonali. Hic autem modus reflexionis non accidit ex proprietate axis, vel puncti in quod cadit, vel foraminis per quod intrat, vel proprietates speculi. Accidit enim in quodlibet foramen quecumque sit lux, et per quamcumque lineam descendat, et in quodcumque speculi punctum cadat. Quoniam quocumque puncto speculi sumpto, si lux in ipsum descendat, cum idem sit ei situs respectu longitudinis speculi, et cuicumque alii erunt similiter idem respectu linearum ab eo

105 *post* probandi *add.* hic *C1* 106 sunt: fit *O*; sint *R*/ortogonali: ortogonaliter *E*/ortogonali . . . superficiem *om. O*/super *om. P1* 107 *ante* eiusdem *add.* quod *R*/sunt: sint *R*/descensus *corr. ex* descendsus *F* 108 reflexionem: rationem *O* 109 perpendicularem: perpendiculares *OC1* 110 *post* speculo *scr. et del.* sperico *S* 111 medius: medium *R*/punctus: punctum *OR* 112 erit in eo: in eo erit (in eo *inter.*) *L3* 113 quod: que *S*; qui *ER*/*ante* in *scr. et del.* in superficie regule *S* 115 speculo sperico *transp. E*/determinatum: declaratum *R* 116 descendit: descendet *E*; descendat *R* 117 transit: transeat *R*/eadem *corr. ex* edem *O* 118 superficie: specie *O*; *corr. ex* specie *L3*/puncto[2] *corr. ex* predicto *O* 119 medie *corr. ex* medium *F* 120 *post* in *add.* eadem *R* 121 *post* contingentem *scr. et del.* secundum *O*; *scr. et del.* per *C1*; *add.* per *L3*/illo puncto *transp. O*/medio *inter. a. m. E*/equidistantem *corr. ex* equidistans *F* 122 et *inter. a. m. C1*/ut: quod *E*; que *R* 123 palam: planum *OL3C1ER*/speculum eorum: horum speculorum *R* 124 cadit: ceciderit *R*/et *om. E* 126 intrat: intravit *O*/proprietates: proprietate *ER* 127 accidet enim *om. E*/quodlibet foramen: quolibet foramine *OR*/*post* foramen *add.* in quolibet foramine *E*/et: vel *E* 128 *post* quacumque *scr. et del.* planum ergo quod omnis lux in quodcumque speculum eorum cadit reflexio et descensus sunt in eadem superficie ortogonali *S* 129 puncto speculi *transp. C1*/si . . . descendat (130) *om. P1* 130 ei *corr. ex* eis *C1*/*post* speculi *add.* qui *O* 131 cuicumque *corr. ex* quicumque *L3*; quicumque *E*

protractarum, que eiusdem sunt declinationis cum lineis a puncto priori intellectis, sicut et puncto priori, vel cuicumque alii.

[3.87] Et generaliter idem est situs cuiuslibet puncto in quod cadit lux qui et in priori sumpto et respectu axis, et respectu acuminis tabule enee. Et eadem in omnibus probatio, et similis demonstratio, unde certum non esse hoc ex proprietate lucis vel figure alicuius speculi sed ex quadam proprietate communi omni rei polite et cuilibet luci. Si autem per diversa in quodcumque punctum descendit lux foramina, videbitur reflexio diversa et angulorum diversitas suo descensui consona, et sic in omnibus.

[3.88] Manifestum ex superioribus quod, si corpus politum opponatur corpori luminoso, cadit in quodlibet punctum eius lux a quolibet luminosi puncto, unde super quodlibet politi punctum cadit piramis cuius acumen in eo, et superficies luminosi basis. Et a quolibet puncto luminosi procedit piramis cuius acumen in eo et basis superficies politi.

[3.89] Si autem inter luminosum et politum intelligatur punctum aliquod, veniet quidem ad illud punctum lux luminosi in modum piramidis cuius acumen in puncto, et latera huius piramidis procedentia usque dum cadant in superficiem politi piramidem efficiunt. Unde in puncto intellecto erunt acumina duarum piramidum quarum bases sunt superficies luminosi et superficies politi, et si ad punctum quodcumque intermedium intelligatur piramis cuius basis superficies politi, et procedant huiusmodi piramidis linee, illud quod occupabunt ex superficie luminosi hoc est a quo procedebat lux ad politum secundum duas piramides quarum acumina in puncto intellecto.

[3.90] Et quod procedit lucis in hiis duabus piramidalibus procedit et includitur in duabus primis piramidalibus, et a luminoso secundum

132 protractarum *corr. ex* postrarum *a. m. E* / priori: priore *R* 133 intellectis *corr. ex* et intellectus *C1* / et: a *C1*; *om. R* / puncto *corr. ex* predicto *O* 134 cuiuslibet: cuilibet *R* / *post* cuiuslibet *add.* in *C1* / *post* quod *scr. et del.* in quod *F* 135 qui et *transp. E* / in *om. O* / priori: priore *R* / respectu *mg. a. m. C1* / *ante* axis *add.* ita *C1* 136 *post* et[1] *add.* in *FP1*; *inter.* est *a. m. E* / similis *alter. ex* sillabica *in* simillima *O*; *alter. ex* similima *in* sillogistica *a. m. C1*; similia *L3* / *post* unde *add.* est *R* 137 figure: figura *R* 138 quadam proprietate *transp. R* / omni *inter. a. m. C1*; *om. ER* / cuilibet *corr. ex* cuiuslibet *C1* 139 in *corr. ex* et *O* / descendit: descenderit *R* 140 descensui: descensu *E* 142 *post* manifestum *add.* autem *R* 143 cadit: cadet *R* / quodlibet punctum: quolibet puncto *E* 144 luminosi puncto *transp. ER* / puncto *corr. ex punctum O* 145 cuius *inter. a. m. E* / et[1] *om. S* / *post* luminosi *add.* est *R* 146 cuius *corr. ex* eius *O* 148 *post* autem *scr. et del.* inter *F* / inter: intra *L3* 149 veniet: tenet *F*; tendit *P1* 150 piramidis cuius acumen *corr. ex* cuius acumen piramidis *E* / cuius . . . piramidis[2] *om. S* 151 cadant: cadat *O*; caderent *L3* / in puncto (152) *mg. a. m. C1* 153 superficies[2] *om. R* / superficies politi *transp. E* / quodcumque: quodlibet *P1* 154 intelligatur: intelligitur *L3C1* / cuius *corr. ex* eius *O* / et *om. FP1SOL3* 155 huiusmodi: huius *L3R* 156 procedebat *corr. ex* procedat *C1* / *post* politum *add.* erit *R* 157 *post* acumina *add.* sunt *R* 158 quod: quicquid *R* / piramidalibus *corr. ex* pyramidibus *P1*; pyramidibus *R* 159 primis *om. O* / *post* primis *scr. et del.* politis *L3*; *add.* pollitis *E* / piramidalibus: pyramidibus *L3R* / *post* secundum *scr. et del.* lineis *P1*

lineas equidistantes procedit lux ad speculum, sed hee linee includuntur in duabus primis piramidibus. Et per quascumque lineas moveatur lux ad speculum, observant linee reflexionis eundem penitus situm quem habebant linee motus lucis; unde si moveatur lux per equidistantes, reflectitur per equidistantes, et lux cadens in modum politi ad modum piramidalis reflectitur observans modum eiusdem piramidis.

[3.91] Cum descendit lux a corpore luminoso per foramen aliquod ad corpus politum, si in superficie foraminis ex parte illuminosi intelligatur punctus a quo puncto intelligantur due piramides basis unius in luminosa alterius in polito, a sola base piramidis cuius luminosum basis venit lux ad politum super illud punctum. Similiter, si in superficie foraminis ex parte politi intelligatur punctum in quo acumina duarum piramidum unius ad speculum alterius ad luminosum, a sola base piramidis que basis est in luminoso accedit lux ad speculum super hoc punctum.

[3.92] Et a parte luminosi hiis duabus piramidalibus communi accidit lux ad partem speculi commune duabus piramidalibus. Venit etiam lux a luminoso ad speculum per lineas equidistantes, sed per quascumque accedat, fit reflexio modo predicto. Et quelibet linee reflexionis conservant situm linearum descensus lucis eas respicientium, et in omni reflexione observatur ydemptitas forme lucis que fuerit in polito corpore, et hoc deinceps explanabimus explanatione evidenti.

[3.93] Amplius, patuit quod lux quanto plus ab ortu suo elongatur plus debilitatur. Patuit etiam quod lux continua fortior disgregata. Cum igitur ab aliquo puncto luminosi procedit lux ad superficiem speculi in modum piramidis, quanto magis elongatur a puncto illo tanto maior est eius debilitatio duplici de causa: et propter elongationem ab ortu suo, et propter disgregationem. Cum autem ab aliquo speculi puncto

160 sed: et *L3* 161 piramidibus: pyramidalibus *P1O*/per *om. FP1* 163 quem: quam *O*/ moveatur: movetur *O* 164 modum: modicum *C1*; *om. R*/politi: politum *R* 165 *post* modum[1] *add.* politi *C1*/piramidalis: piramidis *P1R* 167 *ante* cum *add.* et *R*/a *om. L3*; *inter. a. m. C1*/aliquod *inter. OE* 168 si *inter. a. m. C1*/*post* parte *add.* illius *O*/illuminosi: luminosi *OR* 169 punctus *corr. ex* pes *O*; punctum *R*/quo *inter. a. m. E* 170 luminosa: luminoso *OR*/base: basi *R*/cuius: eius *L3C1*; *corr. ex* eius *OE* 171 *post* luminosum *add.* est *a. m. C1* 172 si *inter. a. m. E*/*post* quo *inter.* sunt *O* 173 duarum: duorum *L3C1* 174 base: basi *R*/que *corr. ex* qui *E*/in luminoso *corr. ex* illuminoso *O*/accedit: accidit *OL3* 176 piramidalibus communi: pyramidibus communis *R* 177 accidit: accedit *R*/speculi: speculum *FP1O*/commune *corr. ex* tot? *O*; communem *R*/piramidalibus: pyramidibus *R* 178 a *corr. ex* ad *S*/*ante* speculum *scr. et del.* punctum *S* 179 *post* quelibet *scr. et del.* in *C1* 180 conservant: observant *ER*/ respicientium *corr. ex* respiciendum *O* 181 lucis *mg. a. m. E* 182 hoc: hec *C1R* 183 quod ... etiam (184) *mg. a. m. S* (plus ab ortu suo: ab ortu suo plus)/quanto *corr. ex* quam *a. m. E*/*post* elongatur *add.* tanto *ER* 184 *post* fortior *inter.* it *F*; *add.* est *P1R* 185 *post* ad *scr. et del.* speculum *P1*/speculi *inter. a. m. S* 186 a puncto illo: ab illo puncto *R*/illo *inter. a. m. E* 187 est: erit *ER*/eius *corr. ex* enim *O*/debilitatio: debilitas *R* 188 aliquo *corr. ex* liquo *O*

reflectitur lux, ista fit debilior tripliciter: et propter reflexionem que debilitat, et propter elongationem a loco reflexionis, et propter disgregationem.

[3.94] Si vero lux reflexa a speculo agregetur in punctum aliquod, fiet quidem fortior propter agregationem, sed debilitatur per reflexionem et elongationem. Si igitur agregatio lucis tantum redit ei fortitudinis quantum subtrahunt reflexio et elongatio, erit lux reflexa agregata eiusdem fortitudinis cuius est in superficie speculi. Si vero agregatio minus addat fortitudinis quam diminuant illa duo, erit debilior, et si plus addat, erit fortior.

[3.95] Similiter, si a superficie luminosi procedat piramis ad aliquod punctum speculi, erit lux procedens secundum hanc piramidalitatem debilior propter elongationem sed fortior propter agregationem. Si autem agregatio potest super elongationem, erit lux in puncto speculi agregata fortior luce unica a luminoso veniente per lineam unam. Unica dico, quia ad quodlibet punctum linee ex illis sumpte venit etiam piramis a luminoso, que quidem piramis cum similibus excluditur in hac consideratione.

[3.96] Si vero elongatio ponderet super agregationem, erit lux puncti politi minor luce sola unius linee sumpta, et si agregatio plus ponderet elongatione, erit fortior. Luces autem que a luminoso ad speculum accedunt super lineas equidistantes erunt debiliores quam modo alio accedentes, quoniam debilitate propter elongationem non agregentur in speculum, et in reflexione per lineas equidistantes moventur. Unde per reflexionem et elongationem debilitantur. Et si agregentur in reflexione, conferetur eis fortitudo comparata ad fortitudinem quam habuerint in speculo secundum posse agregans super reflexionem et elongationem.

[3.97] Amplius, omnis linea per quam movetur lux a corpore luminoso ad corpus oppositum est linea sensualis, non sine latitudine,

189 tripliciter *corr. ex* multipliciter *C1* 193 *post* propter *scr. et del.* e *P1; add.* per *O* / debilitatur per: debilitabitur propter *R* 194 reddit: impendit *O* 195 quantum: quam *E* / et *inter. O* / *post* reflexa *add.* et *L3C1* 196 *post* vero *scr. et del.* in *P1* 197 minus: unius *FSL3C1; alter. in* unius *O* / addat: addit *O* / fortitudinis: fortitudinem *FP1S* / diminuant: diminuunt *R* 199 aliquod *corr. ex* aliqud *O* 200 hanc *inter. a. m. S* 201 *post* agregationem *scr. et del.* si autem aggregatio potest super elongationem et propter fortior agregationem *S* 202 agregatio: aggregationem *O* 203 agregata: aggregati *O* / unica: unicam *O*; *corr. ex* unicam *L3C1* 204 quia *corr. ex* quod *a. m. E* / etiam: et *E* 205 quidem *inter. O* 207 *post* vero *scr. et del.* aggregatio *O* / ponderet: penderet *O* 208 politi: polita *O*; *corr. ex* polita *L3*; *corr. ex* positi *a. m. E* / *post* minor *scr. et del.* line *F* / *post* sola *scr. et del.* luce *L3* / unius: minus *P1* / ponderet: penderet *O* 211 agregentur *corr. ex* agregeretur *L3E*; aggregantur *R* 212 et: sed *FP1S* / reflexione: flexione *F*; inreflexione *E* 213 agregentur: agregerentur *E* 214 comparata *corr. ex* comparatata *O* 215 habuerint: habuerunt *R* / agregans *alter. in* agregationis *C1* 217 *post* omnis *add.* autem *O* 218 sensualis *corr. ex* sensuasensualis *F*

lux enim non procedit nisi a corpore, quoniam non est nisi in corpore. Sed in minori luce que sumi possit est latitudo, et in linea processus eius est latitudo. Et in medio illius linee sensualis est linea intellectualis, et alie eius linee sunt equidistantes illic. Et si dividatur minor ex lucibus, neutra eius pars erit lux, sed utraque extinguetur nec apparebit. Si autem lux minima duplicetur, aut amplius multiplicetur per equalia, et compacta dividatur, erit lux utraque eius pars. Si vero per inequalia fiat divisio, erit altera pars eius lux, altera minime.

[3.98] Lux autem minima procedit ad minimam corporis partem quam lux occupare possit, et processus eius est secundum lineam intellectualem linee sensualis mediam, et extremitates ei equidistantes. Et cadit lux minima non in punctum corporis intelligibilem, sed sensibilem, et refertur per lineam sensibilem cuius latitudo est equalis latitudini linee sensibilis venientis. Et si intelligatur in linea sensibili linea reflexa intellectualis media, eundem habet situm super reflexionis locum quem habet linea intelligibilis media linee sensibilis venientis, et quelibet linea intellectualis in linea reflexa sensibili eundem penitus observat situm cum linea intelligibili alterius sensibilis ipsam respiciente. Observatur ergo in omni luce reflexio linearum et punctorum intellectorum, licet ab eis aut per ipsas non procedat lux, et in hunc modum erit reflexio lucis.

[3.99] Amplius, quare ex politis corporibus non ex asperis fiat reflexio est quoniam lux, ut diximus, non accedit ad corpus nisi per motum citissimum, et cum pervenit ad politum, eicit eum politum a se. Corpus vero asperum nec potest eam eicere, quoniam in corpore aspero sunt pori quos lux subintrat; in politis autem poros non invenit. Nec accidit hec eiectio propter corporis fortitudinem vel duriciem, quia videmus in aqua reflexionem; sed est hec repulsio propria politure,

219 in *inter. L3* 220 minori: minore *R*/que: qui *P1*/possit: potest *ER*/est *inter. a. m. E* 222 illic *alter. in* illi *F*; illi *R*/si *corr. ex* non *O* 223 sed: si *P1* 224 multiplicetur *inter. E*/per . . . compacta (225): et compacta per equalia *L3ER*/equalia: inequalia *inter. a. m. E*/equalia et compacta (225): et compacta equalia *OC1* 225 eius pars *transp. ER* (pars *inter. a. m. E*)/ inequalia *corr. ex* inqualia *O*; equalia *E* 227 ad *corr. ex* in *a. m. E*; in *R* 230 intelligibilem: intelligible *R*/sed sensibilem (231) *mg. a. m. F* 231 sensibilem: sensibile *R*/et . . . sensibilem² *om. O*/refertur: reflectitur *R*/latitudo *corr. ex* altitudo *L3*; altitudo *E*/est *om. SL3C1E*; *inter. O*/*post* equalis *scr. et del.* a *L3* 232 si *inter. F*/sensibili: sinsibili *R* 233 habet *inter. O*; habebit *R* 234 quem: quoniam *FP1O*/*post* linea *scr. et del.* intellu *E*/linee: linea *P1*; *corr. ex* linae *F* 235 intellectualis in linea *om. S* 236 observat *corr. ex* observant *O*/*post* observat *scr. et del.* eundem *L3*/intelligibili *corr. ex* intelligibilis *L3*/*post* alterius *add.* linee *R*/ipsam *corr. ex* illam *a. m. E* 237 ergo *om. E*/reflexio: ratio *R*/*post* linearum *add.* intellectuarum *O* 238 intellectorum: intellectuum *O*; *corr. ex* intellectum *L3*/licet *inter. a. m. E* 239 in *inter. O* 240 non *corr. ex* nec *a. m. C1* 241 accedit: accedet *OL3*; *alter. ex* accidet *in* accedet *E* 242 pervenit *corr. ex* pervenerit *S*; *corr. ex* venit *E*; pervenerit *R*/eum *alter. in* eam *C1*; eam *R* 243 nec: non *C1ER* 244 poros non invenit: non invenit poros *ER* 245 hec eiectio *transp. ER*/eiectio *corr. ex* ectio *O* 246 aqua *corr. ex* qua *S*/repulsio: repulsi *FP1S*/politure *corr. ex* poture *F*; poture *S*

sicut de natura accidit quod aliquod honerosum cadens ab alto super lapidem durum revertitur in altum, et quanto minor fuerit duricies lapidis in quam ceciderit, regressio cadentis debilior erit. Et semper regredietur cadens versus partem a qua processit. Verum in arena, propter eius mollitiem, non fit regressio que quidem accidit in corpore duro.

[3.100] Si autem in poris asperi corporis sit politio, tamen lux intrans per poros non reflectitur, et si eam reflecti accidit, dispergitur, et propter dispersionem a visu non percipitur. Pari modo, si in aspero corpore partes elatiores fuerint polite, fiet reflexa dispersio, et ob hoc occultabitur visui. Si vero eminentia partium adeo sit modica, ut sit eius quasi idem situs cum depressis, tunc comprehendetur eius reflexio tamquam in polito non aspero, licet minus perfecte.

[3.101] Quare autem fiet reflexio lucis secundum lineam eiusdem situs cum linea per quam accedet ad speculum ipsa lux est quoniam lux motu citissimo movetur, et quando cadet in speculum, non recipitur; sed ei fixio in corpore illo negatur. Et cum in ea perseveret adhuc prioris motus vis et natura, reflectitur ad partem a qua processit, et secundum lineas eundem situm cum prioribus habentes.

[3.102] Huius autem rei simile in naturalibus motibus videre possumus et etiam accidentalibus. Si corpus spericum ponderosum ab aliqua altitudine descendere permittamus perpendiculariter super politum corpus, videbimus ipsum super perpendicularem reflecti per quam descenderat. In accidentali motu, si elevetur speculum secundum aliquam altitudinem hominis, et firmiter in pariete figuratur, et in acumine sagitte consolidetur corpus spericum, et proiciatur sagitta per arcum in speculum hoc modo ut elevatio sagitte sit equalis elevationi speculi (et sit sagitta equidistans orizonti), planum quod super perpendicularem accedit sagitta ad speculum, et videbis super eandem

247 sicut *corr. ex* sitis *O* / aliquod: aliquid *C1ER* / honerosum: porosum *R* 248 quanto: quando *L3C1* 249 quam: quem *ER* / ceciderit *corr. ex* cecideret *O* / *post* ceciderit *add.* tanto *R* / cadentis . . . regressio (251) *mg. a. m. S* 250 regredietur: regreditur *ER* / processit: procedit *L3C1* 251 fit *inter. a. m. E* / quidem *om. R* 253 sit: fit *FP1SOL3C1* 254 per *om. OER* / accidit: acciderit *ER* 255 in *inter. E* 256 fiet *corr. ex* fiat *E* 257 sit eius *transp. E* / sit eius quasi: eius quasi sit *R* 258 depressis: depressivi *C1* / tunc: et *FP1SOL3E*; *om. R* 259 *post* non *inter.* in *a. m. L3* 260 fiet: fit *C1*; fiat *R* 261 accedet: accidit *SOL3*; *corr. ex* accidit *a. m. C1*; accedit *R* 262 motu citissimo *transp. L3C1* / cadet: cadit *R* 263 *post* ei *scr. et del.* eo *P1* / et *om. O* / ea: eo *P1*; *corr. ex* eo *F* 264 et[1] *inter. F* / ad: in *O* 267 *post* etiam *add.* in *R* / accidentalibus *corr. ex* accidentatalibus *O*; *corr. ex* accidenta *E* / *post* si *inter.* enim *C1* 269 perpendicularem: perpendiculari *FP1* 270 *post* elevetur *scr. et del.* ad *L3*; *add.* ad *E*; *add.* aliquod *R* 271 *post* altitudinem *add.* velut altitudinem *OL3*; veluti altitudinem *mg. a. m. C1* / figuratur: figatur *OL3ER* 272 *post* spericum *scr. et del.* et proiciatur corpus spericum *S* / et *inter. a. m. E* 273 ut *mg. a. m. FC1*; *om. P1SL3E* / *ante* elevatio *mg.* scilicet *a. m. F* 275 accedit: accidit *O* / *post* et *add.* modo *L3*; *scr. et del.* modo *C1*; videbis: videbit *O*; videbitur *L3R*; *corr. ex* videbitur *E*

perpendicularem eius regressum. Si vero motus sagitte ad speculum fuerit super lineam declinatam in ipsum, videbitur reflecti non per lineam per quam venerat sed per aliam non equidistantem orizonti, sicut et alia erat, et eiusdem situs respectu speculi cum ea et respectu perpendicularis in speculo. Quod autem ex prohibitione politi corporis accidat luci motus reflexionis palam quia, cum fortior fuerit repulsio vel prohibitio, fortior erit lucis reflexio.

[3.103] Quare autem accidit idem motus reflexionis et eius accessus hec est ratio. Cum descendit corpus ponderosum super perpendicularem, reflexio corporis politi et motus descendentis ponderosi directe sibi sunt oppositi, nec est ibi motus nisi perpendicularis. Et prohibitio fit per perpendicularem, quare repellitur corpus secundum perpendicularem, unde perpendiculariter regreditur. Cum vero descendat corpus super lineam declinatam, cadit quidem linea descensus inter perpendicularem superficiei politi per ipsum politum transeuntem et lineam superficiei eius ortogonalem super hanc perpendicularem.

[3.104] Et si penetraret motus ultra punctum in quem cadit, ut liberum inveniret transitum, caderet quidem hec linea inter perpendicularem transeuntem et lineam superficiei ortogonalem super perpendicularem. Et observaret mensuram situs respectu perpendicularis transeuntis et respectu linee alterius que ortogonalis est super illam perpendicularem. Compacta enim est mensura situs huius motus ex situ ad perpendicularem et situ ad ortogonalem.

[3.105] Repulsio vero per perpendicularem incedens, cum non possit repellere motum secundum mensuram quam habet ad perpendicularem transeuntem, quia nec modicum intrat, repellit ergo secundum mensuram situs ad perpendicularem quam habet ad ortogonalem. Et quando motus regressio eadem fuerit mensura situs ad ortogonalem

276 regressus: regressum *R* / ad speculum *om. R* 278 aliam: lineam *ER* / non *inter. a. m. L3; om. OER* 279 situs *inter. a. m. C1* 280 politi corporis (281) *transp. R* 281 quia *corr. ex* quod *a. m. O* / fuerit: fuit *FP1SE* 282 *post* lucis *scr. et del.* erit *E* 283 accidit: accidat *OR* / motus *alter. in* modus *L3*; modus *C1*; *scr. et del. E* / *post* reflexionis *add.* motus *OC1E*; *scr. et del.* motus *L3* 285 reflexio: repulsio *OR*; *corr. ex* repulsio *a. m. C1* 286 *post* directe *scr. et del.* sibi *E* / est *rep. F* / ibi: in *scr. et del. O* / ibi *om. O*; in *L3* 287 per *inter. SE; om. L3* / quare: quia *FP1* 288 vero *corr. ex* quo *L3* 289 descendat *corr. ex* descendet *E*; descenderit *R* 293 *post* penetraret *scr. et del.* locus *S* / quem *corr. ex* quod *a. m. E*; quod *R* / *post* ut *scr. et del.* punctum *E* 294 inveniret *corr. ex* inveniet *C1E* / caderet: cadet *P1* / inter *inter. a. m. S* 295 *post* transeuntem *add.* per politum *R* / et: etiam *O* 296 *post* super *mg.* hanc *a. m. C1* / observaret: observare *L3* 298 illam perpendicularem *transp. R* / compacta . . . perpendicularem (299) *mg. a. m. S* / enim est *transp. C1ER* / est *inter. OL3C1E* (*a. m. C1*); *om. S* 299 situ: situs *OE*; *corr. ex* situs *C1* 300 per *om. SC1; inter. O* 1 quam: quem *P1* 2 *post* transeuntem *add.* per politum *R* / modicum *corr. ex* modum *S* / secundum (3) *om. L3; mg. a. m. C1* 3 *post* habet *scr. et del.* quam *F* / ad *inter. a. m. F*; quam *S* 4 ad *om. L3*

que fuit prius ad eandem ex alia parte, erit similiter ei eadem mensura situs ad perpendicularem transeuntem que fuit prius.

[3.106] Sed ponderosum corpus in regressu, cum finitur repulsionis motus, ex natura sua descendit et ad centrum tendit. Lux autem eandem habens reflectendi naturam, cum ei naturale non sit ascendere aut descendere, movetur in reflexione secundum lineam inceptam usque ad obstaculum quod sistere faciat motum, et hec est causa reflexionis.

[3.107] Patet etiam ex superioribus quod colores simul moventur cum lucibus, unde erit reflexio coloris sicut et lucis. Et si probationem eius videre volueris secundum modum in parte secunda assignatum, poteris iterum per instrumentum. Ad hanc denotandam reflexionem non plene videbis propter debilitatem coloris, debilitatur enim color per elongationem, per reflexionem, per foramen in quod intrat. Quod autem foramen debilitat planum per hoc quod lux apparet maior post foramen magnum quam parvum. Pari modo, cum foramina stricta sint, color post reflexionem aut nullus apparebit aut valde modicus. Tamen, si in predicto instrumento videre volueris, facias speculum argenteum, in ferreo enim speculo color apparet debilior, quoniam in reflexione misceretur cum luce reflexa mixta ex luce descendente et luce speculi ferrei modica, et color ferreus colori reflexo mixtus debilitaret ipsum.

[3.108] Iterum in domo unici foraminis tantum habeatur instrumentum predictum cui domui paries albus opponatur. Et instrumentum foramini domus aptetur cuius foraminis latitudo sit ut duo instrumenti foramina occupare possit per quorum alterum inspiciatur paries albus domui oppositus. Et parti comprehense parietis opponatur corpus coloris fortis, et per aliud instrumenti foramini

5 *post* que *scr. et del.* fiat *F* / erit *inter. a. m. C1* / *post* erit *scr. et del.* ei *O* / *post* similiter *scr. et del.* erit *E* / ei *om. E* 7 ponderosum *corr. ex* pondero *F* / regressu *corr. ex* essu *O*; *corr. ex* gressu *L3* 9 *post* habens *scr. et del.* reflexionem *S* 10 movetur *corr. ex* moveatur *E* / secundum *corr. ex* super *L3*; super *C1* 12 etiam *om. O* 13 unde *corr. ex* ut *O* 14 eius *inter. a. m. E* / assignatum: signatum *L3C1* 15 iterum: verum *OL3R*; *corr. ex* verum *C1*; scire *E* / reflexionem: reflexionis *O* / *post* reflexionem *add.* factum *OR*; *mg.* factum *a. m. C1* 16 non *inter. L3*; *mg. a. m. C1* / propter *inter. a. m. E* 17 in: per *ER* / intrat: intravit *O* / *post* intrat *add.* vel intravit *E* / quod autem (18) *corr. ex* autem quod *L3* 18 debilitat: debilitet *SOC1ER* / quod: quia *R* / post: per *E* 19 *post* quam *inter.* per *a. m. E*; post *R* 20 sint: sunt *R* / valde *corr. ex* vade *a. m. E* / modicus . . . speculum (21) *mg. S* 22 ferreo *corr. ex* eo *O* 23 *post* reflexione *mg.* coloris *a. m. C1* / misceretur: miscetur *R* / ex *corr. ex* cum *a. m. C1* 24 *post* et *scr. et del.* fo *C1* / color *corr. ex* calor *P1* / *post* colori *inter.* in *O* / reflexo: reflexione *O* 25 debilitaret: debilitat *FP1R* / ipsum: ipsam *O* 26 unici: unicum *O* / foraminis: foramine *L3* / *post* tantum *scr. et del.* enim *C1* 27 instrumentum predictum *scr. et del. O* / predictum . . . instrumentum (28) *mg. a. m. E* / domui *alter. in* domus *E* / albus opponatur *transp. ER* / opponatur: apponatur *O* 28 foramini *corr. ex* foraminis *L3* 29 quorum alterum: quarem altereum *O* 30 parietis *corr. ex* parieti *C1* 31 opponatur: apponatur *C1ER* / aliud *corr. ex* illud *a. m. E* / instrumenti *corr. ex* instrumentum *P1* / *post* instrumenti *scr. et del.* videbitur color reflecti *S* / foramini: foraminum *E*; foramen *R*

videatur pars parietis. Cum ergo lux intraverit per foramina instrumenti, videbitur color reflecti per foramen illud respiciens, quod est oppositum corpori colorato, per aliud minime. Et ita accidet quocumque opposito corpori foramine, et que dicta sunt in reflexione lucis considerari poterunt in reflexione coloris. Occupavit autem latitudo foraminis parietis duo instrumenti foramina ei adhibita ut maior descendat in speculum lux et melior apparet color reflexus. Et quoniam color debilitatur per foramen directus, et similiter reflexus, cum in corpus ceciderit visui oppositus percipietur secundus, unde, si post reflexionem cadat in corpus album foraminis colorationis adhibitum, forsan propter debilitatem non comprehendet eum visus. Adhibito autem secundo visu foramini colorationis, forsan comprehendetur, quoniam primus non secundus videbitur.

[CAPITULUM 4]

Pars quarta: quod comprehensio forme in corporibus fit per reflexionem

[4.1] Super modum comprehensionis forme in politis corporibus dissentiunt plurimi. Unde quidam eorum radios a visu exire ad speculum, et a speculo redire, et formam rei in reditu comprehendere. Alii affirmant formam corporis speculo ei opposito imprimi, et proinde in eo videri sicut in corporibus fit comprehensio formarum naturalium eius.

[4.2] Verum quod aliter sit palam per hoc: quoniam si quis se viderit in aliqua speculi parte motum in partem aliam, non videbit se in parte prima, sed in secunda, quod non accideret si in parte prima infixa esset eius forma. Pari modo, si ad tertiam mutetur partem, mutabitur locus apparentie forme, nec apparebit in prima vel in secunda parte.

32 ergo: igitur *FP1* 33 *post* videbitur *scr. et del.* color *C1* / respiciens: inspiciens *E* / est oppositum (34) *transp. ER* 34 corpori colorato *transp. OC1ER* / colorato *inter. L3E* (*a. m. E*) 35 corpori: corpore *L3* 36 poterunt: poterit *SL3C1; corr. ex* poterit *O* / coloris: corporis *FP1SOL3E* 37 foraminis: foramine *O; corr. ex* foramine *L3* 38 melior apparet: melius appareat *R* / apparet *alter. in* appareat *a. m. E* / *post* reflexus *scr. et del.* cum in *S* 40 oppositus: oppositum *R* / *post* percipietur *scr. et del.* sus *P1* / secundus *corr. ex* secunde *O* / si *inter. a. m. C1* 41 album *inter. a. m. E* / *post* album *scr. et del.* co *F* / foramini *alter. ex* color *in* ?? *F*; foramini *R* / colorationis: reflexionis *C1; corr. ex* reflexionis *L3; alter. ex* remotionis *in* reflexionis *a. m. E* 42 comprehendet: comprehendit *C1; corr. ex* comprehendat *E* 43 visu foramini *transp. E* / colorationis: reflexionis *L3C1; alter. ex* remotionis *in* reflexionis *E* / comprehendetur: comprehenditur *C1* 45 pars . . . reflexionem (46) *om. FP1S* / quod *corr. ex* ut *a. m. E* 46 per *om. E* 48 unde: vident *O* 49 formam *corr. ex* foramina *F*; foramina *P1* / *post* comprehendere *add.* existimant *R* 50 formam *inter. O* / *post* corporis *add.* in *C1* / imprimi *corr. ex* et primi *O* 54 motum: motus *R* / non *inter. O* 55 sed . . . prima[2] *mg. a. m. E* / parte prima *transp. C1*

[4.3] Amplius, viso corpore aliquo, et vidente ab eo situ remoto, poterit accidere quod non videat corpus illud in speculo illo, licet videat totam speculi superficiem, quod quidem non esset si imprimeretur forma in speculo, cum videatur speculum et non mutetur locum, et corpus similiter sit immotum, et forma eius inficiat speculum, sicut et prius.

[4.4] Ut plane appareat non accidere hoc ex comprehensione forme, obturetur medietas foraminum instrumenti, et in aliquo obturatorum sit scriptura aliqua. Si inspiciatur speculum regule per foramen scripturam respiciens, comprehendetur in speculo scriptura, per quodcumque aliud minime. Quod si scripture forma speculo esset impressa, per quodcumque foramen instrumenti posset percipi. Simili modo, in speculis columpnaribus per foramen respiciens tantum comprehendetur scripture situs. Verum in speculis piramidalibus et spericis situs et magnitudo scripture mutabitur.

[4.5] Amplius, speculo columpnari extracto, regula super bases suas directe sita apparebit facies hominis in eo directa. Si vero erigatur regula aut multum declinetur, videbitur distorta. Palam ergo quod non accidet comprehensio ex forma fixa in speculo, cum non comprehendatur res visa in speculis nisi fuerit visus in situ reflexionis. Palam etiam quod distortio faciei apparentis non est ex forma rei sed dispositione speculi.

[4.6] Amplius, viso corpore in speculo et post elongato, comprehendetur corpus magis intra speculum quam prius, quod non erit si forma corporis in superficie speculi sit et ibi comprehendatur. Comprehensionem igitur forme in speculo efficit reflexio.

59 quod: ut *R*/videat[1] *corr. ex* accidat *O*; viderat *L3*; *corr. ex* vidat *C1*/videat[2] *corr. ex* videa *F*; viderat *OL3*; *alter. ex* viderit *in* viderat *C1* 60 quidem: autem *L3C1* 61 in *om. OC1*/*post* in *scr. et del.* etiam *C1*; *add.* etiam *E*/et[1] *inter. P1*/mutetur *alter. in* mutet *F*; mutet *R* 63 prius *corr. ex* primus *S* 64 *ante* ut *add.* et *P1ER*; et *mg. a. m. F*/hoc: huiusmodi *E*/forme *inter. O* 66 aliqua *corr. ex* alia *a. m. E*/si *inter. a. m. C1* 67 *ante* respiciens *scr. et del.* in *O*/respiciens: inrespiciens *E*/*post* scriptura *add.* prima *E* 70 respiciens *alter. ex* inspicione *in* inspiciens *O*/*post* tantum *add.* foramen obturatum in quo est scriptura *R* 71 verum: et erit *OL3*/et *corr. ex* in *O* 75 declinetur: inclinetur *R*/accidet: accidit *R* 76 ex: et *O*/comprehendatur *corr. ex* comprehendantur *P1* 77 fuerit visus *transp. FP1*/etiam: ergo *L3*; *corr. ex* ergo *a. m. C1* 78 *post* est *add.* lux *P1* 79 *post* in *scr. et del.* cor *P1*/*post* post *scr. et del.* elongationem *P1* 80 erit: esset *R* 81 sit: esset *R*/ibi: hic *E*/comprehendatur *corr. ex* comprehendantur *P1*; comprehenderetur *R* 82 reflexio *om. P1*

[CAPITULUM 5]

Pars quinta: in modo comprehensionis formarum in corporibus politis

[5.1] Iam patuit in parte superiori quod, si opponatur speculo corpus coloratum lucidum, a quolibet eius puncto procedit lux cum colore ad totam speculi superficiem, et reflectitur per lineas reflexionis proprias. Igitur a puncto sumpto in corpore opposito speculo procedit lux cum colore ad speculum in modum piramidis continue, cuius basis est superficies speculi, et forma illa reflectitur per lineas eiusdem situs cum lineis accessus, et erit post reflexionem continuitas sicut in accessu. Et si lineis reflexis occurrat superficies corporis, propter continuitatem earum tota occupabitur ut nichil intersit vacuum. Si ergo forma illius corporis moveatur ad speculum per lineas illas (scilicet per reflexas) et ad basem piramidis pervenerit, quoniam linee piramidis eiusdem sunt situs cum lineis reflexis, reflectitur forma per lineas piramidis, et agregabitur tota in puncto sumpto.

[5.2] Quotiens ergo forma alicuius corporis per lineas aliquas ad speculum venerit, si linee ille eiusdem sunt situs cum lineis piramidis a puncto sumpto ad speculum intellecte, cum eas respicientibus, movebitur forma per piramidem illam ad punctum sumptum. Et si in puncto sumpto fuerit visus, videbit corpus cuius est forma illa. Et superius determinatum est quod in situ determinato fit adquisitio forme in speculo. Situs igitur proprius et naturalis adquisitio visus per reflexionem hic est ut linee accessus forme ad speculum eundem habeant situm cum lineis piramidis a centro visus ad capita illarum linearum intellecte unaqueque cum ea respiciente, nec accidit forme reflexe comprehensio nisi in situ isto.

83 pars . . . politis (84) *om. FP1S* 85 *post* iam *add.* autem *L3C1* / *post* superiori *add.* libri *C1* / opponatur *corr. ex* ponatur *O* 86 eius puncto *transp. C1* 88 proprias: prias *E* 91 continuitas *corr. ex* continuitatis *P1* 92 *post* accessu *add.* fuit *C1* 93 *post* continuitatem *scr. et del.* sicut in accessu et si *O* / occupabitur *corr. ex* occupatur *mg. a. m. E* / ut *inter. O* / ergo: modo *O* 94 scilicet *om. FOL3; inter. a. m. S* / scilicet per (95) *om. P1ER* 95 per *inter. L3* / ad *om. O* / basem: basim *R* / *post* linee *scr. et del.* perve *F* / piramidis: piramidum *O* 96 reflectitur: reflectetur *R* 98 quotiens *corr. ex* quosiens *O*; quoties *R* / ergo: igitur *FP1* / *post* corporis *add.* ad speculum venerit *R* / lineas aliquas *transp. R* / ad . . . venerit (99) *om. R* 99 ille: iste *ER* / sunt: sint *ER* / piramidis *corr. ex* piramidalibus *a. m. E* 100 intellecte: intellige *OL3ER* / cum *om. O; corr. ex* tamen *L3*; tamen *ER* 101 illam: illa *O* 102 *post* illa *scr. et del.* in precedenti capitulo *E* 103 determinatum: declaratum *R* / fit: fiat *R* 104 acquisitio: acquisitionis *R* 105 *post* est *add.* dicendus *E* / *post* ut *scr. et del.* sit *E* 106 habeant: habent *P1C1*; habuerint *E* / illarum: earum *FP1* 107 intellecte: intellige *OE*; scilicet *R* / unaqueque: unaquaque *OL3C1E* / ea: eam *C1*; sua *R* / accidit: accit *P1* 108 situ isto *transp. R*

[5.3] Palam ergo quod secundum hanc dispositionem linearum tantum fit comprehensio formarum. Et palam quod ex corpore colorato luminoso procedit lux cum colore ad speculum et reflectitur, nec procedit aliquid ex corpore preter colorem et lucem. Patet igitur quod ex luce et colore tantum huiusmodi forma comprehenditur, et cum moveatur forma ex colore et luce compacta secundum predictam situs observationem, superfluum est dicere quod ab oculo exeant radii ad speculum et reflectantur super situm predictum, sicut a pluribus dictum est. Hic igitur est reflexionis modus geometrarum doctrine non adversus sed consonus, cum in eo geometrice radiorum exeuntium opinione observetur situs, et hic modus michi soli usque nunc patuit.

[5.4] Verum cum a corpore luminoso procedat forma ad speculum secundum varietatem situum propter lineas a quolibet puncto corporis ad totam speculi superficiem intellectas, erit forme eiusdem reflexio per diversas piramides quarum capita sunt diversa puncta et bases speculi superficies situm linearum motus forme observantes. Ob hoc accidit quod eadem hora speculo fixo eadem percipitur corporis forma a diversis super quorum intuitis cadunt capita piramidum reflexarum. Similiter, si idem visus moveatur super illa piramidum capita, apparebit ei speculo immoto a locis diversis eadem forma. Sed diversis in speculo eandem formam comprehendentibus in diversa speculi loca cadunt eorum intuitus, quoniam ab eodem speculi puncto diversorum punctorum corporis formas comprehendere eandem non possunt.

[5.5] Iam dictum est quod a quolibet corporis puncto procedit lux ad quodlibet punctum speculi, unde super quodlibet punctum corporis est acumen piramidis cuius superficies speculi basis. Et quodlibet superficiei speculi punctum est acumen piramidis cuius basis superficies corporis. Tota ergo forma corporis erit in quolibet speculi puncto per lineas procedentes in partes diversas, nec concurrere possibiles. Et

109 ergo: igitur *FP1* 110 tantum fit *transp. C1* / fit: fiat *R* 111 procedit: procedat *R* / reflectitur: reflectatur *R* 112 procedit: procedat *R* / colorem et lucem: lucem et colorem *R* / igitur: ergo *R* 113 tantum: *scr. et del.* tamen *O* / huiusmodi *alter. ex* hominis *in* omnis *O* / *post* et^2 *scr. et del.* con *F* 114 et luce *om. O; inter. a. m. E* / secundum *inter. O* 116 super: secundum *ER* 117 igitur est *transp. ER* / modus: modum *L3* 118 in *inter. L3* / *post* geometrice *add.* situm *O* 120 cum: non *L3* 123 per *alter. in* ad *a. m. E* / sunt *inter. O; om. L3C1E* 124 situm: situum *E* / observantes *inter. a. m. E* 125 quod: ut *R* / speculo fixo *transp. R* / percipitur: percipietur *R* 126 intuitis: intuitus *SOL3C1E* 128 ei: se *C1* 129 formam *om. C1* 131 corporis *inter. L3; om. R* / comprehendere *corr. ex* comprehendunt *L3*; comprehendent *E* / *post* comprehendere *scr. et del.* ut *O; add.* vel comprehenditur et *C1* / eandem: eadem *SOE; corr. ex* et *L3*; easdem *R* 132 *ante* iam *add.* et *R* / iam *corr. ex* item *a. m. E* / est *inter. a. m. E* / corporis puncto *transp. R* 133 unde *corr. ex* unum *S* / punctum corporis *transp. OL3C1ER* 134 *post* superficies *add.* est *C1R* 135 *post* basis *inter.* est *a. m. S; add.* est *C1* 136 tota ergo forma *inter. a. m. E* / quolibet speculi *transp. FP1* / speculi *om. S* / speculi puncto *transp. R* 137 lineas: linea *FP1SO* / et *inter. O*

forma a corpore ad quodcumque speculi punctum accedens per piramidem reflectetur per piramidem. Et licet in speculi superficie super numerum multiplicetur eadem iteratio forme, cum concurrat forma totalis cum qualibet parte et in quolibet puncto. Et non sit in formis illis discretio, sed continuitas inseparabilis in reflexione. Tamen, quia forma totalis non cadit in diversas speculi partes, secundum ydemptitatem situs dirigitur ad loca diversa in quibus eam comprehendit visus.

[5.6] Cum ergo similis sibi fuerit forma speculi figure corporis, erit in speculo complementum forme corporis et figure. Quoniam in speculo eiusdem figure cum corpore forma puncti primi dirigitur ad primum punctum speculi, secundi ad secundum, et sic in omnibus se respicientibus. Et ita erit in speculi superficie figura totalis figure, quod non accidit in speculo alterius figure. Similiter, sumpta quacumque speculi parte cui eadem cum corpore figura, erit complementum figure corporis in ea. Et cum infinite sint tales speculi partes, infinite erunt forme corporis reflexionis sed ad puncta diversa procedentes ex quibus formam comprehendit visus.

[5.7] Cum igitur secundum hanc linearum dispositionem fiat forme comprehensio, non erit forme procedentis a corpore in speculi superficie fixio. Et in hunc modum accidit in omnibus speculis, sed in planis certius; in aliis autem accidit quedam diversitas ex errore visus secundum modum predictum. Et quilibet visus secundum modum predictum ab uno speculi puncto non percipit nisi unum corporis punctum, nec a duobus percipitur in eodem speculi puncto idem corporis punctus.

[5.8] Amplius, si opponatur speculum visui, et intelligatur a centro visus ad speculi superficiem piramis, et basis illius piramidis si sumatur punctum, et intelligatur linea piramidis a centro visus ad illud punctum, cum a puncto illo infinite possunt produci linee, si aliqua earum cum latere piramidis eundem habeat situm et equalem cum perpendiculari teneat angulum, et ita accidat quolibet puncto speculi

138 speculi punctum *transp. OL3C1* / *post* punctum *scr. et del.* punctum *L3* 141 totalis: talis *C1* / parte *inter. L3* / non *inter. a. m. E* 144 comprehendit: comprehenderit *L3* 145 ergo: igitur *FP1E* / sibi *om. P1R* / fuerit: fuerint *OE* / *post* speculi *scr. et del.* specu *F; add.* et *O* / figure: forme *R.* / *post* figure *scr. et del.* cum corpore forma *O* 146 forme corporis *transp. L3* 147 forma *mg. a. m. C1* / puncti primi *ER* 148 punctum speculi *transp. C1* (punctum *mg.*) / secundi *om. O* / se *inter. a. m. C1* 149 *post* figure *mg.* res *F*; res *inter. a. m. S* / quod . . . figure (150) *inter. L3* 151 *post* speculi *scr. et del.* figura *C1* / eadem: eidem *L3C1* 153 ex: a *R* 157 fixio *corr. ex* fixo *O* 158 certius: circulis *FP1* 159 et *om. O; inter. L3* / et . . . predictum (160) *scr. et del. S* / quilibet: quibuslibet *O* / secundum . . . predictum (160) *om. OC1* 160 uno: uni *FP1L3* 161 *post* a *add.* visibus *ER* 162 punctus: punctum *R* 163 et *om. L3; inter. C1* / *post* centro *scr. et del.* c *F* 164 speculi superficiem *transp. ER* (speculi *inter. a. m. E*) / si: et *R* 166 illo *corr. ex* illius *L3*; illius *C1* / possunt: possint *R* 168 *post* accidat *add.* quod *alter. in* in *S; add.* quod a *C1E*(a *inter. a. m. C1*; a *scr. et del. E*) / quolibet . . . planum (169) *mg. L3*

sumpto, planum quod a quolibet puncto speculi potest fieri reflexio. Dico ergo quod inter lineas a puncto sumpto productas est linea eundem habens situm cum latere piramidis, et equalem tenet angulum cum perpendiculari super illud punctum. Et est illa latus piramidis intellecte a puncto illo superficiei rei occurrentis, et quod super terminum illius linee ceciderit, cum per eam ad punctum sumptum venerit, reflectetur ad visum per latus piramidis eius iam dictum. Et hoc piramidis latus cum linea a puncto illo producta erit in eadem superficie ortogonali super superficiem speculum in illo puncto tangentem. Et hoc dico, cum lateris piramidis super punctum sumptum fuerit declinatio. Si enim ortogonaliter cadat super superficiem speculum in puncto sumpto tangentem latus piramidis productum a centro visus, reflectetur in se et redibit in visum ad originem sui motus.

[5.9] In speculo plano planum est quod diximus, quoniam in quodcumque punctum superficiei plane cadat radius a puncto illo potest erigi linea ortogonalis super superficiem illam, et a centro visus potest intelligi linea perpendiculariter cadens in superficiem planam predicte continuam, aut in eandem. Et hee due perpendiculares erunt in superficie eadem, quoniam sunt equidistantes, et linea a termino unius usque ad terminum alterius protracta in superficie plana tenebit angulum acutum cum utraque, et erit in eadem superficie cum utraque. Et radius qui a linea illa elevatur tenebit acutum angulum cum perpendiculari speculi, et similiter cum perpendiculari visus. Et intelligatur linea in alteram partem superficiei plane transiens ortogonaliter per terminos perpendicularium. Tenebit ex parte alia cum perpendiculari speculi angulum rectum, unde ex illo recto poterit

169 sumpto . . . speculi *mg. a. m. C1E / post* sumpto *scr. et del.* planum *S* / a *inter. L3* 170 ergo: igitur *FR* / quod *om. E* / *post* linea *add.* que *R* / eundem (171) *corr. ex* eadem *O* 171 habens: habet *R* 172 *post* punctum *scr. et del.* est *L3* / est illa latus: illa linea est latus *C1ER*; est *mg. a. m. E* 173 intellecte *corr. ex* intellige *O* / *post* illo *inter.* ad *OL3*; *add.* ad *C1* / superficiei: superficiem *L3C1* / occurentis *corr. ex* currentis *O* 175 per: et *O* / *post* latus *scr. et del.* u *F* / piramidis: piramidum *FP1SC1* / eius *om. R* / *post* dictum *scr. et del.* est *O* 176 hoc: huius *R* 177 super *om. F*; *inter. P1O* / *post* superficiem *add.* tangentem *R* / illo puncto *transp. P1* / puncto: loco *O*; *om. L3C1* 178 tangentem: contingentem *FP1*; *om. R* / sumptum: positum *FP1* 179 *post* superficiem *add.* tangentem *R* 180 tangentem *om. R* 182 quoniam in quodcumque (183): in quodcumque quoniam *C1* 183 superficiei *corr. ex* superficie *O* / cadat: ceciderit *R* 184 visus: unius *FP1* 185 linea *inter. a. m. S* 186 aut *mg. a. m. C1* / eandem: eadem *FP1C1* 187 superficie eadem *transp. R* 189 acutum *om. OC1ER*; *inter. a. m. L3* / *post* utraque *add.* ectum *alter. in* rectum *O*; *scr. et del.* et super *E* / superficie *inter. O*; *inter. a. m. C1*; *om. L3E* 190 linea illa: aliena linea *FP1*; *corr. ex* aliena linea *S* / cum *inter. a. m. L3* 191 *post* speculi *scr. et del.* et similiter cum perpendiculari speculi *S* / similiter *om. FP1*; simpliciter *L3C1* / perpendiculari: particulari *FP1* / *post* perpendiculari *scr. et del.* est *O* / *post* et[2] *inter.* si *O*; *add.* si *R* 192 intelligatur: intelligitur *O* / linea *om. OL3C1ER* / alterem partem *transp. ER* / *post* partem *add.* linea *OL3C1E*; *add.* produci linea *R* / plane *alter. in* planum *L3* 193 per: super *R* / perpendicularium: perpendicularem *O*; *corr. ex* perpendiculariter *L3E* (*a. m. E*) / *post* perpendicularium *inter.* que *a. m. C1* / alia: altera *ER*; *scr. et del.* altera *C1* 194 unde *inter. E*

abscindi angulus acutus equalis angulo acuto quem cum eadem perpendiculari tenet radius. Et hii duo anguli in eadem superficie, quare radius exiens et reflexus in eadem superficie et in superficie perpendicularium dictarum. Inspecto autem alio puncto, idem situs accidet radiorum cum perpendicularibus quarum una linea a puncto viso, alia a centro visus.

[5.10] In omni ergo superficie reflexionis accidit quatuor punctorum concursus—scilicet centrum visus, et punctus comprehensus, et terminus perpendicularis a centro visus, et punctus reflexus. Et omnes reflexionis superficies secant se in perpendiculari a puncto reflexionis intellecta, et ipsa est communis omnibus superficiebus reflexionis. Et cum idem accidat quolibet puncto superficiei plane inspecto, erit ex omnibus punctis similis reflexio, et eodem modo.

[5.11] In speculis autem spericis palam erit quod diximus. Opposito visui speculo sperico—et est oppositio ut visus non sit in superficie illius sperici aut in superficie continua et sperica—et inspecto hoc speculo, pars eius comprehensa erit pars spere circulo inclusa quam efficit motu suo radius tangens superficiem spere, si per girum moveatur contingendo speram donec redeat ad punctum primum a quo sumpsit motus principium. Et si intelligantur superficies se secantes super dyametrum spere a polo circuli predicti intellectum, quilibet arcuum superficiei spere et hiis superficiebus communium a polo circuli ad ipsum circulum intellectorum erit minor quarta circuli magni, quoniam linea a centro spere ad terminum radii speram contingentis protracta—et est ad circulum predictum—tenet cum radio angulum rectum ratione contingentie. Tenet ergo angulum acutum cum semidyametro a polo circuli producto, et hunc angulum respicit arcus interiacens polum

195 quem: quam *S*; quoniam *L3* 196 *post* radius *scr. et del.* exiens *S* / *post* anguli *inter.* sunt *a. m. O; add.* sunt *R* / *post* superficie *add.* sunt *C1* 197 *post* eadem *add.* sunt *R* / *post* superficie *add.* sunt *C1* 198 autem: aliquo *E* / *post* idem *scr. et del.* punctus *P1* 199 una *inter. a. m. L3* / linea *om. O* / linea . . . alia (200) *om. R* 200 *post* visus *add.* alia a puncto viso *R* 202 scilicet: que sunt *R* / punctus comprehensus: punctum apprehensum *R* / comprehensus . . . punctus (203) *mg. a. m. S* / terminus (203): tertius *FP1SL3C1* 203 *post* perpendicularis *inter.* ducte *S* / a centro visus *om. C1*; a centro visus ducte *inter. L3* / *post* visus *add.* ducte *ER* / punctus reflexus: punctum reflexionis *R* 204 se *inter. E* / *post* reflexionis[2] *scr. et del.* et cum idem accidat quolibet *S* 205 ipsa est *transp. ER*; est *inter. a. m. E* / *post* est *add.* sectio *C1* 206 idem *rep. C1* / plane *alter. ex* plan *in* plani *O* / *post* erit *add.* et *FS* 207 similis *corr. ex* simul *C1E* (*a. m. E*) / reflexio: reflexo *E* 208 spericis *corr. ex* speris *O* / *post* diximus *add.* hoc modo *C1* 209 speculo *corr. ex* specula *O* 210 sperici: speculi *R* / *post* superficie *add.* ei *R* / et sperica *om. R* / hoc *corr. ex* hic *E* 211 *post* eius *add.* a visu *R* / *post* circulo *add.* minore *R* / quam: quem *R* 212 *post* efficit *add.* in *C1* / *post* superficiem *scr. et del.* speculi *P1* 213 redeat: redderat *O*; *corr. ex* rederat *L3* 214 motus: motum *FP1* / et: quia *R* / intelligantur *corr. ex* intelligeantur *O* / *post* superficies *scr. et del.* se similiter *O* 215 intellectum *alter. in* intellecto *O*; intellectam *R* / arcuum *corr. ex* acumen *O* 216 *post* et *scr. et del.* in *O* / *post* communium scr. et del. terminum *O* / *post* circuli *scr. et del.* ad ipsum circulo *C1* 218 a *mg. L3* 219 et *inter. a. m. C1*; que *R* / est *inter. a. m. E* 221 producto: producta *R*

circuli et circulum, quare quilibet horum arcuum erit minor quarta circuli.

[5.12] Dico ergo quod a quolibet huius portionis puncto poterit fieri reflexio, quoniam, sumpto aliquo eius puncto, dyameter spere ab illo puncto intellectus erit perpendicularis super superficiem planam tangentem speram in puncto illo. Et huius rei probatio est: Intellectis duabus superficiebus speram super dyametrum a puncto sumpto intellectum secantibus, linee communes superficiei spere et hiis superficiebus sunt circuli spere transeuntes per punctum sumptum. Et intellectis duabus lineis tangentibus hos circulos in puncto sumpto, erit dyameter perpendicularis super utramque lineam, quare super superficiem in qua sunt ille linee. Et cum descenderit radius super punctum sumptum, erit in eadem superficie cum dyametro spere cuius terminus est punctus sumptus, et linea a centro visus ad centrum spere intellecta, que quidem transit per polum circuli, et est radius ortogonaliter cadens super superficiem spere. Et ex hiis tribus lineis erit triangulus, et radius super punctum sumptum incidens tenet acutum angulum cum dyametro spere ab exteriori parte, quoniam, cum elatior sit iste radius radio speram contingente, secabit speram cum productus intelligitur. Et superficies tangens speram in puncto sumpto dimissior erit hoc radio, et secabit inter speram et visum dyametrum—id est lineam a centro visus ad centrum spere intellectam per polum circuli transeuntem.

[5.13] Unde cum dyameter spere sit ortogonalis in superficie punctum tangente, tenebit angulum recto maiorem ex interiori parte cum radio in punctum descendente, unde in exteriori parte tenebit cum eo angulum minorem recto. Et productus ortogonalis erit super superficiem contingentem exterius, quare ex angulo recto quem tenebit

222 quare . . . circuli (223) *om. R* 224 ergo: igitur *R* / huius *inter. L3* / huius portionis puncto: puncto huius portionis *R* 226 intellectus: intellecta *R* / super superficiem *corr. ex* superficiem *O* 227 est *om. OL3; inter. a. m. C1* / intellectis *corr. ex* intellectas *a. m. E* 228 sumpto intellectum (229): sumptam intellectam *R* 231 intellectis *corr. ex* intellectus *L3* 232 perpendicularis: perpendiculariter *L3* / *ante* utramque *scr. et del.* ut *C1* 233 *post* sunt *inter.* site *L3* / et *inter. a. m. E* / *post* et *scr. et del.* de *P1* / descenderit: descenderet *L3* 235 est *inter. OL3; om. C1R* / est punctus *transp. E* / punctus sumptus: punctum est sumptum *R* / *post* sumptus *add.* erit *C1* 236 *post* quidem *add.* linea *O* / et *scr. et del. O* 237 *post* spere *add.* est similiter in eadem superficie *R* 238 triangulus: triangulum *R* 239 acutum angulum *transp. FP1* / *post* spere *inter.* et *a. m. E* / ab: et *OL3C1* / *post* ab *mg.* ex *a. m. C1* / cum *inter. a. m. E* 240 elatior *corr. ex* elatio *F* / speram *corr. ex* speras *O* / contingente *corr. ex* continginte *F* / cum: si *inter. OL3* (*a. m. O*); *inter. a. m. E* 241 productus: produci *OC1; corr. ex* productum *L3;* producta *R* 242 dimissior *corr. ex* demissior *OR* / hoc *corr. ex* ob *O* / *post* radio *scr. et del.* et eius cui centro *OL3E* / visum *rep. FP1SL3E* (*mg. L3*) / *post* visum *add.* visam *R* 245 ortogonalis: ortogonaliter *O* 246 recto: rectum *E* / interiori parte *transp. R* 247 descendente: descendentem *FP1L3E; corr. ex* descendentem *S* / unde in exteriori *corr. ex* quoniam in maiori *a. m. E* / in[2] *om. O* / exteriori *corr. ex* ratiori *O* 248 productus: producta *R* 249 quem: quam *S* / *post* tenebit *add.* diameter *C1*

cum superficie ex alia radii parte poterit abscindi acutus equalis ei quem includit radius cum illo dyametro. Et erunt linee tres hos duos angulos includentes in eadem superficie, quare a puncto portionis sumpto potest produci linea in eadem superficie cum radio in punctum illud cadente et linea ortogonali in superficie punctum contingente et ad paritatem angulorum cum perpendiculari illa. Et illi linee occurret forma puncti mota ad superficiem speculi per radium illum. Igitur eiusdem est situs cum linea que poterit reflecti, et erit superficies in qua sunt hee linee ortogonalis super superficiem speram in puncto contingentem, et ita in quolibet puncto portionis intelligendum.

[5.14] Ergo in omni superficie reflexionis erunt centrum visus, centrum spere, punctus reflexionis, et punctus reflexus, et omnes hee superficies secabunt se super lineam a centro visus ad centrum spere protractam. Et cuilibet reflexionis superficiei et superficiei spere communis linea erit circulus spere, et omnes circuli secabunt se super punctum spere in quem cadit dyametrum visus, et est super circuli portionis polum. Cum autem radius ceciderit in speculum ortogonaliter super superficiem in punctum in quem cadit radius speram tangentem—et est radius ille dyameter visus per polum circuli portionis ad centrum spere—fiet reflexio ad visum per eundem radium ad motus radii ortum.

[5.15] In speculis autem columpnaribus patebit quod diximus. Opponatur speculum columpnare exterius politum oculo—et est oppositio ut non sit visus in superficie columpne aut superficie ei continua—et intelligemus superficiem a centro visus ad columpne superficiem secantem columpnam super circulum equidistantem basibus columpne. Et in hac superficie sumantur due linee tangentes circulum sectionis in duobus punctis oppositis. Ab utroque illorum punctorum producatur linea secundum longitudinem columpne, et intelligantur due superficies in quibus sunt hee due linee longitudinis

251 illo: illa *R* / linee tres *transp. F* / duos angulos *transp. R* 254 *post* et[1] *inter.* cum *O*; cum *mg. a. m. C1* / paritatem: parietem *P1* 255 cum: in *OE*; *corr. ex* in *L3* / occuret *corr. ex* occurat *O* 257 que *om. O* / hee linee (258) *om. O* 258 super *inter. O* 259 puncto portionis *transp. R* 261 punctus[1]: punctum *R* / punctus reflexus: punctum reflexum *R* 262 se *om. P1* 263 et superficiei *om. P1* 264 et . . . cadit (265) *mg. a. m. S* (et: ??; spere:??; quem: quam) / *post* circuli *scr. et del.* linee *O* 265 *ante* spere *add.* super punctum *FO* / *post* spere *add.* in quantum; quantum *del. F* / quem: quam *OC1*; quod *R* / dyametrum: dyameter *ER* 266 speculum *corr. ex* polum *L3E* (*a. m. E*) / *post* speculum *scr. et del.* speculum *O* / ortogonaliter *corr. ex* ortogonale *O* 267 punctum: puncto *OR* / quem: quod *R* / cadit radius *transp. R* / tangentem: contingentem *FP1* 269 fiet *corr. ex* et *O* / eundem: eum *FP1* 270 autem: et *E* 272 aut . . . columpne (273) *mg. a. m. S* / *post* aut *inter.* in *a. m. O* 273 intelligemus: intelligamus *R* 274 *post* superficiem *scr. et del.* seq *P1* / equidistantem *alter. in* equidistanter *a. m. E* 275 hac *alter. ex* alia *in* illa *O* 276 *post* circulum *scr. et del.* equidistantem *S* / oppositis *om. FP1* 277 punctorum *mg. a. m. C1* / *post* punctorum *add.* sumatur *L3*; *scr. et del.* sumatur *C1* / *ante* producatur *inter.* et *L3* / longitudinem: longum *C1* 278 longitudinis *om. P1* / longitudinis . . . linee (279) *mg. a. m. S*

et due linee a centro visus ducte contingentes circulum sectionis. Dico quod hee superficies tangent columpnam.

[5.16] Si enim dicatur quod altera secat illam, planum quod sectio erit super lineam longitudinis columpne in quam superficies cadit, et similiter erit sectio super lineam longitudinis columpne huic oppositam. Et circulus sectionis transit per has duas lineas longitudinis. Et linea contingens circulum sectionis, cum sit in superficie aliqua, secat columpnam super aliquas longitudinis lineas sibi invicem equidistantes, et si transit per unam earum, transibit per alteram, et ad paritatem angulorum. Cum ergo transeat per punctum in quo circulus sectionis secat primam longitudinis lineam, transibit etiam per punctum in quo alia longitudinis linea tangit hunc circulum. Et ita secat circulum, quare non erit contingens, quod est contra ypothesim. Palam ergo quod ille due superficies contingunt speculum, et quod inter illas cadit ex superficie speculi est quod apparet visui.

[5.17] Cum autem illarum duarum superficierum sit concursus in centro visus, secabunt se, et linea sectionis communis transibit per centrum visus, et est equidistans axi columpne, quoniam axis columpne ortogonalis est super eundem circulum sectionis. Et linee longitudinis columpne ortogonales super eundem circulum, et superficies tangentes columpnam secundum lineas has sunt ortogonales super circulum eundem. Quare super superficiem secantem columpnam in illo circulo, quare linea communis harum superficierum est ortogonalis super eandem superficiem, quare equidistans axi columpne.

[5.18] Dico ergo quod quocumque puncto in sectione speculi apparente sumpto, linea a centro visus ad punctum producta secabit speculum. Quoniam intellecta linea longitudinis columpne a puncto sumpto, transibit per circulum sectionis, et tanget ipsum in puncto ad

279 *post* linee *scr. et del.* a centro et due linee *O* 280 *post* hee *add.* due *O* 281 *post* altera *add.* illarum *O*; illarum *mg. a. m. C1* / illam: eam *O* / *post* planum *add.* est *R* 282 erit: est *ER* / in . . . columpne (283) *om. P1* 285 sectionis *om. O* / *post* sit *scr. et del.* cum sit *E* / secat: secans *O* 286 super *corr. ex* per *O* / aliquas: has duas *O*; *alter. in* alias *a. m. C1* / longitudinis *corr. ex* longitudines *P1* / longitudinis lineas *transp. O* / *post* lineas *scr. et del.* et linea contingens circulum cum sit in superficie aliqua longitudinis super lineas *O* / invicem *om. O* 287 si *om. O* / transit: transeat *O* / unam *corr. ex* unum *O*; unum *L3* / paritatem *corr. ex* parietatem *O* 288 ergo: igitur *FP1*; *om. S* 289 etiam *corr. ex* et *a. m. C1* 290 *post* linea *scr. et del.* cont *F* 291 erit: est *inter. a. m. E* / erit contingens: contingit *R* / contra: circa *L3* / ypothesim *corr. ex* ypothasim *O* / ergo: igitur *F*; *om. P1* / ille due (292) *transp. R* 292 et *inter. OE* (*a. m. E*) 294 illarum duarum *transp. R* 295 secabunt: secabant *L3* 296 est: erit *R* / quoniam axis columpne *inter. a. m. L3E*; quoniam: quem *L3* / *post* quoniam *add.* enim *R* / columpne *om. FP1S* 297 eundem *om. R* / *post* longitudinis *scr. et del.* or *C1* 298 eundem *om. FP1S* / circulum *om. L3* / et: etiam *R* / et . . . eundem (300) *mg. a. m.*; circulum eundem *transp. L3* 299 secundum: in *O* / lineas *alter. in* lineis *O* / has: illis *O* / sunt *om. R* / sunt ortogonales *transp. E* / *post* ortogonales *add.* erunt *R* / eundem (300) *om. O* 300 *ante* quare *add.* circulum *L3* / quare *inter. E*; ergo et *R* / super *om. P1* 2 *post* quare *add.* est *OC1* 5 a *inter. E* 6 *post* ipsum *add.* aut sectionis *S*

quem punctum, si ducatur linea a centro visus, secabit speculum quod cadit inter lineas contingentes hunc circulum. Et superficies a centro procedens in qua fuerit hec linea secabit speculum. Cum ergo in eadem superficie fuerit linea hec et linea a centro ad punctum sumptum ducta, secabit linea illa speculum, et ita quelibet linea a centro visus ad portionem speculi intellecta secat speculum. Eodem modo quelibet linea a linea communi per centrum visus intellecta ad hanc portionem ducta secat speculum, unde quelibet superficies tangens speculum in aliqua portionis apparentis linea secat superficies que contingunt portionis extremitates. Et nulla omnium superficierum portionem tangentium pervenit ad visus centra; sed inter visum extendetur et speculum.

[5.19] Dico ergo quod a quolibet puncto portionis huius potest fieri reflexio lucis. Dato enim puncto, fiat super ipsum circulus equidistans columpne basibus. Si ergo superficies a centro visus procedens et columpne superficiem equidistantem basibus secans, secet eam super hunc circulum, et linea a centro visus ad circuli centrum ducta transeat per punctum datum. Fiet reflexio forme illius puncti per eandem lineam ad linee ortum, quia linea illa est axis visus super axem columpne perpendicularis. Sumpto autem quocumque per quem transeat axis perpendiculariter super axem columpne, fiet reflexio illius puncti per eundem axim.

[5.20] Si vero pretereat axim punctus sumptus, quecumque sit linea a centro circuli super ipsum equidistantis basibus punctum ducti, ad superficiem in linea longitudinis columpne per punctum illud transeuntis contingentem erit ortogonalis super axem, quare super lineam longitudinis per punctum illud transeuntem. Et quoniam visus est altior superficie punctum contingente, linea a centro visus ad punctum sumptum ducta tenebit acutum angulum cum perpendiculari illa

7 quem: quod *R* / quod: quia *R* 8 *post* circulum *add.* ergo *R* / *post* centro *scr. et del.* visus secabit *O*; *add.* visus *C1R* 9 ergo: igitur *FP1* 10 fuerit: sit *R* / linea hec *transp. O* / hec et linea *om. R* / *post* centro *add.* visus *R* 12 secat: secabit *C1* / quelibet: queque *FP1* 13 a linea *corr. ex* alia *C1* 15 *post* aliqua *add.* linea *C1* / apparentis *om. O* / linea *scr. et del. C1* 16 omnium: omni *P1* 17 centra: centrum *R* / extendetur: extenditur *R* 20 dato *mg. O* / enim puncto *transp. C1* 21 *post* columpne *scr. et del.* et *O* / ergo *om. OL3* 22 equidistantem: equidistanter *OR*; *alter. ex* equidistante *in* equidistanter *C1* / basibus: visui *FP1SL3E*; basi *R* 23 hunc *om. FP1* / linea *corr. ex* lenea *C1* / centrum: centra *FP1SL3E*; *corr. ex* centra *O* / *post* ducta *add.* et *OL3*; et *scr. et del. C1* 25 ad *inter. a. m. E* / linea illa *transp. E* / illa est *corr. ex* est illa *O* 26 *post* autem *add.* puncto *R* / quem: quod *R* 27 perpendiculariter: perpendicularis *R* / axem: axim *L3* / illius *rep. L3* / axim: axem *ER* 29 axim: axem *C1ER* / punctus sumptus: punctum sumptum *R* 30 super ipsum *scr. et del. O*; *om. C1R* / super ipsum equidistantis: equidistantis super ipsum *S* / equidistantis: equidistante *O*; equidistantem *L3E* / *post* basibus *add.* super ipsum *C1*; *add.* per ipsum *R* / punctum *om. SO*; puncti *L3E* / ducti *corr. ex* perdicti *O*; *om. L3E* 31 *post* superficiem *add.* punctum *S* / in *inter. L3* 32 transeuntis: transeunte *O* / ortogonalis super axem: super axem ortogonalis *ER*; ortogonalis *corr. ex* ortogonaliter *a.m. E* 33 quoniam *alter. ex* qui *in* quia *O*

a puncto ad centra circuli ducta. Et hoc ex parte exteriori, quia obtusum ex interiori. Et ex angulo recto quem illa perpendicularis tenet cum linea superficiei contingente circulum poterit abscidi acutus huic equalis. Et perpendicularis illa cum centro visus in eadem superficie, quare cum linea a centro ad punctum ducta. Et erit linea reflexa in eadem superficie, et erit hec superficies ortogonalis super superficiem contingentem speculum in puncto illo, quoniam perpendicularis ortogonaliter cadit super hanc superficiem. Et huiusmodi erit reflexionis superficies.

[5.21] Est autem diversitas inter lineas superficiebus reflexionis et superficiei columpne communes, cum enim reflexio erit per eundem radium, cadet idem radius ille ortogonaliter super axem. Et linea communis superficiei columpne et superficiei reflexionis erit linea recta—scilicet latus columpne—cum in superficie reflexionis sit dyameter columpne. Et hoc planum, quoniam columpne compositio est ex motu superficiei equidistantium laterum super unum latus immotum. Unde superficiei columpnam secanti in qua sit axis—id est latus immotum—communis linea ei et superficiei columpne erit latus motum. Et dico quod ex omnibus reflexionis superficiebus una sola est cui et columpne superficiei sit linea communis recta, quoniam unica potest intelligi superficies in qua sit axis columpne et centrum visus, et non plures.

[5.22] Si vero superficies reflexionis sit equidistans basibus columpne, erit linea communis circulus, et hec sola est superficies que cum columpne superficie lineam communem habeant circularem, quoniam in omni reflexione perpendicularis super superficiem contingentem punctum reflexionis est dyameter circuli basibus columpne equidistantis. Et non potest esse in columpne superficie nisi unus circulus equidistans basibus qui cum centro visus sit in eadem

36 centra: centrum *ER*/hoc: hic *R*/*post* hoc *add.* est *ER*/exteriori: exteriore *R*/*post* obtusum *add.* habet *R* 37 ex interiori *corr. ex* exteriori *a. m. E*/interiori: interiore *R*/*post* recto *scr. et del.* videtur *P1*/*post* quem *add.* linea *C1* 38 linea: illa *C1*/contingente: contingentis *R*/abscidi: abscindi *SOL3C1ER*/huic equalis (39) *transp. C1* 39 *post* visus *add.* est *C1R*/superficie: quare *E* 40 quare . . . superficie (41) *om. OL3C1E*/*post* quare *add.* etiam *R*/et . . . superficie (41) *om. P1* 41 superficie *corr. ex* superficum *S*/*post* superficie *add.* quare cum linea a centro ad punctum ducta *R*/*post* hec *scr. et del.* li *P1* 43 huiusmodi *corr. ex* huius *S* 47 *post* radium *scr. et del.* non *P1*/cadet: cadit *P1*/idem *om. O* 49 in *inter. L3*/dyameter: dyametrum *FP1SO* 50 et . . . columpne *mg. a. m. S*/hoc *om. E*/hoc planum *transp. R*/*ante* quoniam *add.* est *R* 51 unde: unum *S* 53 communis . . . columpne: et superficiei columpne communis linea *R* 54 *post* reflexionis *add.* scilicet *E*/cui *inter. OL3* 55 communis . . . plures (56) *inter. O*/superficies (56) *corr. ex* ipsi *L3* 56 in *om. E*/qua: que *E* 58 *post* est *add.* reflexionis *OC1;* scilicet reflexionis *inter. a. m. E* 59 communem *om. FP1*/habeant *alter. in* habeat *L3;* habeat *C1* 60 *post* reflexione *scr. et del.* que *E*/super superficiem *corr. ex* superficiem *O* 62 equidistantis *corr. ex* equidistantie *a. m. E*/et *inter. OL3*/et non *mg. a. m. E*/*post* esse *scr. et del.* co *P1* 63 unus: unius *FP1SOL3C1*

superficie. Omnes alie reflexionis superficies secant columpnam et axem columpne, quoniam perpendicularis ducta a puncto reflexionis secat axem columpne, et linee communes hiis superficiebus et superficiei columpne sunt sectiones quas in columpnis et piramidalibus assignant geometre.

[5.23] Cum superficiebus columpne et reflexionis linea recta fuerit communis, quodcumque punctum illius linee intueatur visus, fit reflexio in superficie eadem in qua scilicet est axis, quoniam est superficies unica contingens columpnam in linea illa longitudinis. Et quocumque puncto huius linee sumpto, perpendicularis ab eo ad axem ducta erit in eadem superficie cum axe, et hec longitudinis linea ortogonalis est super superficiem contingentem superficiem columpne. Sed centrum visus est in superficie ortogonali, ut super eandem, et sit in ea axis columpne et linea communis, et una sola est superficies ortogonalis super illam superficiem in eadem, quare omnes reflexiones a punctis huius linee facte sunt in eadem reflexionis superficie.

[5.24] Verum cum linea communis superficiei reflexionis et columpne fuerit circulus, quocumque puncto illius circuli viso, fiet in una et eadem superficie reflexio, quoniam quecumque perpendicularis a puncto viso ducta erit dyameter huius circuli, quare in superficie huius circuli, et punctum visus similiter. Et superficies huius ortogonalis est super superficiem quodcumque punctum huius circuli sumptum contingentem, quare in hac sola superficie erit cuiuslibet puncti predicti circuli reflexio. Quacumque vero alia linea communi sumpta, non fiet in eadem reflexionis superficie reflexio nisi ex uno tantum huius linee puncto, quoniam perpendicularis ducta a puncto reflexionis ortogonalis est super lineam longitudinis columpne per punctum illud transeuntis, quare et super axem. Et perpendicularis illa est dyameter circuli equidistantis basibus columpne. Et superficies reflexionis et circulus ille secant se, et linea eis communis est dyameter illius circuli, et est illa

64 *post* omnes *add.* autem *R* / reflexionis *corr. ex* reflexiones *E* / reflexionis superficies *transp. R* 67 superficiei *corr. ex* superficiebus *a. m. E*; superficiebus *R* / et *om. E* / piramidalibus: pyramidibus *R* 69 et *inter. O* 70 fit: fiet *R* 71 eadem *corr. ex* eaidem *F*; earumdem *P1* / scilicet *om. C1R* / est axis *transp. E* / unica *om. P1* 73 *post* ducta *scr. et del.* erit *S* 74 longitudinis *om. R* / *post* linea *add.* erit *R* / est *om. OL3C1ER* 75 *post* contingentem *scr. et del.* super *L3E* 76 ut *om. R* / eandem: eadem *FP1* / *post* eandem *add.* superficiem quia in una superficie est centrum visus et *R* / et . . . et[1] (77) *om. R* 77 *post* communis *add.* et axis columpne *R* / superficies *inter. a. m. O* 78 in eadem *om. R* 79 facte *corr. ex* factis *a. m. E* / sunt *om. P1* 80 communis: omnis *E* 81 *post* columpne *scr. et del.* et *L3E* / fuerit: fuit *O* 82 quecumque: quacumque *FP1* 83 viso: reflexionis *R* / dyameter: dyametrum *FP1SOL3C1* / huius[2]: huiusmodi *L3* 84 *post* circuli *add.* est *R* / et[1] *corr. ex* est *SO* (*a. m. S*); est *C1E* / huius *inter. O*; hec *R* 85 *ante* super *scr. et del.* supp *P1* / super superficiem *corr. ex* superficiem *O* 87 vero: non *FP1* / linea communi *transp. O* / non: nec *E* 88 huius *om. FP1S* 90 *post* super *scr. et del.* superficiem *P1* / lineam: lineas *L3*; *corr. ex* lineas *C1* 91 illa *inter. a. m. C1E* / illa est *transp. FP1*; *corr. ex* est illa *L3* 93 secant se *transp. P1* / se *om. F*; *inter. a. m. E* / *post* linea *add.* est *C1* / eis: iis *R* / est[1] *inter. a. m. C1*

perpendicularis, et superficies reflexionis secans est, et est declinata super ipsum. Et in superficie super lineam aliquam declinatam non potest intelligi nisi una linea ortogonaliter cadens in illam. Si a duobus reflexionis superficiei punctis fieret reflexio in eadem superficie, essent due linee illius superficiei ortogonales super axem, quod esse non potest, cum ipsa sit delinata super eum.

[5.25] Amplius, perpendicularis a puncto reflexionis cadit in circulum equidistantem basibus columpne et in puncto axis communis circuli et superficiei reflexionis. Si ergo ab alio linee communis puncto in eadem superficie fieret reflexio, alia perpendicularis ab alio puncto ducta esset dyameter alterius circuli columpne huic equidistantis, et caderet in punctum axis in quod non cadit superficies reflexionis. Et ita in omnibus reflexionis superficiebus est intelligendum quod ab uno tantum puncto linee communis fiat reflexio in eadem superficie respectu eiusdem visus. Quoniam respectu duorum visuum potest fieri a duobus punctis circuli dyametri terminus, id est perpendicularis. Respectu vero unius visus non accidit, quoniam illa duo puncta non simul ab eodem visu possunt comprehendi, semper enim necesse est partem columpne medietate minorem videri.

[5.26] Palam ex predictis perpendicularem super punctum reflexionis intellectam exterius intus transeuntem dyametrum circuli efficere, quia, si non, cum constet dyametrum circuli super punctum illud transeuntem perpendicularem esse super superficiem contingentem columpnam in puncto illo, et perpendicularem exterius similiter, erit continuitas inter has perpendiculares, et unam efficient

94 perpendicularis . . . superficie (95): diameter perpendicularis super superficiem columnam in illo puncto contingentem et superficies reflexionis secat illam lineam longitudinis columne super quam fit contingentia et est declinata super ipsam ergo et super axem erit illa superficies reflexionis declinata et in superficie plana *R* 96 illam: illa *P1* / *post* illam *add.* sed *ER* / si: scilicet *O* 97 reflexionis superficiei *transp. R* / punctis *corr. ex* punctus *O* / fieret . . . superficie *rep. L3* 98 illius superficiei *transp. C1* 99 *post* cum *add.* superficies *R* / ipsa: illa *R* / eum: eam *L3* 100 amplius: nam *R* / *post* perpendicularis *scr. et del.* a *S* 101 circulum equidistantem: circulo equidistante *O* / puncto: punctum *R* / *post* axis *add.* et est sectio *R* 102 *ante* circuli *scr. et del.* axis *OL3; add.* superficiei *R* / *post* et *scr. et del.* in *S* / ab *inter. O; inter. a. m. L3E* / *post* ab *scr. et del.* illo *FP1* 103 superficie *corr. ex* superficiei *L3* / fieret: fiet *FSOL3*; fiat *E* / reflexio *rep. P1* 104 *post* ducta *scr. et del.* ut *C1* / alterius: alius *L3* / *post* circuli *scr. et del.* licet *E* / columpne *corr. ex* columpna *O* / huic *alter. in* basibus *O* 105 punctum *corr. ex* puncto *P1* / cadit *corr. ex* caderet *C1* 106 reflexionis superficiebus *transp. R* 107 tantum puncto *transp. ER*; puncto *inter. a. m. E* 108 *post* duorum *scr. et del.* visus *P1* 109 *post* potest *add.* reflexio *R* / duobus *corr. ex* duabus *O* / *post* punctis *add.* superficiei speculi ut *R* / circuli dyametri *transp. O* / dyametri: dyameter *FP1C1E* / terminus id: terminis que *R* 110 *post* perpendicularis *add.* super ipsam sectionem *R* 112 medietate *corr. ex* mediate *O*; mediate *R* 114 exterius intus transeuntem: extra et intra produci *R* / *post* exterius *inter.* et *a. m. C1* / transeuntem *corr. ex* transeundtem *F* 115 si *om. FP1* / cum *om. FP1SL3* / circuli *om. O* 116 super superficiem *corr. ex* superficiem *OL3* 117 puncto illo *transp. R* / exterius: extra *R* 118 *post* has *inter.* duos *L3; add.* et *C1* / efficient: efficiet *O; corr. ex* efficit *a. m. E*

lineam. Quia, si non est quod dyametrum extra productum perpendiculare sit super illam superficiem, accidet ex eodem superficiei puncto duas erigi perpendiculares. In omni ergo superficie reflexionum patet quatuor punctorum concursus: centri visus, puncti axis in quem cadit perpendicularis, puncti visi in speculo, puncti a quo forma corporis procedit.

[5.27] In speculis piramidalibus super bases suas ortogonalibus politis exterius est oppositio visus ut non sit visus in superficie speculi aut in ei continua, et secundum visus situm respectu speculi piramidalis erit quantitas comprehense in eo partis.

[5.28] Igitur, si radius ab oculi centro ad terminum axis piramidis—id est ad acumen intellectus—faciat cum axe angulum acutum ex parte piramidis, intelligemus a centro visus superficiem secantem piramidem super circulum equidistantem basi piramidis. Et intelligemus duas lineas a centro quidem visus tangentes illum circulum in punctis oppositis, a quibus punctis protrahemus lineas secundum longitudinem piramidis. Superficies ergo ex una harum linearum longitudinis et altera contingentium circulum continget piramidem, si enim secaverit, continget aliud punctum quam punctum contingentie circuli. Super illud punctum producatur linea longitudinis piramidis, et illud punctum et acumen piramidis sunt sicut in hac superficie, quare illa linea erit in hac superficie et transibit per aliquod punctum circuli. Illud ergo punctum est in hac superficie et in circulo, quare est in linea communi circulo et superficiei. Sed illa est contingens circulum, quare contingens transit per duo puncta circuli quem contingit, quod est impossibile. Restat ergo quod superficies illa tangat piramidem.

[5.29] Et generaliter omnis superficies reflexionis in qua concurrunt linea tangens aliquod punctum piramidis et linea longitudinis per illum

119 *post* non *scr. et del.* esset *C1* / quod *om. O* / dyametrum: diameter *R* / productum: producta *R* / perpendiculare (120) *corr. ex* perpendiculariter *P1*; perpendicularis *OR*; *corr. ex* perpendicularis *L3* 121 perpendiculares *corr. ex* perpendicularis *F* / superficie: superficiei *FP1SL3* / *post* superficie *scr. et del.* puncto duos erigi *S* / reflexionum: reflexionis *R* 122 quatuor: quod *OC1*; *corr. ex* quod *L3* / *post* quatuor *inter.* est *O* / quem: quod *R* 123 puncti . . . speculo *mg. a. m. S* / visi: reflexionis *R* 126 politis: positis *FP1SO* / est: et *FP1SL3* 127 ei *corr. ex* eo *a. m. E* / ei continua *transp. R* / piramidalis: piramidis *OC1* 130 ad *om. FP1S* / faciat *corr. ex* fiat *a. m. C1E* 131 secantem *corr. ex* seguntem *O* 132 super . . . piramidis *mg. a. m. C1* 133 quidem *om. O* / *post* illum *scr. et del.* punctum *L3* 134 punctis *om. R* 135 ergo: igitur *O* / *post* longitudinis *scr. et del.* et *P1* 136 secaverit: secuerit *R* 137 continget *corr. ex* contingens *L3*; contingens *C1E* 138 piramidis *om. R* 139 *post* piramidis *add.* simul *R* / sunt *corr. ex* fiunt *a. m. E* / *post* sunt *add.* simul *P1* / sicut *om. R* / quare *inter. O* 141 ergo: igitur *R* / est in hac superficie: in hac superficie est *ER* / hac superficie *transp. L3C1* / in^3 *inter. P1* 142 circulo *inter. a. m. E* / est contingens: contingit *R* / circulum . . . contingens (143) *om. O* 143 *post* contingens *mg.* et *O*; *add.* circulum tangit; *del.* tangit *C1* / quem *mg. a. m. C1* 144 ergo: igitur *FP1R*; *om. E* / quod: ut *R* / superficies illa *transp. R* / tangat *corr. ex* tangant *S* 145 omnis: omnes *O* / reflexionis *inter. a. m. E*; *om. R* 146 linea tangens: linee tangentes *E* / *post* et *add.* illa *E* / linea longitudinis *transp. ER* / illum: illud *OER* / illum punctum (147) *transp. R*

punctum transiens tangit piramidem super lineam longitudinis. Habemus ergo duas superficies ab oculi centro procedentes piramidem contingentes inter quas est portio piramidis apparens visui in hoc situ, et est minor medietatum piramidis, quoniam linee contingentes circulum includunt eius partem medietatum minorem.

[5.30] Si vero linea a centro ad acumen piramidis ducta tenet angulum rectum cum axe, intelligatur circulus secans piramidem equidistans basi. Linea communis huic circulo et superficiei in qua sunt axis piramidis et centrum visus erit ortogonalis super axem piramidis, quoniam axis est ortogonalis super superficiem circuli. Et super lineam communem protrahatur per centrum circuli dyameter ortogonaliter super hanc lineam, et a terminis huius dyametri protrahantur due contingentes circulum, et etiam due linee usque ad acumen piramidis. Due superficies in quibus erunt hee due linee cum contingentibus contingunt piramidem secundum modum predictum. Et quoniam linea communis circulo et superficiei in qua sunt centrum visus et axis piramidis est equidistans linee a centro visus ad terminum axis producte, et huic linee communi sunt equidistantes linee circulum in predictis punctis contingentes, erunt ille contingentes equidistantes linee a centro visus ad terminum axis ducte, quare erunt in eadem superficie cum illa. Igitur utraque superficierum circulum contingentium transit per centra visus, et communis illarum superficierum sectio est linea a centro visus ad terminum axis ducta. Et quod inter illas superficies cadit ex piramide apparet visui, et est medietas piramidis, quoniam lineas has contingentes circulum interiacet medietas circuli. Et ita palam quod in hoc situ apparet medietas piramidalis speculi.

[5.31] Verum si linea a centro visus ducta ad terminum axis piramidis teneat cum axe angulum obtusum ex parte superiori apparenter, et fiat circulus secans piramidem equidistantem basi, linea communis huic

148 duas superficies *transp. C1* / procedentes: cedentes *L3E* 149 apparens: apparentis *ER* 150 medietatum *corr. ex* medietum *O*; medietate *L3C1R* / piramidis: pyramidum *L3* / contingentes: tangentes *FP1R*; contingens *E* 151 includunt . . . tenet (152) *mg. O* / medietatum *corr. ex* miedietatum *F*; medietate *C1R* / minorem *corr. ex* minorum *F*; minore *P1* 152 *post* centro *add.* visus *ER* / tenet: teneat *L3C1ER* 153 *ante* angulum *scr. et del.* in *O* / cum *corr. ex* cur *L3* / *post* axe *add.* et *R* / secans *inter. E* 154 equidistans: equidistanter *R* / communis *corr. ex* longitudini *O* / et *inter. L3* 155 centrum: centra *FP1; corr. ex* centra *L3* / erit *om. L3; mg. a. m. C1* 157 et . . . circuli *mg. a. m. E* / centrum: centra *FP1SL3; corr. ex* centra *C1* 158 ortogonaliter: ortogonalis *R* / et *inter. L3* / *post* dyametri *add.* ortogonaliter *E; add.* ortogonalis *R* 160 hee *mg. a. m. C1* / cum *inter. L3* 161 contingunt: contingent *R* / piramidem *corr. ex* piramides *E* 163 *post* centro *add.* illius *ER* 164 producte *om. O* / et *om. S* 165 *post* predictis *scr. et del.* d *S* / *post* ille *add.* linee *R* / contingentes *inter. a. m. E; om. R* 166 axis *inter. a. m. E* 169 linea *inter. a. m. C1* / et *om. O* / *post* inter *add.* illa *C1* / illas *corr. ex* illa *E* 170 piramide *corr. ex* piramidis *E* / medietas . . . interiacet (171) *mg. a. m. S* 171 has: habens *S; om. O* 172 medietas *alter. ex* radius *in* radiis *O* / speculi *inter. a. m. S* 174 teneat *corr. ex* tenet *C1* / apparenter: apparente *R* 175 piramidem *corr. ex* piramides *C1* / equidistantem *alter. in* equidistanter *L3*; equidistanter *C1ER*

circulo et superficiei in qua est centrum visus et axis est perpendicularis super axem piramidis. Et hec linea communis extra producta concurret cum linea a centro visus ad terminum axis ducta propter angulum acutum quem facit hec linea cum axe ex inferiori parte. A puncto concursus linearum protrahantur due linee contingentes circulum in duobus punctis oppositis, et producantur linee ab hiis punctis ad acumen piramidis. Superficies in quibus sunt linee contingentes cum hiis longitudinis lineis contingunt piramidem, et in utraque harum superficierum sunt duo puncta linee a centro visus ad terminum axis ducte—scilicet terminus axis et terminus perpendicularis in quo scilicet concurrunt linea illa et perpendicularis—quare linea illa est in utraque superficie. Igitur utraque superficies transit per centrum visus, et includunt hee superficies ex interiori in inferiori parte minorem partem piramidis, quia linee contingentes circulum includunt partem eius minorem medietatem. Unde ex parte superiori interiacet superficies piramidem contingentes pars medietatum maior, et illa est que apparet visui, quare in hoc situ comprehendit visus piramidis partem medietatum maiorem.

[5.32] Si autem linea a centro visus ad terminum axis producta cadat super latus piramidis, ut ex ea et latere unum efficiatur continuum, dico quod non latebit visum ex hac piramide preter lineam quandam intellectualem, quoniam omnis superficies in qua est linea a centro visus ad terminum axis ducta et secundum lateris longitudinem prolongata secat piramidem una tantum excepta que contingit piramidem in latere quod est pars linee. Et hoc solum latus intellectuale in tota piramidis superficie super hoc situ visum preterit.

[5.33] Et huius rei veritas patet ex hoc quod, quocumque superficiei piramidis puncto sumpto, si ad ipsum ducatur linea a centro visus et

176 et[1] *inter. P1* / *post* axis *scr. et del.* perpendicularis *O* 177 concurret *corr. ex* concurrat *OE*; concurrat *L3* 179 *post* puncto *add.* igitur *R* 180 *post* linee *scr. et del.* contingentes *O* / *post* circulum *scr. et del.* in duobus *L3* 181 oppositis *corr. ex* oppositionis *a. m. E* 182 sunt *om. FSOL3E* / sunt linee *inter. a. m. C1* 184 duo puncta *transp. C1* 186 et . . . illa[2] *inter. L3* / illa[2] *inter. a. m. E* / *post* illa[2] *add.* que ducitur a centro visus per terminum axis *R* 188 includunt: includuntur *FP1* / *post* hee *inter.* scilicet *a. m. E* / superficies *inter. a. m. E* / interiori in *om. R* / in *om. SOC1*; *inter. a. m. E* 189 *post* piramidis *add.* medietate *R* 190 medietatem: medietate *OC1ER* 191 medietatum: medietate *OC1R* 192 piramidis: piramidum *P1* / piramidis partem *transp. R* 193 medietatum: medietate *OC1R* 194 cadat: cadit *P1ER* 195 piramidis: piramidum *P1* / efficiatur *corr. ex* ficiatur *O* / *post* continuum *add.* latus *R* 196 non *inter. a. m. S* / *post* preter *scr. et del.* lumen *P1* 197 omnis: omnes *F*; *corr. ex* omnes *P1OL3* 198 *post* lateris *add.* venientis ad centrum visus *P1* 199 contingit: contingerit *P1* / piramidem[2]: piramidis *O* / *post* piramidem[2] *add.* superficiem *O* 200 *post* pars *add.* illius contingentis *C1* / et: in *P1* / latus *om. O* / intellectuale *corr. ex* intellectualem *S* 201 super: sub *L3C1ER* / situ *corr. ex* situm *P1*; sui *S* 202 ex hoc *mg. a. m. E* / quocumque *corr. ex* quodcumque *L3* / *post* quocumque *inter.* in *O* / superficiei *om. R* 203 piramidis puncto *transp. L3C1E* / puncto: puncti *FP1* / *post* sumpto *add.* extra latus intellectuale *R*

ab eo linea longitudinis piramidis ad terminum axis, efficient hee due linee triangulum cum linea lateri applicata. Et est triangulus in superficie a centro intellecta piramidem secante, et ex lineis huius superficiei non nisi due cadunt in superficie piramidis—scilicet linea lateris et linea longitudinis a puncto sumpto ad acumen piramidis. Et linea a centro ad punctum sumptum ducta secat lineam longitudinis reflexionis in puncto sumpto et lineam lateris in centro, quare huic linee non accidet concursus de centro cum aliqua harum linearum. Cum igitur non posset sumi punctus alius ad quem linea a centro accedat et in hoc punctum transeat, non occultatur punctus iste ab alio puncto. Et ita apparet visui, cum ei et visui non intercidat corporis solidi obiectio. Et eadem probatio in quolibet superficiei piramidis puncto.

[5.34] Et si linea a centro visus in terminum axis cadens intret piramidem, dico quod nullus occultatur visui punctus in tota piramidis superficie. Sumpto enim quocumque puncto in piramidis superficie, intelligatur ad ipsum linea a centro et alia ab eo usque ad acumen piramidis. Hee due includunt superficiem triangularem cum linea a centro visus ad terminum axis ducta piramidem intrante, et est iste triangulus in superficie piramidem secante. Cum omnis superficies in qua fuerit linea intrans piramidem secet eam, linea a centro ad punctum sumptum ducta secat in illo puncto lineam longitudinis ab eo ad acumen piramidis ductam. Et ex lineis superficiei in qua sunt hee due linee non sunt nisi due linee in superficie piramidis—scilicet hec linea longitudinis a puncto ad acumen ducta et alia opposita secans angulum quem includit hec cum linea piramidis intrante. Igitur linea illa opposita extra piramidem producta secat lineam a centro ad punctum sumptum

204 eo: ea *FP1S* / *post* piramidis *add.* ducatur *C1* / *post* ad *scr. et del.* centrum *E* / *post* terminum *mg.* B *a. m. E* / axis *mg. a. m. E* / hee *inter. a. m. E* 205 et *corr. ex* quod *L3* / *post* et *scr. et del.* quod *S* / est triangulus: erit triangulum *R* 206 *post* centro *add.* visus *R* / *post* ex *add.* hiis *E* 207 due cadunt *transp. E* / *ante* in *add.* linee *E* / superficie: superficiem *R* / linea lateris et (208) *om. R* 208 lateris: lateri *FP1*; *corr. ex* latitudinis *L3*; latitudinis *inter. a. m. E* / longitudinis *corr. ex* longitudins *P1* / *post* piramidis *add.* et linea opposita huic ex altera parte *R* 209 *post* centro *add.* visus *R* 210 reflexionis: ei eius *O*; *om. R* / reflexionis in puncto: in puncto reflexionis *C1* / lateris: lateri *P1*; *corr. ex* latitudinis *OL3* / *post* lateris *add.* continuati cum visu *R* / *post* centro *add.* visus *R* / *post* quare *scr. et del.* hunc *P1* / *post* linee *add.* a centro visus *R* 211 accidet: accidit *OL3C1* / de centro *scr. et del. O* / de . . . posset (212) *om. R* / *post* centro *inter.* inter ipsum et centrum *a. m. O*; *add.* visus *C1* 212 posset: possit *C1*; *alter. in* possit *E* / *post* posset *scr. et del.* po *F* / punctus alius ad quem: punctum aliud ad quod *R* / *post* centro *add.* visus *R* 213 occultatur *corr. ex* occultatum *O* / punctus iste *alter. in* punctum hoc *a. m. E*; punctum istud *R* 214 et[1] . . . ei: quod non perveniat ad centrum visus quare apparet visui cum inter ipsum *R* / et[2] *inter. a. m. C1* / visui: visum *R* / obiectio: abiectio *FP1* 215 *post* eadem *add.* est *OC1* / *post* probatio *add.* est *R* / in *inter. L3*; de *R* 216 et: quod *O* 217 nullus: nullum *R* / punctus: punctum *R* 218 *post* superficie[2] *add.* et *C1* 219 *post* centro *add.* visus *C1R* 220 *post* due *add.* linee *R* 221 *post* centro *scr. et del.* videtur *P1* / visus *om. O* / intrante: intrantis *O* / iste: istud *R* 222 triangulus: triangulum *R* / *post* in *scr. et del.* parte *P1* / secante *corr. ex* secantem *L3* 223 secet *corr. ex* secent *O* / *post* linea[2] *add.* vero *R* / *post* centro *add.* visus *R* 225 due *om. O* 226 sunt *inter. O* 227 alia: illa *P1* 228 piramidis: piramidum *P1*; pyramidem *R* 229 secat *corr. ex* secans *O*

ductam; quare linea hec secat duas lineas que sole ex lineis huius superficiei sunt in piramidis superficie, unam extra piramidem aliam in puncto sumpto, quare producta in infinitum non concurret cum altera illarum linearum. Unde non occultatur visui punctum sumptum secundum modum supradictum.

[5.35] In hoc ergo situ nulla superficierum piramidis tangentium transibit per centrum visus, sed quelibet secabit lineam visus super terminum axis piramidem intrantis inter visum et piramidem, et est in termino axis. Cum vero linea visus linee longitudinis piramidis applicatur, nulla superficierum piramidis tangentium perveniet ad centrum preter illam que in predicta linea contingit piramidem. Et omnes superficies contingentes secabunt lineam illam inter visum et piramidem.

[5.36] Similiter, in situ in quo superficies due contingentes piramidem per centrum transeunt, quelibet superficies tangens piramidem in portione piramidis apparente que duas contingentes interiacet a centro visus divertit. Super quodcumque punctum illius portionis cadat linea visualis, secabit piramidem, cum intercidat duas contingentes visuales. Et superficies in qua fuerit hec visualis et linea longitudinis piramidis secabit piramidem, et erit hec visualis superficies cuicumque superficiei piramidis in hac portione contingat, quare et visus.

[5.37] Dico ergo quod in quolibet situ a quolibet puncto potest fieri reflexio. Sumatur enim punctus, et intelligatur circulus per punctum transiens basi piramidis equidistans. Dyameter huius circuli ab hoc puncto incipiens erit perpendicularis super axem, cum axis sit perpendicularis super circuli superficiem, quare linea longitudinis a puncto ad acumen piramidis ducta tenet angulum acutum cum dyametro et acutum cum axis termino in eadem superficie. Sit linea visualis super punctum cadens in superficie in qua est linea longitudinis

232 in² *om. C1; inter. E* / concurret *corr. ex* concurrit *F* / altera: aliqua *R* 233 illarum: aliarum *C1* / unde *corr. ex* unum *S* / punctum sumptum *transp. R* 235 ergo *om. FP1; inter. L3; mg. a. m. E* / ergo situ *transp. R* / piramidis: piramidum *P1*; piramidem *ER* / *post* tangentium *scr. et del.* et *O* 236 visus: a visu *R* 237 piramidem¹: piramidis *FP1* / intrantis: intrantem *OC1; corr. ex* intrantem *L3* 238 vero: ergo *L3C1* / piramidis: piramidum *P1* 239 piramidis: piramidem *C1ER* / perveniet: pertinet *R* 240 *ante* preter *add.* visus *R* 242 piramidem: verticem pyramidis *R* 243 situ: visu *FP1; corr. ex* visu *OL3; alter. ex* usu *in* visu *S* / superficies due *transp. R* 244 centrum: centra *FP1SOL3* / *post* centrum *add.* visus *R* 245 in portione piramidis *om. FP1* / apparente: apparentem *P1* 246 divertit *scr. et del. O* / *post* divertit *inter.* et *L3*; et *inter. a. m. C1E; add.* et *R* 247 *post* intercidat *add.* inter *R* 248 *post* fuerit *add.* linea *R* / *post* visualis *add.* linea *C1* / et² . . . visualis (249) *mg. a. m. E* 250 contingat *corr. ex* contingant *OL3*; continua *R* 252 *post* quod *scr. et del.* visus *F* 253 sumatur: similiter *S* / punctus: punctum *R* 254 piramidis: pyramidi *R* / *post* dyameter *add.* igitur *R* 256 circuli superficiem *transp. L3C1* 259 in¹ . . . qua *om. O* / est: et *inter. O*

et axis, in qua superficie ducatur perpendicularis super lineam longitudinis in puncto illo. Concurret hec quidem perpendicularis cum axe, et ex ea, et axe, et linea longitudinis efficietur triangulus. Super punctum illud intelligatur linea contingens, et super dyametrum circuli quem fecimus intelligatur dyameter alius ortogonalis super ipsum, qui erit ortogonalis super ipsum axem, et ita super superficiem in qua axis et dyameter primus. Et hic dyameter secundus est equidistans contingenti, quoniam contingens est perpendicularis super dyametrum primum. Et ita linea contingens ortogonalis est super superficiem in qua axis et dyameter primus, quare erit ortogonalis super perpendicularem quem primo fecimus. Et ita illa perpendicularis ortogonaliter cadit super superficiem contingentem piramidem in qua punctus sumptus.

[5.38] Igitur, si linea visualis cadens in punctum sumptum transeat secundum processum perpendicularis, erit quidem ortogonalis super superficiem piramidis illam in puncto contingentem, et fiet reflexio forme per eandem lineam. Si autem deviet a processu perpendicularis, faciet quidem angulum cum perpendiculari acutum in puncto sumpto. Et poterit produci in superficie huius linee visualis alia linea a puncto illo que equalem angulum huic teneat cum perpendiculari, cum perpendicularis ortogonalis sit super superficiem contingentem. Linea autem quacumque super superficiem contingentem in puncto sumpto ortogonaliter cadente, transit ad axem. Et si ab axe ducatur ortogonalis ad hanc superficiem, efficient perpendicularis interior et exterior lineam unam. Quod si non cum perpendicularis interior extra producta sit etiam perpendicularis super superficiem, accidet ab eodem puncto super illam superficiem erigi duas perpendiculares in eandem partem.

[5.39] Palam igitur quod quocumque puncto superficiei piramidis viso potest fieri reflexio ad paritatem angulorum. Et cum linea

260 ducatur: deducatur *ER* 261 concurret: concurrit *FP1*/quidem: quod *O* 262 et[2] *inter. a. m. C1*/triangulus: triangulum *R* 264 intelligatur: intelligigatur *S*/dyameter *corr. ex* diametrum *O*/alius: alia *R*/ipsum qui (265): ipsam que *R* 265 axem *corr. ex* ?? *O*/ita *om. R*/super[2] *inter. L3*/*post* superficiem *add.* est *C1*/*post* qua *add.* est *R* 266 primus: prima *R*/hic: hec *R*/secundus: secundum *L3*; secunda *R* 267 est *inter. O; om. L3*/est perpendicularis *ER* 268 primum: primam *R*/*post* primum *scr. et del.* quare erit ortogonalis super perpendicularem *C1*/*post* ita *add.* etiam *E*/super *om. O* 269 *post* qua *add.* sunt *R*/primus: prima *R*/ortogonalis: perpendicularis *R* 270 quem: quam *R*/*post* illa *add.* prima *R* 271 cadit *om. O*/punctus sumptus (272): punctum est sumptum *R*/*post* punctus *add.* est *E* 272 *post* sumptus *add.* est *C1*; est *inter. a. m. S* 275 piramidis: pyramidum *L3*; piramidem *ER*/fiet *corr. ex* fiat *O* 276 *post* a *scr. et del.* perp *P1* 277 faciet *corr. ex* fiet *O*/*post* in *scr. et del.* punctum *L3* 278 huius: eius *R* 279 cum perpendiculari *inter. a. m. E*/cum perpendicularis (280) *om. L3*/*post* cum[2] *add.* illa *C1* 281 quacumque: quecumque *E*/super *inter. a. m. E* 282 cadente: cadens *R*/transit: transeat *L3*; transibit *C1*/si *corr. ex* sic *L3* 283 perpendicularis *alter. in* perpendiculares *E*; perpendiculares *R* 285 etiam: et *P1O; om. L3; inter. a. m. E*/accidet *corr. ex* accidit *E* 286 illam: aliam *L3C1; alter. ex* aliam *in* aliquam *a. m. E;* aliquam *R*/erigi duas *transp. O* 287 *post* quod *add.* a *R* 288 paritatem: partem *FP1*

reflexionis occurrerit forma, veniet ad speculum super lineam hanc et reflectetur ad visum super aliam, et sunt hee due linee in eadem superficie ortogonali super superficiem contingentem piramidem in puncto reflexionis. Et hec est superficies reflexionis in qua semper fit comprehensio quatuor punctorum: scilicet centri visus, puncti visi, puncti reflexionis, terminus perpendicularis.

[5.40] Diversificantur autem linee communes superficiebus reflexionis et superficiei piramidis. Cum enim radius visualis continuus fuerit axi piramidis—scilicet cum in qualibet superficie reflexionis sit totus axis et perpendicularis ad axem transiens—erit cuilibet superficiei reflexionis et superficiei piramidis communis linea linea longitudinis in hoc situ. Quoniam quelibet superficies in qua est totus axis hanc habet lineam communem cum superficie piramidis.

[5.41] Et in omni alio situ unica longitudinis piramidis linea erit communis illa—scilicet que fuerit in superficie centrum visus et axem continente. Et quia centrum visus non erit in directo axis, una tantum erit superficies talis, et omnis alia communis linea erit sectio piramidalis non circulus. Si enim fuerit circulus, erit superficies illius circuli in superficie reflexionis, et quia axis est ortogonalis super illum circulum (cum quilibet circulus piramidis sit equidistans basi), erunt latera piramidis declinata super circulum, et ita super superficiem reflexionis. Quare in superficie illa non potest duci perpendicularis super lineam longitudinis piramidis. Sed perpendicularis ducta super superficiem contingentem locum reflexionis est in superficie reflexionis, et perpendicularis super lineam longitudinis, cum quelibet superficies tangens piramidem tangat in linea longitudinis.

[5.42] Accidit igitur impossibile, quare restat omnes alias communes reflexionis lineas sectiones piramidales esse, et cum fuerit linea com-

289 reflexionis: declinata *R* / occurrerit: occurit *E* / *post* occurrerit *add.* alicui *C1* / lineam hanc *transp.* *O* 290 reflectetur: reflectet *O* / due *om. OL3C1E* 291 contingentem *inter. L3* / in *om. O* 292 et . . . reflexionis *om. P1* / fit: sit *S* 293 puncti visi *mg. a. m. E* 294 terminus: terminum *O*; termini *R* / terminus perpendicularis *om. FP1*; termini perpendicularis *mg. L3* / *post* perpendicularis *add.* et puncto axis *O* 295 communes *om. L3* / *post* communes *scr. et del.* a *E* / superficiebus: superficiei *R* 296 superficiei: superficie *FP1S* 297 piramidis: piramidum *P1* / qualibet *om. R* / sit *alter. ex* licet *in* fuerit *O*; fuerit *R* 298 cuilibet *om. R* 299 *post* superficiei *scr. et del.* pima *F* / linea[2] *om. O*; *mg. a. m. C1* 300 est *inter. SO*; *om. L3E* / est totus *transp.*; est *inter. a. m. C1* 1 habet: habeat *L3* / communem *om. P1* / cum *inter. a. m. C1* 2 et *inter. O* / unica *corr. ex* iuncta *a. m. E* 3 centrum: centra *FP1SL3C1* / centrum visus *transp. ER* 4 quia: quando *R* 5 talis: taliter *C1* 6 si enim *om. P1* / fuerit circulus *transp. FS* / circulus *om. P1* 7 axis est *om. O* / est *inter. S*; *om. L3C1* / est ortogonalis *ER* 8 quilibet *alter. ex* quibus *mg. F*; quibus *P1* / piramidis *corr. ex* piramidalis *a. m. E* / *post* sit *scr. et del.* equalis *C1* / latera: lata *F* 9 super *inter. a. m. E* 10 super: per *FP1S* 11 longitudinis: longitudini *P1* / super *corr. ex* superficiem *F* 12 *post* reflexionis[1] *add.* apud locum reflexionis *SC1*; *mg. a. m. E* / reflexionis[2] *om. L3* / reflexionis et *om. O* 13 perpendicularis: perpendiculari *O* / *post* lineam *scr. et del.* longidinis *P1* 14 piramidem *om. R* 15 *post* restat *add.* ut *FP1* 16 sectiones *corr. ex* sectionis *a. m. E* / piramidales *alter. in* piramidis *O* / et *om. OL3E*

munis linea longitudinis, ex quocumque puncto illius linee fiat reflexio erit in eadem superficie cum cuiusque alterius puncti reflexione. Quoniam a quolibet huius linee puncto ducta perpendiculari continget axem; et erunt in superficie reflexionis centrum visus, et punctum reflexionis, et punctum axis; et huius reflexionis est superficies in qua sunt linea longitudinis et axis, quare in hac superficie fit reflexio a quocumque puncto.

[5.43] Si vero communis linea non fuerit linea longitudinis, dico quod vel ab uno communis linee puncto in eadem superficie fiat reflexio, vel a duobus tantum. Quoniam ducta perpendiculari a puncto reflexionis, perveniet ad axem, et cadet in aliquod punctum eius. Intellecto circulo super punctum reflexionis, ortogonaliter secabit circulus axem, et cum perpendicularis secat axem equidistans basi, erit perpendicularis declinata super circulum. Et circumquaque ducta, semper erit equalis, unde fiet piramis cuius basis circulus, acumen punctus axis in quem cadit perpendicularis. Igitur superficies reflexionis aut tanget hanc piramidem aut secabit.

[5.44] Si tangat, dico quod a puncto reflexionis sumpto possit tantum fieri in eadem superficie reflexio. Planum quod superficies reflexionis continget hanc piramidem super perpendicularem, que est linea ortogonalis in superficie reflexionis, et si ab acumine totalis piramidis ducantur linee ad sectionem communem superficiei reflexionis et piramidi magne prius, cadent in circulum qui est basis piramidis intellecte quam in sectione preter unam que in punctum reflexionis cadit. Si ergo ab alio sectionis communis puncto fieret reflexio, linea ab illo puncto ad acumen intellecte ducta erit perpendicularis super lineam longitudinis piramidis per punctum illud transeuntem. Sed linea ab acumine piramidis intellecte ad punctum circuli per quem tran-

17 ex *om. O* / quocumque: cuicumque *O*; quecunque *L3* / fiat: fiet *L3C1* 18 in *inter. L3* / cum *inter. L3*; *inter. a. m. E* / cuiusque *corr. ex* cuiuscumque *P1*; cuiuslibet *C1*; cuiuscumque *ER* 19 ducta: producta *C1* / perpendiculari: perpendicularis *R* / continget *corr. ex* contingit *E* 20 et punctum reflexionis (21) *mg. a. m. C1* / punctum: punctus *P1* 21 et[2] . . . axis (22) *om. R* 22 sunt *om. FP1*; *inter. SOL3*; est *mg. a. m. C1*; est *E* 23 *post* puncto *add.* eius *C1* 25 quod *inter. a. m. E* / fiat *alter. in* fiet *O*; fiet *C1* 26 quoniam *inter. O* 27 cadet *alter. ex* cadat *in* cadit *E* / eius *om. O* 28 *ante* intellecto *add.* et *R* / ortogonaliter: ortogonalis *FP1S* / secabit *corr. ex* secat *E* 29 *post* axem[1] *add.* in termino *FP1*; *scr. et del.* in terminum perpendicularis in terminum axis cum *O* / et . . . axem[2] *om. O* / cum: quia *R* / perpendicularis secat *transp. L3* / secat . . . basi: tenet angulum acutum cum axe *R* / *post* secat *inter.* scilicet axem *a. m. E* / equidistans *corr. ex* equidistantem *C1* / erit perpendicularis (30) *om. C1* 31 erit *inter. a. m. C1* / punctus: punctum *R* 32 quem: quod *R* / aut *mg. a. m. C1* 33 tanget *alter. in* continget *O* 34 a *om. FP1* 35 *post* planum *add.* enim *R* 36 linea ortogonalis (37) *transp.*; ortogonalis *mg. L3* 37 ortogonalis in superficie *om. O*; *mg. a. m. E* / in superficie *inter. L3* 38 communem: commune *FP1* 39 piramidi magne: pyramidis totalis *R* / est *inter. a. m. E* 40 *post* intellecte *add.* quam *P1* / sectione: sectionem *R* 41 ergo: igitur *O* / alio: illo *E* / sectionis communis puncto: puncto communis sectionis *R* 42 *post* intellecte *add.* pyramidis *C1R* 43 sed *corr. ex* et *a. m. C1* 44 quem: quam *P1*; quod *R*

sit illa linea longitudinis absque dubio est perpendicularis super eam, quare alia angulum tenet acutum cum hac linea, non rectum.

[5.45] Si vero superficies reflexionis secet intellectualem piramidem, secabit circulum qui est eius basis in duobus punctis. Dico quod hec sola sunt puncta in tota sectione communi a quibus fieri possit reflexio in eadem superficie, quoniam ab utroque istorum punctorum linea ducta ad acumen intellecte piramidis est perpendicularis super lineam longitudinis super punctum suum transeuntem. A quocumque enim sectionis alio puncto ducatur linea ad acumen illius piramidis, tenebit angulum acutum cum linea longitudinis per ipsum transeuntem, cum perpendicularis cum eadem longitudinis linea angulum rectum teneat in circulo. Et linee ducte ab acumine intellecte piramidis ad puncta sectionis que intercidunt speculi acumen et circulum facient angulos obtusos cum lineis longitudinis versus partem acuminis piramidis totalis. Et que ducuntur ad puncta circulum et basem speculi interiacentia faciunt cum linea longitudinis angulos acutos ex parte acuminis speculi, obtusos ex parte basis.

[5.46] In speculis spericis concavis, si fuerit intra concavitatem speculi tota speculi superficies apparebit ei. Quod si extra fuerit visus, poterit comprehendere portionem eius maiorem medietatum quam scilicet fecerit circulus spere quem contingunt duo radii a centro visus ducti.

[5.47] Visu autem in centro huius speculi existente, non fiet ab aliquo puncto speculi reflexio nisi in se, quoniam quelibet linea a centro spere ad speram ducta perpendicularis est super superficiem speram in puncto illo tangentem. Unde in hoc situ non comprehendet visus per reflexionem nisi se tantum.

[5.48] Si vero statuatur visus extra centrum spere, poterit fieri reflexio in aliud corpus a quocumque speculi puncto preter quam ab eo in quem cadit dyametrum a centro visus ad speram per centrum spere ductus,

46 acutum *corr. ex* acumen *L3*; acumen *C1* / hac *om. P1* 48 eius *om. FP1R* / hec *rep. P1* 49 tota *om. P1* / fieri possit *transp. FP1* / possit: posset *L3C1* 51 perpendicularis: perpendiculariter *L3C1E* 53 alio puncto *transp. ER* / ad . . . linea (54) *mg. a. m. S* / *post* illius *scr. et del.* puncti *P1* 54 acutum *om. FP1SL3C1E* / transeuntem: transeunte *R* 55 perpendicularis: perpendiculari *P1*; *corr. ex* perpendiculariter *E* 56 intellecte piramidis *transp. R* / piramidis: piramidum *P1* 57 *post* intercidunt *add.* inter *R* / speculi *corr. ex* speculum *P1* / circulum: circulus *FP1S*; *corr. ex* circuli *O* 59 *post* puncta *add.* inter *R* / basem: basim *R* 60 faciunt: facient *R* 61 *post* basis *add.* ergo a nullo istorum punctorum potest fieri reflexio *R* 62 *post* si *add.* visus *R* 63 si *inter. a. m. E* / extra fuerit: sit extra *O* / visus *om. R* 64 medietatum: medietate *OC1R* 65 fecerit: fecit *R* / quem: quam *P1L3C1E*; *alter. ex* quam *in* quod *a. m. S* 67 *post* in *scr. et del.* co *F* 68 *post* quoniam *add.* enim *R* 69 ducta: ductam *FP1L3* / ducta . . . speram *mg. a. m. E* / superficiem *corr. ex* superficies *L3* 70 unde: ergo *R* / non *inter. OL3* / comprehendet: comprehendit *FP1S*; *corr. ex* comprehendat *O* 71 *post* nisi *scr. et del.* in *E* 73 in aliud corpus: alterius rei visibilis *R* / preter *inter. O* / preter quam: preterquam *R* / ab *corr. ex* in *O* / quem: quod *R* 74 dyametrum: diameter *ER* / ductus: ducta *R*

quoniam dyameter cadit super superficiem contingentem speram ortogonaliter. Sumpto autem alio puncto, ducatur ad ipsum dyameter a centro spere et linea a centro visus. Ex hiis ergo lineis acutus includetur angulus, quoniam linea visualis cadit inter dyametrum et superficiem contingentem punctum, que scilicet est extra speram. Et sive sit oculus intra speculum sive extra, cadit hec visualis linea intra speculum, quia cadit inter lineas visuales contingentes circulum portionis spere cum visus fuerit extra.

[5.49] Oculo cadente intra, planum quod intra cadit linea. Cum igitur dyameter angulum rectum teneat cum contingente, secetur ex eo acutus equalis predicto in eadem superficie. Dico ergo quod linea reflexionis cadit intra speculum, quoniam linea communis superficiei speculi et superficiei reflexionis est circulus tenens cum dyametro angulum acutum maiorem omni rectilineo acuto, et in singulis punctis erit hic modus reflexionis.

[5.50] Palam ex hiis quod in omni superficie reflexionis erunt centrum visus, centrum speculi, punctus reflexionis, punctus visus, terminus dyametri a centro visus per centrum spere ad speram ducti. Et communis omnium linea cum superficie speculi est circulus, et a quolibet linee communis puncto potest fieri in eadem superficie reflexio.

[5.51] In speculis columpnaribus concavis potest totum comprehendi speculum si fuerit visus intra ipsum. Sed eo extra sito, videbitur maior medietatum speculi, portio que scilicet interiacet duas superficies a centro visus procedentes columpnam contingentes.

[5.52] Intelligemus autem superficiem a centro visus procedentem basibus columpne equidistantem. Hec superficies aut cadet in columpnam aut non. Si ciciderit, linea communis huic superficiei et columpne erit circulus, et linea visualis transiens per centrum huius circuli cadet ortogonaliter super superficiem contingentem columpnam

75 dyameter *corr. ex* dyametrum *F*/*post* dyameter *scr. et del.* su *F*/*post* superficiem *scr. et del.* diamet *O*/*post* contingentem *scr. et del.* sumpto *O*; scilicet speram *inter. a. m. E*/speram *om. O* 76 sumpto *inter. O*/dyameter: diametrum *L3C1* 77 hiis *inter. O* 78 angulus: triangulus *C1E* 79 est *om. P1*/et *mg. a. m. C1*/sit: sumpsit *S* 80 *ante* intra[1] *scr. et del.* e *C1*/sive . . . speculum *mg. a. m.*; visualis linea *transp. L3*/hec *om. L3ER*/intra: inter *E* 81 cadit *om. C1*/*post* visuales *add.* spere *L3C1E* 82 visus . . . cum (83) *om. R*/extra *om. FP1*/*post* extra *add.* et *O* 83 *post* oculo *mg.* existente *a. m. C1*/cadente: existente *O* 85 acutus *corr. ex* actus *O* 86 linea communis *transp. ER*/*post* communis *scr. et del.* spe *S*/superficiei *om. R* 91 *post* visus[1] *scr. et del.* centrum visus *E*/punctus[1]: punctum *R*/punctus visus: punctum visum *R* 92 ad speram *om. R*/ducti: ducte *R*/et: palam quod *C1*/*post* et *add.* quod *R* 93 *post* omnium *add.* superficierum reflexionis *R*/est *om. L3E*/est circulus *transp.*; est *inter. a. m. S*/*post* et *add.* quod *R* 94 *post* quolibet *scr. et del.* et *O* 95 totum *om. E*/totum comprehendi (96) *transp. R* 96 intra: inter *L3E*/*post* intra *add.* speculum *P1*/*post* sed *add.* ex *L3*; *scr. et del.* ex *C1E*/sito: scito *P1* 97 medietatum: medietate *OR* 98 *post* procedentes *scr. et del.* a centro visus *F* 99 procedentem *corr. ex* procedentes *P1* 100 cadet: cadit *OE* 101 ciciderit *corr. ex* cideret *O*

in puncto in quem cadit linea. Et fiet reflexio per eandem lineam ad eius originem.

[5.53] Quicumque alius sumatur punctus. Linea perpendiculariter ab hoc puncto ducta cadet in axem, et linea visualis in punctum illud cadens faciet angulum acutum cum linea perpendiculari, cum sit inter perpendicularem et contingentem. Et quia hec linea cadet intra speculum, planum ex hoc quod cadit inter superficies portionem apparentem contingentes. Poterimus igitur in eadem reflexionis superficie ex angulo quem facit perpendicularis cum contingente excipere angulum acutum equalem angulo predicto acuto. Et cadet linea reflexionis hunc angulum continens intra columpnam, quoniam cadet inter perpendicularem et lineam longitudinis per terminum perpendiculariter transeuntem. Erunt igitur in superficie reflexionis centrum visus, punctum reflexionis, punctum visum, punctum axis in quem cadit perpendicularis.

[5.54] Et si hoc modo statuatur visus ut communis linea superficiei reflexionis et superficie columpne sit linea longitudinis, a quocumque puncto communis linee fiat reflexio. In una et determinata erit superficie omnibus hiis reflexionibus communi—ea scilicet in qua centrum visus et axis columpne totus—sicut dictum est superius in columpnari speculo non concavo.

[5.55] Similiter, si linea communis fuerit circulus, omnes reflexiones a punctis illius circuli facte procedent in eadem superficie, sicut in aliis circulis patuit.

[5.56] Et si sectio columpnaris fuerit linea communis, a duobus quidem eius punctis tantum fiet reflexio in eadem superficie, licet in superioribus columpnis circulus tantum ab uno puncto in unica superficie fieret reflexio, unico visu adhibito, quoniam supra latebant visum puncta sectionis se respicientia per que scilicet transit circulus columpne equidistans basibus. Viso enim uno latebat alius propter

104 quem: quod *R* / fiet *corr. ex* fiat *a. m. E* 106 quicumque alius: quodcumque aliud *R* / punctus: punctum *R* / perpendiculariter: perpendicularis *S* 108 faciet *corr. ex* faciens *P1* 109 quia: quod *C1ER* / cadet: cadat *R* / *post* cadet *add.* in punctum *P1* / intra: inter *OL3E* 110 *ante* planum *add.* est *C1* / *post* planum *add.* est *R* / apparentem *om. R* 111 poterimus: perimus *O* / ex *mg. a. m. C1* 113 angulo *om. O* / predicto acuto *transp. R* 114 cadet *corr. ex* cadit *E* / inter: intra *E* 115 perpendiculariter *alter. in* perpendicularis *a. m. S;* perpendicularis *R* 117 quem: quam *O;* quod *R* / perpendicularis: perpendiculariter *L3* 119 superficie: superficiei *C1ER* / sit: fit *FP1S* 120 *post* una *scr. et del.* in una *F* / et *inter. E; om. R* / superficie *corr. ex* superficiei *L3* 121 *post* communi *inter.* in *O* / *post* qua *inter.* est *a. m. S; inter. L3* 122 superius *om. P1* 123 non *mg. a. m. L3; inter. a. m. E* 124 si *inter. L3; inter. a. m. E* 125 *post* circuli *scr. et del.* facte *F* 126 circulis *alter. in* speculis *OL3* 128 punctis *corr. ex* punctus *O* 129 circulus *om. R* 130 fieret *corr. ex* fierit *E* / reflexio *corr. ex* reflecto *O* / *post* adhibito *add.* quod *C1* / quoniam: quem *FL3;* qui *S; corr. ex* quod *E* / supra: illic *R* / *post* supra *add.* quem supra *FP1O* / latebant *corr. ex* latebunt *E* 131 puncta sectionis *corr. ex* punctionis *O* / respicientia *corr. ex* respicienda *O* / scilicet *om. FP1C1* 132 equidistans basibus *transp. R* / *post* uno *add.* illorum punctorum *R* / latebat *corr. ex* latebit *E* / alius *om. P1; alter. ex* visus *E;* aliud *R*

minoris columpne portionis apparentiam, sed in hiis apparet maior columpne portio, unde ab unico visu percipiuntur puncta circuli equidistantis basibus et sectionis communis.

[5.57] In speculis piramidalibus concavis, si fuerit visus intra speculum, videbit ipsum totum. Si vero extra, et linea a centro visus ad acumen piramidis ducta intret piramidem aut applicetur linee longitudinis piramidis, nichil videbitur ex speculo. Quoniam quecumque alia linea ab oculo ad piramidem ducta cadet in piramidis superficiem exteriorem, unde occultabitur interior superficies.

[5.58] Si autem auferatur portio a piramide, poterit videri pars piramidis cadens inter contingentes superficies a centro ductas, scilicet maior, et si linea a centro visus sit perpendicularis super superficiem contingentem piramidem et continuetur axi, erunt linee communes, sicut dictum est in aliis piramidalibus, aut linee longitudinis piramidis aut sectiones. Et in hiis a duobus punctis sectionis poterit reflexio in eadem superficie respectu eiusdem visus; et in superficie reflexionis erunt centrum visus, punctus visus, punctus reflexionis, punctus axis.

[5.59] Sed speculum piramidale integrum, si apponatur visui, et sit visus ex parte basis, non percipiet nisi hoc quod fuerit intra speculum, quoniam perpendicularis tenet angulum acutum cum linea ab oculo ad ipsam ducta ex parte basis. Unde fit reflexio ex parte acuminis, et cadent omnes linee reflexe intra piramidem, et videri poterit quod intra piramidem positum sit.

[5.60] Si autem auferatur ex eo portio secundum longitudinem, poterunt quidem comprehendi exteriora, cum pateat exitus lineis reflexionis. Similiter, si secetur piramis ad modum anuli ut auferatur conus, liberum habebunt linee gressum, et exteriora apparebunt. Et si fuerit visus ex parte concavi, plura poterit comprehendere exteriora quam ex parte basis, quia latior reflexis lineis datur ad egrediendum via.

134 unico: uno *ER*/percipiuntur *alter. ex* percipiantur *in* percipientur *O*/*post* puncta *add.* terminantia diametrum *R*/circuli equidistantis (135) *corr. ex* circulo equidistantibus *O*; circulo equidistanti *L3C1E* 135 basibus: ba *P1*/*post* basibus *add.* columpne *R*/sectionis communis: sectioni communi *L3C1E*; *om. R* 139 *post* ex *inter.* hoc *L3* 140 ad *corr. ex* et *a. m. C1* 144 sit *om. O* 146 est *om. FS*/piramidalibus: piramidibus *SOL3*/piramidis: piramidum *P1R* 147 sectiones *corr. ex* sectionis *E*/*post* sectiones *add.* pyramidales *C1*/a *scr. et del. E*/*post* poterit *add.* fieri *R*/*post* reflexio *add.* fieri *O* 148 reflexionis *om. E* 149 erunt . . . punctus[2] *mg. a. m. E*/centrum: centra *FP1SOL3C1*/punctus visus punctus: punctum visum punctum *R*/*post* visus[2] *scr. et del.* et in superficie reflexionis erunt centra visus punctus visus *S*; *add.* et *E*/punctus[3]: punctum *R*/*post* axis *add.* in quod cadit perpendicularis *R* 150 integrum si *transp.*; integrum *mg. a. m. E*/apponatur *alter. in* opponatur *E*; opponatur *R* 151 hoc *om. O*/fuerit: sit *L3*/*post* speculum *scr. et del.* quoniam perpendicularis nisi quod fuerit intra speculum *O* 155 sit: est *R* 157 quidem *om. O* 159 conus *alter. in* capud *O*; vertex *R*/linee *om. P1*/gressum: ingressum *R* 160 concavi *corr. ex* coni *L3C1* (*a. m. C1*); superficiei concavitatis speculi *R* 161 *post* latior *scr. et del.* est *L3*; *add.* est *C1*/reflexis: reflexionis *O*; incidentibus *R*/lineis datur *transp. ER*/ad egrediendum *om. R*

[5.61] Amplius, sumpto uniuscuiusque speculi puncto, non est possibile in eo percipi formam nisi formam unius puncti ab eodem visu. Quoniam super perpendicularem et centrum visus unica transit superficies, et una sola est linea a centro visus ad punctum, et unicus angulus ex linea et perpendicularis acutus, et unicus angulus in eadem superficie acutus equalis huic, unde unica linea angulum equalem huic cum perpendiculari faciens. Et cum linea pervenit ad punctum corporis, non potest forma alterius puncti per ipsam vehi, cum punctum precedens occultet postpositum. Sed duobus visibus possunt in eodem speculi puncto comprehendi due punctales forme, quoniam infinite possunt sumi superficies super perpendicularem secantes in quarum qualibet circa perpendicularem sumi poterunt duo anguli equales acuti.

[5.62] Iam ergo proprietatem reflexionis declaravimus, et similiter cuiuslibet speculi proprium. Visus, cum per reflexionem formas comprehendit, non advertit quod hec adquisitio per reflexionem sit. Non enim accidit ex proprietate visus reflexio, quoniam, visu remoto, procedit non minus forma a corpore ad speculum et reflectetur secundum modum predictum. Et si accidit visum esse in loco in quem linearum reflexarum fit agregatio, comprehendet visus formam illam in capitibus harum linearum, et est in speculo tamquam non adveniens sed naturalis esset forma in speculo. Amplius, aliquando adquirit visus formas in speculis in sola superficie, aliquando intra speculum, aliquando ultra. Et erit apparens locus forme secundum figuram speculi et secundum situm rei vise, et semper comprehenditur forma in loco proprio, mutato situ visus et speculi. Et erit diversitas elongationis loci forme ad speculi superficiem secundum diversitatem figure speculi. Et locus forme dicitur locus ymaginis, et forma dicitur ymago. Visus autem comprehendit rem visam in loco ymaginis, et nos dicemus locum illum et eius proprium in quolibet speculorum, que numerabimus et

162 uniuscuiusque: cuiuscumque *O* / speculi *inter. E* 163 formam[1] *om. C1* / nisi formam *om. L3* 164 super: enim per *R* 165 ad punctum *inter. L3; inter. a. m. C1* 166 *post* angulus[1] *scr. et del.* in eadem superficie *S* / et[1] *inter. P1; om. R* / perpendicularis: perpendiculari *R* / *post* acutus *add.* est *C1* / et[2] *inter. P1; corr. ex* est *L3* 167 unde *inter. a. m. E;* ergo est *R* / *post* linea *add.* que *R* / *post* huic *scr. et del.* per *S* 168 faciens: facit *R* / pervenit *alter. ex* venit *C1;* pervenerit *R* / punctum: partem *R* 170 precedens: procedens *FP1SO* 171 due *mg. a. m. C1* / punctales: pictales *F;* punctuales *R* 172 perpendicularem: superficiem *P1* / *post* perpendicularem *add.* se *R* / *post* secantes *add.* se *O* 173 qualibet: quelibet *E* 174 ergo: igitur *FP1* / *post* reflexionis *scr. et del.* ut *L3C1* 175 *post* speculi *inter.* est *a. m. O* / *post* proprium *add.* est *C1; inter.* est *a. m. E* / *post* visus *inter.* est *L3;* est *mg. a. m. S; add.* autem *R* / cum: tamen *P1L3* / cum . . . visus (177) *om.O* 176 comprehendit: comprehenderit *E* / advertit: animadvertit *R* 177 remoto *corr. ex* remotu *a. m. C1* 178 reflectetur: reflectitur *R* 179 accidit: accidet *R* / in loco in quem: aliquam *O* 180 *post* linearum *scr. et del.* reflarum *P1* / fit: sit *L3C1* 182 in *om. OC1ER; inter. L3* 185 secundum *om. R* / comprehenditur: comprehendetur *R* 186 *post* proprio *add.* et immutabili et *O* / mutato: mutatu *FS;* immutato *O; corr. ex* mutatio *L3; alter. in* immutato *a. m. C1* 187 loci *corr. ex* locum *O* / superficiem *corr. ex* superficiei *O* 189 dicemus *corr. ex* dicamus *O;* dicamus *L3E* / locum illum (190) *transp. R* 190 numerabimus: enumerabimus *E;* enumeravimus *R*

assignabimus causas comprehendi res visas in loco illo, et hoc in sequenti libro, si deus voluerit.

191 *post* causas *add.* propter quas *R* / comprehendi *alter. in* comprehendendi *a. m. E*; comprehendantur *R* / visas: vise *R* / hoc: hee *FS* 192 sequenti: sequente *R* / *post* sequenti *scr. et del.* loco *F*

[QUINTUS TRACTATUS]

Liber iste in duas partes partitus est. Prima pars est prohemium libri; secunda in ymaginibus.

[CAPITULUM 1]

Prima pars

[1.1] Liquet ex libro quarto quod forme rerum visarum reflectuntur ex corporibus politis, et visus adquirit eas in corporibus politis propter reflexionem. Et patuit quomodo fiat adquisitio rerum ex reflexione formarum. Et visus comprehendit rem visam in loco reflexionis determinato et primo cum non fuerit situs rei vise ad visum mutatio. Et forma in corpore polito comprehensa nominatur ymago. Et nos explanabimus in hoc libro loca ymaginum ex corporibus politis, et dicemus quomodo adquiratur horum locorum scientia, et quomodo inveniantur sillogistice, et demonstratur.

[CAPITULUM 2]

Pars secunda: loqui in ymaginibus.

[2.1] Ymaginis cuiuscumque puncti locus est punctus in quo linea reflexionis secat perpendicularem a puncto rei vise intellectam super lineam contingentem lineam communem superficiei speculi et

1 liber . . . ymaginibus (2) *om. FP1* / *post* prima *scr. et del.* est *L3*   3 prima pars *om. FP1O*   4 libro quarto *transp. ER*   5 adquirit: comprehendit *R* / *post* politis[2] *scr. et del.* et visus *S*   6 et *corr. ex* quod *O* / fiat *inter. a. m. F*; fieret *R*   7 reflexionis *om. R*   9 in . . . comprehensa: comprehensa in corpore polito *R*   12 inveniantur: inveniuntur *C1* / sillogistice: ille *O*; sillogismo *L3*; *corr. ex* sillogismo *a. m. S*; *corr. ex* simile *a. m. E* / *post* sillogistice *add.* vel sillogismo *C1* / et demonstratur *om. R* / demonstratur: demonstrantur *F*; *alter. ex* demonstrative *in* demonstrationis *O*   13 pars . . . ymaginibus *om. OR*   14 punctus: punctum *R*   15 a *om. S* / a puncto *mg. F*   16 et . . . reflexionis (17) *om. ER*

superficiei reflexionis aut superficiei speculo continue et superficiei reflexionis. Et nos declarabimus.

[2.2] Sumatur speculum planum, et statuatur equidistans orizonti, et lignum directum et politum ortogonaliter erigatur supra speculum. Et sit speculi quantitas ut totum possit videri lignum, nisi enim totum apparuerit, error inerit. Et signetur in ligno punctum aliquod nigrum. Apparebit quidem visui lignum huic equale ultra speculum huic ligno continuum et ortogonale supra speculum, et in ligno apparenti apparebit punctus signatus tantum distans a superficie speculi quantum ab eadem distat in ligno superiori. Et si declinetur lignum supra speculum, apparebit apparens eadem declinatione declinatum, et punctus signatus in apparenti signato eque remotus a superficie speculi. Et si a puncto signato lignum aliquod erigatur ortogonaliter supra speculum, videbitur hoc etiam lignum a puncto apparenti ortogonaliter supra speculum et huic ortogonali continuum. Idem accidit pluribus punctis in ligno signatis. Idem penitus accidet elevato aut depresso speculo.

[2.3] Planum ergo per hoc quod ymago puncti visi apparet in perpendiculari ducta a puncto viso ad superficiem speculi, et in hoc speculo que perpendicularis est super superficiem speculi est perpendicularis super lineam communem superficiei speculi et reflexionis.

[2.4] Idem patere poterit in piramide super basim ortogonali, cuius basis plana speculo plano ortogonaliter sit adhibita, apparebit enim huic piramis alia continua quarum eadem basis et harum piramidum acumina equaliter a speculo distantia. Et planum quod, si ab acumine

17 aut . . . continue *mg. a. m. E;* aut: et *E;* vel *R* / *post* aut *scr. et del.* sp *F* / et *inter. O* / et superficie *om. FP1*   18 *post* nos *add.* hec *R* / *post* declarabimus *add.* hoc *C1*   19 sumatur *om. O* / statuatur *alter. in* statuantur *P1*   20 ortogonaliter *om. R* / supra: super *FP1R*   21 possit: posset *O* / nisi *inter. L3*   22 apparuerit: appareat *ER* / in *inter. L3C1E* (*a. m. C1*)   23 huic equale *transp. R*   24 ortogonale: ortogonalis *E* / apparenti: apparente *R* / *post* apparenti *scr. et del.* apparenti *E*   25 punctus signatus: punctum signatum *R*   26 ligno *corr. ex* loco *a. m. F* / superiori: superiore *R*   28 punctus signatus: punctum signatum *R* / in *om. O; inter. L3E* / *ante* apparenti *inter.* erit *O* / apparenti: apparente *R* / apparenti signato: apparien scilicet signatus *O* / signato *om. R* / *ante* eque *add.* apparebit *ER* / remotus: remotum *R* / *post* remotus *add.* vel speculum apparens signatus eque remotus *C1*   29 supra: super *FP1*   30 videbitur . . . speculum (31) *om. FP1* / hoc etiam *transp. ER* / etiam lignum: in ligno *C1* / lignum *corr. ex* ligno *OL3* / apparenti: apparente *R*   31 accidit: accidet *R*   32 idem: ? *C1;* idemque *R* / penitus: punctus *S; corr. ex* punctus *OL3*   33 speculo: speculi *FSO*   34 *post* ergo *scr. et del.* quod *F* / per *om. P1* / per hoc *transp. F* / apparet *om. FP1* / in *inter. L3E* / *post* in *inter.* nisi *a. m. O*   35 a *corr. ex* in *O* / ad *corr. ex* a *O* / in *om. FP1*   36 *post* speculo *add.* linea *C1* / est[1] *om. FP1*   37 *post* et *inter.* superficiei *a. m. E; add.* superficiei *R*   39 poterit: potest *R* / in *om. O* / piramide *corr. ex* piramidis *O* / ortogonali *corr. ex* ortogonalis *L3*   40 ortogonaliter sit *transp. R* / sit *inter. SO; om. L3;* est *inter. a. m. C1* / sit adhibita *transp. E* / *post* enim *inter.* in *a. m. E*   41 alia *om. O* / *post* alia *add.* linea *C1* / quarum . . . basis *om. R* / *post* quarum *scr. et del.* ut *E* / *post* et *scr. et del.* et *O* / harum *alter. in* earum *O* / *post* piramidum *add.* basis eadem et *R*   42 acumina *mg. a. m. S* / *post* acumina *add.* ipsarum *R*

ad acumen ducatur linea recta, erit perpendicularis super basim, et ita super speculum, cum eadem sit superficies speculi et basis, quare conus piramidis in perpendiculari videbitur ab eo ad speculum ducta. Similiter a quocumque puncto piramidis ducatur linea ad punctum respiciens ipsum in apparenti piramide. Erit linea ortogonalis super basim et super speculi superficiem, quia ymago cuiuscumque puncti piramidis cadit in perpendiculari intellecta a puncto illo in speculi superficiem.

[2.5] Sed quicumque corporis punctus opponatur speculo plano est intelligere piramidem cuius punctus ille sit conus, que quidem piramis super basim ortogonalis, et etiam super speculi superficiem aut ei continuam. Et est intelligere aliam huic piramidi oppositam quarum basis eadem et super speculum ortogonalis, et perpendicularis a cono ad conum ortogonalis erit supra speculum, quare ymago cuiusque puncti speculo quidem oppositi cadit in perpendiculari a puncto ad speculi superficiem aut ei continuam. Sed planum quod in speculis non accedet comprehensio formarum nisi per lineas reflexionum, quare ymago puncti visi cadit in lineam reflexionis, et quelibet talis linea est recta, quare ymago cuiuscumque puncti cadit in punctum sectionis perpendicularis ab illo puncto in superficiem speculi et linee reflexionis. Et in speculis planis cadit, et unica est linea communis superficiei speculi et superficiei reflexionis cum linea contingente locum reflexionis, quare planum quod in speculis planis proprius ymaginis est locus punctus sectionis perpendicularis a puncto visi super lineam contingentem communem lineam superficiei speculi et superficiei reflexionis et linee reflexionis.

44 speculi *om. FP1SL3; inter. E*/quare: quoniam *P1*/conus (45): vertex *R* 45 in . . . videbitur: videbitur in perpendiculari *O*/ducta . . . quocumque (46) *rep. F* 46 *post* quocumque *scr. et del.* ducta *F; add.* ducta *P1*/piramidis: piramis *O*/*post* linea *add.* equidistans axi cadet *R* 47 ipsum . . . piramidis (49) *mg. a. m. E*/apparenti: apparente *R*/piramide . . . puncti (48) *mg. a. m. L3*/*post* piramide *add.* et *R*/*post* linea *add.* illa *OL3C1ER*/ortogonalis: perpendicularis *ER* 48 basim: basem *O*/quia: quare *OL3C1ER*/cuiuscumque: cuiusque *SE;* cuiuslibet *L3C1* 49 *ante* cadit *scr. et del.* maxime *E*/perpendiculari intellecta: perpendicularem intellectam *OR* 51 quicumque: quodcumque *R*/corporis: corpus *P1; om. R*/punctus: punctum *R*/opponatur: apponatur *SL3E* 52 *post* intelligere *add.* illam *C1*/punctus ille: punctum illud *R*/sit *om. R*/conus: vertex *R* 53 basim *corr. ex* bases *O*/*ante* ortogonalis *add.* est *O*/*post* ortogonalis *add.* est *R* 54 quarum: quare *E; corr. ex* quare *L3C1* 55 basis: basisis *F*/et[1] . . . ortogonalis *om. R*/ortogonalis: ortogonale *FP1;* ortogonalem *E*/cono: vertice *R* 56 conum: verticem *R*/ortogonalis *corr. ex* ortogonale *P1*/supra: super *R*/cuiusque: cuiuscumque *SL3ER*/cuiusque . . . ymago (60) *om. O* 57 quidem *om. R*/perpendiculari: perpendicularem *R*/*ante* a *add.* ductam *R* 58 *post* planum *add.* est *R*/speculis: speculo *FS* 59 accedet: accidet *S;* accidit *L3C1R* 61 *post* ymago *scr. et del.* puncti visi *S*/cuiuscumque puncti *transp. P1* 62 *post* perpendicularis *add.* ductae *R*/illo puncto *transp. P1*/in: ad *R* 63 cadit *scr. et del. OL3E; om. C1*/cadit . . . est *om. R*/*post* communis *scr. et del.* in *C1*/superficiei *corr. ex* superficie *C1* 64 *post* reflexionis[1] *add.* est una linea *R* 65 *post* speculis *scr. et del.* sperus *O*/planis *om. P1*/proprius: propriis *L3; om. R*/est locus *transp. R*/punctus: punctum *R* 66 visi: viso *L3C1ER* 67 et[2] . . . reflexionis (68) *inter. L3; mg. a. m. C1E*

[2.6] In speculis spericis et extra politis patebit quod diximus. Queratur superficies speculi talis magna in qua appareat forma baculi gracilis perpendiculariter erecta super ipsum. Apparebit quidem forma baculi baculo continua, et apparebit in forma baculi punctus signatus distans a superficie speculi secundum distantiam eius ab eodem in baculo. Et si fuerit baculus gracilior ex parte unius capitis quam ex parte alterius, apparebit quidem in hoc speculo forma eius piramidalis, et est error visus quem postea assignabimus.

[2.7] Amplius, fiat piramis cum eo ortogonalis super basim circularem circulatione perfecta, et applicetur etiam huic speculo. Videbitur quidem piramis huic continua super eandem basim erecta, sed minor ista. Quod appareat piramis planum per hoc quod omnes linee ab apparenti ymagine coni ad circulum basis videantur equales, et si declinetur piramis modicum supra speculum a situ in quo tota videtur, ut scilicet aliquid ex eo abscondatur, dum tamen locus reflexionis in speculo visui exponatur, apparebit inde ymago piramidis. Et si elongetur visus a speculo aut accedat dum tamen super lineam a loco reflexionis ad ipsum protractatum cedat, comprehendetur ymago piramidis, sed et accessus vel recessus secundum hanc lineam erit ut notetur locus reflexionis. Et a nota ad locum visus ducatur linea secundum quam processus fiat.

[2.8] Verum quoniam ymago piramidis ortogonalis super basim piramidis, et basis est circulus ex circulis in spera, erit linea a cono piramidis ad conum ymaginis ducta ortogonalis super circulum illum, et transibit per centrum eius. Et erit ortogonalis super speram et transibit per centrum spere, et erit ortogonalis super superficiem speram contingentem in puncto per quem transit hec linea. Et erit similiter

69 et *om. R* 70 *post* forma *scr. et del.* basis *P1* 71 erecta: erecti *R* 72 et *inter. a. m. E* / punctus signatus: punctum signatum *R* 74 si *om. L3E; inter. a. m. C1* / fuerit: fuit *L3* 75 *post* alterius *scr. et del.* sed non *E* 76 et *inter. a. m. C1* / est error: ex toto *O* / est error visus *corr. ex* ex toto visu *L3* 77 cum eo *om. R* / basim *corr. ex* basum *O* 78 *post* perfecta *add.* et concavum *O* / etiam *om. O* 79 basim: basem *OE* / erecta: recta *SC1; corr. ex* recta *O* 80 *post* quod[1] *add.* autem *R* / *post* planum *add.* est *R* / omnes *corr. ex* omnis *P1* 81 apparenti: apparentis *P1; alter. in* apparentis *C1*; apparente *R* / ymagine *alter. in* ymaginis *OC1* / coni *alter. ex* conu *in* cono *O*; verticis *R* / basis *corr. ex* bases *O* 83 eo: ea *P1R* 84 visui: visu *F* / *post* apparebit *add.* etiam *R* / piramidis *corr. ex* paramidis *O* 85 tamen *inter. E* / a loco (86) *om. P1* 86 reflexionis *om. R* / protractatum: protractum *S*; pertractatam *C1* / cedat: incedat *O*; cadat *ER* 87 et *om. FP1R* / erit: exit *P1S*; erat *C1* 88 notetur: vocetur *FP1SL3* 89 processus fiat *transp. OL3C1ER* 90 quoniam: quando *O* / *post* piramidis *inter.* est *O* / ortogonalis *corr. ex* ortogono *L3* / *post* ortogonalis *add.* est *P1R* 91 basis: basi *F* / *post* est *scr. et del.* ? *O* / in *om. O* / spera *alter. in* spere *O* / a: in *FSL3E; mg. a. m. C1* / cono: vertice *R* 92 conum: verticem *R* / ducta ortogonalis *transp. O* / ortogonalis: ortogonaliter *E* 93 et[2] . . . spere (94) *mg. S* / super . . . ortogonalis (94) *mg. a. m. L3* / speram . . . super (94) *mg. a. m. E* / et[3] . . . speram (94) *om. P1* 94 spere: eius *F; corr. ex* speres *C1* / superficiem *om. F* 95 quem: quod *R*

ortogonalis super lineam contingentem circulum spere per punctum illum transeuntem, et hec contingens est linea communis superficiei reflexionis et superficiei contingentis speram in puncto illo, et hec linea est contingens circulo spere communi superficiei spere et superficiei reflexionis. Linea ergo a cono piramidis ad conum ymaginis ducta est perpendicularis super lineam contingentem lineam communem superficiei reflexionis et superficiei speculi, que quidem est circulus.

[2.9] In hac igitur perpendiculari videtur ymago coni, et planum quod ymago coni est in linea reflexionis, quare comprehendetur ymago coni in concursu linee reflexionis et perpendicularis a cono ad speram ducte sive ad contingentem circulum communem superficiei spere et reflexionis. Sumpto autem quocumque puncto huic speculo opposito est intelligere piramidem super superficiem speculi ortogonalem aut super continuam, cuius conus sit punctus sumptus. Et linea ab illo puncto ad ymaginem puncti illius erit in superficie reflexionis et perpendicularis super superficiem speculi vel ei continuam modo predicto, quoniam punctus visus et ymago eius semper sunt in superficie reflexionis, quare et linea a puncto viso ad eius ymaginem ducta.

[2.10] In speculis columpnaribus exterius non apparent que in ligno et piramide diximus, quoniam recta in hiis speculis videtur non recta, et est error visus cuius postea causam assignabimus. Accidit tamen in solo corporis puncto videre locum ymaginis predictum.

[2.11] Verbi gratia, adhibito precedentis libri instrumento, immittatur regula cui sit infixum columpnare speculum ut media portionis speculi linea sit in superficie regule. Et non transeat hec regula tabulam eneam sed super ipsum cadat ortogonaliter ita quod altitudo regule sit super lineam dividentem triangulum tabule enee. Erectio facta in hac tabula,

97 illum: illud *R* / *post* hec *add.* linea *C1* 98 contingentis: contingenti *SOL3C1ER* / *ante* speram *scr. et del.* et *L3* 99 est contingens: contingit *R* / *post* contingens *add.* a *P1* / circulo: circulum *ER* / communi: communem *R* / spere et superficiei *om. FP1* 100 *post* reflexionis *scr. et del.* et *L3C1* / ergo *inter. a. m. E* / *ante* a *add.* erit *E* / cono: verticem *R* 102 quidem: quasi *L3* 103 in *inter. OL3; om. E* / perpendiculari *corr. ex* perpendicularis *L3* / coni: verticis *R* 104 ymago[1] *om. R* / coni: verticis *R* / est *rep. P1* / *post* quare *scr. et del.* consistit *P1* / comprehendetur: comprehenditur *OC1E* 105 coni: verticis *R* / cono: vertice *R* 106 et *inter. L3E* (*a. m. E*) 107 *ante* reflexionis *add.* superficiei *R* / autem: *scr. et del.* a *L3* 109 super *om. OL3E* / *post* continuam *scr. et del.* erit *O; add.* ei *R* / conus: vertex *R* / punctus sumptus: punctum sumptum *R* 110 puncti illius *transp. L3C1* 111 vel ei *corr. ex* in *a. m. S* / ei *om. O; inter. L3* / ei continuam *transp. C1* 112 punctus visus: punctum visum *R* / eius *om. R* / *post* semper *scr. et del.* habuerit fuit *P1* / *post* sunt *inter.* sicut *a. m. E; add.* simul *R* 113 et *inter. P1* / eius *om. O* 114 *post* ducta *add.* erit in superficie reflexionis *C1* 115 exterius: extremis *L3* / *post* exterius *add.* politis *C1R* 117 *post* visus *add.* non communis *FP1; add.* communis *alter. in* non communis *S; scr. et del.* communis *L3C1; add.* communis *ER; scr. et del.* communium *O; inter.* causam *O* / causam *om. O* 118 predictum *inter. a. m. E* 119 verbi gratia: hoc modo *R* 120 ut: in *F* / portionis: proportionis *FP1* 122 ipsum: ipsam *L3ER* / cadat: cadit *L3E; corr. ex* cadit *C1* / quod: ut *R* 123 erectio *alter. in* reflexio *L3E;* erectione *R* / *post* erectio *add.* vel reflexio *C1*

impleatur cera, et inducatur ei planities ut sit in eadem superficie cum tabula, et est ut certior fiat ortogonalis regule directio super tabulam.

[2.12] Deinde queratur regula acuta, et acuatur extremitas, et applicetur huius regule acuitas medie superficiei anuli linee. Et descendat super hanc lineam, et ubi ceciderit super regulam fiat signum. Postea acus descendat super hanc lineam in qua sit infixum modicum corpus album, et hoc in termino, nec descendat acus usque ad regulam.

[2.13] Adhibeatur autem visus ut sit in superficie regule, et claudatur unus visuum. Videbitur quidem ymago corporis super lineam a puncto signato ad acumen acus protractam, que quidem linea perpendicularis est super superficiem regule, que superficies tangit columpnam in linea longitudinis; et est perpendicularis super lineam longitudinis columpne, que est in superficie regule et est linea communis superficiei columpne et superficiei reflexionis. Et in superficie reflexionis sunt linea longitudinis et linea perpendicularis.

[2.14] Et si situs visus mutetur, et circa anuli superficiem visus volvatur, apparebunt sicut prius, et in eadem linea corpus et ymago corporis et acus. Et est linea illa perpendicularis super mediam longitudinis columpne lineam, et est hec perpendicularis in superficie reflexionis, quoniam superficies anuli secat columpnam super circulum equidistantem basi columpne, et in hac superficie est visus. Et nos probabimus postea quod, quando visus et visum corpus fuerint in superficie equidistanti basi columpne, illa est superficies reflexionis. In hoc autem situ linea communis superficiei columpne et superficiei reflexionis est circulus, et perpendicularis in qua videntur ymago et corpus ortogonaliter cadit super lineam super circulum contingentem.

[2.15] Hiis peractis, auferatur acus a loco suo, et ponatur regula acuta super lineam mediam ita quod cadat super mediam longitudinis regule lineam, et adhibeatur regula acuta superficiei anuli cera firmiter.

124 *post* ut *inter.* non *L3* 127 huius: huiusmodi *E* 128 super[1]: secundum *SOL3C1R*/hanc *inter. L3E* (*a. m. E*)/hanc lineam *transp. ER*/super[2]: secundum *L3C1* 129 super: secundum *F*; *scr. et del. C1*/super . . . lineam *om. R*/*post* lineam *add.* secundum *C1*/sit infixum *transp. R* 130 nec: ne *R*/*post* nec *add.* ut *L3C1E* (*inter. L3*)/*post* ad *scr. et del.* terminum *P1* 131 claudatur: clauda *FP1* 132 quidem: quid *O*; *corr. ex* quod *L3*; *alter. in* quod *E*/*ante* ymago *scr. et del.* h *O* 133 perpendicularis est (134) *transp. O* 134 est *inter. OL3C1* (*a. m. OC1*); *om. E* 135 et . . . longitudinis *om. S* 136 superficiei: superficie *FP1*; *corr. ex* superficie *O*/columpne *corr. ex* reflexionis *a. m. E*; regulae *R* 137 superficiei: superficie *FP1*/et[2] . . . sunt *mg. a. m. E* 139 si *om. FP1* 140 et[2]: ht *O* 141 *ante* acus *scr. et del.* est *E*/illa *inter. a. m. E* 142 est hec *transp. ER* 143 *post* super *scr. et del.* super *C1* 144 *post* nos *scr. et del.* cerpu *P1* 145 probabimus: probamus *FP1*/postea *om. C1*/*post* et *scr. et del.* et *C1* 146 equidistanti: equidistante *R*/est *corr. ex* et *O*/*post* reflexionis *scr. et del.* est circulus *E* 147 in . . . reflexionis (148) *mg. a. m. L3*/superficiei[1] . . . reflexionis (148): superficiei reflexionis et superficiei columpne *O* 148 *post* reflexionis *scr. et del.* et superficies reflexionis *C1*/videntur: videtur *R*/et[2] *inter. OC1E* 149 cadit: cadunt *R*/super: hunc *OER* 151 *post* lineam *add.* annuli *R*/quod: ut *R*/*post* quod *add.* non *FP1*/mediam[2] *corr. ex* lineam *a. m. E* 152 regula *inter. a. m. E*

Postea auferatur regula in qua est speculum, et accipiatur regula acuta, et applicetur eius acuitas medie longitudinis regule linee, et secundum processum acuitatis fiat cum incausto super speculum protractio. Post sumatur triangulus cereus modicus cuius unum latus sit equale altitudini regule in qua est speculum, et sit altitudo huius trianguli moderata, et superficies huius trianguli sit plane pro posse. Et adhibeatur columpne regule tabule ceree triangulus firmiter sub base regule, et latus eius equale altitudini regule ponatur super latus basis regule. Cum ita fuerit, erit huius trianguli altitudo super basem columpnalis equalem tabule regule, et ut efficiatur superficies plana ad modum superficiei regule, includatur triangulus inter regulam et superficiem planam, et comprimatur donec sit bene planitus. Et super superficiem huius trianguli ponatur regula acuta, et secetur finis huius trianguli cum acuitate regule, et erit finis eius linea recta. Et erit linea hec basis regule in qua est speculum.

[2.16] Postea ponatur regula super superficiem tabule que est in instrumento, et ponatur finis basis eius, que est in longitudine que est latus trianguli cerei, super lineam que est in longitudine eris, sicut factum est prius. Et erit superficies regule in qua est speculum ortogonalis super tabulam eneam, et hec superficies secat tabulam eneam super lineam que est in longitudine eris, et hec superficies tangit superficiem speculi super lineam que est in superficie speculi. Et hec superficies est superficies regule in qua est speculum, et erit angulus regule acute adherentis in media linea superficiei anuli in qua superficie erit speculum declinatum in partem in qua est caput trianguli, quia regula exaltavit unam partem eius cum corpore trianguli, et alia pars que est

153 acuta *corr. ex* in qua *L3* 154 medie: media *L3; corr. ex* media *O* / regule: recte *L3; corr. ex* recte *a. m. C1* 156 triangulus cereus modicus: triangulum cereum modicum *R* 157 est *inter. OC1* (*a. m. C1*); *om. SL3E* / altitudo: spissitudo *R* / trianguli: anguli *F* 158 sit: sint *OC1ER; alter. ex* situm *in* sint *L3* / pro *corr. ex* post *C1* 159 tabule ceree *om. R* / tabule . . . triangulus *inter. a. m. L3* / triangulus: triangulum *R* / *post* triangulus *add.* regule *O* / base: basi *R* 160 *post* regule[1] *add.* tabule enee triangulus regule firmiter sub base regule *L3; scr. et del. C1* / altitudini *om. FP1* / ponatur . . . regule (161) *mg. a. m. L3* 161 basem: basim *R* 162 columpnalis: columpnas *S;* columpne *R* / tabule *om. R* / tabule regule *mg. a. m. C1* 163 includatur: inclinatur *FP1* / triangulus: triangulum *R* / et *inter. O* 164 planitus: complanatum *R* / et[2] *inter. L3* / super *inter. a. m. OE* 165 *post* finis *scr. et del.* huius *F* 166 trianguli: anguli *O; corr. ex* anguli *L3; corr. ex* anguli *a. m. E* / finis *om. C1* / linea hec (167) *transp. O* 168 in *om. S* / in . . . est[1] (169) *om. L3* 169 basis eius *transp. ER* / in . . . est[2] *om. O* 170 *post* latus *scr. et del.* eius *O* / *post* super *scr. et del.* basim *P1* / *post* longitudine *scr. et del.* erit *F* / eris *om. R* / *ante* sicut *add.* tabule *R* 172 tabulam[1] *corr. ex* lineam *SL3C1* (*a. m. SC1*) / eneam[1]: eream *FSL3; alter. in* eream *P1* / et . . . eneam[2] *om. P1* / eneam[2]: eream *F* 173 longitudine: longitudinem *L3* / eris: eius *R* / tangit: tangitur *FP1* 174 hec superficies *om. R* 175 est superficies *om. FP1* / erit: est *C1* 176 *post* superficie *scr. et del.* et *OL3* 177 declinatum: declinatus *R* / *post* in[1] *scr. et del.* qua *F* / trianguli: triangulum *O* 178 *post* trianguli *add.* quia *FP1* / *post* que *scr. et del.* cum in partem in qua est capud trianguli quia regula exaltavit *O*

post caput trianguli est superficies eris, et erit linea que est in medietate speculi declinata.

[2.17] Et quando fuerit latus trianguli cerei super lineam que est in longitudine eris, movebitur regula in qua est speculum, et latus trianguli in hoc motu, si sit super lineam longitudinis eris. Et procedat vel retrocedat donec concurrat angulus regule acute cum puncto aliquo linee superficiei speculi donec firmetur regula acuta, et auferatur linea in speculo cum incausto facta. Et fiat punctus in superficie speculi in directo capitis regule acute. Et auferatur regula acuta, et apponatur acus, et sit acus super lineam mediam superficiei anuli, et adherere cogatur cum cera. Et erit linea intellectualis ab acu in punctum signatum in superficie speculi perpendicularis super superficiem regule que tangit superficiem speculi super punctum signatum et perpendiculariter super quamlibet lineam ab illo puncto protractam in superficiem contingentem speculum. Erit igitur perpendicularis super lineam rectam contingentem lineam communem superficiei alte anuli et superficiei.

[2.18] Ponatur autem visus in superficie anuli in capite eius et videbit in speculo donec comprehendat formam corporis parvi quod est in acu, et tunc percipiet corpus illud, et punctum in speculo signatum, et ymaginem illius corporis. Et linea transiens per corpus parvum et per punctum signatum in superficie est perpendicularis super superficiem contingentem speculi superficiem super punctum signatum. Et hec superficies anuli est ex superficiebus reflexionis, et corpus parvum et centra visus sunt in hac superficie, et punctus reflexionis est in hac superficie, et hoc deinceps probabimus. Et ymago corporis parvi in hoc situ erit super lineam rectam a corpore parvo protractam rectam super superficiem contingentem superficiem speculi, et cum hec linea

179 *post* superficies *add.* tabule enee *R* / eris *om. R* / *post* est *scr. et del.* ortogonalis *P1* / medietate (180): medietatum *O*   180 declinata *alter. in* declinati *a. m. E*   182 eris *om. R* / *ante* movebitur *add.* enee tabule *R* / *post* regula *scr. et del.* est *E*   183 eris *om. R* / *post* longitudinis *add.* tabule enee *R*   184 concurrat: occurat *C1*   186 punctus: punctum *R*   187 directo: directa *FP1* / *post* auferatur *scr. et del.* linea in speculo *S* / apponatur: ponatur *FP1*   189 cera: cerea *E; corr. ex* cere *O; corr. ex* cerea *L3C1* / et *om. R*   190 super *inter. a. m. E* / regule . . . superficiem (191) *rep. C1* / *post* tangit *scr. et del.* superficiem regule que tangit *E*   191 *post* superficiem *scr. et del.* super *O* / super *inter. O*; ad *L3* / *post* punctum *add.* illud *O* / *post* signatum *scr. et del.* in superficie speculi perpendiculari super *O* / et *inter. O* / perpendiculariter: perpendicularis *OC1ER*   192 quamlibet *corr. ex* libet *a. m. C1* / protractam: pertractam *S*   193 contingentem *corr. ex* contingentis *O*; contingentis *L3E*; contingente *C1*   194 alte *corr. ex* ate *F; corr. ex* arte *L3C1; corr. ex* aree *a. m. E* / *post* et *scr. et del.* in *S*   195 *post* superficiei *add.* speculi *R*   196 autem *corr. ex* atem *S* / *post* autem *scr. et del.* ei *C1* / videbit *corr. ex* vibit *F*   197 in acu *corr. ex* natu *C1* / acu: actu *FP1E; corr. ex* actu *O*   198 corpus *corr. ex* corporis *C1* / illud *corr. ex* illum *F*   200 signatum *inter. E* / signatum in superficie: in superficie signatum *R* / *post* superficie *scr. et del.* e et *O; scr. et del.* est et *L3; add.* reflexionis et *C1; add.* et *E*   201 speculi superficiem *transp. O* / *post* hec *scr. et del.* in *E*   203 centra: centrum *C1ER* / et . . . superficie (204) *om. FP1*   204 hoc: hec *R* / probabimus *corr. ex* probamus *C1*   205 protractam: pertractam *S* / rectam² *scr. et del. E; om. R*   206 cum: est *R* / hec: hoc *FL3; inter. a. m. E* / hec linea *transp. E*

perpendicularis super lineam rectam contingentem lineam communem superficiei speculi et superficiei reflexionis que est superficies anuli, et superficies reflexionis est ex superficiebus declinantibus secantibus columpnam inter lineas longitudinis columpne et circulos eius equidistantes basibus, quia regula et speculum quod est in ea sunt declinata, linea ergo communis huic superficiei et superficiei speculi est ex sectionibus columpnaribus. Et ita explanabimus locum ymaginis si mutetur situs regule in qua est speculum et declinetur super superficiem eius alia declinatione minori vel maiori.

[2.19] Palam ergo ex hiis quod ymago percipitur ubi perpendicularis a viso puncto ad speculi superficiem ducta concurrit cum linea reflexionis, et hic est situs predictus. Si a puncto viso ad speculi superficiem ducantur linee ad speculi superficiem, que perpendicularis est minor qualibet alia, quoniam quelibet alia prius secat lineam communem superficiei contingenti speculum in qua ortogonaliter cadit perpendicularis et huic superficiei reflexionis quam veniat ad speculum, et quelibet linea a puncto viso in hac superficie ad hanc lineam communem ducta est maior perpendiculari, quia maiorem respicit angulum, quare propositum.

[2.20] Eadem poterit adhibi comparatio in speculo piramidali exteriori, et idem patebit sive sint ymagines rerum visarum in sectionibus piramidalibus, sive in eis que fuerint secundum lineas longitudinis.

[2.21] In speculis spericis concavis comprehenduntur ymagines quedam ultra speculum, quedam in superficie, quedam citra superficiem, et harum quedam comprehenduntur in veritate, quedam preter veritatem.

[2.22] Omnes quarum comprehenditur veritas apparent in loco sectionis perpendicularis et linee reflexionis, quod sic patebit. Fiat

207 *post* perpendicularis *add.* sit *C1* 208 que . . . reflexionis (209) *om. FP1; inter. a. m. E* 209 declinantibus: declinatis *O; alter. in* declinatis *L3* 210 circulos: circulus *O* 212 et superficiei *om. FP1* 213 est: erit *FP1* 214 si *corr. ex* ut *a. m. E;* ut *R* / super *inter. a. m. E* 215 alia: aliqua *R* / minori vel maiori: maiore vel minore *R* 216 *post* quod *scr. et del.* h *O* 217 concurrit *corr. ex* concurrat *P1* 218 est *om. C1* / predictus: productus *E* / *post* puncto *add.* a *FP1* / viso *om. C1* 219 ad . . . superficiem: a . . . superficiem *scr. et del. O; om. R* 220 est² *om. L3; inter. E* / est minor *transp. R* / qualibet *corr. ex* quamlibet *O* / *post* quoniam *scr. et del.* quem *P1* / *post* secat *scr. et del.* o *O* / lineam *om. FP1SL3E; corr. ex* lineis *O* / lineam communem (221) *transp. C1R* 221 *post* superficiei *inter.* linea *a. m. E* / contingenti: contingentis *R* / in qua: in quam *inter. a. m. F* / qua: quam *R* 222 huic *om. R* / *post* reflexionis *inter.* autem *a. m. E* / quam: antequam *R* 223 et *inter. O* / a *inter. OL3; om. E* / *post* hanc *scr. et del.* superficiem *L3* 224 maiorem: maiore *O* 225 *post* angulum *add.* scilicet rectum *O* / quare *alter. in* quod erat *O* / *post* quare *add.* patet *R* 226 eadem . . . longitudinis (228) *transp. ad* 218 *post* predictus *R* / adhibi: adhiberi *OL3C1ER* / comparatio: operatio *E* / piramidali: piramidis *P1SL3C1E* / exteriori (227): exteriore *R* 227 *post* sint *scr. et del.* exteriores *P1* / *post* visarum *add.* sive *FP1SL3; scr. et del.* sive sint *E* 228 eis: iis *R* / fuerint: fiunt *R* 229 spericis *corr. ex* speciricis *F* 230 citra: circa *FP1C1; corr. ex* circa *L3; corr. ex* cera *O* 231 *post* veritate *add.* et *O* 233 comprehenditur *corr. ex* comprehenduntur *L3* / apparent *corr. ex* appareant *F*

piramis, et ea ortogonalis super basem, et dyameter basis sit minor medietate dyametri spere, et linea longitudinis piramidis sit maior eadem semidyametro. Et secetur ex parte basis ad quantitatem eius, et fiat super sectionem circulus, et secetur piramis super hunc circulum. Postea in medio speculi fiat circulus ad quantitatem basis piramidis remanentis, et aptetur huic circulo piramis, et firmetur cum cera.

[2.23] Deinde statuatur visus in situ in quo ymaginem piramidis possit comprehendere, et adhibeatur lux ut certior fiat comprehensio. Non videbis quidem piramidem huic coniunctam, sed comprehendes hanc ultra speculum extensam, unde apparebit piramis quedam continua cuius basis ultra speculum et pars eius piramis cerea. Et si in hac piramide signetur linea longitudinis cum incausto, videbitur hec linea protendi super superficiem piramidis apparentis, et quoniam conus piramidis est centrum spere, linea a cono secundum longitudinem piramidis ducta erit perpendicularis super contingentem cuiuslibet circuli spere per caput linee transeuntis.

[2.24] Quare quelibet linea longitudinis piramidis apparentis est perpendicularis super lineam contingentem lineam communem superficiei reflexionis et superficiei spere, que quidem linea est communis, et est circulus. Et quilibet punctus piramidis in hac videtur perpendiculari, et quelibet perpendicularis est in superficie reflexionis, quoniam punctus visus et ymago eius sunt in perpendiculari et in hac superficie reflexionis. Et omnis ymago comprehenditur in linea reflexionis, quare ymago cuiuscumque puncti piramidis erit in puncto sectionis perpendicularis et linee reflexionis.

[2.25] Puncta autem quorum ymagines citra speculum comprehenduntur, hoc est inter visum et speculum sunt, cum a quolibet eorum linea ducta ad centrum speculi secet latitudinem vie visum et speculum interiacentis. Et ut videatur hoc, auferatur piramis a medio speculi,

235 et ea: cum *OL3;* cerea *mg. a. m. C1; corr. ex* cum *SE*/ea: eius *R*/*ante* ortogonalis *add.* axis sit *R*/basem: basim *R*/sit *inter. a. m. F* 236 medietate *corr. ex* mediante *O*/eadem (237) *inter. L3* 237 *post* eius *add.* scilicet semidyametri *R* 238 et . . . circulus (239) *mg. a. m. E*/piramis: piramidis *O* 242 ut *corr. ex* et *a. m. E* 243 quidem: quod *FP1* 244 unde: inde *FP1* 245 *post* speculum *add.* est *R*/*post* eius *add.* est *C1*/cerea: cera *FP1* 246 incausto *corr. ex* tantum *a. m. C1* 247 *post* protendi *add.* secundum *L3*/super: secundum *C1*/quoniam: quia *S*/conus: vertex *R* 248 piramidis *corr. ex* piramis *P1*/centrum: centra *O*/cono: vertice *R*/longitudinem *corr. ex* longitudinis *O* 249 *post* super *add.* lineam *R*/cuiuslibet circuli (250): quemlibet circulum *R* 250 *post* caput *add.* cuiuslibet *C1*/transeuntis: transeuntem *R* 253 reflexionis et superficiei *rep. F*/est communis (254) *transp. R* 254 et est *om. R*/quilibet punctus: quodlibet punctum *R*/piramidis *om. P1* 255 perpendiculari: perpendicularis *FC1;* ? *OL3*/perpendiculari et quelibet *om. P1*/et . . . perpendicularis *mg. a. m. C1*/*post* reflexionis *add.* que quelibet perpendicularis *C1* 256 *post* quoniam *scr. et del.* a *S*/punctus visus: punctum visum *R* 257 reflexionis *inter. a. m. S; om. OL3C1ER*/*post* ymago *scr. et del.* non *O* 260 citra: circa *S; corr. ex* cita *O* 261 cum *inter. E*/eorum *inter. L3;* horum *C1* 262 secet: secat *C1R*/vie *inter. O*/*post* vie *add.* inter *OR* 263 videatur *corr. ex* videtur *O*/speculi *corr. ex* speculo *E*

et collocetur in parte. Erit conus centrum speculi, et remotio visus a speculi superficie sit maior semidyametro spere. Deinde sumatur lignum gracile album, et statuatur in speculo ut sit centrum speculi directe medium inter caput ligni et centrum visus, et dirigatur intuitus in punctum speculi a quo linea ad conum piramidis ducta sit inter caput ligni et visum. Et inspiciatur speculum donec non appareat caput ligni et lignum, et apparebit forma capitis ligni citra speculum et propinquior visui cono piramidis. Et erunt in eadem linea recta conus piramidis, et caput ligni, et ymago capitis, et hec linea est perpendicularis super lineam contingentem lineam communem superficiei speculi et superficiei reflexionis. Quoniam superficies reflexionis transit per centrum et punctum visus, et linea transiens per hec duo puncta est in superficie reflexionis, et linea communis est circulus. Et hec linea huic circulo erit dyametrum, quoniam centrum illius circuli est centrum spere, quare erit hec linea perpendicularis super lineam contingentem circulum in capite huius linee, et hec linea transit per punctum visum et per eius ymaginem. Et ita quodlibet punctum citra speculum visum comprehenditur in eadem linea cum centro et cum ymagine eius, et quodlibet punctum videtur in linea reflexionis, quare in loco sectionis perpendicularis et linee reflexionis.

[2.26] Et ea quorum veritas comprehenditur in hiis speculis sunt quorum ymagines apparent ultra speculum vel citra superficiem eius, et preter hec nulla sunt que in hoc speculo in veritate comprehendat visus, ipsa enim prohibent ymagines suas veras apparere. Ymagines que apparent in superficie speculi huius sunt ex ultima partitione, et hoc explanabimus, cum aderit sermo in erroribus visus. Quodlibet ergo punctum in veritate in hoc speculo comprehensum apparet in concursu perpendicularis et linee reflexionis, que quidem perpendicularis transit a puncto viso ad centrum spere et cadit ortogonaliter in contingentem lineam communem.

264 *post* parte *inter.* et *O* / conus: vertex *R* / a . . . superficie (265) *om. R* 265 speculi superficie *transp. L3C1* / semidyametro *corr. ex* semidiametri *P1* 268 conum: verticem *R* / *post* inter *scr. et del.* sit *F* 269 et[1] . . . ligni *om. FP1R; inter. S* / non *om. L3C1E* 270 lignum *alter. in* visum *a. m. E;* visum *R* 271 cono: vertice *R* / conus: vertex *R* 272 *ante* ymago *scr. et del.* conus piramidis *P1* 274 centrum: centra *O* / *post* et *scr. et del.* fluxum *C1* / punctum visus (275) *corr. ex* visus punctum *C1* 275 *post* transiens *add.* et *P1* 276 huic . . . erit: erit huic circulo *C1* / erit *om. L3* / *post* erit *scr. et del.* transiens *O* 277 dyametrum: diameter *R* 278 circulum *inter. a. m. E* 279 per *om. R* 280 citra: circa *FP1L3;* ? *E* / visum: visus *FP1* 281 cum *om. FP1* 282 *post* punctum *scr. et del.* et quilibet punctum *O* / in[2] . . . reflexionis (283) *om. L3; mg. a. m. C1* 284 *post* quorum *add.* est *FP1* / comprehenditur . . . speculis: in hiis speculis comprehenditur *R* 286 hoc *inter. a. m. L3* 288 speculi huius *transp. OL3C1ER* / partitione: apparitione *C1* 289 hoc: hec *R* / aderit: adherit *FP1O;* erit *R* / in: de *R* / ergo: vero *FP1* 291 perpendicularis: perpendiculariter *FP1S; corr. ex* perpendiculariter *L3* 292 viso *corr. ex* suo *a. m. C1*

[2.27] In speculis columpnaribus concavis diversificatur ymago, aliquando enim erit locus eius in superficie speculi, aliquando ultra, aliquando citra. Et in hiis omnibus aliquando in veritate comprehenditur, aliquando non.

[2.28] Cum volueris in hiis locum ymaginis percipere, facias sicut fecisti in columpnis exterioribus. Adhibeatur enim regula in qua sit columpna concava sicut adhibita est superius, et acus similiter, et corpus modicum in summitate acus. Et ponatur visus oppositus in medio circuli et in medio superficiei anuli, et sublevetur visus modicum a superficie anuli, et inspiciat donec ymaginem corporis videat et comprehendat formam corporis, et corpus, et punctum in speculo signatum in eadem linea perpendiculari super superficiem speculi—et hoc per sillogismum sensualem. Et erit ymago ultra speculum, et erit reflexio ex puncto linee que est in medio speculi.

[2.29] Deinde statuatur visus in superficie anuli, sed extra medium, donec videat ymaginem corporis parvi. Videbis quidem eam citra speculum, et videbis corpus, et eius ymaginem, et punctum in speculo signatum in una linea recta perpendiculari super lineam rectam contingentem circulum equidistantem basi speculi super punctum signatum in superficie speculi. Et superficies huius circuli est superficies reflexionis in hoc situ, et est superficies faciei anuli, et punctus reflexionis est punctus illius circuli.

[2.30] Postea adhibeatur cum manu alia acus in cuius summitate sit corpus modicum, et statuatur in superficiem et axem hoc modo ut hoc corpus et punctus signatus sint in eadem linea secundum sensualem sillogismum. Et sit visus in superficie anuli inter caput eius et medium. Videbit quidem ymaginem corporis, et videbit hanc ymaginem, et corpus eius, et punctus signatum in superficie speculi in eadem linea recta.

294 speculis *corr. ex* speculo *P1* / diversificatur: diversatur *FP1*; diversicatur *S* 296 aliquando: aliqua *O* / aliquando citra *mg. a. m. C1*; *om. R* / comprehenditur (297): comprehendetur *E* 297 non *corr. ex* et *a. m. C1* 298 locum ymaginis *transp. E* 299 fecisti: fecistis *C1* / columpnis: columpnaribus *R* 1 summitate: sumpmitate *S*; *corr. ex* sumpmitate *F*; *alter. ex* veritate *in* acuitate *O* 2 sublevetur . . . et[1] (3) *mg. a. m. E* 4 *ante* corpus *scr. et del.* p *L3* / in speculo *rep. E* 5 perpendiculari: perpendicularis *O* 7 *post* linee *add.* recte *SOL3C1ER* 9 videat *corr. ex* videa *F* / videbis: videbit *OL3C1ER* / citra: circa *FP1*; *corr. ex* circa *L3* 10 videbis: videbit *OL3C1ER* 11 una: eadem *O* 12 basi *corr. ex* basis *O* 13 superficie speculi *R* / circuli *om. FP1SR* / est . . . situ (14) *rep. C1* 14 *ante* reflexionis *scr. et del.* faciei anuli et punctus *S* / in . . . reflexionis (15) *rep. C1*; faciei[1,2] *om. C1* / faciei: facie *FP1* / punctus: punctum *R* 15 punctus: punctum *R* 16 in *inter. S* / summitate: sumpmitate *S* 17 superficiem: superficie *FP1* / hoc *om. C1R* 18 punctus signatus: punctum signatum *R* / *post* signatus *add.* in speculo *C1* 19 *post* sillogismum *scr. et del.* sit *F* / et[1] *mg. F* 20 *post* ymaginem[2] *scr. et del.* corporis *F* 21 et[1] . . . superficie *inter. a. m. E* / et[2]: in *C1* / punctus: punctum *C1ER*

[2.31] Si autem declinetur linea recta cum triangulo parvo quem fecimus—et visus sit in medio anuli—videbis ymaginem citra speculum, sed in eadem linea recta cum corpore et puncto signato. Et hec reflexio erit ex sectionibus columpnaribus, quoniam speculum est declinatum, et scimus quod non percipitur ymago nisi in linea reflexionis. Palam ergo quod locus ymaginis est ubi secat perpendicularis predictam lineam reflexionis, cum comprehenditur veritas, et licet non comprehendatur certitudo ymaginis, tamen erit modus harum ymaginum cum veritatis ymaginibus.

[2.32] Pari modo videre poteris in piramidibus concavis in concursu perpendicularis cum lineis reflexionis. Palam ergo quod in omnibus speculis comprehenduntur ymagines in loco predicto, qui quidem locus similiter dicitur ymago.

[2.33] Quare autem comprehendantur res vise per reflexionem in locis ymaginis et quare ymago sit super perpendicularem a re visa in speculi superficiem declarabimus causam. Visus, cum adquirit formam per reflexionem, adquirit eam statim sine certitudine, et adquirit longitudinem per estimationem. Et hanc longitudinem comprehendet forsitan in veritate per diligentiam intuitus adhibitam, forsitan non. Et istud explanavimus in libro secundo, et ibi dictum est quod visus adquirit longitudinem per sillogismum ex magnitudine corporis et angulo aliquo sub quo comprehenditur magnitudo. Et adquisitio rei site ignote manifeste est in hunc modum. Res etiam note comprehenduntur in hunc modum, conferuntur enim rebus cognitis et magnitudinibus vel longitudinibus notis. Cum visus comprehendit rem aliquam per reflexionem, non comprehendit longitudinem ymaginis nisi per estimationem; deinde adhibita diligentia, adquirit longitudinem, et verificat per sillogismum ex magnitudine rei vise et angulo piramidis super quam forma reflectitur ad visum.

23 linea *om. P1* / linea recta *transp. F* / quem: quod *R* 24 visus *om. FP1* / videbis: videbit *OL3C1ER* / videbis ymaginem *transp. C1* 25 sed in: eorum *C1* 26 sectionibus columpnaribus *transp. R* / columpnaribus *om. FP1SOE* 27 quod: quoniam *C1* 28 secat *corr. ex* seca *S* 30 *post* tamen *add.* fueris *O* / erit *corr. ex* eirit *L3* / modus: modum *O*; *corr. ex* modum *L3* 31 veritatis *corr. ex* veritas *O* / ymaginibus *om. O* 32 *post* poteris *add.* imaginem *R* / piramidibus: piramidalibus *R* / *post* concursu *scr. et del.* piramis *P1* 33 lineis: linea *R* 34 comprehenduntur: comprehendantur *P1S*; *alter. ex* comprehendandantur *in* comprehendantur *F* 35 ymago: imaginis locus *R* 37 ymaginis: imaginum *R* 38 *post* visus *scr. et del.* autem *F* / cum *corr. ex* autem *a. m. L3* 40 et *corr. ex* ad *O* / comprehendet: comprehendit *FP1* 42 explanavimus: explanabimus *FP1S*; *corr. ex* explanabimus *O* 43 longitudinem *inter. L3* / ex *alter. in* et *a. m. E* 45 site *alter. in* vise *a. m. L3*; vise *R* / ignote: ceperant *O*; note *R* / res . . . modum (46) *om. P1* / note: ignote *R* 46 conferuntur *corr. ex* confirmitur *O* / *post* enim *add.* in *O*; *scr. et del.* in *L3E* 47 visus *corr. ex* visum *O* / comprehendit: comprehendat *FP1*; *corr. ex* comprehendet *E* 48 comprehendit: comprehendet *E* / ymaginis . . . longitudinem (50) *rep. F* 50 *post* longitudinem *scr. et del.* ymaginis nisi per estimationem *P1* / verificat *corr. ex* verficat *S* 51 quam forma *corr. ex* quem forman *F*

[2.34] Cum ergo res visa ex rebus notis fuerit, visus adquirit eius longitudinem per iam notam longitudinem equalem angulum huic tenentem et huic longitudini similem. Similiter res visa, cum fuerit ignota, conferetur magnitudo magnitudinis eius alii magnitudini rerum notarum, et adquiritur longitudo huius ymaginis per sillogismum mensure anguli quem tenet ymago in centro visus in hora reflexionis. Et locus in quo est forma rei vise comprehensus per reflexionem, forma ab eo directe veniens ad angulum circa oculum, accedet super piramidem ipsam per quam forma reflectitur ad visum, et eadem piramis occupabit totam formam que fuerit in loco ymaginis. Visus ergo, cum adquirit rem visam per reflexionem, adquirit eam in loco ymaginis, quoniam forma comprehensa in loco ymaginis per reflexionem, quare similis est forme directe comprehense, occupare ab illa piramide, et hec est causa quare comprehendatur in loco ymaginis.

[2.35] Quare autem comprehendetur ymago in perpendicularem dicemus. Scimus quod punctum visui perceptibile non est intellectuale sed sensuale, et forma eius sensualis. Dico igitur in speculis planis quod ymago, cum non apparet in superficie speculi sed ultra, competentius est et rationabilius quod appareat super perpendicularem quam extra eam. Cum enim in loco perpendicularis assignata, fuerit distantia eius a puncto reflexionis speculi—que scilicet est pars linee reflexionis a loco ymaginis ad punctum reflexionis ducte—est equalis distantie puncti visi a puncto reflexionis. Quoniam superficies speculi est ortogonalis super perpendicularem, unde linea a puncto reflexionis ad perpendicularem ducta est latus duobus triangulis commune, et

53 iam . . . longitudinem: longitudinem iam notam *C1* / equalem angulum *transp. R* 54 longitudini: longum *FP1S*; longitudinem *L3C1* / similem *corr. ex* sillogismum *O* / visa: nota *FP1* 55 conferetur *corr. ex* confirmetur *O*; confertur *R* / magnitudinis *om. R* / magnitudinis . . . alii *rep. FP1* 56 *ante* notarum *add.* visarum *R* / adquiritur: adquiretur *O* / huius: eius *R* 57 in[1] *om. FP1* 58 locus: a loco *R* / comprehensus: comprehensa *R* 59 ab eo *om. R* / circa: citra *OE*; *alter. in* citra *L3*; *alter. ex* citra *in* intra *a. m. C1* / *ante* oculum *add.* circulum *O*; *inter.* ad *O* / accedet: accidet *O*; *corr. ex* accidet *E*; accedit *R* / super: supra *FP1*; *corr. ex* semper *O* 60 reflectitur *corr. ex* reflectetur *O* 61 piramis *corr. ex* piramidis *O* / totam *corr. ex* notam *O* 62 *post* reflexionem *scr. et del.* rem *O* 63 *post* ymaginis[1] *scr. et del.* visus ergo cum acquirit rem visam *O* / quoniam . . . ymaginis[2] *rep. P1* / *post* comprehensa *add.* est *ER* 64 occupare: occupatae *R* / illa *corr. ex* alia *O* 65 quare *inter. L3C1E* (*a. m. C1E*) 66 comprehendetur: comprehendatur *R* / perpendicularem: perpendiculari *R* 68 sensuale: densuale *L3* 69 ymago cum *transp. R* / cum non *inter. OL3* / non *om. FP1S*; *inter. a. m. E* / apparet: apparent *R* / *post* sed *scr. et del.* una nec *L3*; *scr. et del.* non citra nec *E* 70 *post* competentius *scr. et del.* enim *L3*; *add.* enim *E* / rationabilius: rationalius *L3E* / quod: ut *R* / super: supra *R* 71 enim: est *S* / assignata: assignato *OL3*; assignatio *C1* 73 est: erit *R* 74 quoniam: quia *ER* / quoniam . . . reflexionis (75) *om. L3* / *post* quoniam *add.* enim *R* 75 unde: et *R* 76 *post* perpendicularem *add.* in superficie speculi *C1* / ducta . . . perpendicularem (77) *om. S* / *post* latus *scr. et del.* i *F* / duobus triangulis: duabus angulis *FP1* / *post* et *add.* angulus lineae accessus est equalis angulo reflexionis quare duo anguli unius trianguli sunt equales duobus angulis alterius trianguli et unum latus commune est quare reliqua latera equalia sunt reliquis lateribus *R*

dividet perpendicularem per duo equalia. Quare duo latera unius trianguli erunt equalia duobus lateribus alterius, et angulus angulo, quia uterque est rectus, quare basis basi.

[2.36] Si ergo ymago in perpendiculari apparuerit, equaliter a puncto et a visu distabit cum corpore a quo procedit, et erit ymagini idem situs respectu puncti reflexionis qui est puncto viso respectu eiusdem, et idem est situs respectu visus, unde in hoc situ apparebit veritas et puncti visi et ymaginis. Si vero ymago fuerit extra perpendicularem, cum sit necesse eam in linea reflexionis esse, aut erit ultra perpendicularem aut citra respectu visus. Si fuerit ultra, erit quidem remotior a puncto reflexionis et a visu quam punctus visus, unde tenebit minorem angulum in oculo quam punctus. Et minorem occupat visus partem, unde, cum sit equalis, videbitur minor eo. Si autem fuerit citra perpendicularem, videbitur maior cum sit propinquior.

[2.37] In speculo sperico exteriori videtur ymago super perpendicularem, aut enim videtur ymago centri visus, aut alterius puncti. Si ymago centri, dico quod dignior est perpendicularis ab oculo ad centrum spere ducta, ut super eam appareat ymago centri quam alia, si enim forma directe procedat secundum hanc perpendicularem usque ad centrum spere, eundem semper servabit situm respectu visus, et ita cuicumque puncto spere opponatur forma, perpendicularis ad centrum mota ydemptitatem situs tenebit respectu visus. Et idem situs erit forme in una perpendiculari qui et in alia, quoniam centrum spere eundem habet situm respectu cuiuslibet puncti spere, et omnes huiusmodi perpendiculares eiusdem sunt situs.

[2.38] Si autem extra perpendicularem ymago moveatur ad quodcumque punctum spere, mutabitur eius situs respectu visus, quoniam alium habebit situm extra perpendicularem quam in perpendiculari, et extra speculum movetur perpendicularis et non intra. Et si extra

77 dividet: dividit *L3C1E* / dividet . . . basi (79) *om. R* 78 *post* trianguli *scr. et del.* e *S* / erunt *corr. ex* sunt *a. m. E* / et *inter. E* 80 puncto: speculo *R* 81 et[1] *om. FP1S; scr. et del. C1* / et a visu *om. R* / *post* idem *add.* est *S* 82 puncti . . . respectu (83) *om. S* / *post* puncti *add.* respectu *FP1OE; scr. et del.* respectu *L3* / *post* est *add.* in *ER* / *post* respectu[2] *add.* puncti *R* / et *mg. F* 83 idem: eidem *FP1* (*mg. F*) / est *inter. OE;* eis *C1* / *post* est *add.* eis *O* 84 visi *om. S* / sit: fuerit *ER* 85 esse: inesse *C1; corr. ex* inesse *L3* 86 aut *corr. ex* et *S* / si *corr. ex* aliquid *O* 87 reflexionis *om. FP1S* / punctus visus: punctum visum *R* / unde: unum *S* 88 punctus: punctum visum *R* / *post* punctus *scr. et del.* visus unum *S* / occupat: occupabit *R* 89 unde: unum *S* / citra: circa *FP1S* 91 exteriori: extra polite *R* 93 *post* centri *add.* visus *R* / dignior *corr. ex* digniorem *O* / ad *om. E* 94 ducta . . . spere (96) *om. FP1* / alia: aliud *E* / si enim (95): sicut *SE;* ? *C1* 95 *post* perpendicularem *scr. et del.* ai *S* 97 centrum: centra *FP1SO; corr. ex* centra *L3* 98 mota: noto *FP1;* nota *SO* / situs erit *transp. R* 99 qui: que *R* 100 huiusmodi: huius *O* 101 *post* perpendiculares *add.* puncti qui respectu spere *O* / eiusdem: eius *O* 102 ymago *inter. O* 103 eius situs *transp. ER* (situs *inter. a. m. E*) 104 alium: aliquando *FP1SO* / quam in perpendiculari *mg. a. m. C1* 105 movetur *alter. in* moverit *a. m. E;* movebitur *R* / non intra et *om. FP1S* / et si *inter. L3* / si *inter. a. m. E* / *post* si *add.* intra *S*

perpendicularem appareat, non servabit situm, et convenientius fuit ut servaret ymago situm quam ut mutaret, ut visus rem visam certius comprehenderet. Ob hoc ymago centri super perpendicularem apparet, et huic ymagini non possumus certum in perpendiculari assignare punctum, quoniam non invenitur dignitas in uno perpendicularis puncto plus quam in alio, ut hec ymago determinate appareat in eo. Sed scimus quod in quocumque huius perpendicularis puncto appareat semper apparet continua cum apparenti oculo, et semper in totali forma apparenti eundem tenet locum et situm.

[2.39] Cuiuscumque puncti ymago preter centrum visus ad speculum accedat, movetur declinate, quare non durat ei similitudo situs respectu puncti visus, et perpendicularis a puncto viso ad speculum ducta cadat supra centrum spere in qua quidem perpendiculari observaret ymago similitudinem situs. Non est ergo punctus in quo comprehensa ymago servet similitudinem situs nisi in perpendiculari illa, et cum oportet ipsam comprehendi in linea reflexionis, comprehendetur in concursu linee huius cum hac perpendiculari. Iam ergo assignavimus causam huius rei, verum rerum naturalium status respicit statum suorum principiorum, et principia rerum naturalium sunt occulta.

[2.40] Idem erit modus probationis in speculo sperico concavo; similiter in piramidali concavo vel exteriori. Et universaliter erit locus ymaginis in perpendiculari in quocumque speculo, quoniam non est locus extra perpendicularem in quo forma observet similitudinem et situs ydemptitatem.

[2.41] Hiis explanatis, restat demonstrative declarare locum ymaginis in qualibet speculorum specie.

106 perpendicularem: speculum *R*/appareat: apparea *S*/situm . . . servaret (107) *om. FP1S*/et *om. O* 107 ymago *corr. ex* ymabigo *F*/ymago situm *transp. R* 108 *post* centri *add.* visus *R* 109 in . . . assignare: assignare in perpendiculari *R*/assignare: assignate *FS* 110 quoniam: quod *O*/non *om. O*/*ante* invenitur *inter.* scilicet *O*/*post* puncto *scr. et del.* post *S* 111 plus: maior *R*/determinate: determinata *FP1*; determinatum *S*/appareat *corr. ex* apparet *E*; appateat *R* 112 huius . . . puncto: puncto huius perpendicularis *R*/semper: sed *FP1* 113 apparet *corr. ex* appareat *O*/apparenti: apparente *R*/oculo: oculi *O*; *alter. in* oculi *L3*/apparenti (114): apparente *R* 115 *post* cuiuscumque *add.* vero *R* 116 accedat *corr. ex* concedat *L3*; accedit *R*/movetur: cogetur *O*/declinate: declinare *P1SO*; *alter. in* declinare *E*/durat: ducat *SL3*; esset *O*/ei *corr. ex* ai *S* 117 puncti *om. R*/speculum *corr. ex* punctum *O* 118 cadat: cadet *OL3C1*; *alter. in* cadet *E*; cadit *R*/supra: super *ER* 119 observaret: observat *R*/*post* ymago *scr. et del.* et *L3*/similitudinem *corr. ex* similitudo *F*/est *om. O*/punctus: punctum *R* 120 *post* comprehensa *add.* est *E* 121 oportet: oporteat *R* 122 linee huius *transp. ER* 123 rerum *corr. ex* tunc *O* 124 statum: statim *S*; situs *ER* 126 erit *corr. ex* rerum *O*/*post* concavo *add.* et *O* 127 piramidali: piramide *FP1SL3*; *corr. ex* piramide *a. m. C1*/exteriori: extra polito *R*/universaliter: vel *FP1*/erit: dicitur *O* 128 in . . . perpendicularem (129) *om. S* 129 et situs (130) *transp. R* 131 explanatis *corr. ex* explatis *F*

[2.42] Dicimus quod quodcumque vel quodlibet punctum comprehensum a visu in speculo plano, quando egressus est a perpendiculari que a centro visus cadit in superficiem speculi plani, quod linea per quam reflectitur forma illius puncti ad visum concurret cum perpendiculari producta ab illo puncto ad superficiem speculi. Et erit punctus concursus, qui est locus ymaginis, intra speculum, et erit longitudo illius a superficie speculi equalis longitudini puncti visi a superficie speculi. Et visus non adquirit ymaginem puncti visi nisi in loco illo, et quodcumque punctum adquirit visus in hoc speculo non apparebit ex eo nisi unica ymago.

[2.43] Quodcumque autem punctum comprehendit visus in speculo [sperico] exteriori, quando egreditur forma perpendicularem ductam a centro visus ad centrum speculi, linea per quam reflectitur ymago ad oculum concurret cum linea producta a puncto illo ad centrum speculi, que linea est perpendicularis ducta a puncto illo ortogonalis super lineam contingentem lineam communem superficiei reflexionis et superficiei speculi. Et situs puncti concursus (qui est locus ymaginis) a superficiei speculi erit secundum situm visus a superficie speculi. Et forsitan erit punctus concursus ultra speculum, forsitan in superficie speculi, forsitan erit intra speculum. Et visus comprehendit ymagines omnes ultra speculum, licet diversa sint eorum loca, et non comprehendit locum cuiuslibet ymaginis sillogistice in superficie speculi. Et quodlibet punctum comprehensum a visu in hoc speculo non pretendit nisi unam ymaginem.

[2.44] In speculo columpnari exteriori, quodcumque punctum comprehendit visus (eadem in speculo piramidali exteriori), cum fuerit extra perpendicularem ductam a centro visus ortogonalem super superficiem contingentem superficiem speculi, linea per quam

133 *post* dicimus *add.* ergo *R* / quodcumque vel *om. O* / vel quodlibet *scr. et del. C1* 134 *post* quando *add.* ipsum *R* / a² *om. FP1E* / perpendiculari: perpendicularis *FP1* 135 quod . . . visum (136) *om. R* 137 ad: ultra *O; alter. in* ultra *L3* 138 punctus: punctum *R* / qui: qua *S* / intra: ultra *C1E* 139 *post* speculi *add.* et *FP1S; scr. et del.* et *L3* / a²: in *FP1SE* 140 visi nisi *corr. ex* nisi visi *E* / nisi *mg. a. m. C1* 144 *ante* exteriori *add.* columpnari *O* / exteriori: extra polito *R* / *post* forma *add.* super *O; add.* a *R* / perpendicularem: perpendicularis *FP1;* perpendiculari *R* / ductam: ducta *FP1E* 145 ad centrum *rep. L3* (centrum^1,2: centra *L3*) / centrum: centra *FP1SE* / ymago *om. S* 146 concurret *om. FP1S; inter. L3E (a. m. E)* / ad . . . illo (147) *om. FP1; rep. S* / *post* speculi *scr. et del.* linea per quam reflectitur ymago ad oculum *O* 147 ortogonalis: ortogonaliter *OR* 150 erit . . . speculi *om. FP1S* 151 punctus: punctum *R* / *post* speculum *add.* et *C1* 152 erit *om. R* / intra *om. FP1SL3; inter. a. m. E* 153 eorum: earum *R* 154 *post* ymaginis *add.* nisi *R* / in *om. FP1SL3* 155 quodlibet: quolibet *S* / comprehensum *corr. ex* comprehend *F* / a visu *inter. L3; om. R* 157 exteriori *mg. a. m. C1* / *post* exteriori *add.* extra polito piramidali extra polito *R* 158 eadem *om. OC1* / eadem . . . exteriori *om. R* / *post* eadem *add.* et *C1E (inter. E)* / piramidali: piramidalis *P1S* / *post* fuerit *add.* ille punctus *C1* 159 ortogonalem: ortogonalis *FP1SOL3* / super *om. FP1; inter. OC1 (a. m. C1)*

reflectitur ad visum forma concurrit cum perpendiculari ducta a puncto illo rectam super lineam contingentem lineam communem superficiei reflexionis et superficiei speculi. Et loca ymaginum horum speculorum quedam sunt ultra superficiem speculi, quedam in superficie, quedam citra. Et visus adquirit omnes ymagines horum speculorum ultra superficiem speculi, et quodcumque punctum comprehendat visus in hiis speculis non efficit nisi unam ymaginem tantum.

[2.45] In speculo sperico concavo, linee per quas reflectuntur forme punctorum visorum quedam concurrunt cum perpendicularibus ductis a punctis illis rectis super lineas contingentes lineas communes superficiei speculi et superficiei reflexionis, et quedam sunt equidistantes hiis perpendicularibus. Et que concurrunt cum perpendicularibus, locus concursus, qui est locus ymaginis, quidam ultra speculum, quidam citra speculum. Qui citra speculum fuerit quidam inter visum et speculum, quidam super centrum ipsum visus, quidam ultra centrum visus. Et adquisitio visus formarum rerum visarum quas adquirit in hiis speculis quasdam comprehendit in loco ymaginis, qui est punctus concursus—et hee sunt quas visus certe comprehendit—quasdam comprehendit extra locum concursus—et est comprehensio sine certitudine. Et res visas quas adquirit visus in hoc speculo quedam unam prefert ymaginem tantum, quedam duas, quedam tres, quedam quatuor, nec potest esse quod plures.

[2.46] In speculo piramidali concavo et columpnari concavo, linee per quas flectuntur forme ad visum quedam concurrunt in perpendicularibus ductis a punctis visis rectis super lineas contingentes lineas communes, et quedam sunt equidistantes perpendicularibus. Que concurrunt cum perpendicularibus, in quibusdam punctis concursus est ultra speculum, in quibusdam citra. Que autem citra fuerint, quedam

161 ad . . . forma: forma ad visum *R* / concurrit: concurret *R* / *post* a *scr. et del.* tro *E* / puncto illo (162) *transp. ER* 162 *ante* rectam *add.* super *C1E*; (*inter. a. m. E*) / rectam: recte *O* / rectam super *transp. R* / super *scr. et del. C1* 163 et[1] *inter. E* / et superficiei *om. FP1SL3* / superficiei *inter. a. m. C1*; *om. ER* 164 quedam: que *FP1SL3*; *corr. ex* que *E* 166 comprehendat: comprehendit *FP1SR* 167 ymaginem: magnitudinem *FP1S* 169 ductis: ductus *L3* 170 illis: vel *L3* / rectis *om. R* 171 et[2] *om. L3ER* 172 *post* et *add.* earum *R* / *post* perpendicularibus *add.* quedam habent *R* / locus (173): locum *R* 173 concursus *corr. ex* con circulus *L3* / quidam[1]: quedam *FP1SL3*; *om. R* / quidam[2]: quedam *FP1L3R* 174 *post* speculum[1] *scr. et del.* fuerit *C1*; *add.* et *R* / qui: que *R* / citra: circa *FP1S* / speculum fuerit *inter. a. m. E* / fuerit *om. P1*; habent *R* / quidam: quedam *FP1SR* 175 quidam[1]: quedam *FP1SR* / centrum ipsum *transp. R* / ipsum *om. FP1C1* / quidam: quedam *R* 176 adquisitio *om. R* / *post* visus *add.* quasdam *R* 177 quasdam *om. R* / qui: que *FP1SL3E* / punctus: punctum *R* 178 certe: recte *L3C1*; certo *R* 179 concursus: cursus *O* / est *om. E* 180 visas: visae *R* / prefert: profert *O*; prae se ferunt *R* 181 duas . . . quedam[3] *om. FP1* 182 *post* quod *add.* una res pretendat *R* 184 flectuntur: reflectuntur *OR* / in: cum *R* 185 rectis *om. R* 186 communes: communem *FS* 187 *post* perpendicularibus *scr. et del.* que concurrat *F* / in . . . est: quedam habent concursum *R* 188 in *om. R* / quibusdam: quedam *R* / fuerint *inter. E*; *om. R*

erunt inter speculum et visum, quedam supra centrum visus, quedam ultra centrum visus. Et comprehensio rerum visarum in hoc speculo per visum, quedam sit in loco ymaginis, qui est locus concursus, quedam extra locum concursus. Et eorum que comprehenduntur aliud pretendit unam ymaginem tantum, aliud duos, aliud tres, aliud quatuor, nec aliquid est quod potest pretendere plus quam quatuor. Et nos declarabimus hec omnia demonstrative.

[2.47] **[PROPOSITIO 1]** Sit A [FIGURE 5.2.1, p. 563] punctus visus, B centrum visus, DGH speculum planum. Et sit G punctus reflexionis, DGH linea communis superficiei reflexionis et superficiei speculi. A puncto G ducatur EG perpendicularis super lineam communem. A puncto A ducatur perpendicularis super speculi superficiem, que sit AH, et producatur ultra speculum. Et AG sit linea per quam accedit forma ad speculum, BG per quam reflectitur ad visum. Igitur BG, EG, AG sunt in superficie reflexionis, et cum EG sit equidistans AH, et BG declinata sit super EG, concurret BG cum AH. Concurrat ergo in puncto Z. Dico quod ZH est equalis HA.

[2.48] Quoniam angulus BGD equalis angulo AGH, et angulus AHG equalis angulo ZHG, et latus HG commune, quare triangulus equalis triangulo; quare ZH equalis AH.

[2.49] Et si voluerimus per perpendicularem invenire locum reflexionis, secetur ex perpendiculari ultra speculum pars equalis parti eius usque ad speculum, et est ut sit ZH equalis AH. Et ducatur linea a centro visus ad punctum Z, que sit BGZ. Dico quod G est punctus reflexionis.

189 erunt *om. R* / inter: intra *FP1* / supra: super *OR* / quedam[2] . . . visus (190) *om. FP1* 190 comprehensio: comprehensionum *FP1S* / rerum *om. FP1S* 191 qui: que *FP1S* / *post* concursus *add.* et *C1* 192 quedam: quidam *FP1SE* / quedam . . . concursus *om. L3* / eorum: eo *FP1* / aliud: alium *FP1* 193 aliud[1]: alium *FP1* / duos *om. FP1S* / *post* duos *add.* et *E* / aliud[2] *inter. a. m. E* 194 aliquid est *transp. O* / potest: possit *R* / plus: plures *R* / quam *om. O* 195 demonstrative: demonstrare *FP1S* 196 punctus visus: punctum visum *R* 197 DGH: CDE *R* / speculum . . . DGH (198) *mg. a. m. E* / G: D *R* 198 DGH . . . reflexionis *om. FP1* / DGH: CDE *R* 199 G: D *R* / EG: DF *R* / perpendicularis: perpendiculariter *O* / *post* communem *add.* et *ER* 201 AH: AC *R* / AG: AD *R* / per *om. O* / accedit: accedat *E* / accedit forma (202) *transp. R* 202 BG[1]: AG *P1S; corr. ex* AG *a. m. F*; BD *R* / BG[2]: BD *R* / EG AG (203) *corr. ex* AG EG *S*; FD AD *R* 203 sunt in *inter. a. m. F; om. P1* / EG: FD *R* / AH: AC *R* / BG: DB *R* 204 sit *om. FP1* / EG: HG *O*; FD *R* / concurret: concurrit *FP1E* / BG: BD *R* / AH: AC *R* 205 Z: G *R* / ZH *corr. ex* HZ *C1*; GC *R* / *post* est *scr. et del.* perpendicularis *P1* / HA: AHA *S*; CA *R* 206 *post* quoniam *add.* enim *R* / BGD: BDE *R* / *post* equalis *add.* est *R* / AGH: ADC *R* / AHG: ACD *R* 207 *post* equalis[1] *add.* est *C1* / ZHG: GCD *R* / HG: CD *R* / triangulus equalis: triangulum equale *R* 208 triangulo: angulo *FP1S* / quare *inter. E* / ZH: GC *R* / AH: AC *R* 209 per *om. FP1SO; inter. a. m. E* / per perpendicularem *inter. a. m. L3* / perpendicularem: perpendicularis *FP1SO* / *ante* invenire *add.* perpendicularis *L3* 210 *post* ex *scr. et del.* pyramide *L3* 211 *post* speculum *scr. et del.* pf pars equalis parti eius usque ad speculum *F; scr. et del.* pars equalis parti *P1* / ZH: GC *R* / AH: AC *R* 212 *post* punctum *scr. et del.* visus *C1* / Z: G *R* / Z . . . BGZ: quod est G *O* / BGZ: BDG *R* / *ante* dico *scr. et del.* et *O* / G: D *R* / punctus: punctum *R*

[2.50] Quoniam AH et HG sunt equalia HG et HZ, et angulus angulo, quare triangulus triangulo. Igitur angulus ZGH equalis angulo HGA. Sed ZGH est equalis angulo DGB. Restat ergo ut angulus BGE sit equalis angulo EGA, et ita G punctus reflexionis. Et ita propositum.

[2.51] **[PROPOSITIO 2]** Amplius, sit A centrum visus [FIGURE 5.2.2, p. 563], et AG perpendicularis super speculum planum, et D secet hanc perpendicularem in superficie oculi. Dico quod in hac perpendiculari non est punctus qui reflectatur ab hoc speculo ad visum preter D.

[2.52] Si enim sumatur ultra visum punctus in hac perpendiculari, et sit H. Non perveniet forma eius ad speculum super perpendicularem propter solidi corporis interpositionem, et ita non reflectitur forma eius super perpendicularem.

[2.53] Et si dicatur quod ab alio puncto speculi potest reflecti, sit illud B. Movebitur quidem forma eius ad punctum B per lineam HB, et reflectitur per lineam BA. Dividatur angulus HBA per equalia per lineam TB. Igitur TB erit perpendicularis super superficiem speculi. Sed TG est perpendicularis super eandem, quare ab eodem puncto est ducere duas perpendiculares ad superficiem speculi, quod est impossibile.

[2.54] Eadem erit probatio quod forma puncti D non potest reflecti ab alio speculi puncto quam a puncto G, quare non reflectitur nisi super perpendicularem. Punctum autem in hac perpendiculari sumptum inter G et D, si dicatur formam per reflexionem ad visum mittere improbatio, quoniam aut erit corpus solidum aut rarum.

[2.55] Si solidum, procedet secundum perpendicularem forma eius ad speculum et regredietur secundum eandem usque ad ipsum, et

214 *post* quoniam *add.* enim *R*/AH: AC *R*/HG: CD *R*/sunt equalia HG *mg. a. m. E*/HG: CD *R*/HZ: CG *R* 215 quare triangulus: ergo triangulum *R*/ZGH: GDC *R*/*ante* equalis *add.* est *OER*/HGA: ADC *R* 216 ZGH: GDC *R*/DGB: DGH *S*; BDE *R*/BGE: BGC *P1*; BGD *O*; BDE *R* 217 EGA: HGA *O*; ADC *R*/*post* ita[1] *add.* est *O*/G *inter. L3*; D *R*/*ante* punctus *add.* est *R*/punctus: punctum *R*/reflexionis: re *FP1*/*post* ita[2] *add.* patet *R* 218 amplius *om. R*/centrum: centro *O* 219 et[1] *om. O*/D *om. O* 220 oculi: circuli *L3*/*post* dico *scr. et del.* in h *F*; *add.* ergo *O* 221 punctus qui: punctum quod *R* 222 si enim: sm *FS*; *alter. in* sive *L3*; sin *R*/*ante* sumatur *add.* autem *P1C1ER*/punctus: punctum *R* 223 et *inter. a. m. C1*/*post* non *add.* iam *R*/perveniet: pervenient *S*/*post* perpendicualrem *add.* AH *R* 224 propter . . . perpendicularem (225) *om. O*/interpositionem: interpositione *P1*/non: si *FP1*/reflectitur *corr. ex* reflectatur *P1*; reflectetur *R* 226 alio: aliquo *FP1S*/potest: possit *R* 227 movebitur . . . B[2] *mg. a. m. E*/quidem: quod *P1* 228 reflectitur: reflectetur *R*/per[1]: quod *S*/*post* per[3] *scr. et del.* d *S* 229 igitur TB *om. L3*; *inter. a. m. C1*/TB *om. R* 230 TG: AG *S*/est[1] *om. FP1*/ab *inter. L3* 234 *post* nisi *scr. et del.* p *L3* 235 *post* perpendicularem *add.* DG *R*/autem: aliquod *O* 236 dicatur: ducatur *O*/formam: forma *OC1*/mittere *om. FP1*; *mg. a. m. C1* 237 *ante* improbatio *scr. et del.* tem *C1*/improbatio *corr. ex* probatio *L3*; improbo *R*/aut *corr. ex* autem *C1*/solidum . . . si (238) *mg. a. m. C1*/aut rarum *corr. ex* ? *O* 238 procedet: procedit *O*; *corr. ex* procedat *F*; *corr. ex* procedit *E* 239 regredietur: regradietur *FP1*; *corr. ex* regreditur *E*

propter soliditatem non poterit transire et ad visum pervenire.

[2.56] Si autem punctum illud fuerit rarum, forma eius regrediens a speculo super perpendicularem miscebitur ei, et adherebit, nec reflectetur ad visum.

[2.57] Quod autem forma cuiuscumque puncti in hac perpendiculari inter G et D sumpti non possit ab alio puncto speculi ad visum reflecti modo supradicto potest probari. Similiter, forma puncti inter A et D sumpti nec reflectitur ad visum per perpendicularem nec per aliam, quoniam puncta centrum visus et superficiem eius interposita sunt valde rara, unde nec mittitur eorum forma nec reflectitur ut sentiatur. Et quoniam quodlibet punctum preter D in superficie visus sumptum opponitur speculo non ad rectum angulum, videbitur quodlibet super perpendicularem ab eo ad speculum ductam, et eius ymago ultra speculum equidistans a superficie speculi, sicut ipsum punctum. Et quoniam D videtur continuum cum aliis superficiei visus punctis, et ymago eius continua cum aliis ymaginibus, videbitur ymago D tantum distans a superficie speculi quantum distat D ab eadem.

[2.58] Palam ergo quod cuiuscumque puncti in speculo visi ymago videbitur super perpendicularem, et elongatio ymaginis et visi corporis a superficie speculi eadem.

[2.59] **[PROPOSITIO 3]** Amplius, forma puncti visi in speculo plano non reflectitur ad eundem visum nisi ab uno puncto tantum. Sit enim A [FIGURE 5.2.3, p. 563] centrum visus, B punctum visum, ZH speculum. Si ergo dicatur quod a duobus punctis speculi reflectitur forma B ad visum, sit unum punctum D, aliud E. Et ducatur linea a puncto viso ad visum, scilicet AB, que quidem linea aut erit perpendicularis supra speculum, aut non.

[2.60] Si non fuerit perpendicularis, scimus quod illa linea est in superficie reflexionis ortogonali super superficiem speculi, et in una

240 soliditatem *corr. ex* liditatem *L3* / et *inter. a. m. E* 241 regrediens: regradiens *P1*; egrediens *O* 242 et adherebit *inter. L3* / adherebit: adhebit *F* 244 quod: quia *O* / *post* forma *scr. et del.* eius *O* 245 *post* sumpti *add.* si *C1* / non: si *FP1S* 247 nec[1]: non *R* / ad visum *om. O* / per[1] *om. FP1SL3*; *inter. OE* (*a. m. E*) / *ante* perpendicularem *add.* hanc *C1* / perpendicularem: perpendicularis *FP1SL3* 248 quoniam: quando *P1O* / *post* puncta *add.* inter *R* 249 mittitur: mittetur *O*; mutatur *E* 250 *post* quoniam *inter.* ad *L3* / quodlibet *alter. in* ad quodlibet *O* 251 speculo: speculationi *FP1S*; speculum *OL3* / quodlibet *inter. a. m. E* 252 eius ymago *transp. R* 253 speculi *om. R* / *post* speculi *scr. et del.* sint *C1* / *post* punctum *add.* E *F* 254 *post* aliis *add.* punctis *C1* / superficiei *inter. a. m. E* / punctis: punctus *FP1S*; *corr. ex* punctus *L3*; *om. C1* 258 super *inter. a. m. E* 259 *post* speculi *add.* est *OR* 260 forma *corr. ex* formarum *L3* 261 enim: est *S* 262 visus: visum *FP1*; *inter. E* 263 speculi reflectitur *corr. ex* reflectitur speculi *C1* / reflectitur: reflectatur *R* 264 *post* visum *add.* A *R* / *post* D *add.* et *FP1* / aliud: alium *FP1* / et *om. O* / ducatur: producatur *O* / *post* puncto *scr. et del.* ni *F* 265 AB: BA *R* 268 ortogonali: ortogonalis *O* / super *om. S* / et in *inter. O* / una: unaquaque *F*

sola tali. Quoniam si in duabus, erit communis duabus superficiebus ortogonalibus, et sumpto in ea puncto, et ducta ab illo linea in alteram superficierum super lineam communem huic superficiei et superficiei speculi, erit quidem hec linea ortogonalis supra speculum. Similiter, ab eo puncto ducatur linea in alia superficie super lineam communem ei et superficiei speculi, et erit hec linea ortogonalis supra speculum, quare ab eodem puncto erit ducere duas perpendiculares.

[2.61] Cum ergo BA sit in una sola superficie ortogonali, et tria puncta A, B, E sint in eadem superficie ortogonali, et erunt AE et EB in illa superficie ortogonali in qua est AB, similiter EB et DB. Quare EA, EB in eadem superficie cum DA, DB. Sed angulus AEH est equalis angulo DEB, et angulus HEA maior angulo ADE, quia extrinsecus, quare BED maior ADE. Sed BDZ equalis ADE, et BDZ maior BED, quare ADE maior BED, et dictum est quod minor. Restat ergo ut a solo puncto fiat reflexio.

[2.62] Si vero AB sit perpendicularis supra speculum, iam dictum est quod unicum est punctum in linea a centro visus ad speculum ortogonaliter ducta cuius forma reflectitur a speculo ad visum. Et iam probatum est quod ymago illius puncti ab uno solo reflectitur puncto, quare propositum.

[2.63] **[PROPOSITIO 4]** Amplius, inspecto aliquo puncto ab utroque visu, una tantum et eadem ymago apparet utrique, et in loco predicto. Verum planum est quod forma puncti non reflectitur ad utrumque visum ab eodem puncto speculi. Si enim linea reflexionis ad unum visum procedens angulum teneat cum perpendiculari erecta super superficiem speculi equalem angulo quem tenet linea accessus forme ad speculum cum eadem perpendiculari, non poterit in eadem superficie sumi linea alia que equalem angulum huic efficiat cum

269 erit . . . duabus *inter. L3* 270 sumpto: sumpta *O* / *post* ea *add.* eo *P1* / puncto: puncta *S* / ducta: ducto *FP1SO* / illo: illa *FP1* / alteram: altera *C1; alter. ex* alterum *in* altera *L3* 272 quidem *om. R* / hec *om. FP1* / linea . . . ducatur (273) *om. P1* / supra: super *L3C1R* / *post* speculum *add.* quare *O* 273 eo: eodem *OL3C1E* / alia: illa *FP1S* / *post* superficie *scr. et del.* superficie *F* 274 ei: huic superficiei *R* / et[1] *om. E* / et[2] *om. R* / hec *om. FP1S* / supra: super *R* 275 *post* perpendiculares *add.* ad superficiem speculi *R* 276 ortogonali *corr. ex* ortogonalis *L3* 277 ortogonali *corr. ex* ortogonalis *L3* / et[1,2] *om. R* / EB *corr. ex* EDB *F*; EDB *SL3C1E* 278 in . . . AB *om. R* / similiter . . . DB *om. O* / EB *alter. in* AD *a. m. F*; AD *P1* / *post* DB *add.* DA *R* / quare EA EB (279) *mg. a. m. C1* 279 *post* EB *add.* sunt *R* / est *om. L3* 280 *post* angulo[1] *scr. et del.* DH *S* / DEB: DEH *L3*; BED *R* / HEA: AEH *R* / extrinsecus: exterior *R* 281 *post* maior *inter.* angulo *E* / equalis . . . BDZ[2] *om. FP1S* / et . . . BED *om. R* 282 ADE: AED *FP1SL3E* / BED *corr. ex* BDE *F*; ADE *E* / *post* quod *scr. et del.* in *O* / restat: constat *O* 284 supra: super *R* 285 *post* est[1] *inter.* in secunda figura *a. m. L3; add. C1* 287 uno: illo *FP1* 288 *post* quare *add.* patet *R* / propositum: per ipsum *O* 289 inspecto: in speculo *SL3* 290 *post* utrique *add.* visui *R* 291 verum: unde *L3C1ER* / planum: palam *O* 292 speculi: speculo *S* / si: quia *R* / *post* enim *add.* una *C1* 293 teneat: tenet *R* / super superficiem (294) *corr. ex* superficiem *O* 294 quem: quam *FP1S* 296 linea alia *transp. R* / que: non *S*

perpendiculari. Unde ab hoc puncto non reflectetur linea aliqua ad alium visum. Oportet igitur ut a diversis speculi punctis fiat reflexio.

[2.64] Sint illa puncta T, Z [FIGURE 5.2.4, p. 564]. Et sit speculum planum QE; A punctus visus; B, G duo visus; AD perpendicularis. Palam ergo quod BT et AT et ET, AD sunt in eadem superficie ortogonali super superficiem speculi. Similiter, AD, AZ, GZ sunt in eadem superficie ortogonali, et DT linea communis superficiei ADTB, et DZ communis superficiei ADZG. Si BT, GZ fuerint in eadem superficie ortogonali, erit TDZ linea una, et perpendicularis AD aut erit inter duas perpendiculares predictas ad superficiem speculi a duobus visibus, aut extra [FIGURE 5.2.4a, p. 564].

[2.65] Utrumlibet sit, linea reflexionis BT secabit ex perpendiculari AD ultra speculum partem equalem parti que est AD. Similiter, GZ secabit ex eadem perpendiculari partem ultra speculum equalem illi parti. Igitur ille due linee reflexionis secabunt perpendicularem ultra speculum in eodem puncto. Ergo ymago puncti A in eodem perpendicularis puncto percipietur ab utroque visu, quare unica tantum erit ymago et eadem, et in eodem loco que esset uno tantum visu adhibito.

[2.66] Si vero puncta T, Z non fuerint in eadem superficie reflexionis ortogonali super speculum [FIGURE 5.2.4b, p. 564], eadem tantum erit probatio, cum utraque linea reflexionis secet ex perpendiculari partem equalem parti superiori, et erit sectio linearum reflexionis cum perpendiculari in eodem puncto, quare propositum.

[2.67] Si vero fuerit punctus A in perpendiculari ducta ab uno visu ad superficiem speculi tantum, secundum eundem visum comprehenditur ultra speculum in puncto perpendicularis tantum elongato a superficie speculi quantum distat A ab eadem, quia forma A videtur continua cum formis aliorum punctorum que quidem videntur in locis

297 unde: unum *S* / reflectetur: reflectitur *O* 298 alium: alterum *R* / ut *om. S* / speculi punctis *transp. ER* 300 A punctus visus: punctum visum A *R* / punctus: punctis *FP1S; corr. ex* puncta *O* / B . . . visus: duo visus B G *R* / AD perpendicularis *transp. R* 1 et[1]: ET *F; om. SR* / AT: AD *O* / et[2] *om. P1O* / et ET *om. R* 2 super *inter. O* / super . . . ortogonali (3) *rep. S; mg. a. m. E* (AD AZ *transp. E*) / speculi *om. FP1S* / AD AZ GZ: GZ AZ AD *R* / AZ: EZ *FP1C1* 3 DT linea *transp. R* / ADTB: ADDB *FP1S* / *post* ADTB *add.* et superficiei speculi *R* / *post* DZ *add.* est linea *C1; add.* linea *R* 4 *post* communis *inter.* linea *a. m. L3* / *post* ADZG *add.* et superficiei speculi *R* / *post* si *add.* iam *R* / eadem: eam *F* 5 *post* una *add.* recta *R* 6 perpendiculares: lineas *O* / predictas: ductas *C1*; productas *R* 8 utrumlibet: utrumque *S*; utramque *O*; utramlibet *E* / sit *om. O* / reflexionis *om. R* 9 speculum *corr. ex* speculi *C1* 11 igitur ille *transp. R* / igitur . . . due: ille due igitur *E* / ille *mg. a. m. C1* 12 *post* A *scr. et del.* pe *C1* / eodem: eadem *FP1S* / perpendicularis (13): particularis *S; corr. ex* particularis *L3* 13 percipietur: participetur *FP1S* 14 visu: viso *FP1SL3C1; corr. ex E* 16 tantum *om. P1*; tamen *R* / *post* erit *scr. et del.* puncta *C1* 17 cum: quod *R* / secet: secat *FP1S* 18 equalem parti *transp. O* 19 *post* quare *add.* patet *R* 20 punctus: punctum *R* / visu: viso *FP1SOL3E* 21 tantum: tamen *O* / comprehenditur (22): comprehendetur *R* 22 perpendicularis *corr. ex* perpendiculari *C1*; perpendiculari *E*

similibus. Et ab alio visu comprehenditur ymago A in eodem perpendicularis puncto, quare et sic utrique visui unica tantum apparet ymago puncti A, et in eodem eiusdem perpendicularis puncto, quod est propositum.

[2.68] **[PROPOSITIO 5]** In speculis spericis exterioribus patebit quod dicimus. Sit A [FIGURE 5.2.5, p. 565] punctus visus, B centrum visus, G punctum reflexionis. Palam quod BG, AG sunt in superficie ortogonali super superficiem contingentem speram in puncto G. Linea communis superficiei reflexionis et superficiei spere est circulus, et sit ZGQ. Linea contingens hunc circulum in puncto reflexionis sit PGE. Perpendicularis super hanc lineam sit HG. Planum quod HG perveniat ad centrum spere. Quod si non, cum linea a centro spere ducta ad punctum G sit etiam perpendicularis super lineam PGE, erit ab eodem puncto in eadem parte ducere duas lineas perpendiculares super unam lineam.

[2.69] Sit autem centrum spere N, et ducatur linea a puncto viso ad centrum spere, scilicet AN, que quidem erit perpendicularis super superficiem contingentem speram in puncto spere per quem transit. Et quoniam planum quod BG secat speram, cum sit inter HG, GP que continent rectum angulum, concurret cum linea AN. Et cum perpendicularis HG sit in superficie reflexionis, erit centrum spere in eadem, et ita AN in eadem superficie cum HG.

[2.70] Sit ergo concursus BG cum AN D. Planum quoniam D erit locus ymaginis, et hec quidem intelligenda sunt quando linea ducta a puncto viso ad centrum visus non fuerit perpendicularis super speculum.

[2.71] **[PROPOSITIO 6]** Amplius, linea PGE [FIGURE 5.2.6, p. 565] secat lineam AN. Sit punctus sectionis E, et dicitur punctus iste finis contingentie. Dico quod in hoc situ linea a centro spere ad locum ymagi-

25 visu: viso *FP1SE*/comprehenditur: comprehendetur *R*/in: ab *FP1SL3; corr. ex* ab *a. m. C1* 26 perpendicularis *corr. ex* perpendiculari *C1*/et *om. O; inter. E*/sic: sit *FP1SL3* 27 perpendicularis *om. FP1L3E* 29 spericis *mg. a. m. F; om. SL3C1E*/exterioribus: extra politis *R* 30 dicimus: diximus *R*/*post* sit *add.* autem *O*/punctus visus: punctum visum *R* 31 *post* in *add.* eadem *R* 32 ortogonali: ortogonalis *P1SL3*/contingentem speram *transp. R* 33 circulus: circumferentia *R* 35 HG[1]: HZ *P1*/planum quod HG *om. S*/perveniat: perveniet *OR; corr. ex* perveniant *E* 36 punctum (37): centrum *FP1; corr. ex* centrum *L3E* (*a. m. E*) 38 eadem parte: partem eadem *S;* eandem partem *L3C1ER*/ducere *inter. L3; mg. a. m. C1*/duas *inter. a. m. O* 39 sit *corr. ex* si *E* 40 su-per *om. F* 41 quem *corr. ex* quam *F;* quod *R* 42 *post* planum *add.* est *R*/HG: GH *FP1* 45 *post* eadem[1] *add.* superficie cum HG (*post* HG *scr. et del.* si ergo) *C1*/HG: BG *SO* 46 *post* AN *add.* in puncto *C1; inter.* punctus *E; add.* punctum *R*/D[1] . . . D[2] *om. O*/*post* D[1] *inter.* punctus *a. m. L3*/quoniam: quod *R* 47 et *inter. O*/hec: hoc *P1*/intelligenda: intelligendum *FP1S*/sunt: est *FP1*/quando: quoniam *FE; corr. ex* quoniam *L3*/linea ducta *transp. FP1* 48 non: si *S* 50 linea *mg. a. m. C1* 51 secat *corr. ex* fiet *O*/punctus[1,2]: punctum *R*/E *om. S*/iste: istud *R* 52 in hoc situ: nihil *P1;* nihil situ *scr. et del. F*

nis ducta maior est linea a loco ymaginis ducta ad locum reflexionis: id est, DN maior DG.

[2.72] Quoniam angulus BGH equalis angulo HGA, sed angulus BGH equalis angulo NGD. Ergo angulus HGA equalis eidem, et EG perpendicularis super HGN, quare angulus AGE equalis est angulo EGD. Igitur proportio AG ad DG sicut AE ad ED.

[2.73] Protrahatur a puncto A equidistans DG, et concurrat cum linea HN in puncto H. Erit igitur angulus NGD equalis angulo GHA. Sed angulus NGD equalis est angulo AGH. Ergo angulus GHA est equalis eidem, quare duo latera AG, HA sunt equalia. Igitur proportio AH ad DG sicut AG ad idem. Sed proportio AH ad DG sicut AN ad DN, quare proportio AN ad DN sicut AG ad DG. Igitur proportio AN ad AG sicut DN ad DG. Sed AN est maior AG, quia respicit angulum maiorem recto in triangulo ANG. Igitur DN maior DG, quod est propositum.

[2.74] **[PROPOSITIO 7]** Amplius, dico quod linea ducta a fine contingentie, qui est E, usque ad speram perpendiculariter, id est pars linee EN, minor est semidyametro.

[2.75] Sit F [FIGURE 5.2.7, p. 566] punctus in quo AN tangit superficiem spere. Dico ergo quod EF minor est NF.

[2.76] Quoniam, ut dictum est, proportio AG ad DG sicut AE ad ED, sed AN ad DN sicut AG ad GD. Igitur AN ad DN sicut AE ad ED. Igitur AN ad AE sicut DN ad DE. Sed AN maior AE, quare DN maior DE, quare DN maior EF, quare NF maior EF, quod est propositum.

[2.77] **[PROPOSITIO 8]** Amplius, sit G [FIGURE 5.2.8, p. 566] centrum visus, D centrum spere, DZG perpendicularis a centro visus ad speram. Dico quod nullius puncti forma reflectitur per hanc perpendicularem nisi puncti eius quod est in superficie visus.

54 *post* DN *add.* est *O* 55 *post* quoniam *add.* enim *R* / *post* BGH *add.* est *ER* / *post* equalis *add.* est *O* / HGA: NGD *E* / sed . . . NGD (56) *scr. et del. E* 56 *post* equalis[1,2] *add.* est *R* / NGD *corr. ex* NDG *FL3*; NDG *SO* / eidem *inter. O* 57 *post* angulus *scr. et del.* HGN *F* / equalis est *transp. L3C1* 58 EGD *corr. ex* EDG *FL3*; EDG *SE* / *post* EGD *add.* est enim equalis angulo BGP qui opponitur EGD *C1* / DG: GD *R* 59 *post* puncto *add.* A linea *mg. F*; *scr. et del.* que *F* / A *om. P1* / *post* equidistans *add.* ipsi *R* 60 NGD *corr. ex* NSDG *F*; NDG *P1*; *corr. ex* NDG *L3* / NGD . . . angulus[1] (61) *om. S* 61 NGD *corr. ex* NGDG *F*; NGDG *S*; *corr. ex* NDG *L3*; NDG *E* / est[1] *om. C1* / *post* AGH *scr. et del.* quare duo latera AG HA *P1* / ergo . . . eidem (62) *mg. a. m. F*; *om. SOL3C1ER* 62 AG HA: AGNHA *S* / *post* equalia *scr. et del.* igitur proportio AG ad DG *F* / igitur . . . DG (64) *om. FP1* 63 idem: eandem *R* / *post* idem *add.* GD *C1* / *post* sicut[2] *scr. et del.* AG ad DG sed proportio AN ad DG est *C1* / *post* AN *scr. et del.* ad *E* 64 proportio[1] *om. R* 66 in *inter. L3* / ANG: AGN *R* / igitur: G *S* 68 contingentie (69): contingente *FP1* 69 qui: que *O* / E *inter. a. m. F* / *post* est[2] *add.* EF *R* 71 punctus: punctum *R* / tangit: secat *R* 72 est *om. S* 74 ad[2] *inter. O* / GD: DG *E* / AN: NA *O* 76 EF[1]: DF *R* / quare . . . EF[2] *mg. a. m. C1*; *inter. a. m. E* (*post* maior *add.* est *E*) 78 DZG: GZD *SO* 79 nullius: nullus *FP1L3* / puncti *om. O* 80 puncti *scr. et del. F*; punctum *O*

[2.78] Punctorum enim forme post centrum visus sumptorum non reflectuntur per eam propter causam supradictam. Similiter nec puncta inter superficiem visus et speculum sumpta. Dico etiam quod nullum punctum huius perpendicularis reflectitur ab alio puncto speculi.

[2.79] Si enim dicatur quod ab alio puncto, sit illud punctum A. Erit linea GA linea reflexionis, et a puncto illo intelligemus lineam ad A, que est linea per quam movetur forma. Et includunt hee due linee angulum super A, quem quidem angulum necessario dividet dyameter DA, cum sit perpendicularis super punctum A, quia perpendicularis dividit angulum ex linea motus forme et linea reflexionis per equa. Et ita dyameter DA concurret cum perpendiculari GD inter punctum sumptum et G. Et ita due linee recte in duobus punctis concurrent et superficiem includent.

[2.80] Restat ergo ut solius puncti qui est in superficie visus forma reflectatur a speculo ad perpendicularem, et videatur in primo ymaginis loco propter eius cum aliis continuitatem.

[2.81] **[PROPOSITIO 9]** Amplius, GA, GB [FIGURE 5.2.9, p. 567] sint linee a centro visus ducte contingentes speram, et signetur circulus super quem superficies hiis lineis inclusa secat speram. Erit AB portio apparens ex hoc circulo. Dico ergo quod loca ymaginum que per reflexiones ab hac portione factas comprehenduntur quedam sunt intra speculum, quedam in superficie speculi, quedam extra speculum. Et singulum horum est determinandum.

[2.82] Ducatur a puncto G linea secans circulum, et pars eius que est corda arcus circuli sit equalis semidyametro circuli. Sit linea illa GHK, et corda equalis semidyametro sit HK. Et producatur a puncto H perpendicularis, que sit DHM. Dico quod forma reflexa a puncto H locus eius erit intra speram.

81 forme: forma *FP1L3E* / post centrum: posterius *O* / *ante* visus *add.* quoniam *O* 82 per eam: preter eum *O* 83 inter: intra *E* / etiam: ergo *L3* / nullum: nullius *O* 84 huius *rep. P1; inter. E* / *post* perpendicularis *add.* huius *E* / speculi *om. O* 85 si . . . puncto *mg.* (dicatur: reflectatur; quod *om.*) *O* / illud: illum *P1* 86 GA *corr. ex* GH *F*; GHA *S*; *corr. ex* GHA *L3*; *corr. ex* G *a. m. C1* / linea *om. P1* / intelligemus: intelligamus *R* 87 A *corr. ex* DA *L3* / includunt *corr. ex* includuntur *S* / due linee *transp. FL3E* 88 quem *corr. ex* quoniam *F*; que *P1* / necessario *om. O* / *post* dividet *add.* per equalia *R* 89 *post* perpendicularis[1] *scr. et del.* super perpendicularem *F; add. L3; scr. et del.* p *S* / A *inter. O* 90 et[2] *inter. a. m. C1* 91 GD: G *P1*; DG *O* 92 recte *om. P1* 93 includent: inducent *FP1SE*; *corr. ex* inducent *L3* 94 solius *corr. ex* solus *P1* / puncti *om. P1* 95 ad: et *S*; per *inter. O*; per *R* / primo: proprio *R* 96 *post* aliis *add.* punctis *R* 98 sint *corr. ex* sunt *C1* / *post* sint *add.* due *P1* / *post* centro *scr. et del.* a centro *F* 99 superficies: superficiem *S* 100 apparens: apparentque *FP* 101 reflexiones *corr. ex* reflexionis *F*; *corr. ex* flexiones *a. m. C1* 102 quedam[1] . . . speculum[2] *mg. a. m. F* 103 *ante* et *scr. et del.* intra *F* / singulum: unumquodque *R* / horum *inter. a. m. C1* / *post* horum *scr. et del.* h *O* 105 sit equalis *transp. S* / *post* semidyametro *add.* magni *C1* / circuli[2] . . . semidyametro (106) *inter. a. m. L3* 106 producatur: protrahatur *FP1* 107 quod *om. S* / forma reflexa: forme reflexe *R* 108 eius *inter. a. m. C1*; *om. R* / erit: est *R* / erit . . . linea (109) *mg. a. m. E*

[2.83] Ducatur a puncto H linea equalem tenens angulum cum MH angulo MHG, et sit OH. Reflectentur quidem puncta huius linee a puncto H ad visum, et non alterius. Sumatur ergo aliquod eius punctum, et sit O, et ducatur ab eo linea ad centrum spere, que sit OD. Erit quidem OD perpendicularis super superficiem contingentem speram super punctum eius per quod transit OD. Verum angulus OHM equalis est angulo, ex ypothesi, MHG, quare similiter equalis est angulo contraposito, scilicet KHD. Sed KHD est equalis KDH, quoniam respiciunt equalia latera.

[2.84] Igitur angulus OHM equalis est angulo KDM, quare linee KD, OH sunt equidistantes. Ergo in infinitum producte numquam concurrent. Et linea OD secabit lineam interiacentem KD, OH, et ita quodcumque punctum sumatur in linea OH, linea ducta ab illo puncto ad punctum D secabit lineam reflexionis intra speram, que quidem linea erit perpendicularis super speram, sicut est OD. Quare ymago cuiuscumque puncti linee OH apparebit intra speram.

[2.85] Amplius, arcus circuli interiacens punctum H et punctum per quem transit perpendicularis a centro visus ducta est HZ. Dico quod, a quocumque puncto huius arcus fiat reflexio, locus ymaginis erit intra speram.

[2.86] Probatio: sit I [FIGURE 5.2.9a, p. 567] punctus sumptus, et ducatur linea a centro visus secans circulum super punctum illum, que sit GIS, et ducatur perpendicularis a puncto hoc que sit DIT. Et fiat linea PI equalem tenens angulum cum IT angulo TIG. Palam quod sola puncta linee PI reflectuntur a puncto I ad visum. Palam etiam quod linea IS maior est linea KH, quare maior SD. Igitur angulus SDI maior est angulo SID, quare est maior angulo GIT, quare maior angulo TIP.

110 OH: PH *R* 111 *post* visum *add.* G *R* 112 O: P *R*/sit: si *FP1*/OD: PD *R*/erit . . . OD (113) *inter. L3* 113 quidem *om. R*/OD: PD *R*/perpendicularis: perpendiculariter *FP1; corr. ex* perpendiculariter *a. m. E* 114 OD: PD *R*/*ante* verum *add.* et coniungatur DK *R*/OHM: PHM *R*/equalis est (115) *transp. R* 115 ex ypothesi *om. SOC1R*/contraposito (116): circa posito *S* 116 scilicet *om. R*/scilicet KHD *om. P1*/*post* KHD[1] *scr. et del.* est equalis HDB *F*/sed KHD *om. S; inter. L3E* (*a. m.* E)/sed . . . KDH (KDH: KDM) *mg. a. m. F*/KDH *corr. ex* DH *a. m. C1*/quoniam: quia *SOC1* 118 OHM: PHM *R*/equalis est *transp. L3*/est *om. E*/KDM: KDH *O* 119 OH *corr. ex* OG *FL3E;* HO *C1;* PH *R*/sunt . . . OH (120) *om. S*/ergo *inter. O*/in *om. FP1L3; inter. a. m. E*/*post* producte *scr. et del.* in *E*/numquam: numquid *FP1* 120 OD: PD *R*/*post* interiacentem *add.* inter *R*/*post* KD *add.* et *R*/OH: PH *R* 121 quodcumque *corr. ex* quodcum *a. m. C1*/OH: PH *R* 122 intra: inter *FP1* 123 erit perpendicularis *transp. ER*/OD: PD *R* 124 OH: PH *R* 125 interiacens: interiacentis *FP1L3*/*post* interiacens *add.* inter *R*/H et punctum *inter. L3* 126 quem: quod *R*/est: esto *R* 127 quocumque: quolibet *O*/huius: huiusmodi *C1*/*post* reflexio *scr. et del.* huius *O* 128 intra: inter *FP1OL3* 129 probatio *om. R*/punctus sumptus: punctum sumptum *R* 130 illum: illud *R* 131 GIS: GI *R*/perpendicularis: perpendiculariter *C1*/a puncto: per punctum *R* 132 tenens *om. P1*/*post* IT *add.* et *P1*/TIG *corr. ex* TG *E* 133 *post* I *scr. et del.* palam *E*/etiam *inter. O* 134 quare *mg. F*/quare . . . SID (135) *om. S*/*post* quare *add.* est *OC1*/*post* maior *add.* angulo (*post* angulo *scr. et del.* quare maior angulo) *L3*/SD: ID *P1E; corr. ex* GSD *L3* 135 SID: SIT *E*/*post* quare[2] *add.* est *R*/*post* maior[2] *inter.* est *a. m. E*

[2.87] Igitur linea PI et SD numquam concurrent, et linea ducta a puncto quocumque PI linee ad punctum D secabit lineam SI intra speram, que SI est linea reflexionis. Et omnis linea ducta a quocumque puncto PI linee erit perpendicularis super speram, sicut est PD. Et cum locus ymaginis sit in concursu perpendicularis a puncto viso et linee reflexionis, erit ymago cuiuslibet puncti linee PI intra speram. Palam ergo quod omnium ymaginum arcus HZ locus proprius erit intra speculum, quod est propositum.

[2.88] Amplius, sumpto quocumque puncto arcus HB, dico quod quedam eius ymago erit intra speculum, quedam in superficie speculi, quedam extra speculum.

[2.89] Sumatur aliquod eius punctum, et sit N [FIGURE 5.2.9b, p. 567], et ducatur linea a puncto G secans circulum, que sit GNQ. Et ducatur perpendicularis DNF, et protrahatur linea equalem angulum tenens cum perpendiculari angulo FNG, et sit EN. Quoniam linea NQ minor est linea KH, est etiam minor linea QD, et ita angulus QDN minor est angulo DNQ, quare minor angulo GNF, quare etiam minor angulo ENF. Igitur linea EN et DQ concurrent. Sit ergo concursus in puncto E. Palam quod linea EQD est perpendicularis super speram, et secat lineam GNQ, que est linea reflexionis, in puncto Q, qui est punctus spere. Quare ymago puncti E, cum fuerit reflexio super punctum N, apparebit in puncto Q, et est in superficie spere.

[2.90] Si vero in linea EN sumatur punctum ultra E, utpote R, perpendicularis ducta ab eo ad centrum spere, sicut RD, secabit lineam reflexionis, que est GNQ, ultra punctum Q. Et est extra speram, quare ymago cuiuslibet puncti linee EN ultra E sumpti erit extra superficiem speculi.

136 SD: DS *O*/concurrent: concurrunt *FP1L3E* 137 puncto *corr. ex* puncta *F*/puncto quocumque *transp. L3*/secabit: secat *ER*/lineam: linea *L3*/SI: sumptam *O*; *om. C1* 138 que SI *transp. O*/*post* linea[1] *scr. et del.* ducta *F*/a *corr. ex* ei *a. m. F* 139 *post* linee *add.* ad punctum D *R*/erit perpendicularis: erunt perpendiculares *FP1* 140 a: cum *E* 141 intra speram *mg. O* 142 ergo *om. O*/quod *om. FP1* 147 N: enim *S* 148 ducatur: deducatur *L3* 149 *post* linea *inter.* tenens *a. m. O*/*post* angulum *scr. et del.* G *L3* 150 tenens *om. O*/*post* perpendiculari *add.* FN *mg. a. m. F*; *add. P1*/*post* sit *inter.* linea illa *L3*/*post* quoniam *add.* ergo *O* 151 minor *corr. ex* maior *FL3* (*mg. a. m. F*)/linea[1] *om. R*/angulus QDN *transp. R* 152 DNQ: QND *FP1*/DNQ . . . angulo (153) *mg. a. m. F*/*post* minor[1] *add.* est *FP1*/*post* GNF *add.* contraposito *FP1*/quare[2]: ergo *FP1*/*ante* etiam *add.* erit *FP1* 153 ENF *corr. ex* GNF *S*/linea *om. R* 155 secat *alter. in* secabit *a. m. F*/linea *inter. L3*/*post* reflexionis *scr. et del.* que est linea *F*/qui est punctus: quod est punctum *R*/qui . . . Q (157) *om. S* 156 E: G *O*/N *corr. ex* enim *L3* 157 apparebit: apparet *FP1*/et est *om. C1* 158 EN: NE *R*/R: K *FP1*; *corr. ex* ? *O* 159 sicut: que sit *R*/RD: KD *FP1*/*post* lineam *add.* GNQ *R* 160 que est GNQ *om. R*/est[1] *inter. a. m. E* 161 puncti: punctum *FP1*

[2.91] Si vero in linea EN citra punctum E sumatur aliquod punctum, perpendicularis ab eo ducta ad speculum secabit lineam GNQ intra speram, quoniam in puncto quod sit inter N et Q. Quare ymago cuiuslibet puncti linee EN inter E et N sumpti apparebit intra speram.

[2.92] Eadem penitus erit probatio, sumpto quocumque alio arcus BH puncto. Et ita ymago cuiuslibet puncti arcus BH una sola est in superficie speculi; aliarum quedam in speculo, quedam extra. Et quod demonstratum est in arcu ZB eodem modo potest patere in arcu ZA, et eadem penitus erit demonstratio, cuiuscumque circuli spere sumatur portio visui opposita et perpendiculari GD equaliter divisa.

[2.93] Unde visu immoto et perpendiculari GD manente, si moveatur equidistans perpendiculari linea GHK, secabit ex spera motu suo portionem circularem, et cuiuslibet puncti huius portionis ymago apparebit intra speram.

[2.94] Si vero linea contingens GB moveatur equidistanter perpendiculari visus, secabit ex spera portionem predictam maiorem, et a quolibet puncto excrementi unius super aliam refertur ymago cuius locus erit in superficie spere; et aliarum quedam intra speram, quedam extra.

[2.95] Scimus ex hiis quod in hoc speculo quelibet ymago apparet in dyametro spere, aut intra, aut extra, aut in superficie. Et omnis dyameter in quo appareat ymago aliqua in superficie spere aut extra dimissior est puncto spere quem tangit linea contingens a centro visus ducta, id est ultimo puncto portionis apparentis. Scimus quod quelibet linea reflexionis secat speram in duobus punctis: in puncto reflexionis et in alio.

163 citra: intra *O*; circa *E* 164 *ante* perpendicularis *add.* utpote C *mg. a. m. F; add. P1* / perpendicularis: perpendiculariter *C1* / ducta *corr. ex* ducatur *E* / *post* secabit *add.* sicut GD *mg. a. m. F; inter. P1* / *post* lineam *scr. et del.* reflexionis *S* 165 intra *corr. ex* ultra *mg. a. m. F* / quoniam *om. FP1* / *post* puncto *inter.* V *a. m. F; add. P1* 166 EN *om. O* / inter: citra *P1* / inter E *mg. a. m.* (inter: citra) *F* / E: Q *O* / et N *om. FP1* / N *corr. ex* EN *L3* 167 *post* sumpto *add.* a *L3* / alio *om. P1* 168 puncto . . . BH *inter. a. m. E* / ita *om. O* / ymago: ymaginem *S*; ymaginum *O*; *om. C1* / *post* est *add.* imago *R* 170 ZB *corr. ex* EB *O* / potest patere *transp. C1* / ZA: CA *O* 171 *post* penitus *scr. et del.* ZA et eadem penitus *F* 172 et: a *P1*; *alter. in a. m. F* / perpendiculari: perpendicularem *S*; per axem *O*; perpendicularis *C1* / GD: G et D *S*; GZD *O*; *alter. in* GZD *C1* / equaliter: earum *O* 173 immoto: in moto *S*; in motu *L3* / et *om. O* / GD *corr. ex* GGD *F*; G et D *S*; GZD *OR*; *corr. ex* GSD *L3* 174 equidistans: equidistantem *F*; equidistanter *OR* / *post* perpendiculari *add.* visus *R* / GHK: GH *P1R*; *alter. in F* 175 huius *om. P1S*; huiusmodi *O* / ymago *inter. O* 177 contingens GB *transp. R* / *post* contingens *inter.* scilicet *FO* 178 predictam: predicta *R* 179 unius *corr. ex* minus *O* / *post* unius *add.* portionis *R* / refertur: reflectitur *R* 180 superficie *om. O* / spere: spera *O* / quedam: quodam *F* 182 speculo: spero *S* / *post* ymago *scr. et del.* cuius locuius erit in superficie *S* 183 in[1] *inter. O* / dyametro *corr. ex* dextro *FL3* (*mg. a. m. F*); dextro *SC1* / *post* intra *add.* speram *R* 184 quo appareat: qua apparet *R* / ymago aliqua *transp. C1* / in[2] *inter. O* 185 dimissior: demissior *SR* / quem: quod *R* / linea *corr. ex* lineam *O* 186 id . . . puncto: in ultimum punctum *R* / *post* est *inter.* dimissior *a. m. F* / *post* scimus *add.* etiam *OR* (*inter. O*) / quelibet *corr ex* quecumque *a. m. C1*

[2.96] **[PROPOSITIO 10]** Restat ut loca ymaginum certius determinemus. Dico quod, sumpto dyametro, si ad ipsum ducatur linea secans speram a centro visus cuius pars interiacens punctum sectionis spere et punctum dyametri quem attingit est equalis parti dyametri interiacenti punctum illud et centrum, punctus ille non est locus alicuius ymaginis.

[2.97] Verbi gratia, sit AG [FIGURE 5.2.10, p. 568] circulus spere, H visus, ED dyameter spere, sive perpendicularis. Et HZ sit linea secans speram super punctum F et concurrens cum ED in puncto Z, et sit ZF equalis ZD. Dico quod Z non est locus alicuius ymaginis.

[2.98] Palam enim quod non est locus ymaginis alterius quam alicuius puncti ED, quoniam ymago cuiuslibet puncti est super dyametrum ab eo ad centrum spere ductum. Et quod locus ymaginis alicuius puncti ED non sit in Z sic constabit.

[2.99] Ducatur perpendicularis a puncto D super punctum F, que sit DFN, et super punctum F fiat angulus equalis angulo NFH, et sit QFN. Palam igitur quod angulus QFN est equalis angulo ZFD. Sed ZFD est equalis angulo ZDF, quia respiciunt equalia latera. Igitur QFN est equalis angulo ZDN, quare linea FQ est equidistans linee ED.

[2.100] Igitur in infinitum producte numquam concurrent. Igitur nullius puncti ED forma movebitur ad punctum F per QF, et non potest esse locus ymaginis alicuius puncti in puncto Z nisi forma eius moveatur ad F per lineam QF. Eadem erit improbatio, sumpto quocumque dyametro, quare propositum.

[2.101] Amplius, dico quod nullus punctus linee ZD potest esse locus alicuius ymaginis.

189 *post* restat *add.* iam *R* 190 sumpto: sumpta *R* / dyametro *corr. ex* dextro *O* / ipsum: ipsam *R* / ducatur *corr. ex* dicatur *L3* 191 *post* interiacens *scr. et del.* inter *C1* 192 dyametri[1]: dextri *FP1S*; *corr. ex* dextri *OL3C1* (*a. m. C1*) / quem: quam *R* / attingit *corr. ex* contingit *S* 193 *post* interiacenti *add.* inter *R* / punctus ille: punctum illud *R* / locus *corr. ex* locuius *mg. a. m. F* 196 dyameter: semidyameter *R* / et *om. FP1* 198 equalis: vel *FL3* / *post* equalis *add.* vel *P1* / ZD: EZD *FP1L3* 199 est *inter. a. m. E* / *post* est *scr. et del.* e *S* / alterius quam *om. S*; *mg. a. m. C1* 200 *post* alicuius *scr. et del.* quoniam alterius *C1* / puncti ED *om. S* / *post* puncti *add.* linee *R* / *post* ED *scr. et del.* quoniam ymago cuiuslibet puncti ED *L3* / quoniam . . . ED (202) *om. FP1* / cuiuslibet: cuiuscumque *S*; cuiusque *OC1* 201 dyametrum: dextrum *SL3*; *corr. ex* dextrum *O* / ductum: ductam *R* 203 a puncto D *scr. et del.* (D: B) *F*; *om. P1* / que *inter. a. m. F*; *om. L3*; et *R* / que . . . F (204) *om. SC1* 204 DFN: DFNA *P1L3*; *corr. ex F* / angulo *corr. ex* alio *a. m. C1* / sit[2] *inter. a. m. E* 205 quod *inter. O* / angulus: angelus *FE* / *post* QFN[2] *add.* non *S* / est equalis *transp. L3R* / est . . . angulo: equalis angulo est *FP1E* / angulo: A *S* / *post* angulo *add.* nec *S* / sed ZFD (206) *om. L3* 206 ZDF: ZDFQ *F* / quia . . . latera *om. R* / *post* equalia *scr. et del.* linea *O* 208 in *om. FSC1* / producte *corr. ex* producere *L3* 209 nullius *corr. ex* nullus *a. m. C1* / et *inter. FO* (*a. m. F*); *om. SL3C1E* / *ante* non *inter.* autem *L3E* (*a. m. E*); *add. C1* / *post* potest *add.* autem *R* 210 ymaginis alicuius *corr. ex* alicuius ymaginis *C1* / alicuius: alius *FP1* / eius *inter. a. m. E* 211 F *inter. C1* / erit *om. C1* / *post* erit *inter.* etiam *a. m. E* / improbatio . . . quocumque: probatio sumpta quacumque *R* / dyametro (212) *corr. ex* dextro *O* 212 *post* quare *add.* est *OR* (*inter. O*) 213 nullus punctus: nullum punctum *R* / locus . . . ymaginis (214): alicuius ymaginis locus *C1*

[2.102] Sumatur enim punctus P, et ducatur linea HP secans speram in puncto B. Et ducatur perpendicularis DBM, et angulo MBH fiat angulus equalis, qui sit TBM. Palam quod TBM est equalis PBD, et palam quod angulus DPH est maior angulo PZF, quia extrinsecus. Igitur duo alii anguli trianguli DPB sunt minores duobus aliis angulis trianguli DZF. Sed angulus PDB maior angulo ZDF. Restat ergo ut angulus DBP sit minor angulo DFZ. Sed angulus DFZ est equalis angulo ZDF, quare angulus DPB minor est angulo ZDF. Igitur multo minor angulo PDB, quare angulus TBM minor est angulo PDB. Igitur linee TB, ED numquam concurrent, et ita nulla ymago puncti B refertur ad punctum P. Similiter, nec ymago alterius puncti, et similiter de quolibet puncto linee ZD. Restat ergo ut tota ZD sit vacua a locis ymaginum.

[2.103] **[PROPOSITIO 11]** Amplius, sumpto quocumque dyametro inter lineas contingentie a visu ad speram ductas, preter dyametrum a centro visus ad centrum spere intellectum, et determinato in eo puncto quem diximus, qui est meta locorum ymaginum, dico quod in punctis tantum illius dyametri qui sunt inter superficiem spere et metam predictam sunt loca ymaginum punctorum illius dyametri.

[2.104] Verbi gratia, sint BZ, BE [FIGURE 5.2.11, p. 568] linee contingentie, B centrum visus, A centrum spere, BHA dyameter visualis, DA dyameter sumptus cum meta G punctus spere. Dico quod in sola puncta G et T interiacentia cadunt ymagines punctorum DA.

215 punctus: punctum *R* 216 *post* angulo *scr. et del.* B *C1* 217 TBM[1]: LH *FP1*; TBH *S*; *corr. ex* TB *L3*/TBM[2] *alter. ex* TBH *in* LBM *a. m. F*; BM *P1*; BH *S*; *corr. ex* TBH *L3*; MBH *C1E*/*post* palam *add.* quoniam *P1* 215 punctus: punctum *R* 216 *post* angulo *scr. et del.* B *C1* 217 TBM[1]: LH *FP1*; TBH *S*; *corr. ex* TB *L3*/TBM[2] *alter. ex* TBH *in* LBM *a. m. F*; BM *P1*; BH *S*; *corr. ex* TBH *L3*; MBH *C1E*/*post* palam *add.* quoniam *P1* 218 DPH: DPB *SC1*/PZF *alter. ex* ZF *in* DZF *O*/PZF . . . angulo (220) *om. FP1*/extrinsecus: exterior *R* 219 alii anguli *transp. E*/trianguli[1,2] *om. S* 220 *ante* DZF *add.* NBI *S*/DZF: ZDF *R*/angulus[1] *om. R*/PDB: DPB *C1*/*post* PDB *add.* est *ER*/angulo *corr. ex* angulos *C1* 221 DBP: DPB *FP1C1*; *corr. ex* DPB *S*/sed . . . ZDF *mg. a. m. E*/DFZ *corr. ex* DEF *S* 222 DPB: DBP *P1SOR*; *alter. in* DBP *L3* 223 PDB[1] . . . angulo *scr. et del. E*/quare: ergo *R*/angulus *om. R*/angulo *om. R*/*post* PDB[2] *scr. et del.* quare angulus *S*/linee *om. R*/*post* TB *inter.* et *O*/ED *inter. O* 224 concurrent: concurrunt *FP1L3E*/ymago . . . refertur: forma a puncto B reflectetur *R* 225 P *corr. ex* B *F*; B *P1*; H *R*/*ante* similiter[1] *add.* ut P sit locus imaginis *R*/*post* similiter[2] *scr. et del.* et *F* 226 *post* tota *add.* linea *ER* (*inter. E*) 227 sumpto quocumque: sumpta quacumque *R*/*post* quocumque *add.* primo *FP1*; *add.* puncto *L3*; *scr. et del.* puncto *E*/dyametro: dextro *FP1S*; *corr. ex* dextro *OL3* 228 preter *corr. ex* propter *P1* 229 intellectum *corr. ex* intelli *F*; intellectam *R*/eo: ea *R* 230 quem: quod *R*/qui: quod *R* 231 tantum: centrum *O*/qui: que *R*/metam *corr. ex* metram *F*; metram *P1* 232 *post* sunt *scr. et del.* sicut *P1* 233 sint: sunt *FP1L3*/contingentie (234): contingentes *R* 234 centrum[2] *corr. ex* centro *FL3* (*mg. a. m. F*); centro *P1O* 235 dyameter sumptus: deinde sumetur ymaginis *O*/*post* dyameter *scr. et del.* vis *P1*/sumptus cum: sumpta cuius *R*/cum meta *transp. O*/*post* meta *add.* sit *R*/G *om. O*; T *R*/*post* G *scr. et del.* dico *L3*; *scr.et del.* dico quod *E*/punctus: punctum *R*/punctus . . . dico: dico punctus spere *FS*/punctus . . . quod: dico quod punctus spere *P1*/*post* spere *add.* in quo diameter secat speram *R*/*post* puncta *add.* inter *R* 236 et *corr. ex* Z *S*; *om. R*/cadunt: sunt *inter. O*/*post* punctorum *add.* recte *R*

[2.105] Quod enim non cadent in puncto G vel extra superficiem spere palam per hoc quod supradictum est dyametrum in quo locus ymaginis erit in superficie spere aut extra demissiorem esse puncto contingentie; et cum dyameter DA sit inter lineas contingentie, non erit in eo locus ymaginis, aut in superficie spere aut extra. Quod autem in quodlibet punctum inter G et T sumptum cadat ymago constabit.

[2.106] Sumatur punctum, et sit Q, et ducatur linea BQ secans speram in puncto C. Et ducatur perpendicularis ACL, et angulo LCB fiat angulus equalis DCL, et educatur linea BT secans speram in puncto F, et ducatur perpendicularis AF. Igitur triangulus ACB continet triangulum AFB, quare angulus AFB maior angulo ACB. Restat ergo ut angulus AFT sit minor ACQ. Sed angulus AFT est equalis angulo FAT, quia equalia latera respiciunt. Igitur ACQ maior erit angulo CAQ, quare LCB maior CAQ, unde DCL maior CAQ. Igitur CD, AQ concurrent. Sit D concursus. Forma igitur puncti D reflectatur in puncto C per lineam CB, et locus ymaginis eius est Q. Et eadem est probatio, sumpto quocumque puncto inter G et T.

[2.107] **[PROPOSITIO 12]** Restat ut assignemus loca ymaginum in sectione spere occulta visui.

[2.108] Sint ergo AC, AG [FIGURE 5.2.12, p. 568] linee contingentes portionem apparentem, A centrum visus, B centrum spere, ADBZ dyameter visualis, et ZCG circulus spere in superficie linearum contingentie. Et protrahatur a centro ad punctum contingentie dyameter BG. Palam quod angulus ZBG est maior recto, cum enim in triangulo BAG angulus BGA sit rectus, erit angulus GBA minor recto, quare ZBG maior.

237 *post* quod *scr. et del.* n *C1* / cadent: caden *S*; cadit *E*; cadant *R* / puncto: punctum *R* / *post* superficiem *scr. et del.* terre *P1* 238 quod *rep. C1* / dyametrum: scilicet locus *O* / quo: qua *R* / *post* quo *add.* est *ER* 239 erit *om. R* / demissiorem *corr. ex* demissionem *F* 240 contingentie[2] *corr. ex* contingentes *a. m. E*; contingentes *R* / non: nunc *E* 241 eo: ea *R* 242 quodlibet *corr. ex* quolibet *O* / punctum *inter. E* / cadat *corr. ex* cadit *E* / *post* ymago *add.* sic *R* 244 C: P *R* / et *om. O* / ducatur: educatur *O* / ACL: APL *R* / angulo: angulus *F*; *corr. ex* angulus *L3* / LCB: LPB *R* 245 angulus equalis *transp. R* / DCL: DCB *FP1S*; DEB *O*; DPL *R* 246 et ducatur *om. O* / triangulus: triangulum *R* / ACB: APB *R* / continet *corr. ex* contineat *E* 247 AFB[1] *corr. ex* AB *L3* / *post* AFB[1] *scr. et del.* quare angulus *E* / *post* AFB[2] *add.* est *P1* / *post* maior *add.* est *R* / ACB: APB *R* 248 *post* minor *add.* angulo *SO* / ACQ: APQ *R* / angulus[2]: angelus *F* / est *om. FSL3* / equalis *corr. ex* inequalis *FL3*; inequalis *S* / *post* equalis *add.* est *mg. a. m. F* 249 FAT: fiat *F* / ACQ: APQ *R* / maior erit *transp. ER* / *ante* angulo *add.* FAT ergo et *R* / angulo CAQ *om. FP1SO*; *inter. L3*; *mg. a. m. C1E* / CAQ: PAQ *R* 250 quare LCB: FAQ quare *O* / LCB: LPB *R* / *ante* maior[1] *add.* erit *O* / *post* maior *add.* est *R* / CAQ[1,2]: PAQ *R* / unde . . . CAQ[2] *mg. a. m. L3* / DCL: DPL *R* / CD: PD *R* 251 D[2] *om. S* / reflectatur *alter. in* reflectetur *O*; reflectetur *R* / in: a *R* / C: P *R* 252 CB: CD *FP1*; PB *R* / eius *inter. a. m. E* 253 T: G *FP1*; D *SO* 254 ymaginum *corr. ex* ymaginis *P1* 256 sint: sunt *FP1L3* / *post* AC *add.* et *SO* 257 *post* centrum[2] *scr. et del.* de *C1* / ADBZ: ABZ *O*; DZ *E* / dyameter (258): dyametrum *FP1* 258 et *om. SC1ER*; *inter. O* / ZCG: ZOG *FP1* 259 et *om. O* 260 ZBG: et BG *P1* / enim *om. S* 261 angulus BGA *om. P1* / BGA: BAG *FSL3*; *corr. ex* BAG *a. m. E* / GBA *corr. ex* GA *O*; BGA *L3* / quare *corr. ex* quando *F* / *post* quare *add.* erit *O* / maior sit (262): minor fit *FP1*

Sit ergo HBG rectus. Erit ergo HB equidistans linee contingentie AG. Igitur producte numquam concurrent, et quilibet dyameter inter H et G concurret cum linea AG.

[2.109] Ducatur a puncto A linea secans speram, que sit AMO, ita quod corda, que est MO, sit equalis semidyametro OB, et concurrat dyametro BO cum linea AG in puncto T. Dico quod in quolibet puncto TO est locus ymaginis, et in nullo alio puncto dyametri TB est locus ymaginis, et sunt O, T termini locorum ymaginis.

[2.110] Sumatur enim punctum, et sit K, et ducatur ANK secans speram in puncto N. Et ducatur perpendicularis BNX, et angulo XNA fiat angulus equalis per lineam FN. Palam quod FN non cadet inter B et T, quoniam aut secans speram aut secat contingentem AP in duobus punctis. Igitur forma puncti F movebitur per FN ad punctum N, et reflectetur ad A per lineam AN, et apparebit ymago eius in puncto K. Et eadem probatio, sumpto quocumque alio puncto.

[2.111] **[PROPOSITIO 13]** Amplius, dico quod in arcu OG, quicumque sumatur dyameter, continebit loca ymaginum et intra speculum, et unam in superficie speculi, et alias extra speculum.

[2.112] Sumatur ergo punctus L [FIGURE 5.2.13, p. 569], et protrahatur dyameter BL usquoque secet AP in puncto E. Et producatur linea AL secans speram in puncto R. Palam quod RL minor est LB, quia est minor MO, que est equalis semidyametro. Si ergo ab A ducatur linea ad dyametrum LB cuius pars interiacens circulum et dyameter sit equalis parti dyametri a puncto in quod cadit ad centrum, cadet quidem

262 ergo[1] *inter. a. m. E* / *post* erit *scr. et del.* perpendicularis *P1* / HB *om. P1* / *post* AG *inter.* HB *L3* 263 concurrent *corr. ex* concurrunt *F* / et[1] *inter. E* / quilibet: quelibet *R* 264 *post* G *add.* tunc *O* / concurret: concurrent *S*; continet *O*; *corr. ex* concurrent *C1* 265 *post* ducatur *add.* ante *S*; *add.* autem *OC1* 266 semidyametro: semicirculo *FP1* / OB: OH *FP1*; *om. O* 267 dyametro *om. O*; *corr. ex* linea *E*; semidiameter *R* / dyametro BO: DIS ab O *FP1* 268 TO: DO *P1* / et *om. FP1SL3*; *inter. OC1* (*a. m. C1*) / et . . . ymaginis[1] (269) *inter. a. m. E* / dyametri: dicit *FP1L3*; diametrum *S*; *alter. ex* dicit *in* diametrum *O* / TB: EH *P1*; OB *SC1E* 269 ymaginis[2]: ymaginum *OC1R* 270 ducatur ANK *transp. ER* (ducatur *inter. a. m. E*) / *post* ducatur *add.* linea *O* / ANK: AK *O* 271 ducatur *corr. ex* dicatur *C1* / angulo *corr. ex* alio *mg. a. m. F* / *post* angulo *add.* alio *P1* / XNA: YNA *FSOL3*; NA *P1* 272 FN[2]: NF *R* / cadet: cadit *FP1* 273 et *om. R* / T *corr. ex* D *a. m. F*; TC *P1*; G *R* / *post* quoniam *add.* sic *R* / *post* aut[1] *inter.* esset *L3* / secans: secaret *OR*; secat *C1*; *alter. ex* secat *in* secaret *E* / secat: secaret *OR*; *alter. in* secaret *a. m. E* / AP: AG *R* 274 per: super *FP1* 275 AN *corr. ex* AM *F* / eius *om. FP1* 276 *post* probatio *add.* est *R* 277 quicumque (278): quecumque *R* 278 *post* continebit *inter.* quedam *O* / et *om. O* / *post* speculum *inter.* quasdam *L3E* (*a. m. E*); *add. R* 279 *post* unam *inter.* tantum *O* / speculi *corr. ex* puncti *FOC1* (*mg. a. m. C1*); puncti *P1S*; circuli *E*; *om. R* 280 ergo *inter. a. m. E* / punctus: punctum *FP1R* 281 BL *inter. O*; RBL *R* / usquoque: usquequo *FP1SL3*; quousque *R* / AP: GP *O*; AT *R* / in: a *FP1*; etiam *S* 282 R: L *O*; *om. S* / RL: YL *inter. O* / LB: TB *R* / quia *corr. ex* quod *a. m. E* 283 MO: LM *O* / que *corr. ex* qui *O* / est *om. SO* / semidyametro: semicirculo *FP1* / *post* ducatur *add.* semidyameter *mg. a. m. F*; *add. P1* 284 ad *inter. O* / LB: BL *R* / *post* interiacens *add.* inter *R* / dyameter: dyametrum *ER* 285 quod: quo *O* / *post* cadit *add.* usque *R* / cadet: cadit *C1* / quidem *om. R*

inter L et B. Si enim inter L et E ceciderit, erit RL maior LB, et omnis linea interiacens centrum et illam equalem erit maior parte dyametri que terminatur, secundum probationem assignatam in explanatione mete ymaginum.

[2.113] Sit ergo punctus in quem linea equalis cadet I. Dico quod in quolibet puncto in EI sumpto est locus ymaginis. Et erit eadem demonstratio que fuit in TO.

[2.114] Igitur quedam ymagines in dyametro EB sortiuntur loca intra speculum, quedam extra speculum, una sola in superficie, scilicet puncto L. Et ita poteris demonstrare in quolibet dyametro partium OG transeunte.

[2.115] **[PROPOSITIO 14]** Amplius, sumpto quocumque dyametro in arcu OH, locus ymaginis in eo erit extra speculum.

[2.116] Sumatur dyameter BQ [FIGURE 5.2.14, p. 570], et concurrat cum contingente in puncto P. Et ducatur linea ANQ secans speram in puncto N. Iam dictum est quod MO est equalis OB, sed NQ est maior MO, quare NQ est maior QB. Et linea ducta ad circumferentiam ad dyametrum PB equalis parti BP ipsam et centrum interiacenti non cadet inter Q et B. Si enim ceciderit, secundum supradictam probationem erit NQ minor QB.

[2.117] Restat ergo ut linea equalis cadat inter P et Q. Et quod non cadat in punctum P palam per hoc quod angulus PGB rectus. Igitur PB maius PG. Cadet ergo citra P.

[2.118] Sit punctus in quem cadit S. Erit igitur S meta locorum ymaginum, et quilibet punctus inter P et S erit locus ymaginum, et est eadem probatio quam supra.

286 inter[1]: intra *FP1SOC1E* / L et B: B et L *L3* / inter[2] *om. C1* / RL: L *O; corr. ex* LR *E* / *ante* maior *inter.* illa *O* / et[3] *om. R* / *post* omnis *add.* enim *R* 287 *post* interiacens *add.* inter *R* / *post* illam *add.* partem linee reflexionis illi parti diametri *R* 288 que: qua *R* / terminatur: terminantur *FP1*; terminat *SO* / explanatione: expletione *FP1*; explatitione *S*; exploratione *O*; *corr. ex* expletione *a. m. C1* 290 sit *corr. ex* si *F* / ergo punctus *transp. FP1* / punctus: punctum *R* / quem: quod *R* / cadet: cadit *R* / I:L *FP1L3* 291 quolibet: quocumque *C1* / in *om. FP1SL3E*; linee *R* / EI *om. S* / sumpto *om. FP1L3ER* / est locus *transp. E* 292 TO: DO *FP1*; SD *O* 293 intra (294): inter *FP1O* 294 *post* scilicet *add.* in *SOC1R* 295 quolibet: qualibet *R* / partium: per puncta arcus *R* 297 sumpto quocumque: sumpta quacumque *R* 298 in arcu OH *corr. ex* m ar cuoh *mg. a. m. F* / *post* OH *add.* m ar cuoh *P1* / locus . . . P (300) *mg. O* 299 dyameter: dyametrum *SC1* 300 in[1] *inter. E* / ANQ: AUQ *OL3C1ER* 1 N: U *R* / NQ *alter. in* UQ *a. m. E*; UQ *R* / est *inter. a. m. E* 2 MO . . . maior *mg. a. m. L3* / NQ: UQ *L3R*; *alter. in* UQ *a. m. E* / *post* maior *add.* OB id est *R* / QB: BQ *R* / ad[1] *alter. in* a *O*; a *R* / circumferentiam: circulo *O*; circumferentia *R* 3 BP: PB *SOL3C1ER* / ipsam . . . interiacenti: interiacenti inter ipsam et centrum *R* / interiacenti: interiacentem *E* 5 NQ: ut *O*; *alter. in* UQ *a. m. E*; UQ *R* / minor *alter. in* maior *O* 6 cadat: cadit *L3* 7 quod *inter. O*; quia *R* / *post* PGB *add.* est *R* / PB *corr. ex* PG *L3* 8 *post* maius *add.* est *R* / cadet *corr. ex* cadat *F* / citra *alter. ex* circa *in* intra *O* / *post* citra *add.* punctum *L3C1ER* / P: PAP *FP1* 9 punctus in quem: punctum in quod *R* / erit igitur S *om. FP1* / S *inter. O* 10 quilibet punctus: quodlibet punctum *R* / ymaginum *corr. ex* ymaginis *P1* / est eadem (11) *transp. R* 11 quam *alter. in* que *a. m. E*; que *R*

[2.119] Palam ex hiis quod ymagines dyametrorum arcus HO omnes sunt extra; ymagines dyametri FB una in superficie, que est in O, alie omnes extra, scilicet in TO; omnes autem ymagines dyametri arcus OG, quedam intra, quedam extra, quedam in superficie.

[2.120] **[PROPOSITIO 15]** Amplius, in arcu HZ [FIGURE 5.2.15, p. 570] non potest sumi dyameter in quo est locus ymaginis, quoniam nullus dyameter ibi sumptus concurrit cum contingente AP.

[2.121] Et a quocumque puncto illius talis dyametri ducatur linea ad speram. Cadet quidem in portione GZC et nulla in portione GDC, nisi secando speram. Ergo nulla forma puncti alicuius talis dyametri veniet ad portionem visui apparentem.

[2.122] Quod autem dictum in arcu GH potest eodem modo demonstrari in parte arcus CZ eam respiciente. Et sumpto arcu citra Z equali HZ, in nullo dyametro illius arcus erit ymaginis locus.

[2.123] Idem est demonstrandi modus in quocumque circulo, quare, si linea HB moveatur, eodem manente angulo HBZ, signabit motu suo portionem spere in dyametris cuius nullus sit ymaginis locus. Si vero HB immota, moveatur OB, describetur portio cuius omnes ymagines extra; verum dyametrum TB una in superficie, alie extra. Moto autem arcu OG, fiet portio cuius quedam ymagines in superficie, quedam extra speculum, quedam intra.

[2.124] Verum visus non comprehendit que ymagines in superficie spere aut que extra, nec certificatur in comprehensione earum nisi quod

12 ymagines *om. O*/dyametrorum: dyametrum *O*/omnes: OS *FP1SO; corr. ex* OG *a. m. E* 13 *post* extra *add.* superficiem speculi *R*/ymagines: imaginum *R*/dyametri: dicit *FP1;* dyametrum *S*/FB: FY *R*/superficie *corr. ex* se *O*/*post* superficie *add.* speculi *R*/in O *corr. ex* LO *a. m. E*/O: L *R*/*post* O *inter.* nec *O*/*post* alie *add.* intra scilicet in IL alie *R* 14 TO: DO *FP1;* LE *R*/omnes . . . ymagines: omnium autem imaginum *R*/*post* ymagines *scr. et del.* extra *P1*/*post* dyametri *add.* et *FP1SO; scr. et del.* et *L3C1* 15 *post* intra *add.* speculum *R* 17 quo: qua *R* 18 nullus: nulla *R*/sumptus: sumpta *R*/sumptus concurrit *corr. ex* concurrit sumptus *C1*/concurrit: concurrunt *S* 19 quocumque puncto *transp. C1*/dyametri ducatur *transp. FP1* (dyametri: dyameter *F*) 20 portione[1,2]: portionem *R*/in[2] *inter. E* 21 ergo: quare *R*/puncti alicuius *transp. R* 23 dictum in arcu: in arcu dictum *L3*/*post* dictum *add.* est *C1ER*/GH: GHZ *R* 24 demonstrari *corr. ex* demonstra *F*/respiciente: respicientie *FP1;* respicientem *O*/et *inter. a. m. E*/arcu: arca *S; om. O*/citra: cui *FL3;* in *P1;* circa *SO;* circu *C1; corr. ex* circa *E*/Z *alter. ex* AZ *in* CZ *mg. F; alter. ex* CD *in* C *P1;* et *S;* AZ *L3* 25 nullo: nulla *R* 26 idem est *transp.* (idem *inter.*) *O*/est *rep. S* 27 si linea: similia *FP1S; corr. ex* similia *L3*/HB moveatur *corr. ex* moveatur HB *C1*/*post* HB *scr. et del.* immota *S*/signabit: significabit *E*/*post* signabit *scr. et del.* n *P1* 28 dyametris cuius *transp. R* 29 immota *alter. in* mota *O; alter. in* in meta *a. m. E*/*ante* moveatur *scr. et del.* mota *F; inter.* non *O*/OB: OH *SER*/portio: proportio *FP1* 30 *post* extra[1] *add.* speculum sunt *R*/verum . . . extra[2] *om. R*/TB: TH *FP1*/alie: alia *O* 31 fiet: fiat *FP1L3E*/*post* ymagines *add.* sunt *R*/in superficie: et sic *FP1*/extra (32): intra *SO* 32 intra: extra *SO*/*post* intra *add.* quedam extra speculum *R* 33 *post* ymagines *add.* sint *OR* (*inter.O*) 34 que *inter. O*

sunt ultra portionem apparentem. Iam ergo determinata sunt in hiis speculis ymaginum loca.

[2.125] **[PROPOSITIO 16]** Amplius, puncti visi forma non potest in hoc speculo ad visum reflecti nisi a solo puncto speculi.

[2.126] Sit enim punctus B [FIGURE 5.2.16, p. 571], A centrum visus, et non sit A in perpendiculari ducta ad centrum. Dico quod B refertur ad A ab uno solo puncto speculi, et unam solam pretendit visui ymaginem in hoc speculo.

[2.127] Palam quod ab aliquo puncto potest reflecti forma eius. Sit illud G, et ducantur BG, AG. Et sit N centrum spere, et ducatur dyameter BN secans superficiem spere in puncto L, et termini portionis opposite visui sint D, E. Et secet linea AG perpendicularem in puncto Q, qui est locus ymaginis.

[2.128] Palam quod A, N, B sunt in eadem superficie ortogonali super speram. Et cum omnes superficies ortogonales super speram in quibus fuerit BN secent se super BN, et non possit superficies in qua linea BN extendi ad punctum A nisi una tantum, palam quod A et B sunt in una superficie ortogonali tantum super speram, non in pluribus. Et cum necesse sit quod punctus visus et A sint in eadem superficie ortogonali super punctum reflexionis, palam quod non fiet reflexio puncti B ad visum nisi in circulo spere qui est in superficie ANB. Sit ergo circulus DGE. Dico iterum quod a nullo puncto huius circuli, preter quam a G, fiet reflexio.

[2.129] Si dicatur quod a puncto L, cum BN sit perpendicularis, et AL non sit perpendicularis, et forma per perpendicularem veniens nec-

35 sunt[1]: sint *FP1* / apparentem *corr. ex* apparens *O* 36 *post* loca *add.* A *C1* 37 *post* visi *add.* in *S* / forma *inter. a. m. E* 38 *post* ad *add.* unum *R* / a: ab *R* / *ante* solo *add.* uno *R* / *post* solo *add.* uno *C1* 39 enim *om. O* / punctus: punctum *R* / *post* punctus *inter.* visus *a. m. O*; *add.* qui erit visus *C1*; *add.* visum *R* 40 et *inter. S* / non: enim *P1* / sit *om. FP1* / A *inter. L3* / perpendiculari: perpendiculariter *FP1* / *post* centrum *add.* spere *ER* (*inter. a. m. E*) / refertur: reflectitur *R* 41 puncto speculi *transp. ER* (puncto *inter. a. m. E*) / speculi *om. O* / pretendit: ostendit *R* / visui *alter. ex ? in* ipsa *O* 43 ab *om. S* 44 N: enim *S*; *corr. ex* enim *L3* 45 termini *corr. ex* TL *mg. F*; *om. P1* / opposite visui (46) *transp. R* 46 linea: lineam *FP1SOR*; *corr. ex* lineam *L3* / qui: quoniam *S*; quod *R* 47 ymaginis: ymaginum *FP1L3ER* 48 sunt: sint *R* 49 *post* speram[2] *scr. et del.* est *O* 50 fuerit: fuit *O*; fuerint *R* / se *om. O*; *inter. a. m. E* / super BN: superficiem BNA *O* / qua linea (51): quo lineam *SO* 51 linea *om. FP1* / linea BN *transp. R* / BN: EBN *P1* / ad punctum *rep. P1* / et *inter. O* / B: BN *OC1*; *alter. in* N *E* / *post* B *add.* et N *L3R* (*inter. L3*) 52 ortogonali *om. FP1* / ortogonali tantum *transp. R* / ortogonali . . . superficie (53) *mg. S* / *post* speram *inter.* et *O* / non *alter. ex* NN *in* N *F*; N *S* 53 quod . . . visus: ut omne punctum visum *R* / *post* quod *inter.* omnis *E* 54 reflexio *corr. ex* refer *mg. F* 55 circulo: circulum *FP1* / qui: quoniam *F*; EM *P1* / ANB: AMB *P1* 56 ergo *om. FP1* / iterum: esse *scr. et del. O*; igitur *R* / a nullo: autemlo *S* 57 a G *om. FP1* / reflexio *corr. ex* reflexionibus *F*; *corr. ex* inflexio *O*; inflexio *L3* / *post* reflexio *add.* a puncto G *P1* 58 *post* si *add.* enim *R* / L: R *FP1* / *post* sit *add.* super superficiem speculi *R* / perpendicularis: particularis *S* 59 perpendicularis: particularis *S* / per *om. FP1*; *inter. L3E* / perpendicularem: perpendicularis *FP1*; partem *S*

essario per perpendicularem reflectatur, planum quod non refertur B ad A a puncto L. Iterum nec ab alio puncto arcus LE. Quia ad quodcumque punctum illius arcus ducatur linea a puncto B, tenebit cum contingente illius puncti angulum obtusum ex parte E, et linea ducta a puncto A ad illud punctum tenebit cum contingente illa angulum acutum ex parte L. Quare si ab illo puncto fieret reflexio, esset angulus acutus equalis obtuso.

[2.130] Iterum a nullo puncto GL potest fieri reflexio. Sumatur enim punctum quodcumque, et sit Z, et ducatur linea AZO secans perpendicularem in puncto O. Et ducatur linea contingens circulum in puncto Z, que cadit necessario inter BG et BL, et sit MZ. Et sit FG linea contingens circulum in puncto G. Palam ex superioribus quod proportio BN ad NQ sicut BF ad FQ. Eodem modo erit proportio BN ad NO sicut proportio BM ad MO. Sed maior est proportio BN ad NQ quam BN ad NO. Igitur maior est proportio BF ad FQ quam BM ad MO, quod plane impossibile, cum BF sit minus BM, et FQ maius MO. Restat ergo ut a puncto Z non fiat reflexio.

[2.131] Verum quod ab aliquo puncto arcus GD non fiat reflexio sic constabit. Sumatur quodcumque punctum, et sit T. Educatur linea BT, et linea ATH secans BN in puncto H. Et ducatur contingens circulum in puncto T, que sit CT. Erit ergo proportio BN ad NH sicut BC ad CH, et BN ad NQ sicut BF ad FQ. Sed maior BN ad NH quam BN ad NQ. Ergo maior BC ad CH quam BF ad FQ, quod plane falsum, cum BF maior BC, et CH maior FQ. Restat ergo ut a nullo puncto arcus GD fiat reflexio puncti B, quare quodlibet punctum ab uno solo puncto spere refertur ad visum. Ergo una sola erit linea reflexionis cuiuslibet puncti visi, quare unica unius puncti ymago.

60 per: propter *F; om. O; corr. ex* propter *a. m. E* / perpendicularem: partem *S*; perpendiculariter *O* / planum: palam *R*; palam *mg. a. m. E* / refertur: reflectetur *R* / B ad A (61): BA DA *FS* 61 iterum nec *om. R* / *ante* ab *add.* planum etiam est quod non reflectetur *R* / alio: aliquo *FP1* 62 *post* puncto *scr. et del.* A ad illud punctum tenebit cum contingente *S* 63 contingente *corr. ex* contingentie *S* / puncti . . . illa (64) *mg. a. m. E* 64 illud: illum *FP1* 67 iterum: item *SC1* / *post* a *scr. et del.* viso *P1* / *post* puncto *add.* arcus *C1R* / potest: post *L3* / enim *corr. ex* EN *F* 68 AZO: AZD *FP1S* 69 perpendicularem: partem *S* 70 Z *corr. ex* C *a. m. F*; C *P1*; *om. S*; *corr. ex* E *L3* / cadit: cadet *SOC1* / sit^2 *om. R* / linea contingens (71) *om. R* 71 circulum *corr. ex* circulo *F* / *post* circulum *add.* contingat *R* / proportio *corr. ex* portio *E* 72 BN^1: DN *E* / erit *om. ER* / BN^2 . . . proportio^2 (73) *om. S* 74 maior est *om. P1* / est *alter. in* erit *a. m. E* 75 BF: LF *F* / minus: minor *R* / maius: maior *R* 77 verum . . . reflexio *om. P1* / ab *inter. O* 78 educatur: et ducatur *R* / BT: DT *C1* 79 ATH: ATB *C1* / BN . . . CH (80) *mg. a. m. C1* / et^2 *om. R* / et ducatur: educatur *O* 80 CT: PT *R* / *post* ergo *add.* ex superioribus *R* / BC: HC *S*; BP *R* / CH *corr. ex* BH *P1*; PH *R* 81 NQ *corr. ex* QN *L3* / maior . . . NH: BN ad NH maior est *R* / ad^3: A *S* / NH *corr. ex* NF *O*; NK *C1* / BN^3 *om. FP1SO; inter. L3C1* 82 *post* maior *add.* est proportio *R* / BC: BP *R* / CH *corr. ex* BH *P1*; PH *R* / ad^2 *om. F* / cum: quoniam *S* / *post* BF^2 *add.* sit *R* 83 BC: BP *R* / CH: PH *R* / *post* CH *scr. et del.* et *L3* 84 spere refertur (85): speculi reflectitur *R* 85 puncti *corr. ex* puncto *O* 86 unica *corr. ex* mututa *mg. a. m. F*; mutata *P1*; linea *L3*

[2.132] Si autem punctus B fuerit in perpendiculari visuali, palam quod reflectitur ab uno solo puncto, quoniam per perpendicularem tantum, et unica erit eius ymago et propter continuitatem aliorum punctorum in loco ymaginis proprio.

[2.133] **[PROPOSITIO 17]** Amplius, si in aliquo dyametro sumantur duo puncta ex eadem parte centri, locus ymaginis puncti centro propinquioris erit remotior a centro spere loco ymaginis puncti remotioris a centro spere. Et locus reflexionis puncti propinquioris centro erit remotior a centro visus quam puncti remotioris a centro spere.

[2.134] Verbi gratia, dico quod locus ymaginis puncti C [FIGURE 5.2.17, p. 572] remotior est a centro loco ymaginis puncti B, et punctus reflexionis puncti C remotior ab A puncto puncto reflexionis puncti B, qui est punctus G. Dico quoniam punctus C non reflectitur nisi ab aliquo puncto arcus GL.

[2.135] Palam enim quod non reflectetur ab aliquo puncto arcus LE, nec a puncto L, nec a puncto G, cum B reflectitur ab eo. Et si dicatur quod ab aliquo puncto arcus GD, sit ille punctus T, et sit CT linea per quam forma movetur ad speculum. Et ducatur perpendicularis NT per que necessario dividet angulum CTA per equalia, et ducatur perpendicularis NGK. Erit angulus NTA maior NGA. Restat ergo angulus PTA minor angulo KGA, quare angulus CTP minor angulo BGK. Sed angulus CTP valet angulum TNC et angulum TCN, quia extrinsecus. Et angulus BGK valet angulum GNB et angulum GBN. Erunt ergo duo anguli TNC, TCN minores duobus angulis GBN, GNB, quod est impossibile, cum angulus TNC contineat GNB tanquam partem et angulus TCN sit maior GBN.

87 si . . . ymago (89) *om. S* / autem punctus *corr. ex* a puncto *O* / punctus: punctum *R* / fuerit *om. FP1* 88 reflectitur: reflectetur *R* / puncto *inter. a. m. E* / quoniam *om. R* / per *om. FP1O; inter. L3E* (*a. m. E*) / *post* per *add.* quod *R* / perpendicularem: perpendicularis *FP1R; corr. ex* perpendicularis *a. m. E* 89 eius ymago *transp. FP1* 91 alio: aliquo *L3*; aliqua *R* / sumantur duo puncta (92): duo puncta sumantur *FP1* 92 eadem . . . centri: parte centri eadem *R* / puncti *om. R* / centro . . . puncti (93) *mg. a. m.* (centro propinquioris (93) *transp.*) *C1* 94 et . . . spere (95) *om. R* 95 centro[1] . . . remotior *inter. L3* / visus . . . centro[3] *rep. P1* / *post* remotioris *scr. et del.* verbi gratia *O* 96 C *inter. O*; P *R* 97 et *inter. O* / punctus: punctum *R* 98 C: P *R* / remotior: remotius *R* / ab: a *L3E* / A *om. E* / A puncto *transp.* (A *inter.*) *L3* / *post* A *add.* visus *C1* / *post* puncto[1] *add.* visus *R* / puncto[2] *om. P1OE* / reflexionis *corr. ex* remotionis *O* 99 qui: quod *R* / punctus[1,2]: punctum *R* / quoniam: quod *ER* / punctus[2]: punctum *FP1* / C *om. FP1*; P *R* / reflectitur: reflectetur *OC1* 100 aliquo: alio *SO* / puncto *inter. L3* / *post* arcus *add.* puncto *L3* 101 reflectetur: reflectitur *R* 102 LE . . . arcus (103) *mg. a. m. E* / nec[1]: nisi *ER* / reflectitur: reflectetur *R* 103 ille *alter. in* illud *O* / ille punctus: illud punctum *R* / CT: TP *R* 104 et ducatur: educatur *O* / NT: PTU *R* 105 per[1] *om. R* / CTA: PTA *R* / et ducatur *corr. ex* educatur *O* 106 NGK: NGH *S* / NGA: NQA *F* 107 PTA: UTA *R* / KGA *corr. ex* GA *P1* / CTP: PTU *R* / BGK *corr. ex* GP *O* 108 CTP: OTP *S*; PTU *R* / valet: valeat *FP1* / TNC: DNC *F*; TCN *O*; TNP *R* / angulum[2]: angulus *FP1*; *corr. ex* angulus *L3*; *om. R* / TCN: TNC *OE*; TPN *R* 109 extrinsecus: exterior *R* 110 TNC: tunc *S*; TNP *R* / TCN *inter. L3*; TPN *R* / *post* GBN *add.* et *C1* 111 TNC: PNT *R* / *post* contineat *add.* angulum *R* 112 TCN *corr. ex* CTN *a. m. F*; CTN *P1*; TPN *R*

[2.136] Restat ergo ut punctus C non reflectatur nisi a punctis G et L intermediis. Et omnes linee a puncto A per hec puncta ducte ad dyametrum BN cadunt in puncta a centro spere remotiora puncto Q, et cadunt in puncta spere a centro visus magis elongata puncto G, et ita propositum.

[2.137] **[PROPOSITIO 18]** Amplius, dato speculo et dato puncto viso, est invenire punctum reflexionis.

[2.138] Sit enim B [FIGURE 5.2.18, p. 572] punctus visus, A centrum visus, et ducantur ab eis due linee ad centrum speculi. Si fuerint ille linee equales, erit facile invenire. Quoniam sumetur circulus spere in superficie illarum linearum, et scimus quod ab uno solo puncto illius circuli fit reflexio. Dividetur ergo angulus quem continent in centro ille due linee per equalia.

[2.139] Et ducatur linea dividens angulum extra speram. Erit quidem perpendicularis super lineam contingentem hunc circulum in puncto per quem transit. Et si ducantur ad illud punctum due linee, una a centro visus alia a puncto viso, efficient cum perpendiculari illa et duabus primis lineis duos triangulos, quorum duo latera duobus lateribus equalia, et angulus angulo. Et ita punctus circuli per quem transit illa perpendicularis est punctus reflexionis, quod est propositum.

[2.140] Si vero linea a puncto viso ad centrum spere ducta fuerit inequalis linee a centro visus ad idem centrum ducte, oportet nos quedam antecedentia proponere, quorum unum est:

[2.141] **[PROPOSITIO 19]** Sumpto circuli dyametro et sumpto in circumferentia puncto, est ducere ab eo ad dyametrum extra productum

113 ut: quod *E*/punctus C: punctum P *R*/*post* punctis *add.* inter *R* 114 ducte: ducta *S* 115 *ante* BN *add.* super *O*/cadunt . . . et *om. R*/*post* remotiora *add.* a *O; scr. et del.* a *E*/puncto *corr. ex* puncta *F*/*post* puncto *add.* L *FP1* 116 elongata *corr. ex* elevata *P1*/*post* elongata *add.* a *FP1L3; scr. et del.* a *E*/G: P *P1*/*post* G *add.* amplius dato PN speculo et dato puncto viso est invenire punctum reflexionis *FP1* (PN *om. F*)/*post* ita *add.* patet *R* 117 *post* propositum *add.* est *P1* 118 amplius . . . reflexionis (119) *om. FP1* 120 enim: N *P1S*/B *inter. a. m. C1*/punctus visus: punctum visum *R*/ 121 et ducantur: educantur *O*/*post* fuerint *add.* due *R*/ille linee (122) *transp. P1* 122 quoniam *corr. ex* quando *L3; corr. ex* qui *a. m. E*/*post* in *scr. et del.* spere *E* 123 *post* superficie *add.* duarum *R*/uno: alio *S;* unico *C1ER;* angulo *L3; alter. ex* angulo *in* aliquo *O*/solo puncto *transp. C1* 124 circuli: cum *scr. et del. O*/*post* fit *add.* unius puncti *R*/dividetur *alter. in* dividatur *a. m. C1;* dividatur *R* 125 ille due *transp. R* 128 quem: quod *R*/illud: illum *FP1* 129 visus *om. S; inter. O*/alia *om. R* 130 duos triangulos: duo triangula *R*/*post* latera *add.* a *P1* 131 punctus: punctum *R*/quem: quam *S;* quod *R* 132 transit *inter. O*/ transit . . . perpendicularis: perpendicularis illa transit *R*/punctus: punctum *R*/quod est propositum *om. FP1R* 133 linea *om. O*/*post* linea *scr. et del.* perpendicularis *F*/*post* puncto *scr. et del.* linea *O*/*post* viso *inter.* linea *O* 134 *post* inequalis *add.* R *S*/*post* a *add.* quod est propositum *FP1* 135 antecedentia: accidentia *FP1S; corr. ex* accidentia *OL3*/praponere: preponere *SR* 136 sumpto[1]: sumpta *R* 137 productum: productam *R*

lineam que a puncto in quo secat circulum usque ad concursum cum dyametro sit equalis linee date.

[2.142] Verbi gratia, sit QE [FIGURE 5.2.19a, p. 573] data linea, GB dyameter circuli ABG, A punctus datus. Dico quod a puncto A ducam lineam que a puncto in quo secaverit circulum usque ad dyametrum GB sit equalis linee QE, quod sic constabit. Ducantur due linee AB, AG que aut erunt equales aut inequales.

[2.143] Sint equales, et adiungatur linee QE linea talis ut illud quod fiet ex ductu totius cum adiuncta in adiunctam sit equale quadrato AG, et sit linea adiuncta EZ. Cum igitur illud quod fiet ex ductu QZ in EZ sit equale ei quod fit ex ductu AG in se, erit QZ maior AG. Si enim EZ fuerit equalis aut maior AG, est impossibile ut ductum QZ in EZ sit equale quadrato AG. Si autem minor, palam quod QZ maior AG.

[2.144] Producatur ergo AG ad equalitatem, et sit AGT. Et posite pede circini super A, fiat circulus secundum quantitatem AGT, qui quidem circulus secabit dyametrum BG, et secet in puncto D. Et ducatur linea AD, que secabit necessario circulum, si enim esset contingens in puncto A, esset equidistans BG, et numquam concurreret cum ea. Secet igitur in puncto H, et ducatur linea GH.

[2.145] Palam, cum ABGH sit quadrangulum intra circulum, duo anguli oppositi, scilicet ABG, AHG, valent duos rectos. Sed AGB est equalis ABG, cum respiciant equalia latera, ex ypothesi. Erit igitur angulus AHG equalis angulo DGA, et angulus HAG communis triangulo totali ADG et partiali AHG. Restat ergo ut angulus HDG sit equalis angulo HGA, et triangulus sit similis triangulo, quare proportio DA ad

138 *post* puncto *scr. et del.* secat *F* / concursum *corr. ex* centrum *P1*; *corr. ex* circulum *C1* 140 sit *om. S* 141 dyameter: diametrus *C1* / punctus *corr. ex* punctis *P1* / punctus datus: punctum datum *R* / ducam: ductam *F*; *corr. ex* ductam *P1L3* 142 secaverit: secavit *O*; secuerit *R* 143 ducantur *corr. ex* dicantur *F* 145 sint: sicut *S* / adiungatur *corr. ex* adiungantur *OC1* / illud: illum *FP1*; istud *C1* 146 fiet: fit *R* / totius: eius *inter. O* 147 illud: illum *F* / fiet: fit *R* / ex . . . fit (148) *om. FP1* 148 QZ: quia *S* / *post* AG[2] *add.* et EZ minor eadem *R* 149 equalis *corr. ex* equales *O* / impossibile: impossibilis *FP1SL3*; *corr. ex* impossibilis *C1* / ductum: ductus *R* 150 equale: equalis *R* / quod *inter. a. m. E* / QZ *corr. ex* QD *P1* / *post* QZ *add.* est *R* 151 ergo AG *mg. a. m. F*; *transp.* (AG *mg. a. m.*) *L3* / AG *om. SOC1E* / *post* AGT *scr. et del.* et posito pede super A fiat circulus secundum quantitatem *F* / et . . . AGT (152) *mg. a. m. F* / posite: posito *R* 52 circini: circuli *P1* / secundum: super *C1* 154 AD: ad A *O* / que *om. O* / secabit necessario *transp. SOC1* / esset contingens: contingeret *R* 155 esset: esse *FP1S* / concurreret: concurrent *S*; *corr. ex* concurrent *P1*; *alter. ex* concurrent *in* concurret *OC1*; concurret *E* 157 palam: planum *C1* / cum *corr. ex* autem *O* / ABGH: ABH *P1* / *post* circulum *add.* ABG AHG *R* / duo *om. R* 158 anguli . . . valent: angulos oppositos valere *R* / est equalis (159) *om. P1* 159 *post* equalis *add.* angulo *R* / ABG *inter. O* / equalia latera *corr. ex* equalater *P1* / *post* ypothesi *scr. et del.* erit igitur angulus ABG equalis angulo DGA et duos rectos seu ABG est equalis ABG cum respiciant equalia latera ex ipothesi *O* 160 AHG: AGH *FP1* / HAG *om. P1*; AHG *L3* / *post* communis *scr. et del.* commune *O* 161 *post* partiali *scr. et del.* H *F* / angulus *corr. ex* triangulus *C1* 162 triangulus sit similis: triangulum simile *R*

AG sicut AG ad AH. Igitur quod fit ex ductu DA in HA est equale quadrato AG. Sed DA equalis TA; igitur est equalis QZ. Et erit AH equalis EZ et DH equalis QE, que est data linea, et ita propositum.

[2.146] Si vero AB et AG non sint equales [FIGURE 5.2.19c, p. 573], protrahatur a puncto G linea equidistans AB que sit GN, et sumatur linea quecumque, que sit ZT, et fiat super punctum Z angulus equalis angulo AGD per lineam ZF. Et ducatur a puncto T linea equidistans ZF, et sit TM, et ex angulo TZF secetur angulus equalis angulo NGD per lineam ZM, quoniam hec linea necessario concurret cum TM, cum sit inter equidistantes. Et sit punctus concursus M. Restat igitur angulus MZF equalis angulo AGN.

[2.147] Et a puncto T ducatur linea equidistans linee ZM, que sit TO, que quidem necessario concurret cum FZ. Et sit concursus in puncto K. Et sumatur linea cuius proportio ad lineam ZT sicut BG ad EQ, lineam datam, et sit I. Deinde fiat super punctum M sectio piramidalis quemadmodum docet Ablonius in libro secundo de piramidalibus, propositione quarta, et sit UCM, que quidem sectio non secet lineas KO, KF. Et in hac sectione ducatur linea equalis linee I, scilicet MC, et producatur usque ad lineas KT, KF, et sint puncta sectionum O, L. Igitur, sicut ibidem probabitur, erit OM equalis CL .

[2.148] Et a puncto T ducatur linea equidistans CM, que sit TF, et super punctum A fiat angulus equalis angulo ZFT per lineam AND. Palam quod hec linea concurret cum GD, cum angulus AGN sit equalis FZM angulo, et angulus GAN angulo ZFT. Igitur AD linea aut erit contingens circulo aut secabit ipsum, quoniam, si non fuerit contingens, et arcus AB fuerit maior arcu AG, secabit arcum AB, et si AB fuerit minor, secabit arcum AG.

163 *post* AG[1] *add.* est *mg. a. m. F; add.* sit *C1* / sicut AG *om. P1; mg. a. m.* (*post* sicut *add.* proportio) *F* / sicut . . . AH *inter. a. m. E* / quod: cum *L3; corr. ex* cum *C1* / HA: AH *FP1ER; corr. ex* AG *L3* 164 *post* DA *add.* est *R* 165 DH: TH *F; alter. in* TH *P1* / *post* ita *add.* quod *C1; add.* est *R* 166 et . . . AB (167): fuerit maior AG et linea ducta ab A fuerit contingens ducatur AG equidistans AB *mg. O* 167 *post* linea *scr. et del.* reflexionis *P1* / GN *corr. ex* NG *a. m. E* 168 que *om. C1R* / fiat . . . Z: super punctum Z fiat *R* 170 ZF: et F *S* / et[1] *inter. L3E* / et[1] . . . TZF *om. P1* / TZF: ZF *S* / angulus *corr. ex* angulos *F;* angulis *C1* / NGD: NDG *S; corr. ex* NDG *L3;* DGN *C1ER* 171 quoniam: que *O; om. R* / hec *om. P1* / *post* hec *add.* igitur *R* / concurret: concurrit *R* 172 et *om. R* / punctus: punctum *R* 174 et *inter. a. m. C1* 175 que *om. F* 176 *post* sicut *scr. et del.* et *C1* / EQ: QE *R* 177 I: L *FP1;* Z *S* / piramidalis: pyramidis *L3* 178 Ablonius: Apollonius *R* / piramidalibus: piramidibus *O* 179 UCM: UC *O* / secet: secat *R* 180 I:L *C1E* / I scilicet *om. FP1L3* / scilicet *inter. OC1* / MC: LMC *FP1L3;* IMO *S* 181 ad *om. FP1SE; inter. O* / *post* et *add.* si *F; add.* sicut *S* / sectionum: sectionis *O* / O: D *C1* 182 sicut ibidem: sunt idem *P1* / probabitur *corr. ex* probatur *E;* probatur *R* 184 ZFT *corr. ex* FT *a. m. C1* / AND *corr. ex* AFT *P1* 186 angulo[1] *om. O* / ZFT: ZFZ *F* / erit contingens (187): tanget *R* 187 circulo: circulum *OR* / ipsum: circulum *C1* / fuerit contingens: tetigerit *R* 188 arcu AG *inter. a. m. E* / arcum . . . arcum (189) *om. P1* / et[2] . . . AG (189) *rep.* (AG[1]: AB) *S; mg. a. m. E*

[2.149] Sit igitur contingens in puncto A. Cum angulus GAN sit equalis angulo ZFT, et angulus AGN angulo FZY, erit tertius tertio equalis, et erit triangulus AGN similis triangulo ZFY. Similiter, cum AGD sit equalis angulo FZT, erit triangulus AGD similis triangulo FZT. Igitur que est proportio AN ad AG ea est proportio FY ad FZ, et que est proportio AG ad GD ea FZ ad ZT, quare que est proportio AN ad GD ea est FY ad ZT.

[2.150] Verum, cum TM sit equidistans FL, et FT equidistans ML, erit FT equalis ML, quare erit equalis CO, cum sit MO equalis LC. Sed MO est equalis YT, cum sit ei equidistans, et YM equidistans TO. Restat ergo FY equalis CM. Sed CM equalis I. Erit igitur FY equalis I. Sed proportio I ad ZT sicut BG ad EQ. Igitur proportio AN ad GD sicut BG ad EQ.

[2.151] Verum angulus GAN est equalis angulo GBA, sicut probat Euclides in tertio. Sed angulus NGD est equalis angulo ABG, cum NG sit equidistans AB. Igitur angulus NGD equalis est angulo NAG, et angulus NDG communis, quare tertius tertio equalis, quare triangulus NDG similis triangulo ADG. Igitur proportio AD ad GD sicut GD ad ND, quare quod fit ex ductu AD in DN est equale quadrangulo DG.

[2.152] Verum quadratum AD est equale ei quod fit ex ductu BD in DG, sicut probat Euclides, et quadratum AD est equale ei quod fit ex ductu AD in DN et ei quod fit ex ductu AD in NA. Et illud quod fit ex ductu BD in DG est equale quadrato DG et ei quod fit ex ductu BG in GD, sicut probat Euclides. Ablatis ergo equalibus, restat ut quod fit ex ductu AD in AN sit equale ei quod fit ex ductu BG in DG. Igitur proportio secundi ad quartum sicut tertii ad primum, quare proportio

190 sit[1] *om. R* / igitur *om. FP1* / igitur contingens: tangat igitur *R* / contingens: congingens *F* / in *om. FP1S; inter. O* / *post* cum *add.* igitur *R* 191 angulo[2] *om. FP1* 192 triangulus: triangulo *O*; triangulum *R* / similis: simul *S*; simile *R* / triangulo *om. O* / *post* cum *add.* angulus *C1* 193 FZT[1] *corr. ex* ZFT *L3* / *post* FZT[1] *scr. et del.* erit tertius tertio *S; add.* et *S* / erit . . . FZT[2] *om. FP1* / triangulus: triangulum *R* / similis: simile *R* 194 AN *mg. C1* / est[2] *om. E* / FY *om. P1; corr. ex* AN *L3* / et *inter. P1* 195 *post* ea *add.* est *R* / que *om. FP1* / AN *corr. ex* NA *E* / ad[3] *inter. O* 196 ea: eadem *P1* / *post* est *add.* proportio *FP1* / FY: SY *F* 197 FL *corr. ex* A *L3* / *post* FT *add.* sit *ER* / ML: LM *R* 198 erit[1]: est *R* / erit[1] . . . ML *om. S* / FT: ET *P1* / ML: LM *R* / *post* ML *scr. et del.* et FT equalis LM ergo et *C1; scr. et del.* equalis ML *E* / sit MO *transp. R* 199 ei *om. FP1; inter. L3*; ipsi *R* / *post* equidistans[1] *scr. et del.* totum sit MO equalis LC sed MO est equalis *S* 200 *post* CM[2] *add.* est *R* / erit *om. R* / igitur: quare *R* / *post* FY[2] *add.* est *R* 201 ZT: TZ *L3* / *post* sicut *scr. et del.* sicut *F* 204 in tertio *om. O* 205 angulus *corr. ex* anguli *F* 206 *post* tertio *add.* est *R* / triangulus: triangulum *R* 207 similis: simile *R* / ADG: AGD *C1* / ad[1] *om. P1S* / GD[1]: DG *R* 208 ND: DN *R* / *post* AD *scr. et del.* ad ZB *O* / DN *alter. ex* CND *in* ND *O* / quadrangulo: quadrato *ER* 209 fit *mg. F* / BD: IBD *E* 210 *post* Euclides *add.* 35a propositione *E; add.* 36 propositione *R* / equale: equalis *S* / ex . . . fit[1] (211) *om. P1* 211 *ante* AD[1] *scr. et del.* AD *O* / *post* ductu[2] *add.* BG in GD sicut probat Euclides et quadratum AD est equalis ei quod fit ex ductu AD in DN *S* / AD[2] . . . DG[2] (212) *om. S* 212 *post* equale *scr. et del.* quadrangulo *C1* 213 GD: DG *FP1* / sicut . . . DG (214) *mg. E* / Euclides *corr. ex* eudocles *C1* / ut *om. L3* 214 in[1]: I *P1* / AN *corr. ex* AM *F* / ductu[2] *om. R* / DG: GD *R* / igitur *alter. ex* L *in* ergo *L3* 215 proportio secundi: prime linee ad secundam *R* / ad quartum *om. R* / quartum: quartam *FP1* / *ante* sicut *add.* est *R* / tertii: tertiae *R* / primum: quartam *R*

AN ad DG sicut BG ad AD. Sed iam dictum est quod proportio AN ad GD sicut BG ad EQ. Igitur EQ equalis AD, quod est propositum.

[2.153] Quod si AD non fuerit contingens circulum sed secans, et fuerit AG maior AB [FIGURE 5.2.19d, p. 574], secabit quidem AG. Secet ergo in puncto H, et ducatur linea AG.

[2.154] Palam quod duo anguli AHG, ABG valent duos rectos. Sed angulus NGD est equalis angulo ABG. Igitur angulus AHG et angulus NGD sunt equales duobus rectis. Quare angulus NGD est equalis angulo NHG, et angulus NDG communis, quare tertius angulus tertio angulo equalis est, et triangulus HGD similis triangulo NDG. Igitur proportio HD ad DG sicut proportio DG ad DN, quare illud quod fit ex ductu HD in DN est equale quadrato GD.

[2.155] Sed quod fit ex ductu AD in DH est equale ei quod fit ex ductu BD in DG, sicut probat Euclides, et illud quod fit ex ductu AD in DH est equale ei quod fit ex ductu DH in DN et DH in AN. Et quod fit ex ductu BD in DG est equale ei quod fit ex ductu BG in GD et quadrato GD. Ablatis igitur equalibus (scilicet quadrato GD et eo quod fit ex ductu DH in DN) restat ut illud quod fit ex ductu DH in AN est equale ei quod fit ex ductu BG in DG, quare proportio secundi ad quartum (id est AN ad GD) sicut tertii ad primum (id est BG ad DH). Sed iam probatum est quod proportio AN ad DG sicut BG ad EQ. Igitur EQ est equalis DH, et ita propositum.

[2.156] Si vero AG erit minus AB (et secet ad arcum AB), sit sectio punctus H [FIGURE 5.2.19e, p. 574], et ducatur linea HG. Palam quod angulus NGD est equalis angulo ABG. Sed anguli ABG, AHG sunt equales, quia cadunt in eundem arcum. Igitur angulus NGD equalis est angulo AHG, et angulus NDG communis, quare tertius tertio equalis, et

216 DG: GD *SOC1R* 217 igitur EQ *om. S; inter. L3/post* EQ² *add.* est *ER* 218 fuerit contingens: tetigerit *R*/secans: secuerit *R* 219 *post* quidem *add.* ad arcum *P1; add.* arcum *R* 220 ergo in puncto: in puncto ergo *C1*/H *corr. ex* AG *O*/et ducatur: educatur *O*/AG: HG *OR* 221 AHG: AH *S*/ABG *inter. O* 222 NGD: NDG *L3*/est equalis *transp. R*/*post* est *scr. et del.* q *P1*/angulo *om. ER*/*post* angulus *add.* NGD /AHG *inter.* (*ante* AHG *inter.* et) *O*/et angulus NGD (223) *om. O* 223 NGD²: NDG *L3* 225 equalis est *transp. OER* (est *inter. O*)/est *om. SL3C1*/triangulus: triangulum *R*/HGD: HDG *SO*/similis: simile *R* 226 ad DG *om. S*/*post* DG¹ *add.* est *R*/*post* quare *add.* et *L3* 227 est . . . GD: valet quadratum DG *E*/GD: DG *R* 228 ductu: conductu *FP1S; corr. ex* conductu *C1*/AD *inter. a. m. E*/DH: HD *R*/*post* equale *scr. et del.* quadrato *S* 229 ductu¹: conductu *L3*/fit *inter. a. m. E* 230 equale: equalis *O*/DH² . . . ductu² (231) *rep.* (ei²: est) *S*/DH³ *corr. ex* H *C1* 232 GD¹˒²: DG *R*/igitur equalibus *corr. ex* equalibus igitur *E*/equalibus *corr. ex* qualibet *O*/GD²: DG *C1R* 233 in¹ . . . BG (234) *om. P1*/ut *inter. a. m. E*/DH² *corr. ex* DG *L3*/est: sit *R* 234 secundi *om. P1;* secunde linee *R* 235 tertii ad primum: tertie ad primam *R* 236 igitur EQ *inter.* (igitur: quare) *O* 237 DH: BH *L3; corr. ex* HD *C1*/*post* ita *add.* est *R* 238 erit minus: sit minor *R*/et . . . AB² *inter. L3*/AB sit *rep. O*/sectio punctus (239): sectionis punctum *R* 239 punctus: puncti *O* 240 sunt: fuit *E* 241 arcum *corr. ex* arcus *O*/NGD: NDG *S; corr. ex* NDG *L3*/equalis est *transp. R* 242 AHG: DHG *O*/communis *om. S*

trianguli similes. Igitur proportio HD ad GD sicut GD ad DN, quare quod fit ex ductu HD in DN est equale quadrato GD.

[2.157] Sed quod fit ex ductu HD in DA est equale ei quod fit ex ductu BD in DG, et quod fit ex ductu HD in DA est equale ei quod fit ex DN in HD et AN in HD. Et ductus BD in DG valet quadratum DG et ductum BG in DG. Igitur, remotis equalibus, erit ductus HD in NA sicut BG in DG. Igitur proportio AN ad DG sicut BG ad HD. Sed iam dictum est quod proportio AN ad DG est sicut BG ad EQ. Igitur EQ est equalis HD, quod est propositum, quia a puncto A dato duximus lineam secantem circulum, et a puncto sectionis ad dyametrum est equalis linee date.

[2.158] **[PROPOSITIO 20]** Amplius, a puncto dato in circulo extra dyametrum eius est ducere lineam per dyametrum ad circulum, ut pars eius a dyametro ad circulum sit equalis linee date.

[2.159] Verbi gratia ABG [FIGURE 5.2.20, p. 575] sit circulus datus, BG dyameter, A punctus datus, HZ linea data. Dico quod a puncto A est ducere lineam transeuntem per dyametrum BG cuius pars a dyametro ad circulum sit equalis linee HZ.

[2.160] Probatio: ducantur linee AB, AG, et super punctum H fiat angulus equalis angulo AGB per lineam MH, et super idem punctum fiat angulus equalis angulo ABG per lineam HL. Et a puncto Z ducatur equidistans linee HM, que sit ZN, que quidem secabit HL, et a puncto Z ducatur linea equidistans HL, que sit ZT, et secet HM in puncto T. Et a puncto T ducatur sectio piramidalis TP, quam assignabit Ablonius in libro piramidis, que quidem sectio non continget aliquam linearum ZN, HL inter quas iacet. Similiter fiat sectio piramidalis ei opposita inter easdem lineas, que sit CU.

[2.161] Cum igitur linea minima ex lineis a puncto T ad sectionem CU ductis fuerit equalis dyametro BG, circulus factus secundum hanc minimam lineam, posito pede circini super punctum T, continget sec-

243 trianguli similes: triangula similia *R* / GD[1,2]: DG *R* 244 GD: DG *R* 245 ex *inter. O* / ex ductu *om. FP1S; inter. L3; mg. a. m. C1* / ductu *om. O* / fit[2] *om. E* 246 BD: DH *C1* / quod[1] *om. S* / ductu[2] *om. SOC1* / ex[2] *mg. a. m. C1* / *post* ex[2] *add.* ductu *R* 247 quadratum: quantum *FP1S; corr. ex* quantum *OL3* 248 ductus: ductio *P1; alter. ex* reductio *in* ductio *F* / NA: HNA *S; corr. ex* HNA *L3E*; AN *R* / *post* NA *add.* equa et ductus *mg. O* 249 sicut *om. O* / *post* sicut *scr. et del.* ad EQ igitur EQ est equalis HD quod est propositum *S* / BG[1] . . . DG[2] *mg. O* / *post* BG[2] *scr. et del.* in DG *S* / *post* ad[2] *scr. et del.* HG *P1* 250 est[2] *om. SOC1* 251 quia: quare *L3ER* / duximus *corr. ex* diximus *L3* 252 equalis: equale *FP1L3* 256 ad: a *O* 257 circulus datus *transp. R* 258 punctus datus: punctum datum *R* / *post* HZ *scr. et del.* et *C1* 261 probatio: preter ea *O; om. R* 262 AGB *corr. ex* ABG *L3* / idem punctum *transp. FP1* 263 AGB *corr. ex* ABG *L3* / HL *corr. ex* HHL *F* / Z: et *S* 265 linea *om. FP1L3E* / ZT: GT *E* / et[2] . . . T (266) *om. P1* 266 piramidalis: piramidis *FP1L3ER* / quam *om. S* / assignabit *alter. in* assignat *OC1; alter. in* assignavit *E*; assignavit *R* / Ablonius: Apollonius *R* / *post* Ablonius *scr. et del.* et *O; scr. et del.* UZ *C1* 267 *post* libro *add.* secundo *O* / piramidis: piramidum *SOR* / linearum: lineam *FP1L3E* 268 piramidalis: piramidis *FP1L3ER* 270 cum *scr. et del. L3* / *post* cum *scr. et del.* sit *C1* / *post* igitur *add.* si *FL3; add.* similia *S* 271 ductis: ductus *FP1* / ductis . . . CU[1] (273) *om. S*

tionem CU. Si vero minima ex lineis a puncto T ad sectionem CU ductis fuerit minor dyametro BG, circulus factus modo predicto secundum quantitatem BG secabit sectionem CU in duobus punctis.

[2.162] Sit ergo CT minima et equalis dyametro BG, que quidem secabit ZQ et HF, cum ducatur ad sectionem que eas interiacet. Et ducatur a puncto Z equidistans huic que quidem secabit HM et HL sicut sua equidistans. Secet ergo in punctis M, L, et sit MZL, et punctus sectionis in quo CT secat ZN sit Q, et super dyametrum GB fiat angulus equalis angulo HLZ, qui sit DGB. Et ducantur due linee AD, BD.

[2.163] Palam cum angulus GAB sit rectus, alii duo anguli trianguli AGB valent rectum, quare angulus LHM est rectus, et est equalis angulo GDB. Et angulus HLM est equalis angulo DGB. Igitur tertius tertio, et triangulus similis triangulo, quare proportio GB ad BD sicut LM ad MH.

[2.164] Sed quoniam angulus ADB equalis est angulo BGA, quia cadunt in eundem arcum, et angulus BGA equalis angulo MHZ, ex ypothesi, erit angulus ADB equalis angulo MHZ. Et iam habemus quod angulus GBD est equalis angulo HMZ. Ergo tertius tertio, et triangulus DEB similis triangulo MHZ. Sit E punctus in quo linea AD secat dyametrum BG. Igitur proportio BD ad DE sicut MH ad HZ. Igitur proportio BG ad DE sicut LM ad HZ.

[2.165] Verum Ablonius probat quod, cum fuerint due sectiones opposite piramidales inter duas lineas, et producetur linea ab una sectione ad aliam, pars eius que interiacet unam sectionem et unam ex lineis est equalis alii parti que interiacet aliam sectionem et aliam lineam, quare QC equalis TF. Sed TQ est equalis MZ, cum sit ei equidistans et inter

273 CU[2]: CP *S* / ductis *corr. ex* ductus *F*; ductus *P1* 274 maior: minor *R* 275 CU *om. R* 276 CT: TC *OL3C1ER* 277 secabit *om. O* / ZQ: ZN *R* / HF: HL *R* / *post* que *scr. et del.* iacet *L3*; *add.* inter *R* / eas *inter. L3* 278 que *inter. L3* / et *om. R* 279 *post* equidistans *scr. et del.* huic que quidem *S*; *add.* TC *R* / *post* et[2] *add.* in *O* / punctus *corr. ex* punctis *P1*; punctis *O*; punctum *R* 280 CT: TC *SL3C1ER* / ZN *corr. ex* ZM *L3* 281 HLZ *alter. in* HLM *F*; HLM *P1R* / DGB *corr. ex* DBG *L3* / due linee *R* / BD: DB *C1* 282 *post* palam *add.* quod *E*; *add.* ergo *R* / trianguli: scilicet *P1* 283 LHM *corr. ex* HHM *O* 284 DGB *corr. ex* GDGB *L3* 285 triangulus similis: triangulum simile *R* / *post* triangulo *scr. et del.* simul *E* / GB: BG *R* / *post* BD *add.* est *R* 287 equalis est *transp. SOC1* / BGA *corr. ex* BGB *S*; *corr. ex* GBA *L3* 288 ex ypothesi (289) *om. R* / ex . . . MHZ (289) *inter. a. m. E* 289 ypothesi: pothesi *S* / erit: est *R* / *ante* angulus *add.* ergo *R* / ADB *om. P1* 290 GBD: DAB *S*; *corr. ex* DAB *L3*; DBG *C1R*; *alter. ex* DAG *in* DAB *E* / tertius tertio *transp. C1* / triangulus: triangulum *R* 291 similis: simile *R* / *post* similis *scr. et del.* tib *S* / *post* sit *add.* autem *R* / punctus: punctum *R* 292 igitur[2] . . . HZ (293) *om. OR*; *mg. a. m.* (proportio BG *transp.*) *L3* 294 Ablonius: optimum *FP1*(vel Ablonius *inter. a. m. F*); *corr. ex* oblonus *O*; OB cum *L3*; obtum *alter. in* obtusum *alter. in* Ablonius *inter. a. m. E* / *ante* probat *add.* 12 *FP1*; *add.* NZ *L3E* / probat *inter. a. m. E* / fuerint: fiunt *L3* / sectiones *corr. ex* sectionis *O* 295 piramidales: piramidum (vel piramidales *mg. a. m.*) *F*; piramidis *P1*; piramides *L3*; piramidi *E* / piramidales . . . lineas *om. R* / inter duas: interpositas *C1* / producetur: producitur *FP1*; producatur *R* / ab *inter. L3*; a *ER* / una *om. L3ER* 296 *post* interiacet *add.* inter *R* 297 alii: alteri *C1* / *post* interiacet *add.* inter *R* / *post* aliam[1] *add.* in *FP1* / *post* lineam *scr. et del.* in sectionem *P1* 298 *post* equalis[1] *add.* et *R* / TF: DF *E* / sed TQ est *inter. a. m. O* / sed . . . TF (299) *om. S* / TQ: TF *P1* / ei: enim *E*; illi *R* / et *inter. a. m. E*; *om. R*

duas equidistantes. Igitur MZ equalis FC, et ZL equalis TF. Igitur ML equalis TC, quare proportio GB ad ED sicut TC ad HZ, et cum TC sit equale BG, erit ED equalis HZ, quod est propositum.

[2.166] Si autem linea a T ad sectionem CU ducta et minima fuerit minor dyametro BG, producatur ultra sectionem donec sit equalis. Et secundum quantitatem eius fiat circulus qui quidem secabit sectionem in duobus punctis a quibus linee ducte ad T erunt equales BG. Et a puncto Z ducatur equidistans utrique, et tunc erit ducere a puncto A, modo predicto, duas lineas equales linee date. Erit idem penitus probandi modus.

[2.167] **[PROPOSITIO 21]** Amplius, dato triangulo ortogonio, et dato puncto in uno laterum angulum rectum continentium, est ducere a puncto illo lineam ad aliud latus continentium rectum lineam secantem tertium oppositum recto, ita quod pars huius linee interiacens punctum sectionis et latus in quo non est punctus datus se habeat ad partem lateris oppositi recto que est a sectione ad latus in quo est punctus datus sicut data linea ad datam lineam.

[2.168] Verbi gratia ABG [FIGURE 5.2.21, p. 576] est triangulus datus, cuius angulus ABG rectus, et in latere GB est punctus datus D extra triangulum aut intra. Dico quod a puncto D est ducere lineam secantem latus AG et concurrentem cum latere AB ita quod pars eius interiacens latera AB AG sit eiusdem proportionis ad partem lateris AG que est ab illa linea usque ad punctum G sicut se habet E ad Z, que sunt date linee.

[2.169] Probatio: sit punctus D in ipso triangulo ABG, et ducatur ab eo linea equidistans AB, que sit DM. Et fiat circulus super tria puncta G, M, D, et protrahatur linea AD. Et quoniam planum quod angulus GMD est equalis angulo GAB, erit maior angulo GAD. Secetur ex eo

299 FC: FZ *P1* / et . . . TF *om. O* 300 *post* TC^{1} *add.* quare proportio TC ad HZ sicut ML ad HZ *R* / ED *corr. ex* OED *F* / HZ *corr. ex* HE *a. m. F*; *corr. ex* HTL *P1* 1 equale: equalis *R* / ED *om. S*; *inter. L3E* (*a. m. E*) / est *mg. E* 2 linea *corr. ex* aliam *S* / a *om. R* / T: TC *R* 4 *post* quidem *add.* circulus *R* 5 duobus punctis *transp. C1* 6 Z: T *S*; *corr. ex* ZT *C1* / *post* utrique *scr. et del.* et *P1* / A . . . predicto (7) *om. O* 7 *ante* duas *inter.* dato *O* / lineas *om. SOC1* / *post* date *inter.* et *O* / erit: eritque *inter. a. m. C1*; eritque *mg. a. m. E*; eritque *R* 8 modus: motus *P1* 11 ad *om. E* / latus: laterum *R* / continentium *corr. ex* continendum *O* / lineam^{2} *om. R* 12 oppositum *alter. in* rectum *a. m. E* / quod: ut *R* / interiacens: interiacentis *FP1L3* 13 punctus datus: punctum datum *R* / habeat *corr. ex* habet *a. m. E* 14 oppositi *corr. ex* oppositum *C1* / a *corr. ex* de *a. m. E*; de *R* 15 punctus datus: punctum datum *R* 16 ABG . . . datus: est triangulum datum ABG *R* 17 *post* angulus *rep. et del.* in (13) . . . lineam (15) *O* / ABG *corr. ex* ATG *O* / est: cum *O* / *ante* punctus *inter.* fuerit *a. m. O* / punctus datus: punctum datum *R* / D *om. O* 19 et: ad *S* / quod: ut *R* / *post* interiacens *add.* inter *R* 20 eiusdem: eius *FP1L3ER* 21 E ad Z *corr. ex* EA DZ *FL3* 23 probatio *om. R* / punctus: punctum *R* 24 sit: sunt *S* / fiat . . . D (25): super tria puncta G M D fiat circulus *R* / super *corr. ex* circa *a. m. C1* 25 M D *corr. ex* D M *C1* / et^{2} *scr. et del. E*; *om. R* / *post* planum *add.* est *R* 26 GAB: GNB *E* / angulo *om. R* / *post* eo *scr. et del.* a *C1*

equalis per lineam MN, et sit DMN, et fit H linea ad quam se habeat AD sicut se habet E ad Z. Et a puncto N, qui est punctus circuli, ducatur linea ad dyametrum GM equalis linee H secundum supradicta, et sit NL, et punctus in quo secat circulum sit C. Et ducatur linea GC, et a puncto D ducatur linea ad punctum C, que, cum cadat inter duas equidistantes, tenens angulum acutum cum altera, si producatur, necessario concurret cum alia. Concurrat igitur, et sit punctus concursus Q.

[2.170] Palam quod angulus GMD est equalis angulo GCD, quia cadunt in eundem arcum, et angulus GMD est equalis angulo GAB. Restat igitur ut angulus GCQ sit equalis angulo GAQ. Sit T punctus in quo DQ secat AG, et angulus GTC est equalis angulo ATQ. Igitur tertius tertio, quare triangulus ATQ similis est triangulo TCG. Igitur proportio QT ad TG sicut AT ad TC.

[2.171] Verum angulus NMD est equalis angulo TAD et angulo NCD, quare NCD equalis TAD. Et angulus CTL communis duobus triangulis, quare tertius tertio, et triangulus similis triangulo, scilicet TLC triangulo TAD. Igitur proportio TA ad CT sicut proportio AD ad LC, quare erit proportio AD ad LC sicut QT ad TG. Sed LC est equalis linee H, et proportio AD ad H sicut E ad Z. Igitur proportio QT ad TG sicut E ad Z, quod est propositum.

[2.172] Si vero D sumatur in illo latere extra triangulum [FIGURES 5.2.21a, 5.2.21b, pp. 576 and 577], producatur a puncto D equidistans AB, et sit DM, et ducatur AG donec concurrat cum DM in puncto M. Et fiat circulus transiens per tria puncta G, D, M, et ducatur linea AD. Erit quidem angulus GAD maior angulo GMD. Fiat ei equalis, et sit NMD, et a puncto N, qui sit punctus circuli, ducatur linea equalis H linee ad quam H se habeat AD sicut E ad Z, et sit NCL, et hoc super dyametrum MG. Et concursus sit L.

27 MN *alter. ex* ANU *in* KM *S; corr. ex* NM *a. m. E* / *post* linea *scr. et del.* equidistans *P1* 28 E ad Z: EA DZ *scr. et del. F; corr. ex* EA DZ *L3*; E ad DZ *P1* / qui est punctus: quod est punctum *R* 29 linee: linea *FP1* / supradicta: supradictam *SOL3* 30 punctus: punctum *R* / C *inter. a. m. E* 31 cadat: cadit *C1* 32 equidistantes: equidem *S* / tenens: tenet *C1* 33 punctus: punctum *R* 34 Q *corr. ex* QI *F* 35 palam: planum *R* 36 cadunt: cadent *L3* / GAB *corr. ex* GABB *C1*; GMD *E* 37 ut *corr. ex* un *P1* / punctus: punctum *R* 38 *post* secat *scr. et del.* se *C1* 39 triangulus: triangulum *R* / ATQ: AQT *S* / similis: simile *R* / est *om. R* 40 QT *corr. ex* QTA *F* / TC *corr. ex* DC *F* 41 NMD: HMD *SL3* / et ... TAD (42) *om. P1* 42 NCD[1]: NOD *FSL3* / NCD[2]: LCT *R* / NCD[2] ... CTL *inter. a. m. E* / *post* NCD[2] *add.* est *C1* / *post* equalis *add.* angulo *O* / CTL: CTA *FP1* / *post* communis *scr. et del.* duo *S* 43 triangulus similis: triangulum simile *R* / scilicet: S *S* 44 TLC *om. FP1* / triangulo *corr. ex* triangulus *F* / TAD *corr. ex* DAD *mg. a. m. F* / *ante* igitur *add.* similis triangulo LCT *mg. a. m. F; add. P1* / TA: AT *C1* 45 quare ... LC[1] *mg. a. m. L3* / erit *om. R* 46 AD *inter. L3* 48 D: DQ *S* / sumatur: sumuatur *F* / *post* sumatur *add.* quidem *O* / in illo *corr. ex* nullo *C1* / triangulum *corr. ex* angulum *E* 49 producatur: producto *SOC1* / *ante* a *add.* ducatur *OC1* / a puncto *om. S* / equidistans: equidistanter *FP1*; equidem *SOL3* 51 AD: et *S* 53 N: K *S* / qui sit punctus: quod est punctum *R* 54 *post* quam *add.* lineam *R* / et hoc *om. R*

[2.173] Cum igitur angulus NMD et angulus NCD valeant duos rectos, et angulus NMD sit equalis angulo TAD, erunt duo trianguli TCL, TAD similes. Et cum duo anguli GCD, GMD sint equales, erunt duo trianguli GCT, TAQ similes, et erit proportio AD ad CL, que est equalis H, sicut QT ad TG, et ita E ad Z sicut QT ad TG, quod est propositum.

[2.174] **[PROPOSITIO 22]** Amplius, duobus punctis datis, scilicet E, D, et dato circulo, est invenire punctum in eo ut angulum contentum a lineis a punctis predictis ad illud punctum ductis dividat per equalia linea circulum contingens in puncto illo.

[2.175] Verbi gratia ducatur a puncto E [FIGURE 5.2.22, p. 578] ad centrum circuli dati linea EG, et producatur usque ad circumferentiam, et sit ES. Deinde ducatur linea GD et sit MI linea in puncto C divisa ut proportio IC ad CM sicut EG ad GD. Et dividatur MI per equalia in puncto N, et ducatur perpendicularis NO. Et super punctum M fiat angulus equalis medietati anguli DGS per lineam MO. Palam quod erit minor recto, et angulus ONM rectus. Igitur MO concurret cum NO. Concurrat autem in O puncto, et a puncto C ducatur linea ad triangulum que sit CKF ita ut proportio KF ad FM sit sicut proportio EG ad GS. Et super punctum G fiat angulus equalis angulo MFK per lineam usque ad circulum ductam, que sit AG, et sit angulus AGE. Et ducantur due linee AG, DG. Dico quod A est punctus quem querimus.

[2.176] Ducatur linea EA. Cum ergo MFK sit equalis angulo AGE, et proportio KF ad FM sicut proportio EG ad GA, cum GA sit equalis GS, erit triangulus AGE similis triangulo MFK. Igitur, angulus FMK est equalis angulo EAG, et angulus AEG equalis angulo MKF.

56 et[1] . . . NMD (57) *om. FP1; mg. a. m. L3* / *post* duos *scr. et del.* DM in puncto *O* 57 duo *inter. a. m. E* / trianguli: triangula *R* 58 similes: similia *R* / sint: sunt *L3* 59 trianguli: triangula *R* / similes: similia *R* / CL *corr. ex* AD *C1* 60 ad[1] *om. O* / et . . . TG[2] *om. S* / *post* ita *add.* est *ER* / E: est *FP1; inter. a. m. C1* / quod: que *F* 63 est . . . punctum: invenire punctum est *C1* / ut angulum *corr. ex* ut A *a. m. O* / *post* contentum *scr. et del.* contingens (65) . . . dati (67) (circuli dati: circumdat) *O* 64 a lineis: alienis *S* / lineis: leneis *F* / illud: illum *FP1;* aliud *O* / *post* ductis *scr. et del.* lineis *P1* 65 puncto illo *transp. R* 68 *post* ducatur *rep. et del.* a (66) . . . EG (67) (circuli dati: circumdati) *S* / linea[1]: line *S* / GD: DG *C1* / MI: IM *C1* / in . . . divisa: divisa in puncto C *R* / *post* divisa *inter.* ita *a. m. F* / *post* ut *add.* sit *R* 69 IC: LC *FP1;* AT *S* / *post* CM *inter.* sit *a. m. F; add. P1* / et *om. R* / MI: IM *C1* / per equalia *om. P1* 71 DGS *corr. ex* DG *S* 72 angulus *om. R* / MO *om. R* 73 O *om. S* / O puncto *transp. OL3C1ER* / a . . . ducatur: ducatur a puncto C *R* 74 CKF: CHF *P1; corr. ex* CKFF *C1* / FM: *corr. ex* MF *E;* MF *R* 75 *post* equalis *scr. et del.* e *C1* 76 ductam: ductum *O* / sit[2] *om. O* / *post* angulus *scr. et del.* ductus *C1* / AGE *corr. ex* AEG *F* / et[2] . . . DG (77) *scr. et del.* (AG: AGA) *F; om. P1* 77 due *om. R* / DG: ADG *S;* AD *C1R* / *ante* dico *add.* igitur *C1* / A *inter. S* / *post* est *scr. et del.* est *S* / punctus quem: punctum quod *R* 78 *ante* MFK *add.* angulus *R* / MFK *corr. ex* FMK *F;* FMK *P1;* MFG *S* 79 FM: MF *R* / proportio[2] *om. R* / EG: GE *R* / GA[1] *corr. ex* HG *L3* / cum GA sit *om. P1* / GA sit *transp. FL3E* 80 GS *corr. ex* GP *F* / triangulus: triangulum *R* / similis: simile *R* / igitur . . . FMK *mg. a. m. F* / FMK *om. P1* 81 MKF *alter. in* MFK *S*

[2.177] Igitur a puncto A ducatur linea tenens cum linea AE angulum equalem angulo NMK, et sit linea AZ, que necessario concurret cum linea GE, quoniam que est proportio KF ad FM ea est EG ad GA, et angulus GAZ equalis angulo FMC. Igitur, sicut linea MO concurret cum FK in puncto F, concurret AZ cum GE. Sit concursus in puncto Z, et producatur AZ usque ad punctum Q ita ut linea AZ se habeat ad ZQ sicut MC ad CI, et ducatur linea EQ.

[2.178] Deinde a puncto A ducatur equidistans EQ, que sit AT. Erit quidem angulus AQE equalis angulo QAT, et quoniam duo anguli ZEA, EAT sunt minores duobus rectis, concurret AT necessario cum EZ. Sit concursus punctus T. Palam quod angulus AEG est equalis angulo MKF. Ducta a puncto E linea perpendiculari super AZ, que sit EL, erit angulus AEL equalis angulo MKN, cum angulus EAL sit equalis angulo KMN, et angulus ALE equalis angulo MNK, quia uterque rectus. Restat ergo ut angulus LEZ sit equalis angulo NKC, et angulus ELZ rectus equalis angulo KNC. Restat ut angulus EZL sit equalis angulo KCN. Igitur angulus EZQ equalis angulo KCI.

[2.179] Palam ergo quod triangulus EAG similis triangulo FMK, et triangulus EAL similis triangulo KMN, et triangulus ELZ similis triangulo KNC, et triangulus EAZ triangulo KMC. Ergo proportio AZ ad ZE sicut MC ad CK, et proportio QZ ad ZA sicut IC ad CM, et proportio QZ ad ZE sicut IC ad CK, quare triangulus QZE similis triangulo ICK, et triangulus QLE similis triangulo IKN. Erit proportio NM ad NI sicut AL ad LQ, et ita AL equalis LQ, et EQ erit equalis EA,

82 igitur: iam *R* 83 sit: fiat *O* 84 GE: EG *R* / que *inter. O* / FM: MF *ER* / ea: eadem *C1* / *post* EG *add.* que *FP1; scr. et del. L3* 85 *post* equalis *add.* est *ER* / MO: MC *C1* / concurret: concurrit *R* 86 F: C *OC1* / *post* F *add.* sic *R* / GE: G *O*; EG *R* 87 *ante* ZQ *scr. et del.* punctum *P1* / ZQ: QZ *C1* 88 CI *alter. ex* CY *in* CU *F*; CU *P1*; Q *S* 89 deinde: item *FP1* / *post* ducatur *scr. et del.* EQ *F* / EQ: EI *SO* / AT: GAT *S; corr. ex* GAT *L3* / *ante* erit *add.* et *S* 91 *post* EAT *scr. et del.* sicut *L3* 92 punctus: punctum *R* / T *inter. O* / quod *alter. in* quoniam *a. m. F*; quoniam *P1* 93 *post* ducta *add.* autem *R* / perpendiculari *om. S* 94 *post* angulus[1] *scr. et del.* equalis angulo *S* / AEL *corr. ex* AL *a. m. F* 95 KMN *corr. ex* MKN *L3* / KMN . . . angulo (96) *om. S* / et . . . MNK: et anguli ADN GZL equales MNK *inter. O* / angulo *om. R* 96 ut angulus *om. R* / LEZ *corr. ex* LEF *a. m. F; corr. ex* LZ *E* / sit *om. R* / angulo *om. P1L3E* / NKC *corr. ex* NKCE *L3* / ELZ *corr. ex* ELS *a. m. F* 97 KNC *alter. ex* MNKC *in* NKC *S* / EZL: ELZ *E* / angulo[2] *om. R* / KCN *corr. ex* KCM *F*; KCM *P1* 98 angulus *om. R* / *post* EZQ *add.* est *L3* / *post* equalis *add.* est *FP1* / KCI: KCU *L3* 99 triangulus: triangulum *R* / EAG: EAT *O* / similis: simile *R* / *post* similis *inter.* est *a. m. E; add. R* / FMK . . . triangulo (100) *inter. a. m. L3* 100 triangulus[1,2]: triangulum *R* / similis[1,2]: simile *R* / *post* triangulo *scr. et del.* KAN ei *F* / KMN *inter. a. m F* / ELZ *corr. ex* EHZ *F* 101 triangulo[1] *om. R* / triangulus: triangulum *R* / triangulus . . . quare (103) *rep.*; et[2] (102): igitur *S* / EAZ *corr. ex* KNC *F; corr. ex* EZ *E* / *post* EAZ *add.* similis *mg. L3* / triangulo[2] *corr. ex* triangulus *F* / AZ *corr. ex* AZT *C1* 102 QZ: AZ *R* / ad ZA *mg. a. m. F* / ZA: ZE *FP1L3; corr. ex* ZE *E*; ZQ *R* / IC: MC *P1R* / CM: CK *FP1L3*; CI *R* / et[2]: igitur *R* / et[2] . . . CK (103) *om. FP1L3; mg. a. m.* (IC *alter. in* AC) *E* 103 ZE: EZ *R* / triangulus: triangulum *R* / QZE: EQZ *S* / similis: simile *R* 104 ICK *corr. ex* ACK *O* / triangulus: triangulum *R* / similis: simile *R* / IKN *corr. ex* IKM *L3*; INK *R* / *post* IKN *inter.* ergo *a. m. F; add. P1* / *post* erit *add.* ergo *R* 105 *post* EQ *inter.* ergo *a. m. F*

et angulus EQZ equalis angulo LAT, et angulus EZQ equalis angulo AZT. Igitur tertius tertio equalis, et triangulus EZQ similis triangulo ZAT, quare proportio QZ ad ZA sicut EZ ad ZT, et sicut EQ ad AT, et sicut AE ad AT. Sed QZ ad ZA sicut EG ad GD. Igitur AE ad AT sicut EG ad GD.

[2.180] Fiat autem supra punctum A angulus equalis angulo GAE, qui sit UAG. Palam quod angulus GAL est medietas anguli UAT, sed est medietas anguli DGU, quare angulus UAT est equalis angulo DGU. Sed anguli TAU, TUA sunt minores duobus rectis, cum AT et UT concurrant, quare duo anguli TUA, DGU sunt minores duobus rectis. Igitur AU concurret cum DG.

[2.181] Dico quod concurret in puncto D, quoniam efficiet cum lineis UG, GD triangulum similem triangulo AUT, habebunt enim angulum AUG communem, et angulus TAU equalis angulo UGD. Igitur proportio AU ad AT sicut UG ad lineam quam secat AU ex GD, et proportio EA ad AU sicut EG ad GU, cum sit angulus UAG equalis angulo GAE.

[2.182] Cum ergo eadem sit proportio EA ad AT sicut EG ad GD, proportio EA ad AT sit compacta ex proportione EA ad AU et AU ad AT. Erit proportio EG ad GD compacta ex eisdem, quare erit compacta ex proportione EG ad GU et GU ad lineam quam secat AU ex GD. Sed est compacta ex proportionibus EG ad GU et GU ad GD. Igitur linea quam secat AU ex GD est linea GD. Igitur AU secat GD in puncto D.

[2.183] Producatur ergo a puncto A contingens que sit AH. Erit ergo GAH rectus. Sed GAL medietas anguli DGU. Igitur angulus LAH est medietas anguli DGE, cum illi duo valeant duos rectos. Sed cum angulus TAU sit equalis angulo DGU, erit angulus TAD equalis DGE.

106 *post* EQZ *add.* est *SOC1*/EZQ: EQZ *ER* 107 AZT: ZAT *R*/et . . . ZAT (108) *om. R* 108 sicut[1] . . . ZA (109) *rep. S*/*post* EZ *rep. et del.* ad . . . EZ *L3* 109 GD: DG *R*/igitur . . . GD (110) *om. L3* 111 autem *inter. F*/supra: super *R*/angulo: angulus *S* 112 UAT . . . anguli (113) *inter. a. m. L3*/sed . . . DGU (113) *inter. a. m. E* 113 anguli *om. ER*/DGU *corr. ex* DGA *O* 114 TAU: UAT *R*/*ante* TUA *add.* et *R*/minores *corr. ex* maiores *FL3*; maiores *P1*/cum . . . rectis (115) *mg. a. m.*; concurrant (115) *inter.*; TUA (115): TAU; *ante* DGU (115) *add.* et *E*/UT: TU *R* 115 TUA *corr. ex* TAU *L3C1*; DUG *O*/*ante* DGU *add.* et *P1L3R*/*post* DGU *scr. et del.* sed anguli TAU *C1* 116 AU *inter. O* 117 *post* cum *scr. et del.* E *F* 118 similem: simile *R*/AUT: aut *P1*; AUD *S*/angulum *corr. ex* angulus *E* 119 *post* TAU *add.* est *R*/*post* equalis *inter.* est *O*/UGD *corr. ex* UDG *L3*; DGU *R* 120 AU[1]: AB *S*/GD: DG *R*/et *inter. a. m. C1* 121 GU *corr. ex* AD *L3* 123 ad[1] *rep. L3*/sicut *om. FP1*; *inter. a. m. L3C1*; et *inter. O*/sicut . . . AT (124) *om. S*/*post* GD *add.* et *OR*; *inter. a. m. C1E* 124 *post* proportio *add.* autem *FP1* (*mg. a. m. F*)/compacta: composita *R*/proportione *corr. ex* proportio *F*/*post* EA[2] *scr. et del.* AU *F*/et AU *om. S* 125 compacta: composita *R*/ex *corr. ex* cum *O*/eisdem: ijsdem *R*/quare . . . ex[1] (126) *om. P1* 126 GD: DG *R*/*post* GD *scr. et del.* e *F* 127 est: EA *S*/et GU *mg. F* 128 ex *inter. F*/GD: DG *R* 129 AH *corr. ex* HA *E*; HA *R* 130 GAH: GHA *E*/*post* GAH *scr. et del.* respectu *P1*/*post* GAL *add.* est *FP1R* (*inter. F*) 131 est *inter. FOE* (*a. m. E*); *om. SL3C1*/medietas *corr. ex* medietatas *L3*/*post* DGE *add.* et *S* 132 TAD *corr. ex* TAG *a. m. F*/TAD . . . anguli (133) *om. S*/DGE *alter. ex* DGDGC *in* DGC *P1*

Igitur angulus LAH est medietas anguli TAD, et angulus EAL medietas anguli EAT. Igitur angulus EAH medietas anguli EAD, quare AH dividit angulum EAD per equalia, quod est propositum.

[2.184] Si vero AU, cum sit angulus super punctum A equalis angulo GAE, non cadat super lineam ES extra circulum vel intra, sit ergo equidistans [FIGURE 5.2.22a, p. 578]. Igitur angulus UAG equalis est angulo AGE. Sed idem est equalis angulo GAE, quare angulus GAE equalis est angulo AGE. Igitur EG equalis AE. Similiter, angulus TAD erit equalis angulo ATG, quia coalternus. Sed iam dictum est quod angulus TAD est equalis angulo DGT. Igitur angulus ATG est equalis angulo DGT, et similiter duo anguli ADG, DGT sunt equales. Igitur duo anguli ADG, TAD sunt equales.

[2.185] Sequetur ergo ex hiis quod linea quam secat AU ex DG sit equalis linee AT. Et iam dictum est quod EG equalis AE. Igitur proportio EG ad lineam quam secat AU ex DG sicut AE ad AT. Sed iam dictum est quod AE ad AT sicut EG ad GD. Igitur linea quam secat AU ex DG est GD, et cum TAD sit equalis angulo DGT, erit LAH medietas anguli TAD, sicut dictum est supra, et EAL medietas EAT. Erit ergo EAH medietas anguli EAD, quod est propositum.

[2.186] **[PROPOSITIO 23]** Amplius, dato circulo cuius G [FIGURE 5.2.23, p. 579] centrum, et dato in eo dyametro GB, et dato E puncto extra circulum, est ducere a puncto E ad dyametrum GB lineam secantem circulum ita quod pars eius a circulo usque ad dyametrum sit equalis parti dyametri interiacentis ipsam et centrum.

[2.187] Verbi gratia, ducatur a puncto E perpendicularis super dyametrum, et sit EC, et ducatur linea EG. Et sumatur linea QT equalis linee EC, et fiat super QT portio circuli ut quilibet angulus cadens in hanc portionem sit equalis angulo EGB, et compleatur circulus. Et a medio puncto QT ducatur perpendicularis ex utraque parte usque ad circulum. Erit quidem dyameter huius circuli. Et a puncto Q ducatur

133 anguli *om. FP1* 134 EAT *mg. a. m. F*/*ante* igitur *add.* erit *FP1; scr. et del.* DGU *S*/EAH: EAG *P1*/*post* EAH *inter.* est *O*/quare . . . EAD (135) *mg. a. m. C1* 135 per *corr. ex* pars *F* 136 super: supra *C1* 137 cadat: cadit *FP1C1; corr. ex* cadit *a. m. E* 138 equalis . . . AGE (139) *rep. P1* 139 AGE: DGE *SO; corr. ex* GE *a. m. E*/sed . . . AGE (140) *rep.* (angulo[1]: angulus; AGE[1]: DGE) *S* 140 equalis est *transp. R* 141 quia coalternus *om. R* 142 DGT *corr. ex* DGN *O; alter. in* DGE *L3*/igitur . . . DGT[1](143) *mg. a. m. E* 143 DGT[2] *om. S* 144 *post* anguli *scr. et del.* in puncto *P1*/TAD: ATG *R* 145 sequetur: sequitur *C1* 146 iam *corr. ex* am *S*/*post* equalis *inter.* est *a. m. E; add.* sit *R* 147 ad[1] *corr. ex* a *O*/*post* DG *add.* est *R*/sicut: sit *O*/sed . . . AT (148) *mg. a. m. C1* 148 quod: ut *R*/sicut: sic *R*/igitur . . . GD (149) *om. P1* 149 *post* est *add.* DG et *S*/GD: DG *C1R*/angulo *om. R* 150 *post* sicut *add.* iam supradictum *O*/dictum *inter. a. m. E*/supra *om. O* 151 *ante* EAD *scr. et del.* EAD *P1* 152 G centrum (153) *transp. R* 153 dato: data *R*/GB: BG *R* 154 GB: BG *R* 155 quod: ut *ER* 156 parti *corr. ex* parte *F*/interiacentis: interiacenti inter *R* 158 EG *corr. ex* AEG *P1* 159 linee *om. ER* 161 perpendicularis . . . parte: ex utraque parte perpendicularis *R*

linea ad hunc dyametrum secans eum in puncto F, et producatur usque ad P punctum circuli ita ut FP sit equalis medietati GB, et ducatur linea PT et linea TF. Et ducatur a puncto P linea equidistans dyametro, que sit PU. Concurrat cum TF in puncto U, et a puncto U ducatur equidistans TQ, que sit UO. Et a puncto T ducatur perpendicularis super PQ, que sit TN, et a puncto T ducatur equidistans PQ, que sit TS, et a puncto U perpendicularis super PQ, que sit UH. Deinde ex angulo BGE secetur angulus equalis angulo QPU, qui sit BGD, et ducatur linea EDZ. Dico quod DZ est equalis ZG.

[2.188] Et ducatur a puncto D perpendicularis super BG, que sit DI, et ducatur a puncto D contingens, que sit DK. Palam, cum dyameter FL sit perpendicularis super QT et super OU, et PU sit equidistans ei, erit angulus OUP rectus. Et cum OU dividatur a dyametro per equalia et ortogonaliter, erit FO equalis FU, quare angulus FOU equalis angulo FUO. Sed, cum duo anguli POU, OPU valeant rectum, erit angulus FUP equalis angulo FPU, quare FP equalis FU, et ita equalis FO. Et ita PO equalis BG, et equalis GD, et ita proportio EC ad GD sicut TQ ad PO.

[2.189] Sed cum angulus KDG sit rectus equalis angulo GID, et angulus IGD communis, erit triangulus IGD similis triangulo KGD, et erit proportio GD ad DI sicut GK ad KD. Sed angulus KGD equalis angulo OPU, et KDG rectus equalis OUP, et ita triangulus KDG similis triangulo OUP, et proportio KG ad KD sicut OP ad OU. Igitur DG ad DI sicut OP ad OU. Ergo proportio EC ad DI sicut QT ad OU.

[2.190] Sed proportio QT ad OU sicut TF ad FU, cum triangulus TFQ sit similis triangulo OFU. Verum angulus UTS equalis angulo HFU, quia coalternus ei, et angulus UST rectus equalis angulo FHU.

163 hunc *om. FP1*; hanc *R* / eum: eam *R* 164 ad *om. FP1* / P *om. O* / P punctum *transp. ER* / *ante* ita *scr. et del.* ita *O* / FP *corr. ex* P *a. m. F* / medietati: mediedietati *F* 166 *post* PU *inter.* que *a. m. F*; *add.* que *P1*; *add.* et *L3C1* (*inter. L3*) / concurrat: concurrent *O*; *alter. in* concurrens *E* / *ante* cum *add.* que *ER* / TF *corr. ex* T *mg. a. m. F* / a *mg. F* 167 T *inter. SO* 168 PQ: PG *P1* 169 U: A *P1* / perpendicularis: perpendiculariter *FP1* / BGE *corr. ex* BG *C1* 170 qui: que *R* / dico *om. S* 172 perpendicularis . . . D (173) *mg. a. m. E* 173 *post* palam *add.* quod *FP1* / cum *corr. ex* quod *L3* 174 sit perpendicularis *transp. C1* / et[1] *inter. F* / sit[2]: sint *FP1* / sit equidistans *transp. C1* / equidistans: equidistantes *FP1* / ei *corr. ex* eis *F*; eis *P1* 175 angulus OUP *transp. L3* / dyametro *inter. a. m. E* 176 quare: et *R* / equalis[2] *corr. ex* sequalis *F* 177 OPU *corr. ex* OPOU *S* / valeant *corr. ex* valent *E* 178 quare . . . FU *om. P1* / *ante* FU *add.* est *R* 179 proportio *om. R* 181 KDG: KDH *P1* / *post* rectus *add.* est *FP1* (*mg. a. m. F*); *inter.* ei *L3* 182 IGD[1,2]: LGD *L3* / triangulus: triangulum *R* / IGD[2] *inter. O* / similis: simile *R* / KGD: KDG *OR*; *corr. ex* KDG *E* 183 erit *om. R* / *post* KD *scr. et del.* sicut OB ad NO *C1* / *post* KGD *add.* est *R* 184 *post* et[1] *add.* angulus *R* / KDG[1] *corr. ex* KGD *L3*; KGD *E* / triangulus: triangulum *R* / KDG[2]: KGD *FP1SL3C1ER* / similis: simile *R* 185 KG: GK *R* / OU: UO *C1* / DG . . . ergo (186) *om. S* 186 DI[1]: D *O*; GL *L3*; GI *E* / ergo . . . FU (187) *mg. a. m. E* / *post* proportio *scr. et del.* que est equalis QT *C1* / EC: hec *S* / QT: TQ *R* 187 sed . . . OU *om. S*; *rep. L3* / *post* FU *scr. et del.* ergo proportio que est equalis QT ad DI sicut QT ad OU sed proportio QE ad OI sicut TF ad UT *E* / triangulus: triangulum *R* 188 similis: simile *R* 189 HFU *corr. ex* FH *C1* / UST: UFT *E* / *post* UST *add.* est *P1*

Erit triangulus UST similis triangulo HUF, et ita proportio TU ad UF sicut SU ad UH, quare proportio TF ad UF sicut SH ad UH. Sed TN equalis SH, cum sit equidistans ei, et sint inter duas equidistantes. Igitur proportio TF ad UF sicut TN ad UH, quare proportio QT ad OU sicut TN ad UH, et EC ad DI sicut TN ad UH.

[2.191] Sed cum angulus GID sit rectus equalis angulo PHU, et angulus IGD equalis angulo HPU, erit triangulus IGD similis triangulo HPU, et proportio ID ad GD sicut HU ad UP, quare proportio EC ad GD sicut TN ad UP. Sed cum angulus CGE sit equalis angulo NPT, et angulus GCE rectus equalis PNT, erit proportio GE ad EC sicut PT ad NT. Igitur proportio GE ad GD sicut PT ad UP.

[2.192] Et angulus DGE equalis angulo UPT. Igitur triangulus DGE similis triangulo UPT. Igitur angulus GDE equalis angulo PUT. Restat ergo angulus GDZ equalis angulo PUF, et angulus DGZ equalis angulo UPF, quare tertius tertio, et proportio DZ ad ZG sicut UF ad FP. Sed UF equalis FP. Ergo DZ equalis ZG, quod est propositum.

[2.193] **[PROPOSITIO 24]** Amplius, dato triangulo ortogonio ABG [FIGURE 5.2.24, p. 580] cuius angulus ABG rectus, et dato in BG vel AB puncto D, est ducere lineam a puncto D ad latus AG concurrentem in puncto qui sit Q et ex alia parte concurrentem cum alio latere ut ipsa totalis se habeat ad GQ sicut E est ad Z.

[2.194] Verbi gratia, ducatur a puncto D equidistans AB, que sit DM, et fiat circulus transiens per tria puncta D, M, G. Erit MG dyameter. Et ducatur linea AD, et sit H linea ad quam se habeat AD sicut E ad Z. Et cum angulus DMG sit equalis angulo BAG, secetur ex eo equalis angulo DAG, et sit CMD. Et ducatur MC usque contingat circulum in puncto

190 triangulus: triangulum *R*/UST: UFT *O*/similis: simile *R*/UF: US *FP1C1* 191 SU: FU *O*/quare . . . UH[2] *om. R*/TN: TQ *FP1* 192 *post* equalis *scr. et del.* sed H *F*/SH: FH *O*/equidistans *corr. ex* equalis *mg. a. m. F*/equidistans ei *transp. R*/sint *corr. ex* sicut *L3* 193 TN: TQ *FP1* 194 et . . . UH[2] *om. S*/UH[2]: HU *E* 195 angulus *corr. ex* proportio *L3*/GID: GLD *L3* 196 HPU *corr. ex* HPK *a. m. F*; HUP *S*/erit . . . HPU (197) *mg. a. m. E*/triangulus: triangulum *R*/similis: simile *R*/triangulo HPU (197) *transp. R* 197 ID: IG *L3*/GD *alter. in* DG *a. m. E* 198 TN *corr. ex* TQ *a. m. F*/angulus *om. R*/*post* angulus *scr. et del.* ultis *E*/CGE: DGE *O*/NPT *alter. ex* ATP *in* APT *P1*; HPT *S* 199 proportio *om. R*/*post* ad[1] *add.* GE *alter. in* GD *P1*/EC *alter. ex* GD *in* ET *mg. F* 200 NT: UP *F*; *corr. ex* PN *P1*; *alter. ex* UP *in* UT *S*; TN *R*/igitur . . . UP *mg. a. m. FE*/proportio *om. FP1R*/GE: EG *FP1*/GD: DG *FP1*/PT: TP *FP1* 201 et: sed *R*/*post* DGE[1] *add.* est *R*/igitur . . . UPT (202) *om. S*/triangulus: triangulum *R* 202 similis: simile *R*/angulus: triangulus *FP1SOL3*; *corr. ex* triangulus *C1*/GDE *corr. ex* DGE *E* 203 PUF *alter. ex* PF *in* UP *F*; FUP *R*/et . . . UPF (204) *scr. et del. E*; *om. R* 204 UPF *corr. ex* PUF *S*/et *inter. a. m. C1*/*post* UF[1] *add.* et *E*/sed . . . FP (205) *om. FP1*; *inter. a. m. L3* 205 *post* equalis[1] *add.* est *ER* 207 AB *alter. ex* AG *in* ABG *F*; ABC *P1*; AG *SO* 208 lineam: lineas *FP1L3E*; *corr. ex* lineas *O*/*post* puncto[2] *scr. et del.* A A *F*/AG *corr. ex* G *P1*/*post* puncto[3] *add.* et *OL3*; *add.* Z *E* 209 qui: quod *R*/et *om. S* 210 sicut: sunt *S*/E *om. P1*/E est *transp. R*/est *om. S* 213 habeat: habet *ER* 214 DMG . . . angulo[1] *mg. F*/angulo[1] *corr. ex* angulus *L3*; *om. R*/BAG . . . angulo[2] *om. S* 215 *post* DAG *rep. et del.* secetur (214) . . . DAG *C1*/usque: quousque *R*

C, a quo ducatur linea ad dyametrum MG et usque ad circulum ita quod LN sit equalis linee H. Et ducatur linea NG, et linea DN concurrens cum AG in puncto Q.

[2.195] Cum igitur angulus DMC sit equalis angulo DNC, quia super eundem arcum, erit angulus QNL equalis angulo DAQ, et angulus NQL equalis angulo DQA, quare triangulus NQL similis triangulo DQA. Ergo proportio AQ ad QN sicut AD ad NL.

[2.196] Verum, cum angulus DMG sit equalis angulo DNG, erit QNG equalis angulo TAQ. Sit T punctus in quo DN concurrit cum AB, et angulus TQA equalis angulo NQG. Erit triangulus TQA similis triangulo NQG, et erit proportio AQ ad QN sicut TQ ad QG. Ergo proportio TQ ad QG sicut AD ad LN. Sed NL equalis H, et AD ad H sicut E ad Z. Igitur TQ ad QG sicut E ad Z, quod est propositum.

[2.197] Potest autem contingere quod a puncto C erit ducere duas lineas similes CN, et tunc erit ducere duas lineas a puncto D similes TQ ut utriusque ad partem quam secat ex AG sit proportio sicut E ad Z, et erit eadem probatio.

[2.198] **[PROPOSITIO 25]** Predictis habitis, dato speculo sperico, erit invenire punctum reflexionis in eo.

[2.199] Verbi gratia, sit A [FIGURE 5.2.25, p. 581] centrum visus, B punctus visus, G centrum spere, et ducantur linee AG, BG. Et sumatur superficies in qua sunt hee due linee, et sumatur circulus communis huic superficiei et speculo. Invenietur ergo punctus reflexionis in hoc circulo.

[2.200] Et sumatur linea alia MK, et dividatur in puncto F ut FM se habeat ad FK sicut BG ad GA. Et dividatur MK per equalia in puncto O, et ducatur a puncto O perpendicularis, que sit CO, et ducatur a puncto K linea ad CO tenens cum CO angulum equalem medietati anguli BGA, que sit KC. Et a puncto F ducatur linea ad CK, que sit FP,

216 ducatur: ducat *L3* / et *om. R* 218 *post* Q *add.* et cum AB in puncto C *L3ER* (*inter. a. m. L3*) 220 angulo *om. FP1* / DAQ *corr. ex* DQA *S* / angulus[2] *om. R* 221 NQL *corr. ex* NQ *a. m. F*; NQ *P1* / triangulus: triangulum *R* / NQL *alter. ex* ACB *in* QNL *a. m. F*; ACB *P1*; AQB *S* / similis: simile *R* 222 proportio *om. R* 223 *post* verum *add.* est *SL3*; *scr. et del. OC1E* / sit . . . DNG *inter. a. m. E* 224 angulo *om. R* / TAQ: TAG *C1* / T: D *SE*; *inter. O* / punctus: punctum *R* / *post* in *scr. et del.* sicut *O* / concurrit: concurret *C1* 225 equalis *corr. ex* similis *a. m. E* / erit . . . NQG (226) *inter. a. m.* (TQA: TQD) *L3* / triangulus: triangulum *R* / similis: simile *R* 226 proportio *om. S* / QG *alter. ex* G *in* GQ *a. m. F*; GQ *P1*; QS *E* / ergo . . . QG (227) *om. S*; *mg. O* 227 LN: NL *R* / NL: LN *E* / H[1] *corr. ex* HI *L3* / et . . . H[2] *mg.* (AD *corr. ex* AT) *F* / AD: A *inter. L3* 228 igitur: G *S* / QG: GQ *C1* / est *om. S* 229 duas lineas (230) *transp. ER* 230 similes[1] . . . lineas[2] *om. S* / *post* similes[1] *scr. et del.* t *F* / CN: CLN *OR*; ELN *L3C1*; ELU *inter. a. m. E* 232 probatio: proportio *FP1* 234 in eo *om. FP1L3ER* 236 punctus visus: punctum visum *R* 237 superficies *corr. ex* res *O*; *mg. L3* / hee *om. FP1* 238 speculo: spere *O* / punctus: punctum *R* 239 linea alia *transp. FP1* / FM: MF *R* 240 dividatur: dividitur *O* 241 O[2] *om. P1* / O[2] . . . puncto (242) *mg. a. m. E* 242 tenens cum CO: contingens *inter. O* / CO: ea *R* 243 KC: HC *SE*

et concurrat cum CO in puncto S ita ut proportio SP ad PK sicut BG ad semidyametrum GD. Et ex angulo BGA secetur angulus equalis angulo SPK, scilicet DGB, et ducantur linee SK, BD.

[2.201] Erit igitur proportio BG ad GD sicut SP ad PK, et ita triangulus SPK similis triangulo BGD, et erit angulus SKP equalis angulo BDG. Sed forsan, secundum predictam, poterimus a puncto F ducere aliam lineam ad CK similem SP ut sit proportio eius ad partem quam secabit ex CK sicut SP ad PK, et tunc a puncto K ad OS ducetur alia linea quam SK alium cum CK angulum tenens maiorem vel minorem angulo CKS. Si maior ex hiis angulis non fuerit maior recto, non erit invenire punctum reflexionis. Sit igitur angulus CKS maior recto, et invenitur punctum sic.

[2.202] Erit angulus BDG maior recto. Ducatur contingens NDY, et cum angulus PKO sit minor recto, secetur ex angulo BDG equalis ei, qui sit QDG. Cum igitur angulus SPK sit equalis angulo QGD, erit triangulus FPK similis triangulo QGD, et erit angulus DQB equalis angulo KFS, et triangulus DQB similis triangulo KFS.

[2.203] Producatur autem DQ, et a puncto B ducatur perpendicularis super ipsam, que sit BZ. Erit igitur angulus BQZ equalis angulo SFO. Et angulus BZQ rectus equalis angulo SOF, et ita triangulus BQZ similis triangulo SFO.

[2.204] Ducatur DZ usque ad punctum I, et sit ZI equalis ZD. Palam ergo quod ZQ ad QB et QB ad QD sicut OF ad FS et FS ad FK, et ex hoc erit ZD ad QD sicut OK ad FK, et ita ID ad QD sicut MK ad FK, et ita IQ ad QD sicut MF ad FK, et IQ ad QD sicut BG ad GA.

244 concurrat *corr. ex* concurrant *C1* / S: C *P1* / proportio: portio *S* / *post* PK *add.* sit *P1R*; *add.* sit *mg. a. m. E* / *post* sicut *scr. et del.* per premissam *C1* 245 GD: GN *S*; G? *O* / ex *rep. S* / angulus *om. R* 246 SPK: FPK *R* / SK *corr. ex* SF *O* 247 *post* ad[1] *scr. et del.* semidyametrum GN BG *S* / PK: PR *E* / *post* PK *add.* et angulus BGD equalis angulo SPK *FP1* (et angulus *inter. F*) 248 triangulus: triangulum *R* / similis: simile *R* / BGD: BDG *E* / SKP: SPK *O* 249 BDG: BGD *O* / predictam: predicta *P1SC1E* 250 similem: simile *P1* / proportio eius *transp. SOC1* / *post* partem *add.* ad *F* 251 ducetur: ducere *SO* 252 alia linea: aliam lineam *O* / *post* linea *inter.* ut EK *L3* / SK: LK *S*; *corr. ex* S *C1* / *ante* alium *add.* ut EK *C1* / alium: aliquantum *O* / vel minorem (253) *inter. L3E* (*a. m. E*) 253 angulo: an *P1* / si *inter. O* / ex . . . maior (254) *om. O* 254 erit: licebit *R* / invenire *om. P1* 255 et *om. O* / et . . . sic *om. R* / invenitur: invenietur *SOC1* / *post* sic *add.* si *O* 256 BDG: BGD *L3* / *post* recto *add.* et invenitur punctum sic *R* 257 cum: quia *R* / sit: est *R* / BDG *corr. ex* HLG *a. m. E* / equalis ei *transp. SOC1* 258 cum . . . QGD *om. ER* / erit: est *R* / *post* erit *add.* igitur *ER* 259 triangulus: triangulum *R* / FPK: SPK *FE* / similis *mg. a. m. C1*; simile *R* / triangulo . . . equalis *rep.* (triangulo[2]: angulo) *S* 260 KFS[1] . . . DQB *mg. a. m. E* / triangulus: triangulum *R* / DQB: DQF *S* / similis: simile *R* / KFS[2]: FKS *FP1*; KSF *O* / *post* KFS *scr. et del.* erit angulus DQB similis triangulo KFL *E* 261 autem: a *scr. et del. L3*; *om. ER* / DQ: QD *C1* / a . . . ducatur: ducatur a puncto B *R* / perpendicularis: particularis *S* 262 super: supra *E* / igitur *om. FP1R* / angulus *corr. ex* triangulus *F* 263 et angulus BZQ *mg. a. m. E* / *post* ita *scr. et del.* circulus *E* / triangulus: triangulum *R* / similis: simile *R* 266 ergo *om. R* / OF: FO *C1* / FS[1,2]: SF *FP1* / ex hoc *mg. a. m. E* 267 erit *om. ER* / *post* erit *add.* etiam *C1* / *post* sicut[1] *scr. et del.* FO ad F *C1* / OK: CK *S* / *post* FK[2] *add.* et MF ad FK sicut ex ypothesi BG ad GA *mg. a. m. F* / et[2] . . . FK (268) *om. FP1* / ita[2] *inter. a. m. E* 268 sicut . . . QD[2] *mg. a. m. E*

[2.205] Ducatur autem linea BI et ei equidistans DL. Erit triangulus LDQ similis triangulo BQI, et proportio IQ ad QD sicut IB ad DL. Et cum IZ sit equalis ZD, et BZ perpendicularis, erit BD equalis BI, quare erit BD ad DL sicut BG ad GA.

[2.206] Ducatur autem a puncto D linea, que sit DH, equalem tenens angulum cum linea LD angulo BGA. Et cum HL et DL concurrant, erunt duo anguli LHD, LDH minores duobus rectis, et ita duo anguli AGH, DHG eis equales sunt minores duobus rectis, quare HD concurret cum GA. Dico quod concurret in puncto A.

[2.207] Palam quod GDN rectus equalis duobus angulis OCK, OKC, et angulus OKC equalis angulo GDQ. Restat angulus QDN equalis angulo OCK, et ita angulus QDN medietas anguli BGA, et ita medietas anguli HDL. Sed angulus QDB est medietas anguli BDL, quoniam proportio BQ ad QL sicut BD ad DL, cum triangulus DLQ sit similis triangulo BQI, et BD equalis BI. Restat igitur ut angulus NDB sit medietas anguli HDB, et ita BDN equalis NDH. Restat autem BDE equalis angulo HDG. Sed angulus HDG equalis angulo EDA contraposito, quare BDE equalis EDA, et ita D est punctus reflexionis. Ita dico si HD concurrat cum AG in puncto A, quod quidem sic patebit.

[2.208] Ducatur linea HT equidistans BD. Palam quod angulus BDE equalis est angulo HDG. Sed BDE est equalis angulo HTD, quare HT erit equalis HD. Sed proportio BD ad HT sicut BG ad GH, sicut probat Euclides. Igitur proportio BD ad DH sicut BG ad GH. Sed HD producta concurret cum GA, et fiet triangulus similis triangulo HDL, cum habeat angulum LHD communem, et angulus HDL sit equalis angulo HGA. Igitur proportio HD ad DL sicut HG ad lineam quam secat HD ex GA. Et proportio BD ad DL constat ex BD ad DH et DH ad DL. Igitur con-

269 DL: DI *S*/triangulus: triangulum *R* 270 LDQ: BOQ *L3*/similis: simile *R*/et[1,2] *inter. a. m. C1* 271 *post* IZ *scr. et del.* ad *L3*/erit *om. S* 273 autem *om. ER*/ 274 cum[1] *mg. a. m. F*/ cum linea *inter. L3*/linea *om. FP1* 275 duo anguli[1] *om. R*/et . . . rectis (276) *mg. a. m. L3* 276 AGH: AGD *P1*/DHG: ABG *S*; AHG *OC1*/sunt *corr. ex* sint *E* 277 concurret *corr. ex* concurrent *F*; concurrent *P1* 278 *post* quod *add.* angulus *SOL3C1ER*/*post* rectus *add.* est *R*/*post* OCK *rep. et del.* OCK *S*; *inter.* et *L3*; *add.* et *ER* /OKC *inter. L3* 279 restat *corr. ex* resta *F*/*post* restat *add.* ergo *S* 280 angulus *inter. a. m. E*; *om. R*/BGA *corr. ex* GBA *C1* 282 triangulus: triangulum *R*/similis: simile *R* 283 BI: BC *S*/igitur *inter. O*/NDB *alter. ex* NT *in* NDH *O* 284 *post* NDH *add.* producatur GD ultra D ad punctum F quia igitur anguli FDN GDN sunt recti ergo *R*/autem *om. FP1R*/BDE: BDF *R* 285 HDG[1] . . . equalis (286) *om. S*/HDG[2] *corr. ex* HD *O*/ EDA: FDA *R* 286 BDE: BDF *F*/EDA: FDA *R*/D *corr. ex* DA *L3*; illud *C1*/punctus: punctum *R* 287 HD: AD *R*/AG: HAG *S*; *corr. ex* HAG *O* 288 BDE *corr. ex* BDEE *F*; BDF *R* 289 equalis est *transp. C1*/sed . . . HTD *mg. a. m. E*/BDE: DBE *S*; BDF *R*/est[2] *om. SOL3C1* 290 BD . . . proportio (291) *om. S*/sicut[2] . . . Euclides (291) *om. R* 291 Euclides *om. FP1*; *corr. ex* equalis *L3*/ igitur: et *C1*/*post* DH *scr. et del.* sicut BG ad DH *O*/GH *corr. ex* DB *O*/HD: HG *O* 292 triangulus similis: triangulum similem *R*/habeat: habeant *R* 293 LHD *alter. ex* HD *in* GHD *F*; GHD *P1*; SHD *S*/HGA *corr. ex* HG *O*; *corr. ex* GA *a. m. C1* 295 DL[1]: LD *C1*/*post* ex *add.* proportione *R*/DH[1]: BH *P1*/et DH ad DL *mg. a. m. F*/ad DL[2] *rep. F*/igitur *inter. E*

stat ex BG ad GH et GH ad lineam quam secat HD ex GA. Sed BD ad DL sicut BG ad GA. Igitur proportio BG ad GA constat ex proportionibus BG ad GH et GH ad lineam quam secat HD ex GA. Sed constat ex proportionibus BG ad GH et GH ad GA. Igitur, GA est linea quam secat HD ex GA, et ita concurret cum ea in puncto A, quod est propositum.

[2.209] Si vero angulus CKS non fuerit maior recto, dico quod non fiet reflexio ab aliquo puncto speculi ad visum.

[2.210] Si enim dicatur quod potest, sit D punctus reflexionis, et producatur linea AD usque ad H punctum in dyametro BG. Et fiat angulus LDH equalis angulo AGB, et producatur contingens NDY, et fiat angulus QDN equalis medietati anguli AGB.

[2.211] Palam quod triangulus HDL similis est triangulo HGA, quare proportio DH ad DL sicut HG ad GA. Sed BD ad DH sicut BG ad GH, quod patebit per HT equidistans BD. Igitur BD ad DL sicut BG ad GA. Sed cum angulus BDE sit equalis angulo HDG, erit angulus BDN medietas anguli BDH. Sed NDQ medietas anguli HDL. Igitur BDQ medietas anguli BDL, quare proportio BQ ad QL sicut BD ad DL.

[2.212] Ducatur a puncto B equidistans DL, et sit BI, et concurrat DQ cum ea in puncto I. Et dividatur DI per equalia in puncto Z, et ducatur BZ. Erit triangulus BQI similis triangulo QDL. Igitur BQ ad QL sicut BI ad DL, et ita BI equalis BD. Et IQ ad QD sicut MF ad FK, et ita ID ad QD sicut MK ad FK, et ita DZ ad QD sicut OK ad FK, et ita ZQ ad QD sicut OF ad FK.

[2.213] Palam quod BZ est perpendicularis. Ducatur donec concurrat cum DG in puncto X, quod quidem possibile est, cum angulus DZX rectus, ZDX minor recto. Et palam quod proportio BG ad GD sicut SP ad PK. Cum ergo angulus CKS dicatur non esse maior recto, dico quod super punctum K fiet maior recto per lineam concurrentem cum CO in puncto a quo ducetur linea ad CK transiens per punctum F retinens proportionem ad partem CK sicut BG ad GD.

296 et GH *inter. a. m. E* / GH² *inter. a. m. C1* / *post* lineam *add.* HD *S* / secat HD *transp. C1* / *post* GA *add.* et proportio *O* / sed *scr. et del. O* 297 igitur . . . GA² *mg. a. m. C1* / ex *inter. E* 298 GH¹ *inter. a. m. E* / et GH *inter. L3* / *post* et *scr. et del.* F *O* / ad² . . . secat: quam secat ad lineam *L3* / quam . . . linea (299) *om. FP1* / HD: BD *L3* / ex . . . GA (300) *inter. a. m.* (sed: sicut; GH¹: BH) *L3* / sed . . . GA (300) *inter. a. m. E* 300 concurret: concurrit *SOC1* 2 quod: quoniam *SOC1* 3 aliquo: alio *OL3; corr. ex* alio *E* / speculi: speciei *L3* 4 *post* enim *scr. et del.* d *C1* / punctus: punctum *R* 5 ad *inter. a. m. E* 6 angulo *rep. S* 7 QDN *corr. ex* qui DN *E* / equalis *inter. O* / medietati anguli *corr. ex* anguli medietati *C1* 8 triangulus: triangulum *R* / HDL: DHL *C1* / similis: simile *R* 9 proportio *om. L3* / DH¹: DG *P1* / *post* BG *scr. et del.* ad *O* 10 HT equidistans: equidistantem HT *R* / *post* HT *scr. et del.* ad *E* / equidistans: equidistantem *E* / *ante* BD *add.* ipsi *R* 12 NDQ *corr. ex* DNDQ *F* / *post* NDQ *add.* est *R* 14 *post* ducatur *scr. et del.* a *F* / equidistans: equidem *S* 15 ea: E A *R* / per: in *R* 16 triangulus: triangulum *R* / similis: simile *R* / *post* igitur *add.* ut *R* / BQ ad (17) *rep. P1* 17 sicut: sic *R* / IQ *alter. ex* QI *in* LQ *F*; Q *S*; *corr. ex* Q *L3* / MF: MI *R* / FK *corr. ex* FE *a. m. E* 18 ID *corr. ex* AD *F* / MK *corr. ex* ME *E* 20 est: cum *FP1* / ducatur: producatur *R* 22 *post* DZX *add.* sit *R* 23 CKS *corr. ex* CES *E* / dicatur *corr. ex* ducatur *L3* / non *corr. ex* M *O* 24 dico . . . recto *om. S* / quod: quoniam *OC1* 25 ducetur: ducatur *C1* 26 CK: PK *R*

[2.214] Verbi gratia, planum, cum angulus QDN sit equalis angulo KCO, erit angulus QDG equalis angulo CKO. Fiat ergo super punctum K angulus equalis BDQ, et ponatur quod linea hunc angulum tenens concurrat cum CO in puncto S, et ducatur SFP. Planum est, cum angulus BZD rectus equalis angulo SOK, erit triangulus BZD similis SOK, et proportio BZ ad BD sicut OS ad SK. Sed QZ ad QD sicut OF ad FK. Erit ergo angulus ZBQ equalis angulo OSF, et angulus QBD equalis angulo FSK, quare triangulus BGD similis triangulo SPK. Igitur proportio SP ad PK sicut BG ad GD, quod est propositum.

[2.215] Amplius, impossibile est quod duo anguli supra MO constitituti sit uterque maior recto. Si enim uterque talium maior fuerit recto, cum supra idem centrum fiat angulus equalis angulo SKM, fiet supra idem centrum alius angulus diversus ab isto quem efficiet supra KM alia linea similis SK. Et ita a puncto D et ab alio puncto illius circuli fiet reflexio, quod est impossibile, cum iam probatum sit quod unus uni visui sit reflexionis punctus, et iam ostensum est quomodo inveniri possit.

[2.216] Duobus autem visibus, licet duo sint puncta reflexionis, tamen unica erit ymago sensuali sillogismo, et unicus ymaginis locus. Et hoc probabimus, quoniam due linee a centris oculorum ad centrum circuli ducte sunt equales.

[2.217] Si autem situs puncti visi respectu utriusque visus sit idem ut linee a puncto viso ad centra oculorum sint equales, facilis erit probatio, quoniam dyametri visuales secant ex circulo arcus reflexionis, et tenent angulos equales cum linea a puncto viso ad centrum spere ducta, et arcus hanc lineam et dyametros visuales interiacentes sunt

27 verbi gratia: quod autem hoc possibile *R* / *post* planum *add.* est *R* / angulo . . . equalis (28) *om. L3* 29 BDQ: BDG *P1* / hunc . . . tenens (30): tenens hunc angulum *SOC1* 30 CO: eo *SOC1* / SFP *corr. ex* FP *a. m. C1* / cum[2] *corr. ex* quod *inter. O* 31 *post* SOK *add.* quod *R* / triangulus: triangulum *R* / similis: simile *R* 32 SOK: SOE *E* / proportio *om. R* / *post* SK *add.* et BZ ad ZD sicut SO ad OK *R* / sed . . . FK (33) *inter. a. m. L3* 33 QBD: QBO *S* 34 FSK: FLK *F*; FOK *E* / triangulus: triangulum *R* / similis: simile *R* / SPK: SPQ *FP1* 35 PK: NK *L3* 36 quod . . . constituti (37): ut duorum angulorum super MO constitutorum *R* / MO *corr. ex* NO *O* 37 sit: sicut *SOL3*; sint *C1* / uterque[1]: utrique *S* / maior[1]: ma *O* / maior fuerit *transp. R* / fuerit recto (38) *transp. P1* 38 cum *om. S* / supra: super *R* / idem *om. OC1* / SKM *alter. ex* SKN *in* SFK *O* 39 supra[1]: super *FP1* / *post* isto *add.* angulo *L3* / efficiet: efficit *SL3R* / supra[2]: super *R* 40 KM: KN *S*; *corr. ex* K? *O* / SK *alter. in* SF *O* / ita: in *S*; *corr. ex* in *L3* / et *om. S*; *inter. L3* / alio: illo *S* / circuli (41) *corr. ex* culi *O* 41 quod[1]: que *S*; *corr. ex* que *L3* / sit: est *FP1* / unus: unum *R* 42 uni visui *transp. FP1* / *post* visui *scr. et del.* c *O* / punctus: punctum *R* / est: sit *R* / inveniri *corr. ex* invenire *F*; invenire *P1* 43 possit: possunt *O* 44 sint: sunt *FP1* / puncta reflexionis *transp. R* 45 tamen: inde *S* / unicus: unus *ER* 46 quoniam: quando *FP1R* 47 ducte: ducere *S* 48 autem: ergo *R* / *post* situs *add.* oculorum *E* / visi: visui *S* / respectu *om. S* 49 centra: centrum *SOC1E* / sint: sunt *L3* / facilis: facit *FP1* / facilis erit *transp.* (erit *inter.*) *O* 50 probatio: proportio *FP1S*; *corr. ex* proportio *L3*; ? *O* 51 a *inter. a. m. E* 52 *post* arcus *add.* inter *R*

equales. Et si sumantur puncta reflexionis, secundum supradictam probationem, arcus circuli interiacentes hec duo puncta et punctum circuli quod est in perpendiculari a puncto viso ducta erunt equales, quod facile patebit iterata superiori probatione.

[2.218] Et hoc sive puncta reflexionis sunt in eadem superficie reflexionis, sive in diversis; erunt tamen arcus illi equales, et linee ducte a centris oculorum ad puncta reflexionum equales, et linee a puncto viso ad eadem puncta equales. Et linee a centris oculorum ad puncta reflexionum procedentes necessario se secabunt, et evidens est probatio quod super idem punctum perpendicularis a puncto viso ducte erit sectio, et in hoc puncto utrique visui apparebit ymago et una sola, quod est propositum.

[2.219] Est autem ordinatio ymaginum sicut ordinatio punctorum visorum. Si enim in re visa sumatur linea a capitibus cuius ducantur due linee ad centrum spere, fiet triangulus in quo continebuntur ymagines omnium punctorum illius linee. Et si sit in linea illa punctus eiusdem situs, ymago puncti remotioris ab eo erit in dyametro remotiori ab eius dyametro, et propinquioris in propinquiori. Et ita observatur pars in ymaginibus sicut fuerit in punctis visis.

[2.220] Sumpta autem linea in qua est punctus eiusdem situs, quodlibet punctum illius linee eiusdem situs erit respectu duorum oculorum, secundum modum predictum, et unicam habebit ymaginem propter equalitatem angulorum illius linee cum lineis visualibus. Si autem sumatur linea que angulum quem continent due linee a centris oculorum ad punctum visum dividat per equalia, situs cuiuslibet puncti illius linee, quantumlibet producte, erit idem utrique visui sicut fuit alterius, et idem probationis modus.

[2.221] Preter has duas lineas non est sumere eundem observantem situm, unde cum punctum visum comprehendatur in perpendiculari, cadet ymago eius in diversis punctis illius perpendicularis, sed

54 probationem: proportionem *FP1SOL3E* / *post* interiacentes *add.* inter *R* / duo *om. R*   55 circuli: eius *FP1*; C *E* / quod *om. FP1* / viso: suo *S* / erunt: erit *S*; *corr. ex* erit *L3*; est *O*   56 superiori: superiore *R* / probatione: proportione *FP1SOL3C1*   57 et hoc sive *inter. a. m. O* / puncta: puncti *O* / sunt: sint *R* / *post* eadem *scr. et del.* linea *P1*   58 erunt: erant *S*   59 *post* a[1] *scr. et del.* punctis *P1* / reflexionum *corr. ex* reflexionis *S*; reflexionis *L3C1* / equales . . . reflexionum (61) *om. S*; *mg. a. m. L3*   60 puncta reflexionum (61) *transp. R*   61 reflexionum: reflexionis *C1* / procedentes: precedentes *SC1* / se *om. SE*; *inter. L3* / probatio: proportio *SL3*   62 quod: non *FP1*; *om. S*; *inter. a. m. L3* / super: supra *E* / ducte: duce *F*   63 *post* sectio *add.* ambarum linearum reflexionis *R*   66 capitibus cuius *transp. R* / *post* ducantur *rep. et del.* ducantur *F*   67 due *om. FP1* / triangulus: triangulum *R*   68 si *om. FP1* / linea illa *transp. C1R* / punctus: punctum *R*   69 *ante* eiusdem *add.* non *R* / *post* situs *add.* respectu ambarum visuum *R* / remotiori: remotiore *R*   70 propinquioris in *om. R* / propinquiori: propinquiore *R*   71 pars: situs partium *R* / fuerit: fuit *R*   72 punctus: punctum *OL3C1ER*   73 situs erit *transp. OC1*   76 *post* autem *scr. et del.* fuerit *F*   77 puncti illius (78) *transp. C1* / puncti . . . linee (78) *corr. ex* linee illius puncti *E*   78 illius: alius *S*; *om. R* / quantumlibet: quantum que *FP1* / erit: erunt *P1*   79 alterius: uni *R* / *post* idem *add.* est *R*   80 *post* duas *add.* erunt *P1* / *post* sumere *add.* aliam *R*

imperceptibiliter a se remotis. Et ymago cuiuslibet puncti a quotcumque videatur oculis semper observat ydemptitatem partis, unde apparet unitas ymaginis, sicut dictum est in visu directo. Quod forme, licet in diversa cadant loca, propter tamen distantiam eorum insensiblem non diversificant apparentiam nisi diversificent partem. Similiter hic, quando remotio puncti ab uno visu modicum maior quam ab alio, erunt loca ymaginum imperceptibiliter remota, unde apparent simul, et ex eis una compacta, que quidem ymaginum loca aliquando non totaliter distant, sed partialiter.

[2.222] In speculis columpnaribus exterioribus, aliquando linea communis superficiei reflexionis et superficiei speculi est linea recta, aliquando circulus, aliquando sectio columpnaris.

[2.223] Cum fuerit linea communis linea recta, erit locus ymaginis in perpendiculari a puncto viso ducta super superficiem speculi tantum distans a linea communi quantum punctum visum ab eadem. Et eadem probatio que dicta est in speculo plano.

[2.224] Cum autem communis linea fuerit circulus, erit aliquando ymaginis locus intra circulum, aliquando extra, aliquando in ipso circulo. Eius rei eadem penitus assignatio que in speculo exteriori sperico.

[2.225] Si vero communis linea fuerit sectio columpnaris, dico quod ymaginum loca quedam intra speculum, quedam in superficie speculi, quedam extra speculum, que in singulari explanabuntur.

[2.226] **[PROPOSITIO 26]** Sit ABG [FIGURE 5.2.26, p. 582] sectio columpnaris, B sit punctus reflexionis, E punctus visus, D centrum visus. Et ducatur a puncto B perpendicularis super superficiem contingentem

83 quotcumque: quocumque *SC1E; corr. ex* quocumque *L3* 84 oculis: oculus *F;* circulus *S; alter. ex* oculus *in* oculorum *a. m. C1* / observat: observabit *FP1* 85 in[2] *om. P1* 86 loca *corr. ex* lata *F* / tamen: in *scr. et del. C1; om. O* / eorum: earum *R* 87 diversificant *corr. ex* diversificet *C1* / diversificant . . . nisi *om. P1* / nisi *alter. ex* etc *in* et *O;* neque *C1* / diversificent *corr. ex* diversificant *P1; corr. ex* diversificentur *L3; alter. in* diversificant *C1* / *ante* partem *add.* speciem *P1* / similiter hic *transp. O* 88 quando: quanto *SL3;* quare *O* / remotio puncti *OC1* / visu: nusquam *L3* / *post* visu *add.* fuerit *R* / modicum: modico *R* / quam: quoniam *S* / ab *om. P1* 89 loca *inter. O* / simul: similis *SL3* / et *inter. P1* 90 *post* una *add.* imago *R* / que: quando *R* / loca *corr. ex* locu *F* 91 partialiter: perpendicularis *O* 93 superficiei[1] *om. FP1SL3E* 94 circulus: cicirculus *F;* circulis *C1* / circulus aliquando *om. L3* 95 cum fuerit *inter. O* 96 super *om. F;* ad *P1* 97 distans a linea: a linea distans *E* / quantum *mg. a. m. F;* quamvis *S; corr. ex* quamvis *L3* / et eadem (98) *om. FP1* 98 *ante* probatio *add.* est *R* 100 intra: intus *S;* inter *L3* / *post* extra *scr. et del.* circulum *F* / ipso circulo (101): ipsa circumferentia *R* 101 eius: cuius *O* / rei eidem *transp. C1* / *post* penitus *add.* est *FP1* / que in speculo: in speculo que *S* / exteriori: exteriore *R* 103 si . . . explanabuntur (105) *om. S* / communis linea *transp. R* / columpnaris *mg. a. m. C1* / quod: quoniam *OL3C1E* 104 loca *om. FP1OL3ER* / *post* quedam[1] *add.* sunt *R* 105 in singulari: singula *OC1* 106 ABG: ABC *R* 107 punctus[1,2]: punctum *R* / E *inter. O* / visus[1]: visum *R* 108 a . . . perpendicularis: perpendicularis a puncto B *L3* / B perpendicularis *transp. S*

speculum in puncto B, que sit TBQ, et ducatur a puncto E perpendicularis super superficiem contingentem speculum in puncto K, que sit EKQ. Et linea contingens speculum in puncto B sit CU; linea contingens speculum in puncto K sit KM. Dico quoniam due perpendiculares TB, EQ concurrent.

[2.227] Ducantur linee EB, DB, et ducatur linea KB. Palam quoniam KM cadet in figura EKB, et linea BC in figura eadem. Igitur BC secabit EK. Secet in puncto C. Palam quoniam angulus TBK est maior recto, et angulus EKB similiter maior recto, quare TB, EK concurrent. Sit concursus punctus Q. Similiter, DBK maior recto. Igitur DB, EK concurrent. Sit concursus punctus H. Igitur H est locus ymaginis. Dico etiam quod proportio EQ ad QH sicut EC ad CH, et etiam quod QH est maior HB.

[2.228] Ducatur HF equidistans EB. Palam quoniam angulus EBC equalis est angulo DBU. Est igitur equalis angulo CBH. Restat EBT equalis angulo HBQ, cum sit TBC rectus, et QBC rectus. Cum igitur CB dividat angulum EBH per equalia, erit proportio EC ad CH sicut EB ad BH.

[2.229] Sed angulus EBT est equalis angulo HFB, quare HF, HB sunt equalia. Sed proportio EB ad HF sicut EQ ad QH. Erit ergo EC ad CH sicut EQ ad QH, quod est propositum. Et ex hoc, cum sit proportio EQ ad QH sicut EB ad BH, et EQ sit maior EB, erit QH maior HB, quod est propositum.

[2.230] Palam ex hoc quod, si supra sectionem GB ducatur perpendicularis super superficiem contingentem sectionem, concurret cum TB. Similiter, quecumque ducatur supra sectionem AB concurret cum TB. Et hec quidem patent cum punctus visus non fuerit in perpendiculari visuali. Palam enim ex superioribus quod unius solius puncti

109 *post* speculum *scr. et del.* que sit *F* / in puncto B *om. L3ER* / in . . . K (110) *om. S* / que . . . sit[1] (111) *mg. a. m. F* / TBQ: QBT *FP1*; GBQ *R* / TBQ . . . sit[1] (111) *mg. a. m. L3E* / ducatur . . . E: a puncto K ducatur *FP1* 110 in puncto K *om. OL3ER* 111 EKQ: GKQ *F* / *post* EKQ *add.* et linea contingens speculum que sit EKQ *L3* / CU *corr. ex* QCU *O* 112 *post* sit *rep. et del.* sit *F* / *post* dico *scr. et del.* q *C1* / quoniam: quod *R* 113 TB *alter. in* TBQ *F*; TBQ *P1*; GBQ *R* / EQ: EKQ *FP1SL3C1ER*; *alter. ex* EB *in* EK *O* 114 KB *corr. ex* EB *E* / quoniam: quod *R* 115 KM *corr. ex* EM *E* / cadet *corr. ex* cadent *C1* / figura[1,2]: figuram *R* / eadem: eandem *R* 116 quoniam: quod *R* / TBK *corr. ex* TBE *E*; GBK *R* / *post* TBK *scr. et del.* DCN *P1* 117 quare: quando *P1* / TB: GB *R* / concurrent *mg. a. m. C1* 118 punctus: punctum *R* / Q: H *FP1* / *ante* similiter *scr. et del.* igitur H est *F* / DBK *corr. ex* DBE *E* / *post* recto *scr. et del.* igitur DBK maior recto *O* 119 punctus: punctum *R* 121 HB *corr. ex* IHB *S*; HD *E* 122 quoniam: quod *R* 123 equalis est *transp. R* / EBT: EBG *R* 124 sit *om. R* / TBC *corr. ex* THC *S*; GBT *R* / rectus et *om. R* / et *om. FP1S*; *inter. OL3C1* (*a. m. C1*) / QBC: TBQ *R* / rectus[2]: sint recti *R* 125 CB: OB *SO*; BC *C1* / proportio *om. R* 127 EBT: EBG *R* / est *om. SOC1* 128 proportio *om. R* / *post* sicut *scr. et del.* EB *F*; *add. P1* / ad[2] *om. FP1* 130 BH: HF *R* 132 *post* palam *inter.* ergo *O* / quod *corr. ex* quo *F* / supra: super *R* / GB: ABC *R* 133 concurret: concurrit *FP1* 134 TB: GB *R* / similiter . . . TB (135) *om. R* / quecumque: quocumque *L3* / supra: super *E* 135 punctus visus: punctum visum *R* / fuerit: fuit *L3* 136 puncti forma (137) *transp. E*

forma per perpendicularem accedit ad speculum et secundum eandem reflectitur, et est punctus perpendicularis existens in superficie visus, punctus enim ultra visum sumptus non potest reflecti super hanc perpendicularem quia non potest accedere ad speculum super perpendicularem propter predictam rationem. Et similiter non poterit reflecti ab alio puncto speculi quam a puncto perpendicularis, quia accideret duas perpendiculares concurrere et effici triangulus cuius duo anguli recti, sicut supra patuit.

[2.231] **[PROPOSITIO 27]** Amplius, sumatur sectio columpnaris [FIGURE 5.2.27, p. 583], et sumatur in ea punctus A, et ducatur contingens sectionem, que sit AT, et sumatur perpendicularis super AT intra speculum, que sit DA.

[2.232] Palam quod AD dividit sectionem in duas partes in quarum utraque est punctus unicus cuius puncti contingens erit equidistans AD. Sit ergo G cuius contingens concurrat cum AD in puncto H, et ducatur perpendicularis super hanc contingentem, que sit QG, et hec quidem necessario concurret cum HD, sicut ostensum est in precedenti figura. Sit concursus in puncto D, et ducatur linea GA usque ad P, et ducatur linea QA. Igitur angulus QAH aut est equalis angulo HAP, aut maior, aut minor.

[2.233] Sit equalis. Procedet igitur forma puncti Q ad A et reflectetur ad P, qui sit visus, et locus ymaginis erit punctus sectionis columpnaris, scilicet G.

[2.234] Si vero supra punctum Q sumatur aliquod punctum, ut punctum F, erit quidem angulus FAH minor angulo HAP. Fiat ei equalis NAH. Concurret quidem NA cum GQ intra columpnam. Sit in puncto

137 per *om. FP1SC1; inter. OL3* / perpendicularem: perpendicularis *FP1* / accedit: accedet *L3*; accedat *C1* / eandem: eundem *R* 138 reflectitur: reflectetur *L3* / et est: si cum *O* / punctus: punctum *R* / perpendicularis: particularis *S* 139 punctus: punctum *R* / sumptus: sumptum *R* / reflecti: reflecte *S* 140 quia: et *FP1*; quam et *R* / accedere: accidere *L3* 141 *post* propter *scr. et del.* punctum *P1* / *post* predictam *add.* ibidem *ER* 142 *post* perpendicularis *add.* huius *C1ER* 143 perpendiculares: partes *S* / *post* perpendicularis *scr. et del.* a *P1* / effici *corr. ex* efficit *L3*; efficere *ER* / triangulus *alter. in* triangulum *L3*; triangulum *ER* / *post* anguli *scr. et del.* sicut *F* 144 recti *inter. L3* 146 punctus: punctum *R* 147 sectionem: sectioni *SO* / AT[1]: ET *SO*; *corr. ex* ET *L3*; *alter. in* EAT *a. m. C1*; EAT *R* / et *om. L3* / AT[2]: ET *SOL3*; *alter. in* EAT *a. m. C1* 148 speculum: sectionem *SOC1* 149 AD: A *FP1SOL3E*; DA *R* / in[2] *inter. L3* 150 punctus unicus: punctum unicum *R* / unicus: unius *O* / *post* puncti *add.* linea *R* 151 sit *corr. ex* si *C1* / *post* ergo *add.* aliud punctum *R* / *post* G *inter.* punctus *OL3* / *post* cum *add.* linea *R* 152 perpendicularis: perpendiculariter *FP1L3* / *post* contingentem *rep. et del.* erit (150) . . . contingens (151) *S* 153 precedenti: precedente *R* 154 figura *inter. O* 155 QAH: QDH *P1*; *corr. ex* QHA *L3* 157 procedet: procedit *FP1L3E* / Q *inter. O* 158 qui: quod *R* / sit visus *transp. R* / punctus: punctum *R* / columpnaris: corporis *O* 159 scilicet: S *S* / *post* G *scr. et del.* si *F* 160 aliquod: aliud *SOC1* 161 FAH *corr. ex* FHA *FL3* 162 quidem *inter. O* / NA: NAM *S* / columpnam: columpna *F* / sit *om. FP1*

K. Palam ergo quod ymago puncti F erit in puncto K, et ymagines omnium punctorum ultra punctum Q intra columpnam.

[2.235] Si vero intra Q et T sumatur punctum aliquod, ut punctum C, erit angulus CAH maior angulo HAP. Fiat ei equalis HAM. Palam quod MA cadet supra GQ, et extra sectionem. Sit in puncto O. Erit igitur ymago C in puncto O, et omnium punctorum T et Q interiacentium ymagines erunt extra sectionem inter T et G.

[2.236] Si autem angulus QAH fuerit minor angulo HAP, secetur ex eo equalis, et sit HAN. Palam quod ymago Q erit in puncto K, et omnium punctorum superiorum ymagines erunt infra sectionem. Si vero inferius sumatur C punctum, ut angulus CAH sit equalis angulo HAP, erit ymago C in sectione, et omnes inter C et Q intra, omnes inter C et T extra.

[2.237] Si vero angulus QAH fuerit maior angulo HAP, fiat ei equalis HAM [FIGURE 5.2.27a, p. 583]. Palam quod MA secabit sectionem. Et secet in puncto B, et ducatur contingens super punctum B, que concurrat cum DH in puncto L. Erit autem angulus DLB acutus, et angulus HLB obtusus, et LB concurrens cum HG faciet cum ea acutum. Ducatur perpendicularis a puncto B super LB, que sit SB. Secabit quidem HG, et faciet angulum acutum cum ea, cui angulus contrapositus similiter erit acutus. Et HG secat QA. Sit punctus sectionis U, et facit acutum angulum cum ea super punctum U, quare SB et QU concurrunt. Sit concursus in puncto Z. Palam ergo quod forma puncti Z movebitur ad speculum per ZA, et refertur per AM, et locus ymaginis B. Et ymagines punctorum linee ZS ultra Z erunt intra sectionem, et punctorum citra Z extra sectionem, quod fuit propositum.

[2.238] **[PROPOSITIO 28]** Amplius, ab uno solo puncto speculi columpnaris fit reflexio ad centrum visus, utpote punctus B [FIGURE 5.2.28, p. 584] reflectitur ad A a puncto G. Dico quod non refertur ad ipsum ab alio puncto speculi quam a puncto G.

163 K *inter. a. m. C1* 164 *post* punctorum *add.* linee QF *R* 165 intra: inter *SOR*/T: D *FP1* 166 C: R *R*/CAH *corr. ex* CHA *C1*; RAH *R* 167 MA *corr. ex* MAS *F*; *inter. O*/*post* supra *add.* lineam *ER*/et *om. L3* 168 C: R *R*/*post* punctorum *add.* inter *R*/et *om. R*/*post* et *scr. et del.* que *F* 169 T: D *FP1*; *inter. O*; O *R* 170 autem: vero *R*/minor *corr. ex* maior *mg. a. m. F* 172 ymagines: ymaginetur *F*/infra: intra *R* 173 C: R *R*/CAH: RAH *R*/HAP *corr. ex* AHP *FL3*; AHP *S*; HPA *P1* 174 C^1 *om. L3C1*; R *R*/sectione *corr. ex* sectionem *F*; sectionem *P1*/et^1 *inter. O*/*post* inter^1 *scr. et del.* se *F*/C^{2,3}: R *R*/intra: inter *S* 177 et *om. SOC1* 178 concurrat: concurret *R* 179 *post* DH *add.* ut *R*/erit autem: eritque *R* 180 obtusus et LB *mg. F*/HG: HD *C1*/cum^2 . . . faciet (182) *mg. F* 181 a *corr. ex* ad *P1*/*post* HG *add.* ut in puncto X *R* 182 cui: cum *FP1*; que *C1*; quoniam *R*/angulus: angulo *SOL3E*; oppositus triangulo *C1*/contrapositus: circa positus *S* 183 *post* acutus *scr. et del.* acutum *C1*/*post* HG *scr. et del.* et faciet angulum *S*/punctus: punctum *R* 184 sit concursus (185) *om. P1* 185 puncto *om. R*/Z *corr. ex* E *L3* 186 refertur: referetur *E*; reflectetur *R* 187 Z *corr. ex* S *a. m. L3* 188 Z extra *P1* 190 punctus: punctum *R*/B: D *S*; *corr. ex* D *L3*; *inter. O* 191 reflectitur: reflectatur *R*/refertur: reflectetur *R*

[2.239] Quoniam si in superficie reflexionis que est ABG sit totus axis speculi, erit linea communis superficiei speculi et superficiei reflexionis linea longitudinis speculi. Et cum in superficie reflexionis sit centrum visus, punctus visus, punctus reflexionis, et punctus axis in quem cadit perpendicularis, una sola superficies sumi potest in qua sit linea illa longitudinis, sive axis, et puncta A, B, quare non potest fieri reflexio ad A nisi ab aliquo puncto linee longitudinis. Sed iam probatum est quod non potest fieri reflexio ad A ab alio puncto linee longitudinis quam a puncto G, quare in hoc situ ab uno solo puncto speculi fit ad A reflexio.

[2.240] Si vero superficies ABG sit equidistans basi columpne, erit linea communis circulus equidistans basi. Et iam patuit quod ab alio puncto illius circuli non potest fieri ad A reflexio. Et si ab alio puncto speculi fiat reflexio, perpendicularis ducta a puncto illo cadet ortogonaliter super axem, et secabit lineam AB in puncto aliquo. A puncto illo ducatur linea ad axem in superficie equidistanti basi columpne. Erit quidem ortogonalis super axem, et ita due perpendiculares efficient cum axe triangulum cuius duo anguli sunt recti, quod est impossibile. Palam ergo quod in hoc situ non refertur B ad A nisi a puncto G.

[2.241] Si vero superficies ABG secet speculum sectione columpnari, dico quod a solo puncto G fit reflexio.

[2.242] Ducatur a puncto A [FIGURE 5.2.28a, p. 584] superficies equidistans basi columpne, que sit EZI, et a puncto G similiter superficies equidistans basi speculi in qua ducatur ab axe linea ad punctum G, que sit TG. Erit quidem perpendicularis super superficiem contingentem speculum in puncto G. Et concurrat cum AB in puncto K, et ducatur a puncto G linea longitudinis speculi, que sit GZ, et sit axis TQ. Et a puncto B perpendicularis ducatur ad superficiem EZI, que sit BH, et ducantur linee AZ, HZ. Et ducatur a puncto Z in superficie illa ad

193 ABG: AGB *S* 194 *post* communis *add.* super superficiem *P1* / speculi[2] *mg. a. m. O* 195 linea *corr. ex* line *P1* / et . . . superficie *inter. O* / *post* reflexionis[2] *rep. et del.* que (193) . . . reflexionis[1] (195); ABG: AGH *S* 196 sit: sicut *SO*; sint *C1* / *post* centrum *add.* A *S* / punctus[1,2,3]: punctum *R* / visus: visum *R* / *post* visus *add.* et *FP1* / et *om. S* 197 quem: quod *R* / sumi: ? *S* 198 sit *mg. a. m. F; inter. OL3; om. SE;* erit *inter. a. m. C1* / sive *om. R* / A *corr. ex* AA *P1* / *post* B *add.* G *R* / non potest *om. L3* 199 nisi *inter. a. m. E* / aliquo: alio *SO* / sed . . . longitudinis (201) *mg. a. m. E* 200 probatum: propositum *SL3C1* / est *om. FL3* / alio: aliquo *FP1L3* / linee longitudinis *om. R* 201 solo puncto *transp. S* 202 ad *mg. a. m. L3E* 204 linea communis: KG omnis *O* 205 illius . . . puncto *mg. a. m. L3* / ad A *inter. O* / *post* si *scr. et del.* ab *F* 206 *ante* speculi *scr. et del.* illius *L3* / *post* ducta *scr. et del.* a *P1* / cadet: cadent *FP1* 207 *post* aliquo *add.* et *S* 208 *post* in *scr. et del.* perpendiculari *P1* / equidistanti: equidistante *R* 211 recti: recta *FP1* / refertur: reflectetur *R* 216 que: qui *FP1L3C1* 217 equidistans: equidem *F* 218 *post* quidem *add.* TG *E* 219 concurrat: concurret *O* / K *corr. ex* E *a. m. E* / *post* ducatur *rep. et del.* et ducatur *C1* 220 G *inter. F* / GZ *corr. ex* GZG *F* / axis *corr. ex* axit *L3* 221 perpendicularis ducatur *transp. P1* / *post* ducatur *add.* EZ *S* / EZI *corr. ex* ZEI *F*; ZEI *P1* 222 AZ HZ *corr. ex* AS HS *L3* / ducatur: ducantur *S*; *corr. ex* ducantur *O* / Z *corr. ex* S *L3*

axem linea que sit ZQ. Erit quidem perpendicularis super axem, cum axis sit perpendicularis super hanc superficiem, et erit perpendicularis super superficiem contingentem speculum in puncto Z. Et concurrat cum linea AH in puncto L. Dico quod forma puncti H refertur ad A a puncto Z.

[2.243] Ducatur a puncto A equidistans linee KG, que sit AM, que quidem concurret cum BG. Sit concursus in puncto M. Palam quoniam GZ equidistans linee BH, cum utraque sit ortogonalis super superfices equidistantes, quare linea BGM est in superficie harum linearum. Igitur tria puncta M, Z, H sunt in hac superficie. Sed iterum AM est equidistans KG, et LZ equidistans KG, quoniam GZ equidistans TQ et inter superficies equidistantes. Igitur LZ equidistans AM, quare sunt in eadem superficie, et in ea est linea AH. Igitur in hac superficie sunt tria puncta M, Z, H. Et iam patuit quod sunt in superficie MBH. Igitur in linea communi sunt hiis duabus superficiebus. Igitur HZM est linea recta.

[2.244] Palam, cum G sit punctus reflexionis, erit angulus AGK equalis angulo KGB, et ita equalis angulo AMG. Sed est equalis MAG, quia coalternus. Igitur AG, MG sunt equales. Sed quoniam GZ est ortogonalis super quamlibet lineam superficiei AZH, erit quadratum MG equale quadratis MZ, GZ. Et similiter quadratum AG equale quadratis AZ, GZ. Erit igitur AZ equalis MZ, quare angulus AMZ est equalis angulo ZAM. Sed est equalis angulo LZH, et angulus ZAM est equalis angulo LZA, quia coalternus. Igitur angulus AZL est equalis angulo LZH, quare forma puncti H accedens ad punctum Z refertur ad punctum A.

223 ZQ *corr. ex* Z *F*; *corr. ex* SQ *L3* / super *om. S* 225 Z *corr. ex* S *L3* 226 AH: HA *C1*; AK *R* / quod: quoniam *SOC1* / refertur: reflectetur *R* 227 Z *corr. ex* S *L3* 229 quoniam: quod *R* 230 *post* GZ *add.* est *C1ER* / equidistans: equidistantem *F*; equidistanter *P1* / *post* equidistans *add.* est *SO* (*inter. O*); *scr. et del.* est *C1* / superficies *corr. ex* superficiem *L3*; superficiem *R* 231 equidistantes: equidistantem *R* / *ante* quare *add.* basibus columne *R* / quare: quia *E* 233 KG[1] *corr. ex* KQ *L3*; *corr. ex* EG *a. m. E* / LZ *corr. ex* LS *L3* / KG[2] *corr. ex* EG *a. m. E* / *post* GZ *add.* est *SO* (*inter. O*) / equidistans[3] *corr. ex* equidistantes *L3* / *post* equidistans[3] *scr. et del.* axi *F*; *scr. et del.* KG *S* 234 LZ *corr. ex* LS *L3* 235 *post* superficie[1] *scr. et del.* sunt tria *C1* 236 Z *corr. ex* S *L3* / patuit *corr. ex* planum *C1* / sunt *om. FP1*; sint *R* / MBH: BMH *R* 237 in . . . sunt: sunt in linea communi *C1R* / igitur *inter. F* / HZM: GHZM *FP1*; HMZ *C1*; *alter. ex* ASM *in* AZM *L3* 239 palam *om. O* / *post* palam *add.* igitur *R* / punctus: punctum *R* 240 KGB *corr. ex* EGB *a. m. E* 241 AG: SLG *S* 242 super: supra *C1* / AZH: ZAH *R* / quadratum *inter. a. m. F* 243 equale *corr. ex* equalis *C1* / quadratis *corr. ex* quadrato *a. m. F* / MZ: AZ *E* / MZ GZ *transp. O*; *corr. ex* MS GS *L3* / et . . . GZ (244) *om. ER* / *post* AG *inter.* est *O* 244 *post* quadratis *scr. et del.* erit igitur *C1* / GZ *corr. ex* GS *L3* / *post* GZ *add.* ? *F* / MZ *corr. ex* MS *L3* / AMZ *corr. ex* AMS *L3* 245 ZAM[1] *corr. ex* SAM *L3*; MAZ *R* / *post* sed *add.* angulus AMZ *R* / est: ? *O* / LZH: KH *S*; *corr. ex* LSH *L3* / ZAM[2] *corr. ex* SAM *L3* 246 angulo *om. R* / LZA *corr. ex* LSA *L3*; *corr. ex* LZH *E* / *post* LZA *rep. et del.* et (245) . . . LZA (246) *E* / igitur angulus *inter. L3* / AZL *corr. ex* ASL *L3* 247 angulo *corr. ex* angulu *F* / LZH *corr. ex* LSH *L3* / Z *corr. ex* S *L3* / refertur: reflectetur *R*

[2.245] Si ergo dicatur quod ab alio puncto quam a puncto G potest forma B reflecti ad A, illud aliud punctum aut erit in linea longitudinis, que est GZ, aut in alia. Si est in ea, ducatur ab eo perpendicularis, que necessario secabit lineam AK, et erit equidistans linee AM. Et linea ducta a puncto B ad illud punctum necessario concurret cum AM, et erit punctus ille et punctus M in eadem superficie.

[2.246] Et linea illa aut cadet super punctum M aut super aliud. Si super punctum M, erit ducere a puncto B ad punctum M duas lineas rectas. Si autem ad aliud punctum linee AM, ducatur a puncto illo linea ad punctum Z, et probatur quod hec linea cum HZ facit lineam rectam, sicut probatum est de linea ZM. Et ita a puncto H erit ducere duas lineas rectas per punctum Z transeuntes in diversa puncta AM cadentes, quod est impossibile.

[2.247] Palam ergo quod a nullo puncto linee GZ nisi a G potest B reflecti ad A. Si dicatur quod a puncto extra hanc lineam sumpto, ducatur super punctum illud linea longitudinis speculi, et a puncto circuli EZI in quem cadit hec linea, probatur H reflecti ad A secundum supradictam probationem. Sed iam probatum est quod H a puncto Z reflectitur ad A, et ita impossibile. Restat ergo ut a solo puncto speculi reflectitur B ad A, quod est propositum.

[2.248] **[PROPOSITIO 29]** Amplius, dato puncto B, quod reflectitur ad A, erit invenire punctum reflexionis, et hoc patebit per revolutionem probationis.

[2.249] Ducatur a puncto A [FIGURE 5.2.28a, p. 584] superficies equidistans basi columpne, que quidem secabit columpnam super circulum, qui sit EZI. Et ducatur a puncto B perpendicularis super hanc superficiem, que est BH, et inveniatur in hac superficie punctus a quo fit reflexio H ad A, qui sit Z. Et a puncto Z ducatur linea longitudinis, que

249 a *om. S; inter. OL3 (a. m. L3)* 250 aliud *corr. ex* ad *L3;* autem *E* 251 GZ aut *corr. ex* GS ut *L3* / ea: linea GZ *R* / *post* ea *add.* GZ *L3E (inter. L3)* / *post* eo *scr. et del.* que est *C1* 252 equidistans: equidem *SO* / linee AM: lineam *S; alter. in* lineam *O* 253 illud: illum *FP1* / AM: autem *S* 254 punctus[1,2]: punctum *R* / ille: illud *R* / et: etiam *E* 255 super[1]: supra *S* / M: AY *O* / super[2]: supra *C1* / aliud: alium *FP1; corr. ex* illud *mg. C1* 256 super: supra *SC1* 257 *post* rectas *add.* quod est impossibile *ER* / ad *om. S* / aliud: alium *FP1* 258 Z *corr. ex* S *L3* / et *om. P1* / probatur: probabitur *R* / *post* probatur *scr. et del.* cum *P1* / HZ *corr. ex* HS *L3* 259 ZM *corr. ex* SM *L3* 260 Z: ZS *FP1; corr. ex* S *L3* / transeuntes: transeuntem *FP1SL3* / *ante* in *add.* per *FP1* / *post* puncta *add.* linee *R* 261 cadentes: cadens *C1* 262 a[1] *om. S* / GZ: GS *L3* 264 illud: illum *FP1* / *post* linea *scr. et del.* LZ *F* 265 EZI *corr. ex* EZ *F;* EZ *S; alter. ex* ES *in* EZ *L3* / quem: quod *R* / probatur: probabitur *R* / H *inter. O* 266 sed: licet *L3* 267 reflectitur: refertur *SOL3* / ad A *inter. a. m. E* / *post* impossibile *add.* est *E* / *post* puncto *rep. et del.* puncto *F* 268 reflectitur: reflectatur *SOR* 269 reflectitur: refertur *SO;* reflectatur *R* 270 revolutionem: reflexionem *FP1; corr. ex* reflexionem *L3* / *post* revolutionem *add.* predicte *R* 272 *ante* ducatur *add.* verbi gratia *SOC1* 274 qui: que *E* / EZI *alter. ex* SI *in* ESI *S; corr. ex* ESI *L3* 275 est: sit *ER* / punctus: punctum *R* / fit *inter. a. m. F; om. SL3E* / fit reflexio (276): refertur *OC1* 276 qui: quod *R* / Z[1]: S *L3* / et . . . Z[2] *om. S* / Z[2] *corr. ex* S *L3*

sit ZG, et a puncto Z perpendicularis ZL et huic equidistans a puncto A, que sit AM. Et linea HZ producatur usque concurrat cum ea, et sit concursus in puncto M. Et a puncto M ducatur linea ad B, que necessario secabit lineam ZG, cum sit in eadem superficie cum ea. Quoniam cum BH sit equidistans GZ, erit HZM in superficie illarum, et ita MB in eadem, que, si secaverit ZG in puncto G, erit G punctus reflexionis, quod quidem si revolvas probationem predictam videre poteris.

[2.250] In speculis exterioribus piramidalibus, si linea communis superficiei reflexionis et superficiei speculi fuerit linea longitudinis speculi, erit locus ymaginis sicut assignatus est in speculis planis, et eadem probatio.

[2.251] Quod autem non possit esse linea communis circulus palam per hoc quod perpendicularis ortogonaliter cadit super superficiem contingentem speculum in puncto reflexionis, et circulus necessario erit equidistans basi. Superficies vero equidistans basi non erit ortogonalis super superficiem contingentem speculum.

[2.252] Si vero communis linea fuerit sectio piramidalis, ymagines quedam erunt in superficie speculi, quedam intra speculum, quedam extra. Et idem assignationis modus qui est in speculo columpnari exteriori, et eadem probatio. Et sicut ostensum est in columpnari exteriori, per perpendicularem visualem non reflectitur forma ad oculum nisi puncti superficiei oculi tantum, et hoc ab uno solo speculi puncto, et locus ymaginis eius continuus locis aliarum ymaginum, sicut patuit superius.

[2.253] Restat in hiis speculis declarare quod ab uno solo puncto eius fiat reflexio, quod sic patebit.

[2.254] **[PROPOSITIO 30]** Sit A [FIGURE 5.2.30, p. 585] visus, B punctus visus, G punctus reflexionis, et ducatur super punctum G superficies equidistans basi, que quidem secabit piramidem super circul-

277 ZG *corr. ex* SG *L3* / Z *corr. ex* S *FL3* / ZL *corr. ex* L *F*; L *P1*; SL *L3* / huic *corr. ex* hec *L3C1* 278 *post* et[1] *add.* etiam *ER* / usque: quousque *R* 279 et . . . M[2] *inter. a. m. L3* 280 lineam *om. O* / ZG *corr. ex* SG *L3* 281 equidistans *corr. ex* equidem *S* / GZ: BG *O* / HZM *corr. ex* HSM *L3* / MB: BM *R* 282 secaverit: secuerit *R* / *post* secaverit *rep. et del.* ZG (280) . . . eadem (280) *S* / ZG *corr. ex* LS *L3* / punctus: punctum *R* 283 revolvas: revolueris *F*; volueris *P1* / predictam *rep. P1* / videre poteris: videbis *O* / poteris: possis *L3* 285 superficiei[2] *om. R* / longitudinis *corr. ex* lineis *L3* 287 *post* eadem *add.* est *R* / *post* probatio *add.* C *E* 288 autem *om. S* / linea *corr. ex* lineam *S* 289 perpendicularis: superficies reflexionis *R* / *post* perpendicularis *add.* perpendicularem *S* / ortogonaliter cadit: ortogonalis est *R* 291 erit[1]: est *R* / vero: ergo hec *R* 293 ymagines: ymaginum *O* 294 erunt: erant *S*; erit *O* 295 *post* idem *add.* est *R* / qui: que *FP1L3E* / est: fuit *R* 296 exteriori: exteriore *R* / ostensum *om. R* / exteriori (297): exteriore *R* 297 per *om. S*; *inter. OL3E* (*a. m. E*) / reflectitur: refertur *SO*; reflectetur *R* / oculum *corr. ex* speculum *P1* 299 *post* eius *add.* erit *R* 1 puncto eius (2) *transp. SOC1* 3 A visus *transp. R* 4 punctus[1,2]: punctum *R* / visus: visum *R*

um, qui sit PG. Et ducantur linee AG, BG, AB, et a puncto G ducatur ad centrum circuli linea, que sit GT. Et conus piramidis sit E, a quo ducatur axis, qui erit ET. Et ducatur perpendicularis super superficiem contingentem speculum in puncto G, que sit HG, que, cum dividat angulum AGB per equalia, cadet super AB. Punctus casus sit Z.

[2.255] Et a cono ducatur linea longitudinis speculi ad punctum G, que sit EG, cui linee ducatur equidistans a puncto A, que necessario secabit superficiem circuli GP. Secet in puncto N, et sit AN. Similiter, a puncto B ducatur equidistans eidem EG, scilicet BM, que secet superficiem PG in puncto M. Et a puncto N ducatur equidistans GT, que sit NF, et ducantur linee NG, MG, NM.

[2.256] Palam quod TG secabit NM. Secet in puncto Q. Palam etiam quod MG secabit NF, cum secet ei equidistantem. Sit punctus sectionis F. Et a puncto A ducatur equidistans HZ, que sit AL. Palam quod BG concurret cum AL. Sit concursus L. Deinceps ducatur linea communis superficiei contingenti speculum in puncto G et superficiei circuli PG, que sit GC. Palam quod erit ortogonalis super GT, et similiter super NF.

[2.257] Sumatur etiam linea communis superficiei contingenti et superficiei reflexionis, que sit GD, que quidem, cum secet GH, secabit AL. Sit punctus sectionis D, et erit ortogonaliter super AL.

[2.258] Palam ex predictis quoniam NF est equidistans GT, et AL equidistans GH. Igitur superficies in qua sunt NF, AL est equidistans superficiei GTH. Sed EG equidistans BM, quare sunt in eadem superficie, que superficies secat predictas equidistantes, unam super lineam EG, aliam super lineam FL, quare FL equidistans EG. Sed AN equidistans eidem. Igitur FL equidistans AN.

[2.259] Verum superficies contingens speculum in puncto G secat superficies easdem equidistantes, unam in linea EG, aliam in linea CD. Igitur CD est equidistans EG. Igitur est equidistans AN et LF, quare erit proportio AD ad DL sicut NC ad CF.

6 qui: que *C1* / PG *corr. ex* BG *P1* / *post* BG *inter.* et *L3* / AB *inter. L3* / ducatur: ducantur *FP1* 7 conus: vertex *R* 9 G *mg. C1* / HG *corr. ex* DH *P1* 10 punctus: punctum *R* / Z *corr. ex* S *L3* 11 cono: vertice piramidis *R* 12 sit *inter. O* / equidistans *corr. ex* equidem *S* 13 circuli: speculi *P1* 14 equidistans: equidem *S* 15 N *om. FP1* / ducatur *corr. ex* due *a. m. L3* / equidistans: equidem *F* / *post* equidistans *add.* ipsi *R* 18 punctus: punctum *R* 20 sit: et *O* / deinceps: deinde *R* 21 contingenti: continti *F* 22 GC: GO *R* / ortogonalis: ortogonaliter *FP1SOL3E* 23 *post* NF *scr. et del.* S quidem *C1* 24 *post* contingenti *add.* speculum *R* 26 *post* AL^1 *scr. et del.* equidem *C1* / punctus: punctum *R* / ortogonaliter: ortogonalis *C1R* 27 *post* palam *scr. et del.* quod *L3* / equidistans: equidem *S* 28 equidistans^1: equidem *S* / superficies: in superficie *P1* 29 sed: ei *O* / *post* sed *add.* linea *R* / EG *corr. ex* G *C1* / equidistans: equidistat *ER* 31 *post* FL^2 *add.* est *ER* / *post* AN *add.* per *L3* 32 equidistans^1 *inter. O*; equidistat *ER* / *post* FL *add.* est *R* 33 secat: secet *S* 34 easdem: eadem *FP1* / unam: una *P1* / aliam: alia *S* / CD: OD *R* / CD igitur CD (35) *inter. L3* 35 CD: OD *R* / *post* CD *scr. et del.* EG *L3* / *post* LF *add.* et a puncto F ducatur linea equidistans LA secans DO in K et AN in I ergo FK equalis LD et KI equalis DA *R* 36 DL: LD *FL3E* / NC: NO *R* / CF: OF *R*

[2.260] Palam etiam quod angulus BGZ equalis est angulo ZGA, et etiam angulo GLA, et etiam angulo GAL, quare GAL, GLA equales. Et GA et GL equales, et GD perpendicularis super AL. Erit AD equalis DL. Erit igitur NC equalis CF, et GC perpendicularis. Erit angulus CFG equalis angulo CNG, Erit igitur angulus NGQ equalis angulo MGQ. Igitur a puncto circuli PG, quod est G, potest punctus M reflecti ad N, non impediente piramide.

[2.261] Dico igitur quod punctum B a solo G refertur ad A. Si enim dicatur quod a puncto alio potest reflecti illud, aut erit in linea longitudinis, que est EG, aut non.

[2.262] Sit in ea, et sit X [FIGURE 5.2.30a, p. 586], et ab eo ducatur perpendicularis super superficiem contingentem speculum in puncto illo, que quidem perpendicularis erit equidistans ZG, et ita equidistans AL. Erit igitur AL in superficie reflexionis huius perpendicularis, et erit similiter in superficie reflexionis perpendicularis ZG. Igitur ille due superficies reflexionis secant se super lineam AL. Sed secant se super punctum B, quod est impossibile, quoniam B non est in linea AL, quod patet per hoc quoniam FL equidistans BM. Restat ergo ut a nullo puncto linee EG preter quam a G possit reflecti B ad A.

[2.263] Si autem ab alio puncto, sit illud U, et ducatur linea longitudinis EUO, et sumatur superficies equidistans basi transiens per punctum U. Palam quoniam AN secabit hanc superficiem. Sit punctus sectionis Y. Similiter BM secabit eandem. Sit punctus sectionis K, et ducantur linee KU, YU, YK. Et cum superficies illa secet piramidem super circulum transeuntem per U, ducatur a puncto U linea ad centrum huius circuli, que sit RU. Et ducantur linee EK, EY, que quidem secabunt superficiem circuli PG, et sint puncta sectionum I, S. Et ducantur linee IO, SO.

37 ZGA *corr. ex* SGA *L3* 38 angulo[1] *alter. in* angulos *a. m. E* / angulo[1] . . . etiam *om. L3* / quare *om. P1* / quare . . . GLA[2] *inter. a. m.* (*post* GAL *add.* et) *L3* / *post* GLA[2] *add.* sunt *R* / et[2] . . . equales (39): et GA equalis GL *mg. a. m. O* 39 et[1] *om. SC1ER* / equales . . . DL (40) *mg. a. m. C1* / erit . . . DL (40) *om. P1* 40 NC: NO *R* / CF: OF *R* / GC: GO *R* / *post* perpendicularis *add.* super NF *R* 41 CFG: OFG *R* / CNG: ONG *R* 42 punctus: penes *P1*; punctum *R* 44 refertur: fertur *L3E*; reflectitur *C1R* 45 a: ab *R* / puncto alio *transp. R* / illud: illum *F* 49 erit *om. FP1* / erit equidistans *transp. C1* / ZG *corr. ex* EG *L3* / ita *mg. S* 50 erit *om. R* / *post* AL[2] *add.* est *R* 51 erit: est *R* / perpendicularis . . . reflexionis (52) *om. S* 52 reflexionis *corr. ex* rereflexionis *F* / super . . . se[2] *om. L3* / AL: AB *E* / se *om. S* 53 in *om. FP1*; *inter. L3* 54 equidistans: equidistant *FP1E*; equidistat *C1R*; *alter. ex* equidistant *in* equidistat *L3* / restat: restant *FP1* / a *inter. O* / *post* nullo *add.* a *S* 55 linee EG *transp. C1* / a G: AG *P1* / ad *inter. L3* 56 si: sed *S* / alio: aliquo *FP1ER*; *corr. ex* aliquo *L3* / *post* puncto *add.* extra lineam EG *R* / ducatur: dicatur *E* 58 quoniam: quod *R* / AN: autem *S* / punctus: punctum *R* 59 sit . . . piramidem (60) *mg. a. m. O* / punctus: punctum *R* 60 YK: AK *S*; *corr. ex* AK *L3* 61 U[2] *om. P1* 62 *post* que[1] *add.* extra circulum producta *R* / RU: YU *S*; *corr. ex* YU *a. m. E* 63 sectionum: sectionis *E* / I S: E S *S*; *corr. ex* Y S *a. m. E*; S I *R* 64 linee *inter. O* / IO SO: IC SC *R*

[2.264] Sicut ergo probatum est de puncto M quod, non impediente piramide, potest reflecti ad N a puncto G, ita probatur de puncto K quod potest reflecti ad punctum Y a puncto U, et est eadem probatio. Et ita angulus RUY est equalis angulo RUK.

[2.265] Palam ergo quoniam BK est equidistans EG, et linea communis superficiei BGEK et superficiei circuli PG est linea MG. Igitur linea EK, cum sit in hac superficie et secet superficiem circuli PG, cadet super lineam communem, que est MG. Erit igitur SMG linea recta.

[2.266] Eodem modo, cum superficies NYEG secet superficiem circuli PG super lineam NG, linea EY concurret cum linea NG. Igitur ING linea est recta. Palam etiam quoniam superficies IOE secat superficiem circuli PG super lineam IO, et secat superficiem huic equidistantem que transit per U super lineam YU. Ergo YU equidistat IO. Similiter superficies SOEK secat superficies illas equidistantes super duas lineas SO, KU. Igitur SO equidistat KU.

[2.267] Similiter, si sumatur superficies secans speculum super lineam longitudinis EO in qua sunt R, U, O, M, secabit illas superficies equidistantes super duas lineas MO, RU. Igitur hee due linee sunt equidistantes. Igitur angulus SOM equalis angulo KUR, et angulus MOI equalis angulo RUY. Sed iam patuit quod angulus KUR equalis est RUY. Igitur angulus SOM equalis est angulo MOI, quare punctus S potest reflecti ad I a puncto O, non impediente piramide.

[2.268] Sed iam probatum est quod punctum M reflecti potest ad I a puncto G, et ita punctum S, quod est in linea SMG, potest reflecti ad I a puncto G. Igitur punctus S reflectitur ad I a duobus punctis circuli PG, quod est impossibile. Restat ergo ut primum sit impossibile, scilicet

66 *post* G *add.* etiam *O*/probatur: probabiliter *P1*; probabitur *FR* 67 ad . . . U: a puncto U ad punctum Y *R*/U *inter. O*/est eadem *transp. R*/probatio: propobatio *O* 68 ita: in *L3*/*post* ita *rep. et del.* ita *F*/RUY: RUI *O*; *corr. ex* UY *a. m. E*/est *om. FP1L3E*; erit *R*/equalis *corr. ex* equalem *L3*/RUK *alter. ex* IUE *in* RUE *a. m. E* 69 ergo *om. SOL3C1ER*/quoniam *om. F*; quod *P1*/BK *alter. in* BE *a. m. E* 71 linea *corr. ex* lineam *L3*/*post* linea *scr. et del.* tamen fit *P1*/et *inter. O* 73 NYEG: NYES *E* 74 lineam: linea *O* 75 ING: IQG *S*/linea est *transp. P1C1*/est *inter. O*/quoniam: quod *R*/IOE: IDE *S*; *corr. ex* IEO *L3*; IEC *R* 76 IO: IC *R*/equidistantem (77): equidistanter *S* 77 ergo YU *inter. L3*/equidistat *corr. ex* equidistant *C1*/IO: YO *S*; IC *R* 78 SOEK: IOEK *L3*; SOE *E*; SEC *R* 79 SO[1,2]: SC *R*/KU igitur SO *om. FP1*/equidistat *corr. ex* equidistant *C1* 80 si *om. FP1*/*post* speculum *rep. et del.* secans speculum *E*/super: et *O*/super . . . longitudinis (81) *mg. P1* 81 lineam: linea *O*/EO: EC *R*/*post* qua *add.* superficie *R*/R *corr. ex* C *L3*; *corr. ex* Y *a. m. E*/*post* U *add.* et *FP1*/O: C *R* 82 equidistantes *corr. ex* equidistans *C1*/RU *corr. ex* CU *L3*; *corr. ex* YU *a. m. E*/hee due linee: due linee hee *C1* 83 SOM: SCM *R*/*post* equalis *add.* est *R* 84 MOI: MCI *R*/RUY *corr. ex* RUT *F*/*post* quod *scr. et del.* L *F* 85 est *om. C1*/RUY *corr. ex* CUY *L3*/SOM: SCM *R*/MOI: MOY *S*; MCI *R*/punctus: punctum *R* 86 O: C *R* 87 est *om. S*; *inter. OC1* (*a. m.* C1) 88 et . . . G (89) *om. R* 89 punctus: punctum *FP1R*/S: F *L3*/reflectitur: reflectetur *SOC1*/*post* circuli *scr. et del.* puncto *C1* 90 sit impossibile *transp. C1*

quod punctus B reflectatur ad A ab aliquo alio puncto speculi quam a G, quod est propositum.

[2.269] **[PROPOSITIO 31]** Amplius, dato speculo piramidali, est invenire punctum reflexionis.

[2.270] Verbi gratia, sit G [FIGURE 5.2.31, p. 587] conus piramidis, et super ipsum fiat superficies equidistans basi piramidis, que sit MNG. A sit punctus visus, B centrum visus. A et B aut erunt citra illam superficiem; aut ultra; aut in ipsa superficie; aut unum citra, aliud ultra; aut unum in superficie, aliud citra vel ultra.

[2.271] Sint ultra superficiem, et a puncto A ducatur superficies secans piramidem equidistans basi, et ducatur a puncto G linea ad punctum B, que producta cadet in superficiem ab A ductam, cum sit inter superficies equidistantes. Punctus in quo cadit hec linea sit H.

[2.272] Probatur autem modo supradicto quoniam A refertur ad H ab aliquo puncto circuli piramidis quem efficit superficies secans ducta a punctis A, H. Et inveniatur in circulo illo punctus reflexionis, et sit E. Et ducatur linea AB, et linea longitudinis piramidis GE, axis piramidis GT.

[2.273] Et ducatur a puncto E linea ad centrum circuli, que quidem cadet super axem, et sit ET, et erit ortogonalis super superficiem contingentem circulum illum in puncto E. Et ductis lineis AE HE, secabit angulum earum per equalia, et dividet lineam AH. Sit punctus divisionis R.

[2.274] Palam quoniam GE, ET efficiunt superficiem secantem lineam AB. Sit punctus sectionis F, et a puncto F ducatur perpendicularis super lineam GE et sit FC, que quidem erit ortogonalis super superficiem contingentem piramidem super lineam GE. Deinde a puncto A

91 quod punctus: ut punctum *R* / reflectatur: reflectitur *E* / A *inter. L3* / alio *inter. O* / alio puncto *transp. R* / *post* puncto *add.* circuli *E* 92 G: AG *P1* / quod *inter. a. m. C1* 95 conus: vertex *R* / piramidis: piramidalis *FP1SER; corr. ex* piramidalis *C1* 96 *ante* et *add.* speculi *R* / ipsum *corr. ex* ipsam *S* / fiat *corr. ex* fias *P1* / piramidis: piramidali *FP1*; piramidalis *SC1* / MNG A sit (97) *om. FP1* 97 punctus visus: punctum visum *R* 98 aut[2] . . . ultra *mg. a. m. F* / *post* aut[2] *scr. et del.* unum *L3* / ipsa *om. L3*; ipsam *E* / *post* superficie *scr. et del.* aliud in ipsa superficie *L3* / aut[3] . . . superficie (99) *om. S* / *post* citra *add.* et *E* / aut[4] *inter. O* 99 unum . . . aliud *inter. a. m. L3* / *post* superficie *add.* et *FP1* / aliud: alium *F* / ultra *corr. ex* extra *P1* 100 sint: sicut *S*; sit *E* / ultra: citra *R* 101 equidistans: equidistanter *FR* 102 ab: ad *P1* / A: ab *S*; *inter. O* 103 equidistantes *corr. ex* equidem *S* / punctus in quo: punctum in quod *R* 104 probatur autem *rep. S* / supradicto: predicto *FP1* / quoniam: quod *R* / refertur: reflectitur *R* 105 aliquo: alio *FP1SOL3E; corr. ex* alio *C1* / piramidis: piramidalis *SOL3C1E*; *om. R* / *post* secans *add.* piramidem *R* 106 punctus: punctum *R* 107 GE: G *S* / *post* GE *add.* et *ER* / axis . . . GT (108) *mg. a. m. L3* 110 et[2] *om. S* / ortogonalis: ortogonaliter *O* / superficiem: lineam *R* 112 punctus: punctum *R* 113 R: P *S; corr. ex* Y *L3; alter. in* Y *E* 114 ET *corr. ex* et *L3*; ER *C1* 115 punctus: punctum *R* 116 FC: FQ *R* / ortogonalis: ortogonaliter *FP1O*

ducatur equidistans linee FC, et sit AL. FC autem concurret cum axe in punto K. Et a puncto A ducatur equidistans linee RT, que sit AS, et ducatur a puncto E linea communis superficiei AEH et superficiei contingenti piramidem in linea GE, que sit EO. Cadet quidem ortogonaliter super AS, cum sit ortogonalis super ER.

[2.275] Et ducatur linea BC, que producta necessario concurrat cum linea AL. Sit concursus in puncto L, et ducatur a puncto C linea communis superficiei contingenti et superficiei ABL, que sit CP. Et ducantur linee LS, PO.

[2.276] Palam quoniam superficies ALS est equidistans superficiei GEK, et linee CE, PO sunt in superficie contingenti, que superficies secat illas superficies equidistantes super duas lineas CE, PO. Igitur CE equidistans PO.

[2.277] Ducatur autem linea HE donec concurrat cum AS in puncto S. Palam quod linea ES est in superficie HEG, et in eadem est linea BL, et hec superficies secat predictas superficies equidistantes in duabus lineis EC, LS. Igitur EC est equidistans LS. Erit igitur PO equidistans LS, quare proportio AO ad OS sicut AP ad PL.

[2.278] Sed palam quod angulus HER est equalis angulo REA. Erit angulus ESA equalis angulo EAS, et EO perpendicularis. Erit AO equalis OS. Erit igitur AP equalis PL. Et CP perpendicularis super AL, cum sit perpendicularis super FCK. Ergo CL equalis CA, et angulus CLA equalis angulo LAC. Erit ergo angulus BCF equalis angulo ACF. Igitur A refertur ad B a puncto C, quod est propositum.

[2.279] Si vero centrum visus et punctus visus fuerit in superficie MGN [FIGURE 5.2.31a, p. 588], sit unum in puncto M, aliud in puncto N, et ducantur linee MG, NG, MN, et dividatur MGN per equalia per

118 FC[1] *corr. ex* FT *a. m. E;* FQ *R*/et . . . FC[2] *inter. a. m.* (AL *rep.*) *L3*/FC[2]: FQ *R*/*post* FC[2] *add.* A *S*/autem: aut *S*/concurret: concurrat *ER* 119 equidistans *corr. ex* equidem *S* 120 superficiei[1] *om. SC1; inter. L3*/*post* superficiei[1] *add.* reflexionis *R*/AEH et superficiei *om. P1* 121 quidem: quod *FP1* 122 ortogonalis: ortogonaliter *FP1L3E*/ER: ET *R* 123 BC: BO *S;* BQ *R*/concurrat: concurret *C1ER* 124 *post* sit *add.* punctus *E; add.* punctum *R*/in puncto *om. ER*/C: Q *R* 125 CP: OP *R* 126 linee *om. ER* 127 ALS: AL *FP1* 128 GEK: GER *C1;* GEF *E*/CE: QE *R*/*post* CE *add.* donec (131) . . . linea[2] (132) *SL3* (concurrat: concurrant *S;* quod: quoniam *L3*)/*ante* PO *add.* BL *FP1SL3*/PO: BPO *S*/*post* in *scr. et del.* super *O*/contingenti: contingente *R* 129 CE: QE *R* 130 CE: QE *R*/equidistans: equidistet *C1ER* 131 HE . . . linea[2] (132) *om. FP1SL3; mg. O*/AS: HS *R* 132 quod: quoniam *C1E*/BL: HE *F;* HC *P1;* H *S;* HL *C1E; alter. ex* H *in* HC *L3* 134 EC[1,2]: EQ *R*/EC[2]: ES *P1*/equidistans[1]: equidem *S* 135 proportio *om. R*/AO: AC *S*/PL *corr. ex* PA *F* 136 quod: quoniam *C1*/HER *corr. ex* HEL *S*/est *om. E*/est equalis *transp. R*/*post* erit *add.* ergo *FP1* 137 angulo . . . CA (139) *mg. O*/EO: EA *E*/*post* EO *add.* est *R*/*post* perpendicularis *add.* super AS *R*/*post* erit *add.* ergo *R*/AO: AC *S* 138 OS *corr. ex* OF *F*/AP *mg. L3*/PL: PB *FP1*/CP: QP *R*/*post* perpendicularis *add.* est *R* 139 FCK: FK *SC1R;* FCFT *O;* FC *E; corr. ex* FK *L3*/CL: QL *R*/CA: AL *S;* AC *L3C1E;* AQ *R* 140 CLA: QLA *R*/LAC: LAQ *R*/BCF: BEL *FP1;* HCF *S;* BQF *R*/ACF: AQF *R* 141 refertur: reflectetur *R*/a *inter. O*/C: Q *R*/quod est: et ita *SC1; om. O* 142 punctus visus: punctum visum *R*/fuerit: sint *C1;* fuerint *R* 143 aliud: alium *F* 144 *post* dividatur *add.* angulus *R*/MGN: MNG *S*

lineam UG. Palam quoniam N a puncto G refertur ad M. Palam etiam quod linea UG et axis piramidis sunt in superficie secante piramidem super lineam longitudinis.

[2.280] A puncto U ducatur ortogonalis super hanc lineam longitudinis, que sit UE. Et super punctum E ducatur superficies equidistans basi, que secabit piramidem super circulum. Linea communis superficiei UEG et huic circulo sit ET. Palam quoniam cadet super axem et super centrum circuli.

[2.281] Deinde a puncto M ducatur equidistans linee GE, que quidem in superficie illius circuli cadat in punctum H. Similiter, a puncto N ducatur equidistans GE, que cadat in punctum A. Et ducatur AH, et ET secet eam in puncto R.

[2.282] Palam quoniam MH equidistans GE est in eadem superficie cum ea, que superficies secat superficiem MGN et superficiem HEA super duas lineas MG, HE. Igitur MG est equidistans HE. Similiter, AN, GE sunt in superficie secante illas equidistantes super NG, AE. Igitur NG equidistans AE. Similiter, superficies UGE secat easdem superficies super duas lineas RE, UG. Igitur UG, MG equidistantes HE, RE, quare angulus MGU equalis angulo HER, et angulus UGN equalis angulo REA, et angulus HER equalis angulo REA. Et ita punctus A potest reflecti ad H a puncto E.

[2.283] Si ergo a puncto A ducatur equidistans UE et alia equidistans RE, et ducatur ME donec concurrat cum linea equidistante UE, et ducantur linee communes, ut prius, et iteretur probatio predicta, patebit quoniam N potest reflecti ad M a puncto E. Erit igitur E punctus reflexionis, quod est propositum.

145 UG: QG *R* / quoniam: quod *R* / refertur: reflectitur *R* 146 quod *corr. ex* quoniam *P1* / UG: QG *R* / piramidis: piramidalis *S* 148 U: Q *R* / *post* ducatur *scr. et del.* linea *C1* / lineam *om. P1* 149 *ante* que *add.* GE *R* / UE: QE *R* / E *inter. a. m. C1* / ducatur: fiat *R* / equidistans: equidistanter *F*; equidem *SOL3* 151 UEG: QEG *R* / ET: et *S* 153 a puncto: ab *inter. O* / ducatur *corr. ex* ducantur *L3* / equidistans: equidem *SO* / GE: EG *R* / *post* GE *scr. et del.* quoniam *P1* 154 cadat: cadet *E* 155 equidistans: equidem *S* / cadat: cadit *C1*; cadet *E* / AH *corr. ex* KH *S*; AB *R* 156 secet: secat *O* / R: A *FP1* 157 MH: in H *S*; MB *R* / equidistans: equidem *S*; equidistat *R* 158 ea: ipsa *R* / superficiem^2: superficies *S*; *corr. ex* superficies *L3C1* / HEA: BEA *R* 159 *ante* super *add.* equidistans *R* / *post* MG *add.* et *C1* / HE^1: HG *E*; BE *R* / est equidistans *transp. R* / HE^2 *corr. ex* EH *S*; BE *R* 160 AN: NA *R* / *post* illas *add.* superficies *R* / equidistantes *alter. ex* quidem *in* equidem *F*; equidem *OL3* / equidistantes . . . NG (161) *om. S* 161 equidistans *alter. ex* equidem *in* equidistat *mg. a. m. C1*; equidistat *ER* / UGE: QGE *R* / *post* easdem *scr. et del.* superficies *F* 162 RE: R *FP1* / UG^{1,2}: QG *R* / *ante* igitur *add.* igitur RE QG equidistant *R* / *ante* MG *add.* et *ER* / equidistantes: equidistant *C1ER* / HE *corr. ex* EH *S*; BE *R* 163 MGU: MGQ *R* / HER *corr. ex* HEY *a. m. F*; BER *R* / UGN: UGA *E*; QGN *R* 164 HER: BER *R* / punctus: punctum *R* 165 *post* ad *add.* punctum B *R* 166 ergo *corr. ex* vero *P1* / UE: QE *R* / equidistans: equidem *S* 167 ME: BE *R* / equidistante: equidistanter *F*; equidem *SOL3*; equidistanti *E* / *post* equidistante *add.* ipsi *R* / UE: QE *R* 168 *post* prius *add.* et ME NE *R* 169 quoniam: quod *R* / N *corr. ex* non *L3* / punctus: punctum *R*

[2.284] Si vero ambo fuerint citra MGN [FIGURE 5.2.31b, p. 588], fiat piramis huic opposita. Et est ut protrahantur linee longitudinis piramidis iam facte, et a puncto A ducatur superficies secans hanc ultimam piramidem, que sit equidistans basi, que quidem secabit piramidem super circulum, que sit YZ.

[2.285] B aut erit in hac superficie, aut non. Si fuerit, fiat operatio a puncto B. Si non, ducatur linea GB usque dum concurrat cum hac superficie. Et sit concursus in puncto D. Palam quoniam A refertur ad D ab aliquo puncto circuli YZ interiori. Inveniatur punctus ille (sicut deinceps probabimus et docebimus, non ex anterioribus), et sit Z. Et ducentur linee DZ, AZ, et linea PZ dividat illum angulum per equalia.

[2.286] Et producatur linea ZG ad aliam piramidem, que quidem perveniet ad superficiem eius, et erit linea longitudinis. Et sit linea ZGE. Palam quoniam superficies PZE secabit lineam AB. Secet in puncto Q, et ducatur a puncto Q perpendicularis super lineam GE, et cadat in punctum E. Et erit perpendicularis super superficiem contingentem piramidem super lineam GE. Et super punctum E fiat superficies equidistans basi, que sit AEH, et ducatur a puncto D linea equidistans ZE, que sit DH, concurrens cum superficie illa in puncto H. Et eidem linee sit equidistans AA.

[2.287] Palam quoniam DH est equidistans ZE, et sunt in eadem superficie, que superficies secat superficies equidistantes super duas lineas DZ, HE. Igitur HE DZ sunt equidistantes. Similiter, AZ, AE equidistantes. Et palam quoniam PZ transit per centrum circuli YZ, similiter RET per centrum alterius circuli super quem superficies AEH secat piramidem. Igitur superficies PZER secat duas superficies equidistantes super duas lineas PZ, RE. Igitur PZ equidistans RE, quare

171 citra: ultra *R* 173 a puncto *mg. F* / hanc ultimam (174) *transp. L3* 174 que[1] . . . piramidem (175) *om. ER* / *post* sit *scr. et del.* QM sit *C1* / que[2] *om. FP1; inter. L3* 175 que: qui *OE* / que sit *om. R* / YZ *corr. ex* YS *L3* 176 aut[1] *corr. ex* autem *P1*; autem *R* / hac: hanc *S* / operatio: comparatio *C1* 177 GB: DH *P1* / concurrat: currat *S* 178 quoniam: quod *R* / refertur: reflectitur *R* 179 aliquo: alio *SO* / interiori: interiore *R* / punctus ille: punctum illud *R* 180 *post* probabimus *add.* et dicemus *FP1OL3* / et docebimus *om. O* / *post* docebimus *scr. et del.* vel dicemus *C1* 181 ducentur: ducantur *R* / DZ AZ *corr. ex* DS AS *L3*; AZ DZ AD *R* / et *om. O* / PZ *corr. ex* PS *L3* / illum *inter. O* / illum angulum *transp. R* 182 *post* et *add.* a puncto G ducatur GZ linea longitudinis et ducatur AB et *R* / que *corr. ex* quasi *P1* 183 superficiem: superficies *O* / linea[2] *om. R* 184 ZGE *corr. ex* GE *O* / palam . . . GE (185) *mg. O* / quoniam: quod *OR* / PZE: PEZ *P1* 185 perpendicularis: perpendiculariter *FP1L3* / cadat: cadet *FP1* 187 *post* GE *rep. et del.* et[2] (185) . . . GE (187) *E* 188 basi *om. L3* / AEH: HAE *O*; FEH *R* / D: B *SO* 189 ZE *corr. ex* SE *L3* / que . . . ZE (191) *om. S* / DH: BH *O*; *corr. ex* ACB *C1* 190 AA *corr. ex* A *a. m. L3*; *corr. ex* AL *a. m. C1*; AF *R* 191 ZE *corr. ex* SE *L3* / et: quod *R* 193 DZ[1,2] *corr. ex* DS *L3* / AZ *corr. ex* AS *L3* / AE: FE *R* 194 *ante* equidistantes *add.* sunt *R* / et *om. FP1* / et palam: similiter *R* / PZ *alter. ex* PS *in* PZS *F*; PZS *P1*; *corr. ex* PS *L3* / centrum: centra *E* / YZ: YS *L3* 196 secat[1] *corr. ex* secet *E* / PZER: per Z et *P1*; *corr. ex* PSER *L3* / *post* superficies[2] *scr. et del.* duas *E* 197 PZ[1]: PSZ *FP1*; *corr. ex* PS *L3* / PZ[2]: PS *L3* / equidistans: equidistat *C1ER*

angulus AZP equalis angulo AER. Et ita erit angulus AER equalis angulo REH, quare A refertur ad H a puncto E.

[2.288] Igitur, si a puncto A protraxerimus equidistantem QE, et aliam equidistantem RE, et lineas communes, sicut supra, et iteraverimus modum probandi predictum, patebit quoniam punctus A refertur ad B a puncto E, quod est propositum.

[2.289] Si vero centrum visus fuerit in superficie equidistante basi que est supra conum, scilicet G, et punctum visus ultra hanc superficiem, erit invenire punctum reflexionis hoc modo.

[2.290] Sit enim centrum visus M [FIGURE 5.2.31c, p. 589], punctus visus A, et sit MGN superficies equidistans basi piramidis. Et a puncto A ducatur superficies equidistans basi piramidis, que secabit piramidem super circulum DEK cuius centrum T. Et a puncto M ducatur perpendicularis super hanc superficiem, que sit MH, et ducatur linea HT. Et a puncto A ducatur ad lineam HT intra circulum linea AEQ ut EQ sit equalis QT, secundum supradicta. Et ducatur linea TEI, et a puncto H ducatur equidistans TE et equalis, que sit HB . Et ducantur linee MB, BE. Palam quoniam superfices GTE secabit lineam AM. Sit punctus sectionis F, et ducatur a puncto F perpendicularis super lineam GE cadens in puncto O, que sit FOC. Et ducantur linee MO, AO. Dico quoniam O est punctus reflexionis.

[2.291] Palam quoniam HB equidistans et equalis TE. Igitur HT equidistans et equalis BE. Sed MH equidistans et equalis GT, cum utraque perpendicularis. Igitur HT equidistans et equalis MG. Igitur MG equidistans et equalis BE, quare MB equidistans et equalis GE.

198 AZP: AP *FP1; corr. ex* ASP *L3* / AER^{1}: AES *FP1*; AEZ *E* / AER^{1,2}: FER *R* / *ante* et *add.* et angulus DZF angulo HER *R*   199 quare . . . H *om. S* / A: F *R* / refertur: referetur *E*; reflectetur *R* / E . . . puncto (200) *mg. O*   200 A: F *R* / equidistantem *corr. ex* equidistans *a. m. C1*   201 equidistantem: equidem *F* / lineas: linea *FP1* / *post* supra *add.* EC *FP1* / iteraverimus (202): iteravimus *FP1L3*   202 quoniam: quod *R* / punctus: punctum *SOC1R*   203 refertur: reflectetur *R* / B: D *SOC1* / *ante* a *add.* et *S*   204 centrum: punctus *FP1SOC1; corr. ex* punctus *L3* / equidistante: equidistanti *E* / basi *om. ER*   205 conum: verticem *R* / et *om. FP1; scr. et del. L3* / punctum: centrum *FP1SOC1; corr. ex* centrum *L3*; punctus *E* / visus: visum *R* / ultra: citra *R*   206 *post* hoc *scr. et del.* i *C1*   207 punctus visus (208): punctum visum *R*   208 A: ? *S; inter. OL3* / et^{1} . . . superficies *om. FP1*   209 que . . . piramidem *om. P1*   210 *post* circulum *add.* que sit *R* / DEK: DEH *O* / centrum *corr. ex* centri *a. m. L3* / *post* centrum *scr. et del.* visus *S*   211 *post* ducatur *add.* axis GT et *R* / *post* HT *add.* et ducatur ab M ad A linea recta MA *R*   212 AEQ *corr. ex* EAQ *L3* / ut: et *R*   213 supradicta: supradictam *FP1L3*   214 TE: DE *FP1* / *post* equalis *add.* ei *FP1* / et^{2} *inter. O*   215 *post* BE *add.* GE *R* / quoniam: quod *R* / secabit: secat *SOC1* / *post* AM *add.* et *C1* / punctus: punctum *R*   216 perpendicularis: perpendiculariter *FP1L3* / GE: EG *R*   217 *ante* cadens *add.* et producatur ad axem *R* / puncto: punctum *R* / sit *om. FP1; inter. L3* / FOC: FOP *R*   218 quoniam: quod *R* / punctus: punctum *R*   220 equidistans^{1}: equidem *FP1L3* / et^{1} *om. S* / BE: EB *R* / equidistans et equalis: equalis et equidistans *R*   221 *post* utraque *add.* sit *R* / perpendicularis: perpendiculari *E* / *post* HT *add.* est *OER* / equidistans et *om. C1* / igitur MG (222) *mg. O*   222 *post* equidistans^{1} *scr. et del.* M *L3* / BE: DE *S*

[2.292] Palam etiam quod angulus QTE equalis angulo QET, et ita equalis angulo AEI. Sed est equalis angulo IEB. Igitur, IEB equalis angulo IEA, quare A refertur ad B a puncto E. Et cum MB equidistans sit GE, si a puncto A ducatur equidistans FOC et equidistans TE, et iteretur figura supradicta et probatio, palam quoniam A refertur ad M a puncto O, et ita propositum.

[2.293] Si vero M sit in superficie, et A citra superficiem, fiet piramis alia huic opposita. Et fiat super A superficies equidistans basi huius piramidis, et invenitur in circulo huius superficiei punctus reflexionis ex punctis interioribus. Et ducatur a puncto illo linea ADG, et producatur. Et invenietur punctus, secundum superiora, et idem probandi modus.

[2.294] Si autem puncta, scilicet centrum visus et punctus visus, ita disponantur ut unum sit citra superficiem coni, aliud ultra, sit unum L [FIGURE 5.2.31e, p. 589], aliud A, superficies coni MGN.

[2.295] Et ducatur a puncto A superficies equidistans basi secans piramidem super circulum DE cuius centrum T, et ducatur linea LG. Concurret quidem cum superficie AED. Sit concursus K, et in circulo DE inveniatur punctus, qui sit E, ita ut contingens ducta a puncto illo, que sit SE, dividat per equalia angulum quem continent linee KE, AE.

[2.296] Et a puncto L ducatur linea equidistans GE, que necessario concurret cum linea KE. Sit concursus B. Palam quod L est in superficie GEK, et LB in eadem superficie equidistans GE. Et ducatur linea TEI. Palam quoniam superficies GTE secat lineam LA. Secet in puncto U, a quo ducatur perpendicularis super superficiem contingentem, que sit UOC. Et ducantur linee AO, LO.

223 QTE: QDE *FP1* / *post* equalis *add* est *R* / QET . . . angulo² (224) *mg. a. m. L3* 224 *post* sed *add.* QTE *R* / IEB¹: ZEB *FP1* / igitur IEB *om. FP1* 225 angulo *om. FP1L3E*; est *R* / refertur: reflectitur *R* / MB: linea BM *R* 226 *post* sit *add.* linee *R* / si: sed *C1* / FOC: FOP *R* / equidistans *inter. O* / TE: IT *P1SOL3C1ER* 227 quoniam: quod *R* / refertur: reflectetur *R* / M: AN *S* 228 O *om. P1* / *post* ita *add.* est *R* 229 superficie *corr. ex* superbie *F* / et . . . superficiem *mg. O* / citra: ultra *R* / superficiem: superficies *S*; *corr. ex* superficies *L3C1* 230 equidistans: equidem *FSOL3* 231 et *om. S* / invenitur: invenietur *SOC1*; inveniatur *R* / punctus: punctum *R* 232 punctis: puncti *S* / illo *om. P1* / ADG *corr. ex* ADADG *F*; *corr. ex* DG *L3* / et² *om. FP1* 233 punctus: punctum *R* / *post* idem *add.* est *R* 235 scilicet *inter. L3* / et punctus visus *om. FP1*; *mg. a. m. E* / punctus visus: punctum visum *R* / *post* visum *add.* et *L3* 236 disponantur: disponatur *L3* / coni: verticis *R* / ultra *corr. ex* extra *E* / sit² *rep. L3* / L: B *R* 237 aliud: ad *FP1* / coni: verticis *R* / MGN *corr. ex* MG *O* 238 secans: secabit *R* 239 piramidem: sperem *SOC1*; *corr. ex* sperem *L3* / super *rep. S*; *corr. ex* circa *L3* / *post* circulum *add.* que sit *R* / cuius centrum: centrum eius sit *R* / centrum *om. FP1* / T *rep. P1* / *post* et *add.* ducatur axis GT et *R* / LG: BG *R* 240 concursus: circulus *S* / et *om. O* 241 inveniatur *corr. ex* invenietur *a. m. C1* / punctus qui: punctum quod *R* / E: extra *R* 242 KE AE *om. P1* / *post* AE *add.* et ducatur linea longitudinis GE *R* 243 L: B *R* / linea *inter. E* 244 cum *inter. L3* / B: H *R* / L: B *OC1*; H *R* 245 LB in: BHM *R* / *post* superficie *inter.* sit *a. m. O*; *add.* quia *R* / equidistans: equidistanter *FP1* / *post* equidistans *add.* est *R* / GE: GK *E* / TEI: DEI *F*; DCI *P1* 246 quoniam: quod *R* / LA: BA *R* 247 quo *inter. a. m. S* 248 UOC *corr. ex* UIC *L3*; UOP *R* / AO: AC *S* / LO: LC *S*; *corr. ex* LG *L3*; BO *R*

[2.297] Palam quoniam AES equalis est angulo SEK, et cum angulus IES est rectus, et SET rectus, erit IEA equalis angulo TEK. Et ita angulus AEI equalis est angulo IEB, quare A refertur ad B a puncto E. Si ergo a puncto A ducantur equidistans UO et equidistans IT, et iteretur probatio, patebit quoniam refertur A a puncto O ad L, et ita propositum.

[2.298] Palam ergo quomodo sit invenire punctum reflexionis, et hec que dicta sunt in unico visu intelligenda sunt. In duplici autem visu, idem accidit, quoniam eadem forma et idem locus forme comprehendetur ab utroque oculo, et sicut dictum est in speculo sperico exteriori, forme a duobus oculis comprehense in hiis speculis propter contiguitatem videntur eadem, et aliquando simul sunt in loco, aliquando commiscentur earum loca in parte, aliquando separantur, sed modicum.

[2.299] Forma autem que secundum perpendicularem in hiis speculis descendit secundum eandem regreditur, sicut supra patuit, et forma illa ab uno oculo super perpendicularem percipitur ab alio oculo secundum lineam reflexionis. Sed loca formarum continua, unde eadem apparet utrique visui forma.

[2.300] **[PROPOSITIO 32]** In speculis spericis concavis, aliquando perpendicularis a puncto viso ducta secat lineam reflexionis, aliquando est equidistans ei. Quando secat, erit locus forme aliquando in speculo, aliquando ultra speculum, aliquando citra. Et cum fuerit locus forme citra speculum, aliquando erit inter visum et speculum, aliquando in centro visus, aliquando citra centrum visus. Et nos hoc demonstrabimus.

[2.301] Sit A [FIGURE 5.2.32, p. 590] centrum visus, D centrum speculi, et fiat superficies super hec puncta, que secabit speculum super circulum, qui circulus sit HBFG. Erit quidem hec superficies superficies reflexionis, quoniam est ortogonalis super quamlibet superficiem

249 quoniam: quod *R*/*ante* AES *scr. et del.* superficies *P1; add.* angulus *R*/est *inter. O* 250 IES: LES *P1*/est: sit *R*/et SET *inter. L3*/SET *corr. ex* ET *O*/rectus *om. L3* 251 est *om. R*/IEB: IEH *R*/refertur: reflectetur *R*/B: H *R* 252 ducantur: ducatur *OC1R*/UO: UC *SO* 253 quoniam: quod *R*/refertur: reflectetur *R*/refertur A *transp. SOC1* (refertur *inter. O*)/L: B *R*/*post* ita *add.* patet *R* 254 quomodo: quoniam *R*/sit: sic *P1*/invenire: invenitur *FP1; corr. ex* invenitur *L3* 255 in: de *R* 257 comprehendetur: comprehenditur *R*/oculo: visu *R*/est *om. SOL3*/speculo sperico *corr. ex* sperico speculo *P1* 258 exteriori: exteriore *R* 259 eadem: una *R*/simul: similis *SO*/in *om. SO*/*post* loco *add.* et *ER* 262 que secundum: que per *mg. a. m. O*/secundum: per *FP1ER;* per *inter. L3*/in hiis *inter. L3* 263 speculis *om. E* 264 oculo[1]: circulo *C1*/super . . . alio *mg. a. m. L3*/percipitur: percuritur *FP1O; ? S;* percurrit et *L3*/oculo[2] *om. SOL3* 265 lineam: lineas *O*/*post* continua *add.* sunt *ER* 266 utrique: utique *FP1; om. ER*/*post* forma *add.* de spericis concavis *L3E* 269 ei: ti *C1* 271 *post* aliquando *add.* autem *L3* 272 hoc: hec *ER* 276 hec superficies *transp. R*/superficies (277) *inter. L3* 277 quamlibet: quelibet *S; corr. ex* quam *L3*

contingentem circulum. Et ducatur linea AD, et a puncto A ducatur linea ad circulum maior AD, que sit AE. Et a puncto D ducatur ad circulum equidistans linee AE, que sit DH, et producatur AD usque in puncta B, I, et ducatur linea DE.

[2.302] Palam quoniam angulus AED est minor recto, quoniam ED dyameter, et quelibet linea in circulo cum dyametro facit angulum acutum. Et super punctum E fiat angulus equalis angulo AED, qui sit DET. Palam quoniam ET cadet intra circulum, et secabit lineam DH. Sit punctus sectionis T. Palam etiam quod angulus ADE maior angulo DET, quia AE maior AD, et ita ET secabit AB. Secet in puncto Z.

[2.303] Deinde a puncto A ducatur ad arcum EH linea que sit AN, et ducatur linea DN, et supra punctum N fiat angulus equalis angulo DNA per lineam NM, que necessario cadet intra circulum et secabit DH. Secet in puncto M. Palam etiam quod AN concurret cum DH extra circulum. Sit concursus L.

[2.304] Ducatur etiam a puncto A linea ad arcum EF, que sit AG, et ducatur DG, et sit angulus AGD equalis angulo DGQ. Palam quod QG secabit DH. Sit punctus sectionis Q. Palam etiam quod AG concurret cum DH ex parte F. Sit concursus O. Quod autem GQ cadat inter D et H palam cum arcus quem secat GO ex circulo sit maior arcu GH. Si enim ducatur linea GH, angulus HGD respiciet maiorem arcum angulo DGA.

[2.305] Item, a puncto A ducatur ad arcum FB linea AC secans DH in puncto S ut sit CS maior SD, et ducatur DC. Palam quod angulus DCA est acutus. Fiat ei equalis, qui sit DCK. Palam, cum angulus CDS sit maior angulo DCS, CK concurret cum DH. Sit concursus in puncto K.

[2.306] Palam secundum supradicta quod punctus T movetur ad E, et refertur ad A. Et perpendicularis a puncto T ducta est TD, que

278 et² . . . ad¹ (279) *om. L3* 279 maior *corr. ex* maiorem *S* / *post* maior *add.* quam *R* 280 linee *om. R* 282 quoniam: quod *R* / *post* ED *add.* est *O* 283 dyameter: semidiameter *R* 284 AED *corr. ex* ED *O* / *post* AED *scr. et del.* maior angulo *S* / qui: que *R* 285 quoniam: quod *R* / ET *corr. ex* et *FL3* 286 punctus: punctum *R* / *post* maior *inter.* est *O; add.* est *R* 287 quia: qua *S* / quia . . . AD *om. R* / ET *mg. L3* / *ante* secabit *add.* FS *L3* / *post* secabit *add.* lineam *C1* 288 *post* sit *scr. et del.* communis *P1* 289 et² *inter. a. m. C1* / supra: super *R* 290 intra: inter *FP1SL3* 291 etiam: est *S* / AN *corr. ex* angulus *L3* / concurret *corr. ex* concurrens *a. m. E* 292 L: ML *R* 293 *post* etiam *scr. et del.* linea *C1* / linea *om. O* / EF: EIF *R* 294 sit: fiat *R* / angulus *om. O* / AGD: QGD *SO;* DGQ *R* / angulo *om. O* / DGQ: DGA *SO;* AGD *R* 295 *post* DH *add.* ut patuit *R* / punctus: punctum *R* / etiam quod *transp. E* / *post* quod *scr. et del.* angulus *P1* 296 cum *inter. a. m. E* / GQ *corr. ex* GA *L3* / cadat: cadit *L3* 298 HGD: BGD *R* / respiciet maiorem *transp. R* / *post* maiorem *scr. et del.* angulum *C1* 299 DGA: AGD *R* 300 item: iterum *FP1E* / AC: AK *R* 1 CS: KS *R* / SD: LD *S* / DC: KD *R* 2 DCA: dicta *P1;* DKA *R* / qui: que *R* / DCK: *alter. in* DCE *deinde corr. ex* DCE *a. m. E;* DKU *R* / *post* palam *add.* quod *R* / CDS: KDS *R* 3 angulo *om. O* / DCS: CDCS *S;* DKS *R* / CK: KU *R* 4 K *alter. in* E *a. m. E;* U *R* 5 *post* palam *add.* ergo *SO* (*inter. O*) / supradicta: predicta *FP1* / punctus: punctum *R* 6 et¹ *inter. a. m. L3* / refertur: reflectitur *R* / TD . . . est (10) *mg. a. m. FL3*

perpendicularis est super superficiem contingentem circulum, et est equidistans linee reflexionis, que est AE, unde non concurret cum ea.

[2.307] Punctum autem Z movetur ad E, et refertur ad A. Et perpendicularis a puncto Z ducta est AZ, que concurrit cum AE in puncto A, unde locus forme puncti Z erit A.

[2.308] Punctum vero M movetur ad N, et refertur ad A. Et perpendicularis ducta a puncto M, que est MD, concurrit cum AN in puncto L, quod est ultra speculum, et locus forme puncti M erit L.

[2.309] Forma vero puncti Q movetur ad G, et refertur ad A, et locus eius erit O, qui est ultra visum.

[2.310] Et forma puncti K movetur ad C, et refertur ad A, et perpendicularis ab eo est KD, et locus ymaginis S.

[2.311] Palam igitur ex predictis quod ymaginum quedam ultra speculum, quedam inter visum et speculum, quedem in ipso visu, quedam citra visum, quod est propositum.

[2.312] Amplius, palam quoniam visus adquirit formas sibi oppositas, unde, cum locus ymaginis fuerit ultra speculum aut inter visum et speculum, comprehenditur veritas illius ymaginis. Cum autem perpendicularis a puncto viso ducta fuerit equidistans linee reflexionis, apparebit quidem ymago in puncto reflexionis. Quoniam, cum punctus ille sit sensualis, sumpto puncto eius intellectuali medio, ymago cuiuscumque partis illius puncti sensualis ultra medium sumpte erit ultra speculum; ymago partis citra medium erit inter visum et speculum, et cum totalis forma ex ulterioribus et citerioribus videatur una continua, necessario forma illius puncti sensualis videbitur in ipso speculo in loco reflexionis.

[2.313] Verum in ymaginibus quarum locus fuerit in centro visus, non comprehenditur veritas earum, unde sepius error accidit in

7 circulum: speculum *R* 8 equidistans: equidem *S* / que: quem *P1* / unde: unum *S* / concurret: concurrit *FP1E*; *corr. ex* curret *S* / ea: illa *FP1* 9 autem: A *P1* / *post* Z *add.* vero *FP1* / refertur: reflectitur *R* 10 a . . . ducta: ducta a puncto Z *R* / *post* ducta *rep. et del.* est (6) . . . ducta (10) *O* / concurrit: concurret *C1* 11 puncti: in puncto *E* / A[2] *inter. a. m. C1* 12 punctum *corr. ex* punctis *L3*; *corr. ex* productum *a. m. E* / movetur: videtur *L3* / refertur: reflectitur *R* 13 ducta *om. O* / AN: autem *S* 14 quod: que *E* 15 vero *inter. a. m. C1* / et[1] *mg. a. m. C1* / refertur: reflectitur *R* / ad[2] *rep. P1* / A *om. S* 17 K: U *R* / C: K *R* / refertur: reflectitur *R* 18 S: C *P1* 19 *post* predictis *scr. et del.* secundum *L3C1* / quedam . . . speculum[1] (20) *om. R* / *post* quedam *add.* sunt *OC1* (*inter. O*) 21 *post* visum *add.* quedam ultra visum apparent *R* / est *om. L3* 22 *post* visus *add.* perfectius *R* 23 unde: unum *S* 24 illius ymaginis *transp. C1* 26 quidem *om. R* / punctus ille (27): punctum illud *R* 27 sensualis: sensuale *R* 28 cuiuscumque: cuiuslibet *L3* / *post* medium *scr. et del.* illius *P1* / erit ultra speculum (29): ultra speculum erit *C1* 29 *post* speculum *add.* et *R* / partis *om. P1* 30 citerioribus *corr. ex* ceterioribus *L3* / *post* citerioribus *add.* partibus *R* / *post* una *add.* et *R* 31 sensualis *om. O* 33 in[1] *inter. a. m. E* / quarum *corr. ex* qualiter *a. m. F*; *corr. ex* quare *L3C1* (*a. m. C1*) / in centro *corr. ex* intro *L3* / *post* visus *add.* eo *S* 34 unde: unum *S*

huiusmodi speculis. Quod autem hoc pateat, erigatur super superficiem speculi lignum perpendicularem minus medietate semidyametri speculi. Et citra caput huius ligni sit centrum visus, et dirigatur visus ad punctum speculi cuius longitudo a ligno maior quam longitudo centri visus a dyametro per lignum transeunte. Videbitur quidem ymago illius ligni ultra visum, nec erit certa comprehensio eius, immo apparebit quasi arcuata, cum non sit. In hiis ergo speculis non comprehenditur veritas ymaginis nisi cuius locus fuerit ultra speculum aut inter visum et speculum. Cum autem fuerit centrum visus in perpendiculari per lignum transeunte, non plene comprehendit formam illius ligni.

[2.314] Si vero visus fuerit in dyametro spere et in centro eius, cum quelibet linea ab eo ducta ad speculum sit perpendicularis super speculum, non comprehendetur forma alicuius puncti nisi puncti portionis circuli interiacentis latera piramidis visualis que a centro circuli intelligitur protendi. Quoniam forma cuiuslibet alterius puncti cadet in speculum super lineam declinatam, et necessario refertur super declinatam, quare linea reflexionis non transibit per centrum, et ita non continget centrum visus.

[2.315] Si vero fuerit visus in dyametro sed non in centro, non comprehendet formam alicuius puncti semidyametri in quo est. Quoniam angulus quem efficient due linee a puncto sumpto in semidyametro et a centro visus in idem speculi punctum non dividetur per perpendicularem ab illo puncto speculi ductam, cum illa perpendicularis tendat ad centrum speculi. Sed formam alicuius puncti alterius semidyametri percipere poterit.

[2.316] **[PROPOSITIO 33]** Amplius, viso puncto in huiusmodi speculo, cum non fuerit perpendicularis equidistans linee reflexionis,

35 huiusmodi *corr. ex* huius *L3;* hiis *R* / quod: ut *R* / pateat: appareat *C1* 36 perpendicularem *alter. ex* perpendiculari *in* perpendiculariter *L3;* perpendiculariter *C1ER* 37 citra: contra *FP1;* circa *OL3ER* / huius: huiusmodi *C1* / ligni *corr. ex* lignus *C1* 38 ad: in *O* / punctum *corr. ex* speculum *L3* / *post* speculi *scr. et del.* u *L3* / longitudo[1] . . . quam *om. S* / a ligno *om. P1* / a . . . longitudo[2] *inter. a. m. L3* / *post* ligno *add.* est *L3; add.* sit *R* 39 transeunte *corr. ex* transeuntem *L3C1;* transeuntem *E* 40 *post* visum *scr. et del* verum *F* / nec erit *mg. a. m. F* 41 quasi *om. R* / arcuata *corr. ex* arcuta *O* 42 *post* comprehenditur *scr. et del.* y *F* / *post* veritas *rep. et del.* veritas *E* / ymaginis: yma *L3* / fuerit: fuit *L3* / ultra *inter. E* 43 aut: ut *S* / fuerit . . . visus: centrum visus fuerit *R* 46 visus *om. FP1* / et *om. P1* / *post* centro *scr. et del.* si *S* 47 quelibet: qualibet *S* / ducta ad speculum: ad speculum ducta *R* / perpendicularis: perpendiculariter *FP1L3* 48 nisi puncti: nisi forma *mg. F* / puncti: forma *P1* 49 circuli[1]: oculi *OR; alter. in* oculi *L3E* (*a. m. E*) / *post* latera *scr. et del.* oi c *P1* / circuli[2]: speculi *R* 50 alterius puncti *transp. P1* 51 refertur: reflectetur *R* 52 et . . . centrum (53) *om. O* 53 *post* centrum *scr. et del.* vestrum *E* 54 sed . . . centro *om. ER* 55 alicuius: alterius *R* / quo: qua *R* 57 semidyametro: dyametro *P1; corr. ex* diametro *a. m. L3* / dividetur: videtur *P1* 58 per *om. FP1; inter. O* / perpendicularem: partem *L3;* pendicularem *E* / illo: alio *FP1SL3C1E; corr. ex* alio *O* 61 huiusmodi *corr. ex* huiusiusmodi *F;* huius *P1* 62 non *om. P1* / equidistans: equidem *FC1*

linea a centro speculi ad punctum visum ducta sic se habebit ad lineam ab eodem centro ad locum ymaginis ductam sicut linea a puncto viso ad punctum quem diximus contingentie se habet ad lineam a puncto contingentie ad locum ymaginis ductam.

[2.317] Verbi gratia, sit E [FIGURE 5.2.33, p. 590] centrum speculi, B punctus visus, A centrum visus, G punctus reflexionis, linea contingentie ZG. ZG aut concurret cum EB, aut erit equidistans ei.

[2.318] Concurrat in puncto T. Linea EB concurret cum AG, et non in puncto G, cum EB, BG sint due linee. Igitur aut concurrent ultra G, aut inter G et A, aut in A, aut citra A. Sit ultra G, et in puncto H. Dico ergo quoniam proportio EB ad EH sicut BT ad TH.

[2.319] Ducatur perpendicularis EG, et a puncto H ducatur equidistans linee BG, que concurret cum EG. Sit concursus L, et a puncto B ducatur equidistans GH, que necessario concurret cum ZT. Sit concursus Q.

[2.320] Palam quoniam angulus BGE est equalis angulo AGE. Sed angulo BGE equalis est angulo GLH, et angulus AGE equalis angulo LGH. Igitur LH equalis est GH. Similiter, angulus BGQ equalis est AGZ, et angulus AGZ equalis angulo GQB, et ita BQ equalis BG, quare proportio BG ad HL sicut BQ ad HG.

[2.321] Sed quoniam angulus GHT equalis est angulo TBQ, erit triangulus TBQ similis triangulo GHT. Igitur proportio QB ad HG sicut BT ad TH, et ita BG ad HL sicut BT ad TH. Sed cum triangulus BGE sit similis triangulo HEL, erit proportio BG ad HL sicut EB ad EH, et ita EB ad EH sicut BT ad TH, quod est propositum.

[2.322] Eadem erit probatio si locus ymaginis fuerit inter A et G [FIGURE 5.2.33a, p. 591], aut in A [FIGURE 5.2.33b, p. 591], aut ultra A [FIGURE 5.2.33c, p. 591].

63 sic *om. SOC1R; corr. ex* S *L3* 64 linea a *corr. ex* linea *a. m. C1* 65 quem: quod *R*/*post* lineam *scr. et del.* a punctum *P1* 68 punctus[1,2]: punctum *R*/visus *inter. a. m. O*; visum *R* 69 *post* ZG[1] *add.* que *FP1* (*mg. a. m. F*)/ZG[2]: GZ *L3*/*post* ZG[2] *scr. et del.* arcus *P1*; *add.* autem *R*/ concurret *corr. ex* curret *F* 70 *post* linea *add.* vero *R*/concurret: concurrat *C1*; concurrit *R*/*post* cum *add.* linea *SO*/et: sed *R* 71 EB BG: BE AG *R*/*post* sint *scr. et del.* linee *P1*/concurrent: concurret *E*; concurrit *R* 72 aut[2] . . . A[3] *om. P1*/aut citra A *mg. O*/citra: ultra *R* 73 quoniam: quod est *R* 74 ducatur[1]: producatur *R* 75 sit: sicut *L3*/et . . . B *om. C1*/B: H *P1*/*post* B *scr. et del.* d *C1* 78 quoniam: quod *R*/est equalis *transp. C1*/est . . . BGE (79) *om. FP1*/angulo *om. R* 79 angulo[1]: angulus *R*/BGE: AGE *C1*/*post* BGE *rep. et del.* equalis (78) . . . BGE (79) *E*/ equalis est *transp. R*/est *om. P1*/GLH: HLG *O*/et . . . LGH (80) *scr. et del. E*/angulus *om. O*/*post* AGE *add.* est *O*/angulo[3] *om. O* 80 *post* LGH *add.* ergo angulus GLH equalis est angulo LGH *R*/est[2] *om. P1*/*post* est[2] *rep.* GH . . . est *S*; *add.* angulo *R* 81 et[1] *corr. ex* equalis *S*/AGZ *corr. ex* AGS *L3*/*post* equalis[1] *add.* est *OC1ER*/*post* equalis[2] *add.* est *R*/quare *corr. ex* qualiter *F*/quare . . . BG (82) *mg. O* 82 HG: HD *S* 83 *post* quoniam *add.* G *FP1L3*/GHT: GHTT *S*/equalis est *transp. R*/TBQ: TQB *P1*/erit . . . TBQ (84) *rep. S*; *mg. a. m. L3* 84 triangulus: triangulum *R*/ similis: simile *R* 85 triangulus: triangulum *R*/BGE . . . triangulo (86) *inter. a. m. L3*/sit: est *L3* 86 similis: simile *R*/BG *corr. ex* EG *P1*/et . . . EH (87) *inter. a. m. L3* 87 est *om. O* 88 erit probatio *transp. C1* 89 ultra: intra *P1*/A[2] *om. R*

[2.323] Si vero linea contingentie ZG sit equidistans perpendiculari, que est BH [FIGURE 5.2.33d, p. 591], ducatur perpendicularis GE, que, cum sit perpendicularis super ZG, erit perpendicularis super BH. Et erit angulus BEG equalis angulo HEG, et angulus BGE equalis est angulo EGH. Restat triangulus BGE similis triangulo EGH. Igitur proportio BE ad EH sicut BG ad GH, quod est propositum, quia in hoc casu non potest sumi aliud punctum contingentie quam punctus G, eo modo quo punctum contingentie supra appellavimus.

[2.324] **[PROPOSITIO 34]** Amplius, sit circulus DGT [FIGURE 5.2.34, p. 592], et A centrum visus intra speculum, E centrum speculi, B punctus visus. Et ducatur dyameter DAG.

[2.325] Si fuerit B in semidymetro EG, poterit esse reflexio ab aliquo puncto semicirculi GTD et ab aliquo puncto semicirculi ei oppositi. Quoniam quocumque puncto semidyametri EG sumpto, si ab eo ducatur linea ad aliquod punctum semicirculi GTD, et a puncto A ad idem punctum ducatur alia linea, ille due linee efficient angulum quem dividet semidyameter ductus a puncto E in illud punctum. Similiter in semicirculo opposito.

[2.326] Si vero B fuerit extra dyametrum DAG [FIGURE 5.2.34a, p. 592], ducatur dyameter transiens per B, qui sit TBQ. Dico quoniam B potest reflecti ad A per arcum interiacentem semidyametros in quibus sunt A et B, et similiter per eius oppositum, id est per arcum TD et per arcum GQ, et non poterit reflecti ab aliquo puncto arcus GT vel arcus QD.

[2.327] Verbi gratia, sumatur punctum in arcu GT, quod sit K, et ducantur linee AK, KB donec KB cadat super dyametrum DG in puncto O. Cum O et A sint ex eadem parte centri circuli quod est E, perpendicularis ducta a puncto K ad E non dividet angulum OKA, et ita B non refertur ad A a puncto K. Similiter, sumpto alio puncto quod sit F,

91 *post* contingentie *add.* scilicet *FP1* / ZG *alter. ex* GC *in* GZ *F*; GZ *P1* / *post* sit *scr. et del.* EG *L3* / perpendiculari *mg. L3* 92 BH: BEH *R* / *post* BH *add.* et *E* / GE: G *S* 93 ZG: GZ *R* / et *inter. O* 94 BEG: BEH *P1* / HEG: HEB *S* 95 triangulus: triangulum *R* / similis: simile *R* 96 EH: HE *R* / quia *om. FP1*; quare *R* 97 punctus: punctum *R* 99 DGT: ABGD *R* 100 A: H *R* / B: Z *R* 101 punctus visus: punctum visum *R* / DAG: BED *R* 102 B: Z *R* / EG: BE *R* / aliquo: alio *SO* 103 GTD *corr. ex* TGD *L3*; DTG *E*; BAD *R* / aliquo: alio *SOL3E* 104 *post* puncto *scr. et del.* semicirculi *C1* / EG: ?G *F*; BE *R* 105 GTD *om. ER* / A: H *R* 106 efficient: efficiunt *C1* 107 dividet: divident *E* / *post* dividet *add.* per equalia *R* / ductus: ducta *R* / E *inter. F* / in[1]: ad *R* / illud: illum *FP1* / *post* punctum *rep. et del.* punctum *E* 109 *post* B *add.* punctum visum *R* 110 qui: que *C1R* / TBQ: TQ *R* / *post* TBQ *add.* et *FP1* / quoniam: quod *R* 111 *post* ad *add.* visum *R* / semidyametros: diametros *R* 112 *post* est *scr. et del.* arcum et *F* / TD . . . arcum (113) *om. FP1* 113 et: sed *O* / GT: GR *R* / QD: GQ *L3* 114 GT: prope T *R* 115 KB cadat *transp. R* / cadat: cadet *P1* 116 *post* cum *add.* igitur *R* / centri circuli *transp. L3*; *corr. ex* circuli centri *E* / E *om. L3* / perpendicularis (117): perpendiculariter *L3* 117 K *corr. ex* KY *F*; *corr. ex* KS *L3* / et . . . AFB (119) *om. E* / B non *transp.*; B *inter. O* 118 refertur: reflectetur *R* / alio: aliquo *FP1* / *ante* puncto[2] *scr. et del.* al *F*

patebit quoniam perpendicularis EF non dividet angulum AFB, et ita non refertur B ad A a puncto F.

[2.328] Quod autem a puncto arcus TD vel arcus GQ possit fieri reflexio palam per hoc. Sit M punctum arcus DT, et ducantur linee AM, MB; fiet quadrangulum AMBE. Igitur perpendicularis EM dividet angulum AMB.

[2.329] Pari modo, sit H punctus arcus GQ. Linea AH secabit dyametrum TQ in puncto C, et linea HB eundem in puncto B. Et sunt hec duo puncta ex diversis partibus centri, quare linea EH dividet illum angulum.

[2.330] Pari modo, si fuerit B in superficie speculi aut extra speculum, dum A sit intra speculum, idem erit modus probandi qui prius. Similiter, si A fuerit in superficie speculi, B intus aut exterius.

[2.331] Verum, si a puncto A ducatur equidistans TE, que sit AP [FIGURE 5.2.34b, p. 593], loca ymaginum reflexarum a punctis arcus TP erunt extra speculum; loca autem ymaginum arcus PD ultra centrum visus, quod est A; loca autem ymaginum arcus QG sunt inter centrum visus et speculum. Et quod supradictum est de locis ymaginum idem intelligendum, ducta AM equidistans linee TQ.

[2.332] Si vero A fuerit extra speculum, B intra [FIGURE 5.2.34c, p. 593], patebit quod diximus. Ducantur a puncto A linee contingentes circulum GTD, que sint AH, AZ, et ducantur duo dyametri AEG, TEQ, et B in dyametro TEQ. Refertur B ad A ab aliquo puncto arcus TD, sed palam quod non ab aliquo puncto arcus ZD. Igitur ab aliquo puncto arcus TZ, et similiter ab aliquo puncto arcus oppositi TD, scilicet arcus GQ. Sed ab arcu TG vel DQ non fiet reflexio secundum supradictum modum.

[2.333] Si vero B sit extra hunc dyametrum et supra alium, qui similiter sit TEQ, fiet reflexio ab arcu TD, et a sola parte eius TZ, et ab arcu opposito, qui est GQ. Sed ab arcu TG vel DQ non fiet reflexio.

119 quoniam: quod *R* / AFB: OKA *S* / *post* ita *add.* B *S* 120 non . . . B: B non refertur *E* / refertur: reflectetur *R* 121 arcus *om. S* 122 *post* hoc *scr. et del.* quod *L3* / DT: TD *R* 123 MB: AB *S* / *ante* fiet *add.* et *S* / AMBE: ABE *O* / dividet: dividit *FP1*; dividat *E* / angulum (124) *om. S* 125 pari: simili *R* / *post* sit *scr. et del.* modo *S* / punctus: punctum *R* / GQ: GA *S* / AH: AD *S* 126 TQ *corr. ex* DQ *F*; QT *C1* / puncto[2] *corr. ex* punctum *P1* / *post* hec *add.* etiam *FP1L3ER* 129 in: ut *P1* 130 dum . . . speculum *inter. a. m. L3* / modus probandi *transp. R* 131 B *alter. in* D *E* / intus: interius *R* 132 verum . . . TQ (137) *transp. ad* 147 *post* reflexio (equidistans [137]: equidistante) *R* / que: qui *E* 133 punctis: punctus *O* / arcus *inter. O* 134 erunt: esset *FP1* / *post* speculum *add.* que ab ipso P erit perpendicularis equidistans linee reflexionis *O* / loca . . . A (135) *mg. a. m. C1* (autem *om.*; *post* PD *add.* sunt) 135 quod . . . visus (136) *om. S* / autem *inter. P1* / sunt *om. C1* 137 *post* idem *add.* erit *L3* 139 *post* ducantur *add.* enim *R* / a . . . linee: linee a puncto A *R* / A *inter. a. m. C1* 140 GTD: DCG *E*; DTG *R* / sint: sit *FP1L3*; sunt *C1* / AZ: DZ *O* / duo: due *R* 141 TEQ *corr. ex* DEQ *F* / refertur: reflectetur *R* / refertur B *transp. L3* / aliquo: alio *SO*; *corr. ex* alio *L3* / TD: DT *S* 142 ab[1] *om. O* / aliquo: liquo *F* / TZ: CG *S* / et *om. O* 143 ab[1] *rep. S*; *inter. C1* / oppositi: opposito *E* / *post* oppositi *add.* ipsi *R* / TD *om. S* / scilicet: id est *inter. O* / GQ: QG *C1* / *post* GQ *add.* reflexio fiet *R* 144 DQ *corr. ex* D *F*; *corr. ex* Q *P1* 145 sit: fuerit *R* / hunc: hanc *R* / supra: super *FP1R*; *corr. ex* super *S* / qui: que *R* 147 *post* arcu *inter.* ei *a. m. L3*

[2.334] **[PROPOSITIO 35]** Amplius, sumpto dyametro circuli in sperico speculo concavo, quilibet punctus illius dyametri quantumcumque producti potest esse locus ymaginum.

[2.335] Verbi gratia, sit AG [FIGURE 5.2.35, p. 594] dyameter circuli AMG, cuius D centrum. Sumatur in hoc dyametro punctus Z, E centrum visus. Dico quod Z potest esse locus ymaginis.

[2.336] Verbi gratia, ducatur linea ETZ, T punctus circuli. Ducatur linea DT. Erit angulus ETD acutus. Fiat ei equalis qui sit DTL. Palam quod L refertur ad E a puncto T, et eius ymago erit Z.

[2.337] Similiter, sumpto puncto L, patebit quod est locus ymaginis. Ducatur linea EL usque in B punctum circuli, et ducatur linea DB. Erit angulus EBD acutus. Fiat ei equalis, qui sit DBC. Refertur quidem punctus C ad E a puncto B, et locus ymaginis eius erit L, et ita sumpto quocumque alio puncto, eadem erit probatio.

[2.338] Amplius, punctorum qui comprehenduntur in hiis speculis quorumdam ymagines quatuor loca sortiuntur, quorumdam tria, quorumdam dua, quorumdam unum. Punctus cuius ymago in quatuor ceciderit loca a quatuor punctis determinatis refertur, non ab aliis, vel pluribus. Punctus cuius ymago tria sibi usurpat loca a tribus punctis speculi refertur, non a pluribus; cuius duo a duobus punctis; cuius autem ymago in unicum cadit locum poterit esse quod ab uno tantum puncto fit eius reflexio, et poterit esse quod a quolibet circuli determinati puncto, non ab alio.

[2.339] **[PROPOSITIO 36]** Verbi gratia sit E [FIGURE 5.2.36, p. 594] centrum visus; H sit punctus visus in eodem dyametro; D sit centrum circuli. Ducatur dyameter ZEHA. Aut ED est equalis DH, aut non.

148 sumpto: sumpta *R* 149 sperico speculo *transp. FP1* / quilibet punctus: quodlibet punctum *R* 150 producti *corr. ex* produci *L3*; producte *R* / ymaginum: ymaginis *O* 152 hoc: hac *ER* / punctus: punctum *R* 154 verbi gratia *om. R* / ETZ: EZT *P1*; GTZ *E* / ETZ T: et ZT *S* / *post* EZT *add.* per *R* / T *om. P1*; *inter. L3* / punctus: punctum *R* / *post* circuli *add.* et *SOR* 155 ETD *corr. ex* et D *F* / *ante* acutus *scr. et del.* et *L3* / *post* fiat *add.* autem *R* / ei *inter. a. m. C1* / qui *corr. ex* que *F*; que *L3ER* / DTL . . . sit (159) *om. S* 156 refertur: reflectetur *R* / Z *corr. ex* S *L3* 157 sumpto puncto *transp. L3*; *corr. ex* puncto sumpto *O* / puncto L *transp. R* 158 *post* ducatur[1] *add.* enim *R* / EL: EB *L3*; LE *R* / *post* usque *add.* ad *L3*; *scr. et del.* ad *C1* / DB: BD *R* 159 EBD: EDB *E* / DBC: DBP *R* / refertur: reflectetur *R* 160 punctus: punctum *R* / C: P *R* / eius *om. R* / sumpto quocumque *transp. C1* 161 eadem *mg. a. m. C1* / eadem erit *transp. R* 162 qui *corr. ex* quidem *C1*; que *R* 164 quorumdam . . . unum: quorumdam unum quorumdam dua *L3* / dua: duo *R* / punctus: punctum *R* / ymago: hymago *O* 165 *post* a *scr. et del.* G *F* / refertur: reflectitur *R* 166 punctus: punctum *R* / ymago: hymago *O* / *post* sibi *scr. et del.* suscip *F* 167 refertur: reflectitur *R* / punctis: punctus *OL3E*; *om. R* / *ante* cuius[2] *add.* puncti autem *R* / cuius autem (168) *corr. ex* autem cuius *F* 168 autem: enim *L3* / poterit: ponit *FP1*; possit *L3E*; *alter. ex* possit *in* potest *C1* 169 eius *om. R* 172 centrum[1] *rep. L3* / punctus visus: punctum visum *R* / eodem: eadem *R* 173 ZEHA *corr. ex* EHZA *F*; HEZA *P1*; *corr. ex* SEHA *a. m. L3* / ED *mg. a. m. E* / est *corr. ex* quod *E*

[2.340] Sit equalis, et super EH ducatur a puncto D perpendicularis dyameter GDB, et ducantur linee HG, GE, HB, BE. Palam quoniam triangulus HGD equalis triangulo EGD, et equalis triangulo HBD et triangulo EBD. Palam quod, cum angulus HGE divisus sit per equalia, H a puncto G refertur ad E, et locus ymaginis eius E. Similiter, H a puncto B refertur ad E, et locus ymaginis eius E.

[2.341] Si igitur dyametro ZEHA immoto moveatur semicirculus AGZ per speram, aut solus triangulus HGE, describet quidem punctus G motu suo circulum, et a quolibet puncto illius circuli refertur H ad E, et locus ymaginis eius semper erit punctus E, et ita propositum.

[2.342] Quod ab alio puncto quam illius circuli non possit fieri reflexio puncti H ad E palam per hoc. Sumatur punctum C. Erit quidem linea EC maior linea EG, et linea HC minor linea HG, quare non erit proportio EC ad HC sicut ED ad DH. Igitur linea DC non dividet angulum ECH per equalia, quare H a puncto C non potest reflecti ad E. Eadem erit improbatio si sumatur C inter G et Z.

[2.343] Si vero ED fuerit maior DH, mutetur figura, et addatur linee EH linea HQ [FIGURE 5.2.36a, p. 595] ut productum ex EQ in QH sit equale quadrato DQ. Erit igitur proportio EQ ad DQ sicut DQ ad HQ, unde EQ ad DQ sicut ED ad DH, sicut probat Euclides.

[2.344] Fiat circulus ad quantitatem semidyametri QD, cuius Q centrum, G, B loca sectionis duorum circulorum, et ducantur linee EG, EB, QG, QB, DG, DB, HG, HB. Palam ergo quod erit proportio EQ ad QG sicut QG ad QH, et angulus GQH communis utrique triangulo EQG,

174 perpendicularis: perpendicularem *SOL3*; perpendiculariter *ER* 175 dyameter: dyametrum *O; alter. in* dyametrum *L3* / quoniam: quod *R* 176 triangulus *corr. ex* angulus *P1*; triangulum *R* / equalis[1,2]: equale *R* / *post* equalis[1] *inter.* est *O* / EGD *corr. ex* GD *L3*; EDG *ER* / et[1] *rep. L3* 178 refertur: reflectetur *R* / E[1] *corr. ex* EE *F* / *post* locus *add.* est *E* / *post* eius *add.* est *R* / similiter . . . E[2] (179) *inter. a. m. L3; mg. a. m. C1E* (B: D *C1*); *scr. et del. S* (eius: est; *post* E[2] *rep. et del.* similiter H a puncto B) 179 refertur: reflectetur *R* 180 si *scr. et del. S* / ZEHA *corr. ex* SEHA *L3* / immoto: immota *R* / moveatur: moveantur *P1* 181 AGZ *corr. ex* AGS *L3* / *post* speram *add.* speculi *R* / aut: autem *P1* / solus *corr. ex* solut *L3*; solum *R* / triangulus: triangulum *R* / describet *corr. ex* describit *F* / punctus: punctum *R* 182 G *om. S; inter. O* / illius *om. R* / *post* illius *scr. et del.* diametri *P1* / refertur: reflectetur *R* 183 punctus: punctum *R* / et ita: quod est *C1* / *post* ita *add.* patet *R* / *post* propositum *scr. et del.* et est polus illus circuli *O* 184 *post* quod *add.* autem *FP1R* / *post* quam *add.* aliquo *R* / *post* illius *inter.* scilicet *a. m. E* / circuli *inter. a. m. E* 185 *post* reflexio *scr. et del.* et est polus illus circuli *O* / C: O *S* / *post* C *add.* et ducatur EC CH *R* 186 linea[1] *om. R* / linea HC minor *om. S* 187 HC: CH *R* / DH *corr. ex* DHC *L3* 188 non *inter. O* / E *corr. ex* ea *F* 189 improbatio: probatio *FP1R* / sumatur *corr. ex* sumiatur *F* 190 *post* figura *scr. et del.* ada *S* 191 EH: DH *R* / productum *corr. ex* punctum *O* 192 DQ[2]: HQ *O* / sicut . . . HQ *mg. O* / *post* DQ[3] *rep. et del.* sicut DQ *E* / ad[2] . . . Q (194) *om. S* 193 unde . . . DH *om. R* / ad DQ *mg. a. m. F* 195 *post* loca *scr. et del.* ymaginum *P1* / EB: EH *S* 196 DG *rep. P1* / HG HB *transp. C1* / HB *inter. O*

HQG. Igitur illi duo trianguli sunt similes. Erit ergo proportio EQ ad QG sicut EG ad GH. Erit igitur proportio ED ad DH sicut EG ad GH, quare linea DG dividet angulum EGH per equalia.

[2.345] Unde punctus H a puncto G refertur ad E, et locus ymaginis eius punctus E. Similiter, H a puncto B refertur ad E, et locus ymaginis est E.

[2.346] Si ergo moveatur triangulus EGH, punctis E, H immotis, punctus G describet in spera circulum a quolibet puncto cuius refertur H ad E, et semper erit locus ymaginis E.

[2.347] Et quod ab alio puncto quam illius circuli non possit H reflecti ad E palam, ut prius. Si enim sumatur C inter G et A, erit EC maior EG, et HC minor HG, nec erit proportio EC ad HC sicut ED ad DH, et ita DC non dividit angulum ECH per equalia. Similiter, si C sumatur inter G et Z poterit improbari.

[2.348] Et ita propositum, notandum tamen quod E est punctus intellectualis, et circulus ille cuius E est polus est circulus intellectualis, et H punctus intellectualis. Unde quod dictum est secundum geometricam demonstrationem est intelligendum non secundum visus probationem, cum intellectualia visum lateant. Sed quoniam forma H continua videtur formis aliorum punctorum, videbitur quidem a visu forma cuius punctus medius H, et locus puncti medii illius forme erit E, et reflectetur hec forma a loco speculi circularis cuius medium erit circulus predictus, et E polus eius.

[2.349] Cum autem ED sit maior DH, et in tantum poterit esse maior quod non refertur H ad E a puncto G, sciendum quod nisi fuerit

198 HQG *om. P1* / illi: illa *R* / trianguli: triangula *R* / similes: similia *R* / erit ergo *transp. C1* / proportio: portio *S* 199 erit . . . GH[2] *om. S* / proportio: portio *O*; probatio *L3*; *om. R* / DH *corr. ex* AH *F* 200 linea *om. P1* / dividet: dividit *SC1*; dividat *L3*; *alter. ex* dividit *in* dividat *O* 201 unde: unum *S* / punctus: punctum *R* / refertur: reflectetur *R* 202 *post* eius *add.* est *FP1O* / punctus: punctum *R* / refertur: reflectetur *R* / ad *om. FP1* 203 est: cum *P1*; eius *E*; *om. SO* / est E *transp. L3* / *post* est *add.* punctus *E*; *add.* punctum *R* / *post* E *rep. et del.* similiter (202) . . . est E (203) (est E *transp.*) *E* 204 triangulus: triangulum *R* / punctis *corr. ex* punctus *L3* / H: Q *P1* / immotis: immotu *P1*; *om. O* 205 punctus: motus *FP1*; *corr. ex* punctis *L3*; punctum *R* / G *mg. a. m. L3*; E *S* / describet: describit *FP1* / quolibet . . . cuius: cuius quolibet puncto *R* / refertur: reflectetur *R* 206 E[1] *om. P1* 207 quam *inter. a. m. E* / *post* quam *add.* aliquo *R* 208 *post* A *add.* et *L3* / EC *corr. ex* hec *S*; *corr. ex* EA *L3* / EG: ED *E* 209 nec: non *L3ER* / *ante* erit *add.* ergo *R* 210 dividit: dividet *R* / equalia: equa *C1* / si C *corr. ex* sic *P1L3* 211 Z *corr. ex* S *L3* 212 *post* ita *add.* est *P1*; *add.* patet *R* / E *inter.* S / est *om. L3* / punctus intellectualis (213): punctum intellectuale *R* 213 *post* intellectualis[1] *add.* est *L3* 214 punctus intellectualis: punctum intellectuale *R* / *post* secundum *add.* quod *P1* 215 geometricam *corr. ex* geometriam *O* / *post* intelligendum *add.* si *L3E*; *add.* sed *C1* 216 intellectualia: intellectu alia *E* 217 continua videtur *inter. O* 218 forma: formam *S* / punctus medius: punctum medium *R* 219 hec: H *R* / hec forma *transp. L3* / *post* speculi *scr. et del.* s *S* / circularis: circuli *FP1*; circulari *OR* 220 predictus: precedens *FP1* 221 sit: fuerit *R* / et *om. R* / in tantum *corr. ex* iterum *FOL3* (*mg. a. m. F*); item *S* 222 quod[1]: ut *R* / refertur: reflectatur *R* / E a: ea *P1* / quod *inter. O* / nisi *alter. in* si *mg. a. m.F*; si *P1*

proportio EA ad AH maior quam ED ad DH, non poterit H reflecti ad E.

[2.350] Si enim potest reflecti, reflectatur a puncto quod sit G. Erit quidem angulus GDH minor recto, cum respiciat sectionem minorem quarta. Ducatur a puncto G contingens, que necessario concurret cum EA. Sit concursus Q. Erit quidem proportio EQ ad QH sicut ED ad DH (ex [33]), sed maior est proportio EA ad AH quam EQ ad QH. Igitur maior est EA ad AH quam ED ad EH, et ita necessario, si H refertur ad E, erit proportio EA ad AH maior ED ad DH. Palam ergo que dicta sunt cum centrum visus et punctus visus fuerint in eodem dyametro.

[2.351] **[PROPOSITION 37]** Amplius, cum punctum visum et centrum visus non fuerint in eodem dyametro, et fuerint extra speculum, non refertur punctus visus ad centrum visus nisi ab uno tantum speculi puncto.

[2.352] Verbi gratia, sit T [FIGURE 5.2.37, p. 595] punctus visus, H centrum visus, D centrum spere, et ducantur linee HD, TD. Superficies quidem HDT secat speram super circulum EBQG.

[2.353] Palam quoniam T non refertur ad H nisi ab aliquo puncto huius circuli. Palam etiam quod non refertur ab arcu QG vel BA, secundum modum predictum. Refertur ergo aut ab arcu GB aut AQ.

[2.354] Dividatur angulus TDH per equalia per lineam LEDZ, et a puncto E ducatur contingens, que sit KEF. Si puncta T, H fuerint super illam contingentem, non reflectetur T ad H ab aliquo puncto arcus BG. Cum enim a puncto T ducetur linea ad aliquem interiorem punctum huius arcus, linea a puncto H ad idem punctum ducta cadet super ipsum exterius, non interius, et ideo non erit reflexio.

223 ad[1] *inter. L3* 226 quidem: quidam *SL3* / angulus *om. R* / *post* GDH *scr. et del.* non poterit H *S* 228 sit . . . ad[2] *mg. C1* / Q: F *R* / EQ: EF *R* / QH: FH *R* 229 *ante* sed *inter.* figura *a. m. O* / est *om. C1* / EQ . . . quam (230) *inter. a. m. L3* / EQ: EF *R* / QH: FH *R* 230 EH: DH *L3C1ER* / refertur: reflectitur *R* 231 *post* maior *add.* quam *R* / *post* DH *add.* p *F* / palam: patent *R* 232 et *inter. C1* / et . . . visus[2] *inter. a. m. O* / punctus visus: punctum visum *R* / fuerint: fuerit *O* / eodem: eadem *R* 234 eodem: eadem *R* / fuerint: fuerit *OL3* 235 refertur . . . visus[1]: reflectetur punctum visum *R* / *post* visus[1] *add.* H *L3* / visus[2] *om. O* 237 T: Z *FP1* / punctus: punctum *ER* / visus: visum *R* 238 *post* TD *add.* DT HT *R* / superficies (239): ses *P1* 239 *post* circulum *add.* qui sit *R* / EBQG: EB *P1* 240 quoniam: quod *R* / refertur: reflectetur *R* / ad: a *O* / aliquo: alio *SO; corr. ex* alio *L3* 241 *post* circuli *add.* producantur ergo HD TD usque ad circumferentiam circuli *R* / etiam *om. ER* / refertur: reflectetur *R* / *post* refertur *scr. et del.* ad H *S* / BA *corr. ex* AB *C1* 242 refertur: reflectetur *R* / aut: autem *P1* / GB: BG *L3* 243 LEDZ: Z *P1; corr. ex* Z *mg. F;* LED *O;* LEZD *SC1; alter. ex* LESD *in* LEZD *L3* 244 fuerint: fuerit *O* 245 aliquo *corr. ex* alio *L3* 246 aliquem: aliquod *R* / *post* aliquem *add.* minorem *F* / interiorem *mg. a. m. F;* interius *R* / *post* interiorem *add.* minorem *P1* 247 H *corr. ex* D *C1* / *post* ipsum *add.* arcum *C1* 248 non interius *transp. R*

[2.355] Et quod ab uno puncto arcus AQ tantum fiat reflexio palam erit ex hoc. Ducantur linee TZ, HZ. Cum angulus TDH divisus sit per equalia, TDZ equalis angulo HDZ.

[2.356] Linee TD HD aut sunt equales, aut non sunt equales. Si sint equales, et DZ communis, erit triangulus TZD equalis triangulo HZD, et angulus TZH divisus per equalia per lineam DZ, et ita T refertur ad H a puncto Z.

[2.357] Quod ab alio puncto non possit sic constabit. Sumatur punctus O, et ducantur linee TO, HO, et linea ODM dividat angulum illum per equalia. Planum quod TZ minor TO, et HO minor HZ, et proportio TZ ad HZ sicut TL ad LH, et erit proportio TO ad HO sicut TM ad MH. Sed minor est proportio HO ad TO quam HZ ad TZ. Ergo minor HM ad MT quam HL ad LT, quod est impossibile.

[2.358] Palam igitur quod, si T et H equaliter distant a centro et fuerint supra contingentem, non refertur T ad H nisi ab uno speculi puncto tantum, et unicus erit ei ymaginis locus.

[2.359] Amplius, BDQ, ADG [FIGURE 5.2.37b, p. 596] sint duo dyametri spere, et dyameter EDZ dividat angulum BDG per equalia, et a puncto E ducantur due perpendiculares super duos dyametros BD, GD, que sunt ET, EH.

[2.360] Palam quod triangulus ETD equalis est triangulo EHD, cum ED sit communis utrique. T igitur refertur ad H a puncto E. Eodem modo, a puncto Z. Et palam quod non refertur ad E ab aliquo puncto

249 arcus AQ tantum: tantum arcus AQ *R* 250 erit ex hoc *corr. ex* ex hoc erit *P1* / *post* ducantur *add.* enim *R* / TZ HZ *corr. ex* TS HS *L3* 251 *post* equalia *add.* erit *OR* (*inter. O*) / TDZ: TDS *F*; *alter. ex* DS *in* TDS *P1*; *corr. ex* TDS *L3* / *post* TDZ *add.* erit *E* / HDZ *corr. ex* HD *a. m. L3* 252 *post* linee *add.* igitur *R* / sunt[1]: sint *E* / aut . . . equales[2] *om. FP1* / sunt equales[2] *om. O* / sint: sunt *SOC1R* 253 DZ: DS *FP1*; *corr. ex* DS *L3* / communis erit *corr. ex* erit communis *C1* / triangulus: triangulum *R* / TZD: TSD *FP1*; *corr. ex* TSD *L3* / equalis: equale *R* 254 TZH *alter. ex* DIH *in* DZH *F*; DZH *P1*; *corr. ex* TSH *L3* / per[1] *om. F* / DZ: DS *F*; *corr. ex* DS *L3* / refertur: reflectetur *R* 255 Z: S *FP1*; *corr. ex* S *L3* 256 *post* quod *add.* autem *R* / *post* possit *scr. et del.* fieri *P1* / sic *inter. O* / punctus (257): punctum *R* 257 ducantur: reducantur *FP1* / *post* ODM *add.* per centrum D *R* / *post* angulum *scr. et del.* illum *L3* 258 *post* equalia *scr. et del.* per lineam *S* / TZ: TS *FP1*; *corr. ex* TS *L3* / *post* TZ *add.* sit *P1*; *inter.* est *O* / *post* minor[1] *add.* est *R* / HZ: HS *FP1*; *corr. ex* HS *L3* 259 TZ: TH *FP1*; *corr. ex* TS *L3* / HZ: HS *FP1*; *corr. ex* HS *L3* / TL: DL *FP1*; D *L3* / proportio *om. FP1* / TO: DO *P1* / HO: O *S* / MH: H *FP1*; *corr. ex* IMH *C1* 260 HO: HZ *SOC1*; *corr. ex* HZ *L3* / TO: TZ *SOC1*; *alter. in* TZ *L3* / quam: quod *P1*; *inter. a. m. E* / HZ *corr. ex* BZ *S*; *corr. ex* HOS *L3*; HOZ *C1* / TZ: TOZ *SC1*; TO *O*; *alter. ex* TOS *in* TOZ *L3* / *post* minor[2] *add.* est proportio *R* 262 si T *om. S* / distant: distent *R* / *post* et[2] *inter.* si *a. m. O* 263 supra: super *R* / refertur: reflectetur *R* / T: D *FP1* / 264 ei: eius *P1R* / locus *om. P1* / *post* locus *add.* si vero TD HD sunt inequales secentur ad equalitatem et fiat demonstratio ut antea *R* 265 BDQ: DBQ *C1* / sint: sunt *P1E* / duo: due *R* 266 *ante* spere *scr. et del.* EDZ *P1* / dyameter *corr. ex* dyametri *L3* / angulum *om. L3* 267 duos: duas *R* / BD: HD *O* / GD *alter. in* GED *deinde corr. ex* GED *F*; DG *R* 268 que: qui *FP1* / sunt: sint *SOR* / ET EH *transp. C1* 269 triangulus: triangulum *R* / equalis: equale *R* / *post* EHD *add.* et angulus TED angulo HED latusque TD lateri HD et latus ET lateri EH *R* 270 communis *corr. ex* cos *C1* / T *corr. ex* TT *C1* / refertur: reflectetur *R* 271 et *inter. O* / *post* quod *add.* T *ER* / *post* non *add.* T *C1* / refertur: reflectetur *R* / ad . . . refertur (272) *mg. a. m. C1* / E *alter. in* B *L3*; H *C1ER*

arcus AB vel arcus GQ, nec refertur ab alio puncto arcus AQ quam a puncto Z, secundum supradictam probationem. Verum quod ab alio puncto arcus BG quam a puncto E non possit reflecti patebit sic.

[2.361] Detur O punctum, et ducantur linee TO, HO DO. Fiat circulus ad quantitatem linee DE transiens per tria puncta T, D, H, cuius quidem circuli linea DE erit dyameter, cum angulus ETD quem respicit sit rectus. Igitur circulus ille transibit per punctum E.

[2.362] Cum igitur E sit communis utrique circulo, et sit super eundem dyametrum, continget circulus minor maiorem in puncto E, sicut probat Euclides. Igitur circulus iste secabit lineam DO.

[2.363] Secet in puncto I, et ducantur linee TI, HI. Iam patet quod TD equalis est DH. Igitur angulus TID equalis angulo DIH, quia super equales arcus. Restat angulus TIO equalis angulo HIO; et angulus IOT est equalis angulo IOH, ex ypotesi, et IO commune. Erit triangulus TIO equalis triangulo HIO, et erit TO equalis HO, quod est impossibile, quoniam HO maior HE, TO minor TE, et TE, sicut prius probatum est, equalis est HE. Restat ergo ut T non reflectatur ad H ab alio puncto quam ab E vel a Z.

[2.364] Iterum, a puncto E ducatur linea super dyametrum TD, qui sit EM, et secetur a linea HD pars equalis MD, que sit ND, et ducantur EN, EM. Palam quod angulus EMD est maior recto. Secetur ex eo equalis recto per lineam CM, que concurret cum DE. Sit concursus punctus C. et ducatur NC, et fiat circulus ad quantitatem CD transiens per tria puncta M, D, N. Cum CMD sit rectus, erit CD dyameter, et transibit circulus per C. Palam ergo quod M refertur ad N a puncto E, et similiter a puncto Z, et non ab aliquo puncto arcus AB vel QG; et palam quod non ab alio puncto arcus AQ quam a puncto Z.

272 nec: non *L3* / refertur: reflectetur *R* / alio: aliquo *FP1*; illo *L3* 273 Z: S *L3* / supradictam: predictam *C1* 275 TO: TD *S*; ED *L3*; OT *C1* / TO HO DO: DO HO TO *FP1ER* (DO: OD *ER*) / *post* fiat *add.* que *R* 277 circuli . . . angulus *mg. O* / quem: quoniam *FP1* / respicit *alter. in* recipit *L3* 279 sit[2]: sint *OC1* / eundem (280): eandem *R* 280 continget: contingens *O* / maiorem *inter. O* / E . . . I (282) *mg. a. m. C1* 281 Euclides: Euclidis *R* / iste: illo *L3* / DO: TO *FP1* 282 I: L *R* / TI *corr. ex* TH *S*; TL *R* / HI: HL *R* / patet: patebit *FP1*; *corr. ex* patebit *L3* 283 TD: TG *FP1*; *corr. ex* TG *L3* / equalis est *transp. R* / DH: HD *R* / TID: TLD *R* / DIH: DH *FP1*; DLH *R* 284 angulus[1] *om. R* / TIO: TLO *R* / angulo *corr. ex* triangulo *C1* / HIO *corr. ex* HYO *S*; *corr. ex* TLO *L3*; HLO *R* / et . . . HIO (286) *mg. a. m. C1* / IOT: LOT *R* 285 est *om. FP1L3ER* / IOH: LOH *R* / IO: LO *R* / *post* commune *add.* latus *R* / triangulus: triangulo *FP1*; triangulum *R* 286 TIO equalis: TLO equale *R* / *post* equalis *add.* tri *O* / HIO: HLO *R* / *post* et *scr. et del.* I *P1* / TO: DO *FP1* / HO: HA *FP1*; OH *O* 287 *post* HE *add.* et *R* / TE[1]: DE *FP1* / et TE *inter. a. m. C1* 288 T: TE *S* / ad *inter. a. m. C1* / alio: ali *L3*; *scr. et del.* p *C1* 289 ab E: alie *F*; alio *P1* / a Z: A *S* 290 iterum: item *SOC1R* / TD *inter. C1* / qui: que *R* 291 et[1] *om. L3*; *corr. ex* item *E* / a *inter. OL3* / HD: HO *F* / que *om. O* / ducantur: ducatur *SOC1* 292 EN EM *transp. R* / EM *om. SO*; *inter. L3* / angulus *om. R* / est maior *transp. R* 293 *post* equalis *scr. et del.* angulo *S* / CM: PM *R* 294 punctus C: punctum P *R* / NC: NP *R* / CD: PD *R* 295 CMD *corr. ex* CM *L3*; PMD *R* / *post* erit *scr. et del.* rectus *S* / CD: PD *R* / et *inter. O* 296 C: P *R* / *post* ergo *add.* per *S* / refertur: reflectetur *R* 297 et[1] *om. SO* / *post* puncto *scr. et del.* to *C1* / Z: E *FP1*; *corr. ex* S *L3* / aliquo: alio *FP1S*; *corr. ex* alio *OL3* / QG: Q *FP1*; GQ *SOL3ER* 298 arcus *mg. a. m. C1* / AQ . . . arcus (299) *mg. a. m. L3* / AQ: BG *FP1* / Z: S *FP1*

[2.365] Et quod non ab alio puncto arcus BG quam a puncto E palam secundum modum predictum. Sumpto enim puncto, et ductis lineis ad punctum illud a punctis T, D, H, et sumpto puncto in quo circulus ultimus secabit dyametrum, et a puncto sectionis ductis lineis ad puncta T, H, eadem erit improbatio que prius.

[2.366] Palam ergo ex predictis quod si angulum contentum a duobus dyametris per equalia dividat tertius dyameter, et a termino illius dyametri ducantur perpendiculares ad illos dyametros, puncta dyametrorum in que cadunt ad se invicem reflectuntur a duobus punctis speculi tantum. Puncta etiam dyametrorum citra hos terminos perpendicularium sumpta, id est, versus centrum, reflectitur quodlibet a duobus punctis tantum, et unum refertur ad illud quod equaliter distat a centro, et omnium talium duplex est ymaginis locus.

[2.367] Amplius, sumptis duobus dyametris BQ, AG, et EZ [FIGURE 5.2.37c, p. 597] dividente angulum eorum per equalia, et sumatur in BD punctus T supra punctum in quem cadit perpendicularis ducta a puncto E, et in DG sumatur DH equalis DT, et ducantur TE, HE. Refertur quidem T ad H a puncto E, et similiter a puncto Z, non ab alio puncto arcus AQ, nec ab aliquo puncto arcus AB vel GQ.

[2.368] Deinceps, a puncto T ducatur perpendicularis super TD, que quidem concurret cum DE extra circulum spere, cum angulus DTE sit acutus. Concurrat ergo in puncto O, et ducantur linee TO, HO. Et fiat circulus transiens per tria puncta T, D, H, qui necessario transibit per punctum O, et erit DO dyameter eius. Et ducantur linee TO, HO, et ducatur linea contingens circulum BZG in puncto E, que sit KE. Palam quoniam ultimus circulus secabit primum, scilicet BZG, in duobus punctis. Sint puncta illa L, M, et ducantur linee TL, HL, LD, TM, DM, HM.

299 BG *corr. ex* AB *L3*/palam *om. R* 300 *post* enim *scr. et del.* su *F* 1 ad . . . illud *R*/ad . . . lineis (2) *om. S*/illud: illum *FP1*/D: B *FP1* 2 puncto *corr. ex* punctis *L3*; punctis *R* 3 erit *om. P1*/improbatio *corr. ex* probatio *F* 4 si *mg. F*/a *om. R* 5 duobus: duabus *R*/tertius: certius *S*; tertia *R*/termino: tertio *S*; *corr. ex* tertio *L3* 6 illos: illas *R* 7 cadunt: cadent *C1* 8 *post* tantum *add.* et *O*/puncta: punctorum *R*/etiam: autem *FP1R*; *om. O*/terminos *corr. ex* diametros *L3*/perpendicularium (9): perpendicularis *O* 9 sumpta: sumptorum *R*/id est: L *FP1*; I *SOL3C1*/quodlibet: quod si *FP1* 10 unum *corr. ex* unde *L3*/refertur: reflectitur *C1R*/illud *alter. in* aliud *L3* 12 duobus: duabus *R*/BQ: BC *FP1*; *corr. ex* BG *L3* 13 eorum: earum *R*/et *om. R* 14 punctus: punctis *FP1*; punctum *R*/T *corr. ex* Z *a. m. E*/quem: quod *R*/cadit: cadet *SC1* 15 TE: DE *F*; DC *P1*/refertur: ref *FP1*; reflectetur *R* 16 E: HE *S*/Z *corr. ex* S *L3*/*post* Z *add.* et *O*/ab . . . nec (17) *om. C1* 17 aliquo *corr. ex* alio *L3*/GQ *corr. ex* GA *L3* 18 deinceps: deinde *R*/perpendicularis *om. S*/TD: TAD *L3* 19 DTE: BDE *R* 20 concurrat: concurret *FP1* 21 qui: que *FP1L3E*/transibit . . . O (22): per punctum O transibit *C1* 22 et[2] . . . HO *om. R* 23 in . . . BZG (24) *mg. a. m.* (scilicet *om.*) *L3*/que . . . BZG (24) *om. FP1*/sit *inter. a. m. E* 24 quoniam: quod *R* 25 puncta illa *transp. ER*/TL: A *S*/HL: YL *E*/LD: DL *SOC1* (*inter. O*)

[2.369] Cum ergo arcus TD sit equalis arcui HD, erit angulus TLD equalis angulo DLH, et ita T refertur ad H a puncto L. Similiter, angulus TMD equalis angulo DMH, et ita T refertur ad H a puncto M. Palam igitur quod T refertur ad H a quatuor punctis, scilicet E, Z, L, M, et quadruplex erit locus ymaginis eius.

[2.370] Et non potest T reflecti ad H ab alio puncto quam aliquo istorum. Detur enim punctus F, et ducantur linee TF, HF, DF, et producatur DF usque concurrat cum contingenti KE, et sit concursus K. Et ducantur linee TK, HK. Igitur angulus TFD equalis angulo DFH, ex ypotesi; restat angulus TFK equalis angulo KFH. Sed angulus TKF equalis est angulo FKH, quia super equales arcus, et FK communis. Erit triangulus equalis triangulo, et ita TK equalis KH, quod est impossibile, quoniam HK maior HO, et TK minor TO, et TO equalis HO.

[2.371] Palam igitur quod non fit reflexio ab aliquo puncto quam a punctis quatuor circuli.

[2.372] Igitur, si in diversis dyametris sumantur duo puncta, scilicet T, H, equaliter a centro distantia, si fuerint super puncta dyametrorum in que cadunt perpendiculares ducte a termino diametri dividentis per equalia angulum duorum dyametrorum, aut fuerint inter centrum et puncta illa, id est citra perpendiculares, dum equaliter distent a centro, reflectetur quidem T ad H a duobus punctis tantum.

[2.373] Si vero T et H fuerint a locis perpendicularium usque in circulum, reflectetur quidem T ad H a quatuor punctis. Si vero fuerint in circulo vel extra, tamen citra contingentem KE, reflectetur quidem T ad H a duobus punctis tantum. Si vero supra contingentem fuerint, reflectetur quidem T ad H ab uno puncto tantum. Et hec quidem accidunt, dum T equaliter distet a centro cum puncto H.

27 TLD: TLA *S* 28 T: D *L3* / refertur: reflectetur *R* / *post* refertur *rep. et del.* ergo (27) . . . equalis (27) *L3* / ad *rep. L3* / angulus . . . M (29) *om. FP1* 29 DMH *corr. ex* TMH *L3* / refertur: reflectetur *R* / a puncto M *om. E* / *post* a *scr. et del.* quatuor punctis scilicet EZLM (scilicet *inter. a. m.*) *C1* 30 refertur: reflectitur *R* / ad . . . punctis: a quatuor punctis ad H *R* / a *inter. E* / punctis: punctus *S* / scilicet *om. L3* 32 *post* quam *add.* ab *C1R* 33 punctus F: F punctum *R* / TF HF *corr. ex* TK HK *S* / producatur (34) *corr. ex* producantur *S* 34 usque: quousque *R* / contingenti: contingente *ER* 37 equalis est *transp. R* 38 triangulus equalis: triangulum equale *R* / *post* triangulus *scr. et del.* eri *F* / equalis triangulo *mg. O* / TK: KT *L3* / *ante* equalis² *add.* est *L3* 39 HK: HE *E* / HO: HC *E* 41 fit *inter. a. m. E*; est *inter. L3*; est *R* / reflexio: refertur *SOC1* / aliquo *alter. in* alio *L3C1* 42 circuli *om. SOL3C1ER* 43 igitur si *transp. FP1* (igitur *mg. F*) / sumantur: sumatur *O* 44 fuerint: fiunt *L3* / *ante* super *add.* a puncto *L3* / puncta: punctis *R* 45 perpendiculares . . . termino *om. FP1* / a termino: ? *C1* / diametri: perpendicularis *FP1S*; *corr. ex* perpendicularis *L3* / dividentis: dens *FP1* 46 duorum: duarum *R* / fuerint: fuerit *L3* 47 *post* et *scr. et del.* a *L3* / dum *corr. ex* cum *L3* / distent: distant *L3* 48 reflectetur: reflectitur *C1* / quidem *om. C1* 49 T *om. S* / T . . . fuerint: fuerint T et H *R* / in: ad *R* 50 a . . . H (52) *om. S* / fuerint: fuerit *O* 51 contingentem *corr. ex* contingentie *L3* 52 si . . . tantum (53) *mg. a. m.* (uno puncto *transp.*) *L3* / *post* vero *rep. et del.* et (49) . . . locis (49) *S* 53 accidunt (54) *corr. ex* cadunt *S* 54 distet: distat *R*

[2.374] **[PROPOSITIO 38]** Amplius, si fuerint T, H in diversis dyametris, et longitudo eorum a centro fuerit inequalis, reflectentur quidem ad se ab uno puncto.

[2.375] Verbi gratia, ducantur dyametri ADG, BDQ [FIGURE 5.2.38, p. 598], et EZ dividat angulum eorum per equalia. Et T propinquior sit centro D quam H. Et sumatur linea LQ, et dividatur in puncto M ut sit proportio QM ad ML sicut HD ad DT. Et dividatur LQ per equalia in punctum N, et a puncto N ducatur perpendicularis NK, et super punctum L fiat angulus equalis medietati anguli ADT per lineam FL. Erit quidem angulus FLQ acutus, quare FL concurret cum NK. Concurrat in puncto F, et a puncto M ducatur linea ad latus FL concurrens cum latere NK in puncto quod sit K. Et secet linea illa latus FL in puncto C ut sit proportio KC ad CL sicut HD ad DZ.

[2.376] Deinceps super punctum D fiat angulus equalis angulo LCM, qui sit IDA, et sit I punctus circuli supra Z, aut infra. Et supra I punctum fiat angulus equalis CLM, qui sit OID, et super hanc lineam OI ducatur perpendicularis a puncto H, que sit HC, et producatur linea CF equalis linee CI, et ducantur linee HF, HI.

[2.377] Palam secundum predicta quod a puncto M non potest linea duci ad latus FL dividens ipsum eo modo quo dividit linea MCK preter hanc solam lineam MCK. Si enim possit, sit MPO. Palam quoniam PO minor erit CK, quod quidem patebit, ducta linea PY equidistans CK, que erit minor CK et maior PO. Et PL maior CL. Igitur non erit proportio PO ad PL sicut KC ad CL, quare non erit PO ad PL sicut HD ad DT. Restat ergo ut a puncto M non ducatur alia quam MCK similis ei.

[2.378] Verum cum ODI sit equalis angulo LCM, et angulus OID equalis angulo CLM, erit triangulus CLM similis triangulo IOD. Igitur angulus IOD erit equalis angulo LMC. Restat angulus COH equalis

55 si fuerint *om. FP1L3* / si . . . H: T H si fuerint *ER* 56 *ante* et *add.* si fuerint *FP1L3* / reflectentur: reflectetur *SOL3E*; reflexio fiet *R* / quidem ad se (57) *om. R* 59 propinquior: propinquius *R* / sit *om. FP1* 60 LQ: LY *R* / sit proportio (61) *transp. S* 61 proportio: propinquior *FP1* / QM: YM *R* / DT: AT *S* / LQ: LY *R* / per: in *R* 62 punctum *alter. in* puncto *C1*; puncto *R* / NK: NB *S* 63 anguli *om. R* / FL: FK *L3* 64 FLQ: FLY *R* / NK: HK *O* / concurrat *corr. ex* concurret *a. m. C1*; concurrant *R* 65 *post* concurrens *add.* igitur *E* 66 NK *corr. ex* NM *P1* / linea illa: lineam illam *O* / illa *om. P1* / FL: FB *SL3* 67 CL *corr. ex* KCL *F* / DZ: DT *FP1SL3*; DA *O* 68 deinceps: deinde *R* 69 I: A *SO* / I punctus *alter. ex* punctus A *in* A punctus *C1* / punctus: punctum *R* / circuli: cui *S* / et[2] *om. L3* / supra: super *R* 70 qui *inter. O* / OID: CID *S* / super: supra *FE* / hanc *om. O* 71 que sit HC *inter. O* / HC: HR *R* / linea *om. ER* 72 CF: RX *R* / CI: RI *R* / HF: HX *R* 73 predicta *corr. ex* predictam *L3*; supradictam *C1* 74 linea: lineam *R* / preter . . . MCK (75) *inter. E* 75 quoniam *om. FP1*; quod *R* 76 CK[1]: T *FP1* / ducta *om. P1* / PY: PQ *R* / equidistans: equidistanter *C1*; equidistanti *E* 77 que: cum *C1* / que . . . CK *mg. O* / maior[2] . . . PL[1] (78) *om. S* 78 CL: KL *C1* / *post* erit *add.* proportio *R* / DT: DA *O* 79 ut *om. S* / puncto *corr. ex* punctum *L3* 80 LCM . . . angulo (81) *om. P1* 81 CLM[1]: DM *S* / triangulus: triangulum *R* / similis: simile *R* 82 angulus[1] *corr. ex* triangulus *C1* / *post* IOD *scr. et del.* igitur *F* / equalis angulo *corr. ex* triangulo equalis *C1* / *post* restat *add.* ergo *S* / COH: ROH *R*

angulo KMN, et angulus HCO rectus equalis angulo KNM. Restat angulus NKM equalis angulo CHO.

[2.379] Ducta autem linea DI donec concurrat cum HC in puncto R, erit angulus RDH equalis angulo KCF. Erit triangulus RDH similis triangulo CKF. Igitur proportio RD ad DH sicut FC ad KC. Sed HD ad DI sicut KC ad CL. Igitur RD ad DI sicut FC ad CL. Igitur RI ad DI sicut FL ad CL. Sed DI ad IO sicut CL ad LM, cum triangulus DIO sit similis triangulo CLM. Igitur RI ad IO sicut FL ad LM. Sed proportio RI ad IC sicut FL ad LN, quoniam triangulus RIC similis est triangulo FLN. Igitur, proportio IO ad IC sicut LM ad LN. Igitur proportio QM ad LM sicut FO ad IO.

[2.380] Ducta autem a puncto I linea UI equidistante linee HF, et producta linea DA donec concurrat cum UI in puncto U, erit triangulus OUI triangulo HOF similis. Erit igitur proportio HO ad OU sicut QM ad ML, et ita HO ad OU sicut HD ad DT. Sed quoniam triangulus HCI equalis est triangulo HCF, cum HC sit perpendicularis, igitur angulus HFC equalis est angulo CIH, et ita CIH equalis est angulo UIO, quare proportio HO ad OU sicut HI ad UI, et ita HI ad UI sicut HD ad DT.

[2.381] Verum angulus UID maior est angulo DIH. Secetur ab eo equalis, et sit PID, et ducatur linea TP. P sit punctus dyametri DA.

[2.382] Palam quod proportio HI ad UI constat ex proportione HI ad IP et IP ad UI, et proportio HI ad IP sicut HD ad DP, quoniam DI dividit angulum PIH per equalia. Igitur proportio HI ad UI, que est HD ad DT, constat ex proportione HD ad DP et PI ad UI. Sed proportio HD ad DT constat ex proportione HD ad DP et DP ad DT. Igitur proportio DP ad DT sicut PI ad UI.

83 HCO: HCD *P1; corr. ex* H *O;* HRO *R* / *post* equalis *add.* erit *ER* (*mg. a. m. E*) 84 *ante* NKM *add.* ergo *FP1* / CHO: RHO *R* 85 *post* autem *scr. et del.* angulo *F* / HC: HR *R* / R: S *R* 86 *post* erit^{1} *add.* equalis *L3* / RDH^{1,2}: SDH *R* / KCF: HCF *E* / *post* KCF *add.* et *R* / erit: et *C1* / triangulus: triangulum *R* / RDH: SDH *R* / similis: simile *R* 87 RD: SD *R* / KC: CK *R* / ad^{3} *rep. S* 88 DI^{1} *corr. ex* TK *L3* / KC: KO *S* / RD: SD *R* / DI^{2} *corr. ex* CH *L3* / FC: KC *SC1; corr. ex* CK *L3* / CL *corr. ex* D *L3* / RI: SI *R* 89 triangulus: triangulum *R* 90 similis *corr. ex* equalis *L3;* simile *R* / RI: SI *R* / FL: LF *C1* 91 RI: SI *R* / IC: IR *R* / FL . . . IO (92) *om. FP1* / LN: LM *E* / triangulus: triangulum *R* / RIC: SIR *R* / similis: simile *R* 92 IC: IR *R* / QM: YM *R* 93 LM: ML *SOL3* / FO: XO *R* 94 I: U *SC1* / linea UI equidistante: equidistante linea UI *R* / HF: HX *R* / UI . . . linea (95) *mg. a. m. C1* 95 concurrat *inter. O* / *post* UI *add.* concurrat *R* / triangulus: triangulum *R* 96 OUI *om. P1;* OIU *O* / triangulo *rep. S* / triangulo HOF similis: similis triangulo HOF *L3* / HOF *corr. ex* OF *O;* HOX *R* / similis: simile *R* / erit igitur *transp. R* / OU *corr. ex* UO *L3* / QM: YM *R* 97 OU: HOU *S* / triangulus: triangulum *R* / HCI: HOI *S;* HRI *R* 98 equalis: equale *R* / HCF: HEH *F;* HEB *P1;* HRX *R* / HC: HR *R* / 99 HFC: HXR *R* / equalis est *transp. C1* / CIH^{1,2}: RIH *R* / et . . . UIO *mg. a. m. E* 100 UI^{1,2}: IN *L3;* IU *R* 102 ducatur linea *transp. O* / TP: DP *E;* PT *R* / *ante* P *add.* et *R* / punctus: punctum *R* / dyametri *om. P1* 104 et^{1} . . . IP^{2} *mg. O; inter. a. m. L3* / IP^{2}: PI *R* / sicut *mg. a. m. C1* / HD: DH *R* 105 PIH: PIK *P1* / *post* igitur *rep. et del.* proportio (104) . . . equalia (105) *L3* / *ante* proportio *add.* igitur *L3* 106 constat *om. O* / proportione: probatione *FP1* / et . . . DP^{1} (107) *mg. a. m. E; om. R* 107 et DP *om. F; inter. L3* 108 DP: TP *FP1L3E;* quod *S*

[2.383] Verum angulus OIH est medietas anguli UIH, sed angulus DIH est medietas anguli PIH. Restat angulus DIO medietas anguli PIU, sed angulus DIO est medietas anguli TDP, quia est equalis angulo FLM. Igitur angulus PIU est equalis angulo TDP, et proportio DP ad DT sicut PI ad UI. Igitur triangulus UIP similis est triangulo TPD, et angulus UPI equalis TPD. Erit igitur TPI linea recta, quia angulus DPT cum angulo TPO valet duos rectos, et ita angulus OPI cum angulo OPT valet duos rectos. Et ita T refertur ad H a puncto I, et hec quidem erit probatio sive sit T extra circulum, sive intra, et similiter, sumpto puncto H extra vel intra, dum inequaliter a centro.

[2.384] **[PROPOSITIO 39]** Amplius, ductis dyametris BQ, AG, et dyametro EZ dividente angulum BDG per equalia, dico quoniam, quicumque punctus sumatur in arcu AQ preter punctum Z, ab illo poterunt reflecti infinita paria punctorum inequaliter a centro distantium.

[2.385] Verbi gratia, sumatur punctus H [FIGURE 5.2.39, p. 599], et sumatur in dyametro GD punctus L. Et a dyametro BD secetur MD equalis LD, et ducantur linee LM, LH, MH, DH. Punctus in quo EZ dividit LM sit F; erit LF equalis FM.

[2.386] Et ducatur HD usque cadat super LM in puncto N. Erit igitur LN minor NM. Verum, cum angulus MDF sit equalis FDL et angulo QDZ, et angulus MDA equalis angulo LDQ, et angulus ADH equalis angulo NDL, erit angulus LDH maior angulo MDH. Igitur LH erit maior MH, cum MD, DH equalia sint LD, DH. Erit ergo angulus DHL minor angulo DHM, si enim esset equalis, esset proportio LH ad MH sicut LN ad NM, quod est impossibile. Si autem fuerit maior, secetur ex eo equalis, et improbetur hoc modo. Igitur est minor.

[2.387] Secetur igitur ab angulo MHD equalis illi, qui sit THD. Igitur T refertur ad L a puncto H, et TD minor LD.

109 est . . . DIH (110) *om. S* 110 est medietas *transp. R*/PIH: PLH *F* 111 TDP *corr. ex* TPD *C1*/quia: quod *S*/*post* est[2] *scr. et del.* angulus *P1* 112 PIU *om. FP1*/TDP *corr. ex* TPQ *S*; *inter. O*/ et . . . UI (113) *om. O* 113 triangulus: triangulum *R*/similis: simile *R*/est *om. FP1L3ER*/TPD: DPD *L3* 114 angulus[1] *corr. ex* angel *F*/UPI: UIP *L3*/igitur *om. FP1*/quia: quare *E*/DPT: TPD *FP1* 115 angulo[2]: angulus *FP1* 116 duos rectos *inter. O*/refertur: reflectetur *R*/a . . . H (118) *om. S*/hec quidem: eadem *R* 117 similiter: simpliciter *C1* 118 inequaliter: LN equaliter *FP1*/*post* inequaliter *add.* distent *R* 119 BQ: BG *P1*; LQ *S* 120 EZ *corr. ex* DZ *C1*/quoniam . . . punctus (121): quod quodcumque punctum *R* 122 poterunt: potu *FP1*; potuit *S*; *corr. ex* poterit *E*/reflecti *inter. O* 124 punctus H: H punctum *R* 125 dyametro: semidiametro *R*/ GD: DG *S*/punctus: punctum *R*/ dyametro: semidiametro *R* 126 punctus: punctum *R* 127 LF: FL *R* 128 et *om. O*/usque: quousque *R* 129 cum: est *FP1*/MDF: FDM *C1* 130 QDZ *corr. ex* QDS *L3* 131 angulo[1] . . . maior *om. E* 132 *post* erit[1] *add.* M *S*/LD: ID *SO*/ ergo angulus *transp. E* 133 esset: esse *FP1* 134 MH: HM *R* 135 et: ut *L3*/improbetur: probetur *S*; improbabitur *R*/hoc: eodem *R* 136 MHD *corr. ex* MDH *L3*/illi: ei *C1* 137 refertur: reflectetur *R*/*post* et *add.* linea *R*/*post* TD *add.* est *R*

[2.388] Similiter, si sumantur in dyametris HD, GD alia puncta quam L, M equaliter a puncto D distantia, probatur similiter quod a puncto H fit reflexio punctorum adinvicem et inequaliter distantium a centro. Et ita de infinitis punctis in hiis dyametris sumptis similis erit probatio, et a quocumque puncto arcus AQ sumpto preter quam a puncto Z.

[2.389] **[PROPOSITIO 40]** Amplius, sumptis T, L [FIGURE 5.2.40, p. 599] in dyametris quorum inequalis sit longitudo a centro, et reflectantur adinvicem a puncto H, non erit reflecti T ad L ab alio puncto arcus AQ quam a puncto H.

[2.390] Si enim ab alio, sit illud K, et ducantur TK, LK, DK, LT, TH, LH, NDH. Et producatur DK usque cadat in LT in puncto C. Palam quoniam proportio LH ad TH sicut LN ad NT.

[2.391] Et similiter, cum angulus TKC sit equalis angulo LKC, ex ypotesi, erit proportio LK ad TK sicut LC ad CT. Sed LH maior LK, et TH minor TK. Igitur maior est proportio LH ad TH quam LK ad TK, quare maior erit proportio LN ad NT quam LC ad CT, quod plane impossibile. Restat ut ab alio puncto arcus AQ quam a puncto H non possit T reflecti ad L. Palam ergo que accidunt in arcu AQ.

[2.392] **[PROPOSITIO 41]** Amplius, sit A [FIGURE 5.2.41, p. 600] centrum visus, B centrum speculi, et ducatur dyameter DABG. Et sumatur superficies in qua sit AB quocumque modo que secabit speram super circulum qui sit DLG. Dico quod a quolibet puncto semicirculi DLG reflectuntur puncta ad A inequalis longitudinis a centro cum eo.

[2.393] Verbi gratia, sumatur punctus E, et ducantur linee EA, EB. Palam quoniam angulus AEB erit acutus, quia cadet in minorem arcum semicirculo. Fiat ei equalis, et sit OEB, et ducatur linea OE quantumlibet. Palam quod quodlibet punctum illius linee refertur ad A a puncto E.

[2.394] Ducta autem a puncto B ad lineam OE perpendiculari, aut erit perpendicularis illa equalis BA, aut maior, aut minor. Si fuerit

138 dyametris: semidiametris *R*/HD: BD *FP1OL3C1ER* 139 probatur: probabitur *R* 140 et *om. SOL3C1ER*/inequaliter: equaliter *FP1* 141 probatio *corr. ex* proportio *a. m. L3;* proportio *E* 143 *post* sumptis *add.* punctis *R*/L: B *P1E* 144 et *om. R* 145 *ante* adinvicem *add.* ipsi *R*/erit: poterit *R*/T: D *F; om. P1O* 147 illud: illum *F*/LT TH *om. S*/TH *om. F* 148 usque cadat in: quousque concurrat cum *R*/LT *corr. ex* LE *L3; corr. ex* LZ *a. m. E*/C: P *R* 149 quoniam: quod *R*/TH: HT *R* 150 TKC: TPK *R*/angulo *om. R*/LKC: LKF *R* 151 ad TK *om. P1*/TK: KT *R*/LC: LP *R*/CT: PT *R* 152 *post* igitur *rep.* maior (151) . . . igitur (152) *S* 153 LC: LP *R*/CT: TC *P1;* OC *L3;* PT *R*/*post* plane *inter.* est *O* 154 ab *om. L3*/puncto[1] . . . quam *om. P1* 155 T *inter. L3*/*post* L *add.* quod est impossibile *O*/que: quod *FP1*/in *inter. C1* 157 dyameter: dyametri *F* 159 qui: que *FP1*/DLG: DG *FP1* 160 longitudinis: lineis *FP1* 161 punctus: punctum *R*/E *inter. O* 162 quoniam: quod *R*/cadet: cadit *P1O; corr. ex* caderet *E*/in *inter. C1*/arcum *corr. ex* acutum *P1* 163 ei *inter. L3*/OEB: PEB *R*/ducatur: producatur *R*/OE: BE *R* 164 refertur: reflectetur *R*/A *inter. O* 165 OE: CE *O; inter. a. m. C1;* PE *R*/perpendiculari: perpendiculariter *L3* 166 *post* equalis *add.* linee *O*/aut minor *mg. O*

equalis, linee omnes ducte a puncto B ad lineam OE, preter illam perpendicularem, erunt maiores linea BA, et ita quodlibet punctum linee OE, uno excepto, inequaliter distabit a centro puncto A.

[2.395] Si vero perpendicularis fuerit maior, omnia puncta linee illius plus distabunt a centro quam A punctum. Si autem perpendicularis fuerit minor, erit ducere a puncto B duas lineas ex diversis partibus perpendicularis equales linee BA, et omnes alie linee aut minores erunt aut maiores. Palam igitur quoniam a puncto E reflectuntur puncta ad A quorum longitudo a centro inequalis est longitudini A ab eodem, quod est propositum.

[2.396] Constat ex hiis quod, si sumatur A extra circulum—et sit H [FIGURE 5.41a, p. 600]—et ducatur dyameter HDBG et due contingentes HT, HQ, a quolibet puncto arcus TG, preter quam a T vel G, poterit fieri reflexio ad H punctorum inequaliter distantium a centro cum puncto H. Et erit eadem probatio.

[2.397] **[PROPOSITIO 42]** Amplius, ex hiis constabit quod, facta reflexione ad A a puncto E vel alio puncto inequaliter distante a centro puncto A, dyameter in quo fuerit punctus reflexus cum dyametro ABG facit duos angulos, unum respicientem angulum reflexionis, alium ei collateralem, qui quidem collateralis aliquando erit maior angulo reflexionis, aliquando minor.

[2.398] Verbi gratia, ducatur perpendicularis FB [FIGURE 5.2.42, p. 600] super EO. BA aut est perpendicularis super ea aut non.

[2.399] Sit perpendicularis. Erit igitur EA equidistans FB, et erunt duo anguli FBA, FEA equales duobus rectis. Ducta autem linea BO, erunt duo anguli OBA, OEA minores duobus rectis.. Igitur erit angulus OBG maior angulo OEA, qui est angulus reflexionis. Et cum

167 OE: OC *P1*; CE *O*; PE *R*/illam: illum *S*; *corr. ex* illum *C1* 168 *post* linea *scr. et del.* fuerit maior omnia puncta linee illius plus distabunt a centro quam ad punctum si A *S*/ita *mg. F*/OE: OC *P1*; PE *R* 169 *post* centro *add.* cum *R* 171 A punctum *transp. O*/punctum: puncto *FP1* 172 minor *corr. ex* maior *L3*/*post* erit *add.* possibile *R* 173 perpendicularis *corr. ex* perpendiculares *L3*; perpendiculares *E*/et *om. O*/alie *inter. O*/alie linee *transp. C1* 174 quoniam: quod *C1R* 175 A[2] *om. FP1*; *inter. O* 177 A: visus *R* 178 HDBG: HBDG *R* 179 TG: TGQ *R*/*post* a[2] *add.* punctis *R*/vel *om. R*/*post* G *add.* Q *R*/poterit: potest *R* 181 erit eadem *transp. R*/probatio: proportio *FP1* 183 vel *corr. ex* a *S*/puncto[2]: puncti *O*/distante: distanti *SE*; distantis *O*/*post* centro *add.* cum *R* 184 puncto *corr. ex* punctus *S*/dyameter *corr. ex* diametri *C1*/quo: qua *R*/punctus reflexus: punctum reflexum *R* 185 angulos *om. S*/unum: unam U *FP1* 186 angulo: alio *L3*/*post* angulo *add.* constante ex angulo incidentie et *R* 188 perpendicularis: per *F*/perpendicularis FB *transp.* (FB: BF) *SOC1*; pendicularis FB *mg. a. m. F*/FB *om. L3*/*post* FB *add.* quod si EC *mg. a. m. F*; *add.* fiet *P1* 189 super[1] . . . non *om. FP1*/EO: EC *O*/BA *corr. ex* KA *S*/est: erit *L3ER* 190 erit . . . et *om. R*/igitur EA *transp.* (EA *inter. a. m.*) *L3*/equidistans: quidem *FP1*; equidem *SO*/FB: FL *S*/*post* erunt *add.* ergo *R* 191 anguli *inter. O* 192 OBA *corr. ex* FBA *F*/rectis *om. FP1SC1*; *inter. OL3* 193 OEA *om. P1*/*post* angulus *add.* constante ex angulo incidentie et *R*

triangulus EBF sit equalis triangulo EBA, erit BF equalis BA, et ita OB maior BA.

[2.400] Ducta autem linea BN, erunt duo anguli NBA, NEA maiores duobus rectis. Erit ergo angulus NBG minor angulo NEA, et NB maior BA, et ita N et O reflectuntur ad A a puncto E. Et inequaliter distant a centro puncto A, et dyameter OB cum dyametro ABG ex parte G facit angulum maiorem angulo reflexionis, et dyameter NB maiorem, et ita propositum.

[2.401] Si vero BA non fuerit perpendicularis super EA, ducatur perpendicularis, que sit BK, que quidem sive cadat supra AB [FIGURE 5.2.42a, p. 600], aut sub [FIGURE 5.2.42b, p. 600]. Eadem erit probatio.

[2.402] Et BF sit perpendicularis super EO, et ducatur FT equalis AK, et ducatur TB. Palam quoniam in triangulo KEB angulus EKB rectus equalis EFB, et angulus KEB equalis angulo FEB. Restat tertius tertio equalis, et cum EB sit latus commune utrique triangulo, erunt trianguli equales, et erit FB equalis KB. Sed AK est equalis FT. Erit AB equalis BT, et angulus ABK equalis angulo FBT.

[2.403] Addito communi angulo FBA, erit KBF equalis TBA. Sed KBF et FEA valent duos rectos, quare TBA, TEA valent duos rectos, et ita TBG equalis est angulo TEA, qui est angulus reflexionis.

[2.404] Si igitur a puncto B ad lineam ET ducatur linea ultra T, faciet angulum cum BG ex parte G minorem angulo reflexionis. Et erit linea illa maior AB, quoniam TB equalis AB.

[2.405] Et quelibet linea a puncto B ducta ad ET et citra T faciet angulum cum BG ex parte G maiorem angulo reflexionis, et erit inequalis AB, et ita propositum.

194 triangulus: triangulum *R*/EBF: EDF *E*/equalis: equale *R*/*post* EBA *add.* et *R*/BF: BG *P1* 196 BN *inter. C1*/*post* maiores *scr. et del.* d *F* 198 et O *om. FP1S; inter. L3*/reflectuntur: refertur *FP1SOC1*/inequaliter: equaliter *FP1* 199 *post* centro *add.* cum *R*/OB: O *FP1SO; corr. ex* O *L3; corr. ex* OBO *E*/G *alter. in* A *OL3* 200 maiorem[1]: minorem *FP1SOL3E; corr. ex* minorem *a. m. C1*/angulo reflexionis *transp. C1*/*post* reflexionis *add.* et incidentie *R*/et *om. S*/NB: N *FP1S; alter. ex* N *in* GN *O; corr. ex* N *a. m. L3; corr. ex* NBN *E*/maiorem[2] *alter. in* minorem *a. m. C1;* minorem *R*/et ita: quod est *C1* 201 *ante* propositum *add.* patet *R* 202 super . . . perpendicularis (203) *om. S*/*post* EA *scr. et del.* ducatur perpendicularis super ea *L3* 204 probatio: pro *P1* 205 *post* perpendicularis *add.* et *O*/EO: EC *O* 206 AK: AB *L3*/TB: BT *R*/palam *om. P1*/quoniam: quod *R*/angulus EKB *inter.* (EKB: KEB) *O*/EKB: EB *FP1* 207 *post* equalis[1] *inter.* est *O; add.* est angulo *R*/EFB: OFB *P1O*/EFB . . . angulo[2] *mg. a. m. C1*/*post* angulo[2] *add.* reflexionis *E* 208 EB sit latus: sit latus EB *R*/triangulo *corr. ex* angulo *a. m. C1*/*post* erunt *scr. et del.* commune *C1* 209 trianguli equales: triangula equalia *R*/equales *rep. S*/FT: T *FP1*/*post* erit[2] *add.* ergo *R* 211 *post* addito *add.* igitur *R* 212 quare . . . rectos[2] *om. S*/TBA: HA *FP1;* THA *C1*/TEA . . . rectos[2] *inter. a. m. L3* 213 equalis est *transp. C1*/*post* angulus *add.* constans ex angulo incidentie et *R* 214 ET *corr. ex* TE *F*/ducatur *corr. ex* ducantur *P1*/faciet: facit *C1*/angulum *om. L3ER* 215 BG: B *FP1*/*ante* ex *add.* et *FP1*/G *inter. a. m. E*/*post* G *add.* angulum *R*/*post* angulo *add.* constante ex angulo incidentie et *R* 216 maior: minor *FP1*/*post* TB *add.* est *O*/*post* equalis *add.* est *R* 217 et[1] *om. O*/ducta ad ET: ad ET ducta *R*/ad ET *om. FP1*/et[2] *om. R* 218 cum *om. R*/BG: TBG *R*/*post* angulo *add.* constante ex angulo incidentie et *R*/inequalis (219): equalis *FP1;* minor *R* 219 *post* ita *add.* est *R*

[2.406] **[PROPOSITIO 43]** Amplius, sit B centrum visus, G centrum spere. Ducatur dyameter ZBGD [FIGURE 5.2.43, p. 601], et sumatur superficies in qua sit dyameter secans speram super circulum ZEH. Dico quoniam, si punctus A refertur ad B ab aliquo puncto circuli, et inequalis est distantia puncti A a centro et puncti B ab eodem, dyameter AG cum dyametro GD ex parte D faciet angulum quem impossibile est esse equalem angulo reflexionis.

[2.407] Sit equalis, et T sit punctus reflexionis, et sit AG inequalis BG. Ducantur linee TA, TG, TB, et fiat circulus transiens per tria puncta A, G, B, qui necessario transibit per punctum T. Si enim extra, ductis lineis a punctis A, B, ad idem punctum illius circuli extra fiet angulus minor angulo ATB. Et probabitur esse equalis.

[2.408] Quoniam, cum angulo AGB valebit duos rectos, et angulus ATB, cum sit equalis angulo AGD, ex ypotesi, cum angulo AGB valet duos rectos, et ita impossibile. Similiter, si circulus citra T ceciderit, eadem erit improbatio.

[2.409] Restat ergo ut transeat per punctum T, et cum angulus ATG sit equalis angulo BTG, erit arcus AG equalis arcui BG, et ita AG erit equalis BG. Et positum est esse inequalem, et ita propositum.

[2.410] **[PROPOSITIO 44]** Amplius, sumptis in duobus dyametris EGH, ZGD [FIGURE 5.2.44, p. 602] duobus punctis A, B ut BG sit maior AG, dico quoniam, si punctus A refertur ad B a duobus punctis arcus EZ, non erit uterque angulus reflexionis minor angulo AGD.

[2.411] Sumantur enim duo puncta T, Q in arcu EZ a quibus A refertur ad B, scilicet T, Q, et ducantur linee BT, GT, AT, BQ, GQ, AQ. Et si angulus ATB minor est angulo AGD, dico quoniam angulus AQB non erit minor angulo AGD.

221 ducatur: ducantur *L3* 222 *post* circulum *add.* qui sit *R* / ZEH: TEH *O*; EZH *R* 223 quoniam: quod *R* / si *inter. a. m. C1* / punctus: punctum *R* / refertur: reflectitur *R* / aliquo: alio *SOE*; *corr. ex* alio *L3* 224 A *om. FP1*; *inter. SL3* / eodem: eadem *FP1OL3* / dyameter *corr. ex* diametro *L3* 225 GD *corr. ex* HG *L3* / quem: quoniam *FP1* 226 *post* angulo *add.* constanti ex angulo incidentie et *R* 227 *post* sit[1] *add.* enim *R* / *post* equalis *scr. et del.* vel *C1* / punctus: punctum *R* / et[2] *om. FP1* / inequalis: in EK *S* 228 *post* BG *add.* et *R* / TG TB *transp. FP1R* (TG *corr. ex* DG *F*) / circulus transiens *transp. FP1* / transiens *om. O* 229 enim: equalis *FP1* / *post* enim *add.* cadit *R* 230 fiet: fiat *L3* 231 minor: maior *O* / angulo . . . equalis (233) *om. S* / ATB: ATH *FP1* 232 *post* rectos *add.* et anguli AGB et AGD valent duos rectos *R* 233 cum[1] *om. R* / sit: est *R* / *post* ypotesi *add.* ergo angulus ATB *R* / angulo[2]: angulus *L3* 234 *post* si *scr. et del.* L *F* 236 et *om. FP1L3ER* / *post* cum *add.* igitur *R* 237 ita *om. P1* 238 positum: propositum *L3* / inequalem: inequales *R* / *post* ita *add.* est *R* 239 duobus: duabus *R* 240 EGH: EGB *O* / ZGD: ZDG *S* / BG: G *FP1* 241 quoniam: quod *R* / punctus: punctum *R* / refertur: reflectitur *L3*; reflectatur *R* / a *inter. L3* / arcus EZ (242) *om. FP1* 242 angulus: angulo *S* / *post* angulus *add.* constans ex angulo incidentie et *R* 244 refertur: reflectitur *L3*; reflectatur *R* / scilicet T Q *scr. et del. O*; *om. C1ER* / BT: DT *F* / BQ GQ: LQ GA *S* 245 si *om. O* / ATB *corr. ex* ABT *F* / dico *mg. a. m. C1* / quoniam: quod *R* / AQB *corr. ex* ABQ *L3* 246 angulo *om. L3ER* / AGD: AGB *S*

[2.412] Sit enim minor, et ducatur linea GN dividens angulum dyametrorum per equalia, et ducatur linea AB quam dividat GN per punctum F. Palam quod proportio BG ad GA sicut BF ad FA. Sed BG maior GA; erit BF maior FA.

[2.413] Dividatur AB per medium in puncto K, et fiat circulus transiens per tria puncta A, B, T, qui quidem circulus non transibit per G, quoniam essent anguli AGB, BTA equales duobus rectis, et palam quod sunt minores, cum angulus BTA sit minor angulo AGD. Igitur transibit supra G.

[2.414] Similiter, non transibit per Q. Quoniam, sumpto puncto circuli in quo linea GQ secat ipsum, scilicet M, esset arcus AM equalis arcui MB, cum respiciant equales angulos supra Q, quod manet impossibile, quoniam, sumpto O puncto in quo linea GT secat hunc circulum, erit arcus AO equalis arcui OB, quia respiciunt equales angulos supra T. Restat ut hic circulus transeat supra Q, si enim infra, eadem erit improbatio.

[2.415] Ducatur autem linea a puncto O ad punctum K, que quidem, cum dividat cordam AB per equalia, et similiter arcum AB, erit perpendicularis super AB. Verum angulus BAG maior angulo ABG, cum BG maior GA. Et angulus BFG valet duos angulos FAG, FGA, et angulus AFG valet duos angulos FBG, FGB.

[2.416] Sed AGF equalis FGB, et FAG maior FBG. Igitur angulus BFG maior est angulo AFG. Igitur AFG minor est recto, quare NFB minor recto. Sed OK supra FB facit angulum rectum. Igitur producta concurret cum GN supra BF, et inferius numquam.

[2.417] Facto autem circulo transeunte per tria puncta A, Q, B, transibit supra G, et GQ dividet arcum eius AB per equalia. Sed K dividit cordam AB per equalia. Ergo KO concurret cum GN infra BF et supra punctum G. Igitur prius concurret cum GN infra FB, et iam improbatum est.

247 sit *corr. ex* si *L3* 249 proportio *corr. ex* probatio *L3* / GA: GD *FP1* / *post* sed *add.* cum *R* 250 *post* maior[1] *add.* sit *R* 251 puncto: punctum *L3* 252 non *inter. L3* / non transibit *inter. a. m.* (transibit: transit) *E* / per *om. FP1* 253 essent . . . BTA: anguli AGB BTA essent *R* / AGB: HGB *L3* / BTA: HTA *S* / equales: equalis *FP1* 254 cum: quam *S* / sit *inter. O* 258 arcui *corr. ex* arcuu *P1* / MB: BM *R* / supra: super *R* / *post* supra *add.* igitur *C1* / manet: manifeste est (est *inter.*) *O* 259 O puncto *transp. R* 260 *post* AO *rep. et del.* erit arcus AO *E* / equalis: equales *L3* / quia *inter. O* 261 supra[1]: super *R* / ut: in *S* / supra[2]: super *FP1SL3* 262 improbatio *corr. ex* probatio *S* 263 *post* K *scr. et del.* Q *F* / que: qui *E* 264 dividat: dividit *C1* / cordam: coram *FP1* 266 *post* BG *add.* sit *R* / GA *corr. ex* BA *P1* / et angulus BFG *om. E* / *post* BFG *scr. et del.* valet duos angulos FAG et angulus AFG *O* / et[2] . . . FGB (267) *om. O* 267 FGB *om. FP1* 268 *post* AGF *add.* est *O* / *post* equalis *add.* est *R* / FBG: FLG *L3* 269 maior[1]: minor *FP1S* / AFG[2]: FG *FP1*; BFG *C1R* / maior: minor *C1* / NFB: GFB *SL3* 270 minor: maior *L3* / *post* minor *add.* est *OR* (*inter. O*) / supra: super *R* 273 et: quia *E* / sed: set *L3* / K *alter. in* KE *E*; KO *R* 274 dividit: dividet *FP1* / ergo: G *S*; igitur *inter. O* / et: id est *F*; L *P1* 275 *post* igitur *add.* KO concurrens cum BA *R* / FB: BF *R* 276 improbatum: probatum *L3* / *post* est *scr. et del.* GN *E*

[2.418] Restat ergo ut angulus AQB non sit minor angulo AGD, aut A non reflectitur ad B a puncto Q. Similis erit improbatio, sumpto quolibet puncto arcus EN.

[2.419] Sumpto autem puncto in arcu NZ, qui sit C , et fiat reflexio puncti A ad B a puncto C, ut angulus reflexionis supra C sit minor angulo AGD, sicut angulus reflexionis supra T minor eodem, improbatur hoc modo.

[2.420] Ducantur AC, BC, GC. Oportet necessario quod GC dividat KO propter arcum AB, quem dividet ex circulo ABT linea GC per equalia, et similiter linea KO. Sit ergo punctus concursus linee GC cum KO punctus L. Ducta linea TC, cum due linee GC, GT sint equales, erunt duo anguli GCT, GTC equales, et uterque acutus.

[2.421] Ducta igitur perpendiculari super GT a puncto T, erit contingens circulo speculi, et producta cadet super terminum dyametri minoris circuli, cum angulus quem efficit cum TG respiciat semicirculum minoris. Et cum TO cadat super KO, et KO producta transeat per centrum minoris circuli, necessario illa perpendicularis cadet super terminum KO producta, et TC est inferior illa perpendiculari, habito respectu ad N.

[2.422] Igitur quecumque linea ducatur ad lineam TC secans dyametrum illius circuli, qui est OK, cadet in punctum linee TC citra illam perpendicularem. Cum igitur GC cadat in C et secet OK, erit C citra perpendicularem et infra arcum illius perpendicularis.

[2.423] Facto igitur circulo transeunte per tria puncta A, B, C, transibit quidem per C, et secabit circulum ABT in duobus punctis A, B. Et cum exeat a puncto B, iterum redeat in punctum C, et cum sit citra illum circulum, necessario secabit illum in tertio puncto, quod est impossibile.

277 *post* aut *add.* quod *R* 278 reflectitur: refertur *L3*; reflectetur *ER* 279 quolibet: quodlibet *S* 280 *post* sit *scr. et del.* se *S*/C: P *R*/et *om. R* 281 C[1,2]: P *R*/*post* angulus *add.* constans ex angulo incidentie et *R* 282 *post* angulus *add.* constans ex angulo incidentie et *R*/*post* minor *add.* est *R*/ eodem: eadem *O*/improbatur (283): improbetur *L3*; improbabitur *ER* 283 *ante* hoc *add.* autem *R* 284 AC BC GC: AP BP GP *R*/*post* oportet *add.* ergo *R*/quod: ut *R*/GC: GO *S*; GP *R* 285 dividet: dividit *R* 286 punctus: punctum *R*/GC: GP *R* 287 punctus: punctum *R*/*post* L *add.* et *R*/ ducta: ducatur *L3ER*/TC: TP *R*/*post* cum *add.* igitur *R*/GC: GO *S*; GP *R* 288 GCT GTC: GOT GES *L3*; GPT GTP *R*/acutus: arcus *P1* 289 GT: GP *FP1*/erit contingens (290): continget *R* 290 circulo: circulum *R*/super: in *C1*/terminum: tantum *FP1*; tertium *SO* 291 angulus *om. S*/ TG: GT *R*/*post* respiciat *add.* arcum *C1ER*/semicirculum (292): semicirculi *C1ER* 292 *post* minoris *add.* circuli *R*/cum: circuli *L3*/super: supra *R*/KO[1]: KC *FP1*/et KO *om. L3*; *mg. a. m. C1*/KO *corr. ex* EO *a. m. E* 293 circuli: cui *L3*/super (294): supra *C1* 294 terminum: tantum *F*/ producta: producti *O*/TC: PT *R*/*post* perpendiculari *scr. et del.* et *C1*/habito: habitu *FP1* 296 *post* ducatur *add.* a puncto G *R*/TC: TP *R* 297 dyametrum: dyametros *S*; *corr. ex* diametro *C1*/qui: que *FP1R*; *inter. O*/*post* in *scr. et del.* p *C1*/*post* punctum *add.* aliquod *R*/TC: TP *R*/TC citra: et intra *FP1* 298 illam: illum *C1*/cum . . . perpendicularem (299) *om. L3*/igitur: G *S*/GC: C *FP1*; *alter. in* GH *C1*; GP *R*/*post* in *scr. et del.* se *P1*/C[1,2]: P *R* 299 citra *inter. a. m. C1*/infra: ita *S* 300 igitur: isto *FP1*/C: P *R* 1 C: se *P1*; L *OR* 2 *post* B[2] *add.* et *FP1*/punctum: puncto *L3E*/C *om. L3*; P *R*/et *om. R*/*ante* cum *add.* inferius puncto T *R*/*post* cum *add.* P *R* 3 *post* illum[1] *add.* silum *L3*

[2.424] Restat igitur ut punctus A non reflectatur ad B a duobus punctis arcus interiacentis eorum dyametros, id est arcus EZ, ut uterque angulus reflexionis sit minor angulo AGD, quod est propositum.

[2.425] **[PROPOSITIO 45]** Amplius, dico quoniam est reflecti duo puncta a se inequalis longitudinis a centro a duobus punctis arcus ipsa respicientis, id est dyametros in quibus sunt puncta illa interiacentis.

[2.426] Verbi gratia, sumptis duobus dyametris in circulo spere, scilicet BD, GD [FIGURE 5.2.45, p. 603], dividatur angulus eorum per equalia per dyametrum ED, et in BD sumatur punctus M supra punctum in quem cadet perpendicularis ducta a puncto E super BD. Et sumatur ND equalis MD, et fiat circulus transiens per tria puncta D, N, M. Necessario circulus ille transibit extra E, si enim per E, fieret quadrangulum a quatuor punctis D, N, E, M, et duo anguli illius quadranguli sibi oppositi sunt equales duobus rectis, quod quidem non esset, cum linea EM sit supra perpendicularem, et angulus EMD acutus.

[2.427] Ei similiter oppositus supra N acutus, quia EN supra perpendicularem. Similis erit improbatio si transeat circulus citra E. Transibit igitur extra, et secabit circulum spere in duobus punctis, sicut T, L.

[2.428] Et ducantur linee MT, DT, NT, ML, DL, NL, et ducatur linea MN secans TD in puncto F, lineam ED in puncto P. Palam cum MD sit equalis ND, et PD commune, et angulus equalis angulo, erit triangulus equalis triangulo, et erit angulus FPD rectus. Igitur angulus PFD acutus.

[2.429] Ducatur a puncto F perpendicularis super TD, que sit KF. Palam quoniam aliquis punctus linee NL erit inferior puncto K. Sumpta inferioritate respectu T, sit ille punctus Z, et ducatur linea TZ usque ad circulum cadens in punctum circuli, qui sit C. Arcus NC aut est minor arcu TL, aut non.

[2.430] Si non fuerit minor, sumatur ex eo arcus minor, et ad terminum illius arcus ducatur linea a puncto T, et erit idem.

5 punctus: punctum *R* 6 EZ: ES *L3*/ut: et *FP1* 7 *post* angulus *add.* constans ex angulo incidentie et *R*/AGD: GED *L3*/quod est propositum *om. L3ER*/est *inter. P1* 8 quoniam est: quod possunt *R* 9 a[1] *corr. ex* ad *L3*; ad *C1ER*/a centro *om. P1* 10 id est: et *L3*/puncta illa *transp. L3* 11 duobus dyametris: duabus semidiametris *R* 12 GD *om. FP1* 13 per *om. E*/ dyametrum: semidiametrum *R*/BD: BG *FP1S*/punctus: punctum *R* 14 quem: quod *R*/BD: BG *FP1S* 15 et *om. L3*/D N M *mg. a. m. C1*/N M *transp. OER* 16 E[1] *inter. L3* 18 oppositi: opposita *FP1* 19 *post* et *add.* id eo *R*/*post* acutus *add.* et *L3R* (*inter. L3*) 20 ei similiter *transp. R*/*post* ei *inter.* et *O*/supra[1]: super *R*/acutus quia EN: quia EN acutus *FP1* 21 *ante* similis *add.* et *FP1*; *add.* est *R*/improbatio *corr. ex* improbatione *F*; *corr. ex* probatio *a. m. C1* 23 NT: ut *F* 24 TD *inter. O*/in[1] . . . ED *om. FP1*; *mg. a. m. C1* 25 commune: communis *R*/*post* angulus *add.* NDP *R*/*post* angulo *add.* MDP *R*/triangulus: triangulum *R* 26 equalis: equale *R*/FPD: SPD *FP1SOL3*/*post* acutus *scr. et del.* quidem *L3* 28 quoniam . . . punctus: quod aliquod punctum *R*/ NL *inter. a. m. E*/inferior: inferius *R*/*post* inferior *add.* in *FP1* 29 T: N *R*/ille punctus: illud punctum *R*/linea TZ *transp. R* 30 qui: quod *R*/C: O *R*/NC: N *FP1*; NO *R*/est minor *transp. R* 31 TL *corr. ex* D *L3* 32 ad *mg. a. m. L3*/terminum (33): certum *FP1*; *corr. ex* terminus *L3* 33 *post* idem *add.* ac si arcus NO esset minor arcu TL *R*

[2.431] Sit igitur NC minor TL. Palam quoniam angulus TNL erit maior angulo CTN, quia respicit maiorem arcum. Secetur ex eo equalis, et sit INZ, et super punctum T fiat angulus equalis angulo CTN, qui sit OTM. Cum angulus TML sit maior angulo MTO, concurret linea TO cum linea LM. Concurrat in puncto O.

[2.432] Cum igitur angulus LMT sit equalis duobus angulis MOT, MTO, et angulus LNT sit equalis angulo LMT, quia super eundem arcum, et angulus INZ sit equalis angulo MTO, erit angulus INT equalis angulo MOT, et ita triangulus MOT similis triangulo INT, et similiter triangulus INZ est similis triangulo TNZ. Et ita proportio NT ad TO sicut NI ad MO, et similiter proportio TN ad TZ sicut IN ad NZ.

[2.433] Sed TZ maior TO, quod sic patet. Sit R punctus in quo TZ secat KF. Angulus TFR est rectus, quare angulus FTR acutus. Igitur angulus OTF ei equalis est acutus. Et KF perpendicularis super TD, quare producta concurret cum TO, et linea ducta a puncto T ad punctum concursus, cuius linee pars est TO, erit equalis linee TR. Et ita TO minor TZ, quare maior est proportio NT ad TO quam NT ad TZ.

[2.434] Igitur maior est proportio IN ad MO quam IN ad NZ, quare MO minor NZ. Secetur ergo ex NZ equalis ei que sit NS.

[2.435] Quoniam angulus LND cum angulo LMD valet duos rectos, erit angulus LND equalis angulo OMD, et SN, ND equalia OM, MD. Igitur OD equalis est SD.

[2.436] Sed ZD maior SD, quoniam angulus LND cum angulo LMD valet duos rectos. Sed angulus LMD acutus, cum angulus EMD sit acutus. Igitur angulus LND maior est recto. Igitur ZD maior SD, quare ZD maior OD.

34 sit *om. FP1S; inter. O;* sic *inter. L3*/NC: NO *R*/TL: LC *S; corr. ex* D *L3*/quoniam *om. R*/angulus *corr. ex* arcus *L3*/TNL: TOL *O* 35 CTN: OTN *R*/arcum *om. P1*/*post* equalis *scr. et del.* linea *C1* 36 *post* T *add.* linee TM *R*/CTN: et N *FP1;* OTN *R*/sit[2]: si *FP1* 37 OTM: ? *O;* OTN *E;* QTM *R*/*post* cum *add.* igitur *R*/angulus *om. S*/TML: CLM *FP1;* LTM *S*/*ante* sit *scr. et del.* sit maior angulo CLM *P1*/MTO: MTQ *R*/TO: TQ *R* 38 O: Q *R* 39 cum . . . et (40) *om. FP1O*/MOT MTO (40): MQT MTQ *R* 40 LNT: LMT *FP1*/angulo *om. L3ER*/LMT: LM *P1*/*post* quia *add.* sunt *R*/*post* super *rep. et del.* quia super *L3* 41 angulo *om. R*/MTO: MTQ *R*/erit . . . MOT[1] (42) *mg. a. m. E*/INT *corr. ex* LMT *L3* 42 MOT[1,2]: MQT *R*/triangulus: triangulum *R*/similis: simile *R* 43 triangulus: triangulum *R*/similis *om. FP1;* simile *R*/TNZ *corr. ex* TNR *P1*/TO: TQ *R* 44 MO: MQ *R*/TN: DN *FP1* 45 TZ[1] *om. P1*/TO: TQ *R*/sic: si *P1;* sit *S; corr. ex* sit *F*/punctus: punctum *R* 46 secat *corr. ex* secatur *L3*/TFR: TF et *FP1; corr. ex* TF *O* 47 OTF: OIF *FP1;* QTF *R*/*post* est *scr. et del.* et *P1*/perpendicularis: perpendiculares *S* 48 *post* quare *add.* KF *R*/TO: DO *FP1;* TD *E;* TQ *R*/*ante* et *add.* sit concursus S *R*/*post* linea *add.* TS *R*/ducta . . . TO[1] (49) *mg. a. m.* (T *corr. ex* D) *E* 49 TO[1] *rep. P1*/TO[1,2]: TQ *R* 50 TZ: TC *FP1SO; corr. ex* TC *L3*/maior *corr. ex* minor *O*/proportio . . . est (51) *om. S*/TO: TQ *R* 51 est *om. P1*/IN *corr. ex* I *O*/MO: MQ *R*/IN: in *P1*/NZ: OZ *FP1* 52 MO: MQ *R*/*post* minor *add.* est *OR* (*inter. O*)*; scr. et del.* a *L3*/NZ: MZ *O*/*post* equalis *scr. et del.* angulo *S*/NS: NX *R* 53 LND *corr. ex* LMD *L3* 54 erit: EF *L3*/erit . . . rectos (57) *mg. O*/angulo *om. R*/OMD: QMD *R*/SN: XN *R*/OM: QM *R*/MD *corr. ex* OMD *P1* 55 OD: QD *R*/est *om. R*/SD: XD *R* 56 SD: XD *R*/LND *om. P1* 57 *post* LMD *inter.* est *O*/EMD: EMT *C1*/EMD . . . igitur[1] (58) *mg. a. m. L3* 58 angulus *om. L3*/ZD *corr. ex* SD *L3*/*post* maior *scr. et del.* est proportio *S*/SD: XD *R* 59 OD: CD *FP1;* QD *R*

[2.437] Igitur O refertur ad Z a duobus punctis T, L, et O et Z sunt inequalis longitudinis a centro, et in diversis dyametris.

[2.438] Et quod non sunt in eodem dyametro palam ex hoc quoniam angulus SDN equalis est angulo ODM. Addito ergo communi angulo SDM, erit angulus NDM equalis angulo SDO. Sed angulus NDM minor duobus rectis, quare angulus ZDO minor duobus rectis. Quare O, Z non sunt in eodem dyametro, sed in diversis.

[2.439] **[PROPOSITIO 46]** Amplius, sumptis duobus punctis, que sint O, K [FIGURE 5.2.46, p. 604], et inequaliter distantibus a centro, reflectetur quidem unum ad aliud a duobus punctis arcus respicientis semidyametros in quibus sunt, sed non ab alio puncto illius arcus quam ab illis duobus.

[2.440] Verbi gratia, D sit centrum; K remotior a D quam O a D; GD, OD dyametri; T punctus unus reflexionis. Palam ex superioribus quod duo anguli reflexionis non erunt minores angulo ODA nec equales. Alter ergo erit maior. Sit angulus reflexionis qui est supra T maior, et ducantur linee OT, DT, KT.

[2.441] Et ex angulo illo reflexionis secetur equalis angulo ODA, qui sit OTF, et dividatur angulus FTK per equalia per lineam TE. Et a puncto K ducatur equidistans TF, que quidem concurret cum TE. Concurrat in puncto Z, et ducatur linea OK, et dividatur angulus ODK per equalia per lineam DU secantem lineam OK in puncto C, et sit KD maior OD. Cum igitur sit proportio KD ad DO sicut KC ad CO, erit KC maior CO. Item, linea DT secet lineam OK in puncto N. Dico quoniam C cadit inter N et K non inter N et O, quod sic patebit.

[2.442] Angulus KCD valet duos angulos CDO, COD, et angulus OCD valet duos angulos CKD, CDK. Sed angulus CDO equalis angulo

60 O[1,2]: Q *R* / refertur: reflectitur *R* / Z: AN *FP1*; N *S*; *corr. ex* N *L3* / T L: C Z *FP1S* 61 inequalis *corr. ex* equalis *E* 62 sunt: sint *SR* / eodem: eadem *R* / ex: E *S* / ex hoc *om. ER* 63 SDN: XDN *R* / ODM *corr. ex* OND *F*; ODN *P1*; QDM *R* 64 SDM: secundum *S*; XDM *R* / equalis . . . NDM[2] *om. E* / SDO: XDQ *R* / sed: et *R* / angulus NDM[2] *om. R* 65 *post* quare *scr. et del.* OZ non sunt in eodem diametro *C1*; *add.* magis *R* / ZDO: ZDQ *R* / O: Q *R* 66 *ante* Z *add.* et *R* / eodem: eadem *R* 68 sint: sunt *FP1L3E* 69 reflectetur: reflectitur *C1* / aliud: alium *FP1* 70 sunt: fuerint *C1* 72 remotior: remotius *R* / a D: ADG *FP1SL3C1*; ATG *O* / *ante* GD *add.* circulus *O* / GD *om. C1* 73 OD dyametri: BD semidiametri *R* / T *inter. L3* / punctus unus: punctum unum *R* 74 duo anguli *transp. C1*; uterque angulus *R* / *ante* reflexionis *add.* constans ex angulo incidentie et *R* / erunt minores: erit minor *R* / equales: equalis *R* 75 sit *corr. ex* si *O* / *post* angulus *add.* constante ex angulo incidentie et *R* / supra: super *R* / *post* maior *add.* angulo ODA *ER* / et . . . ODA (77) *mg. a. m. E* 77 reflexionis *om. R* / *post* secetur *add.* angulus *ER* 79 equidistans: equidem *FP1* 80 angulus *om. P1* 81 per equalia *inter. a. m. L3* / C: P *R* / sit: est *R* 82 KC[1,2]: KP *R* / CO: PO *R* / *post* CO *add.* erit KC ad CO *P1* 83 CO: PO *R* / quoniam: quod *R* 84 C: cum *S*; P *R* / cadit: cadat *L3* / inter . . . non *om. FP1* / non: N *S* 85 KCD: KPD *R* / CDO COD: PDO POD *R* / et . . . CDK (86) *mg. O* 86 OCD: OPD *R* / CKD: OKD *FP1SL3* / CKD CDK: PKD PDK *R* / CDK: CDH *S* / CDO: PDO *R* / *post* equalis *add.* est *OR* (*inter. O*)

CDK, et angulus KOD maior angulo OKD. Igitur angulus KCD maior angulo OCD, quare angulus KCD maior recto. Et angulus KND acutus, quod sic constabit.

[2.443] Si fiat circulus per tria puncta O, T, K, transibit infra D, quoniam, cum angulus OTK sit maior angulo ODA, erunt duo anguli OTK, ODK maiores duobus rectis, et linea ND dividet arcum illius circuli, qui est OK, per equalia infra D.

[2.444] Si a puncto divisionis ducatur linea ad medium punctum linee OK, que est corda illius arcus, erit linea illa perpendicularis super OK, et cadet inter C et K, cum CK sit maior CO. Et angulus supra N a parte illius perpendicularis et ex parte C erit acutus, et angulus supra C ex parte O est acutus. Si ergo C cadat inter N et O, impossibile erit perpendicularem illam cadere inter N et C, quia secaret DC, et fieret triangulus cuius unus angulus rectus, alius obtusus.

[2.445] Cadet ergo inter N et K, et erit angulus N ex parte perpendicularis acutus; igitur ex parte C obtusus, et ita erit triangulus cuius duo anguli obtusi.

[2.446] Palam quoniam angulus KTD est medietas anguli KTO, sed KTE medietas anguli KTF. Restat ETD medietas anguli FTO, sed FTO equalis est angulo ODA. Igitur ETD medietas anguli ODA.

[2.447] Sed angulus ODA cum angulo ODF valet duos rectos, et tres anguli trianguli ETD duos rectos. Ablato EDT communi, restat angulus TED equalis medietati anguli ODA et angulo ODN. Sed angulus ODC cum medietate anguli ODA est rectus. Igitur angulus TED est acutus, quare ei contrapositus est acutus.

[2.448] Igitur, si a puncto K ducatur perpendicularis ad TZ, cadet inter E et Z. Si enim supra E ceciderit, cum angulus TEK sit obtusus, accidet triangulum habere duos angulos rectum et obtusum. Sit ergo perpendicularis KQ. Dico quoniam KT se habet ad TF sicut KD ad DO.

87 CDK: PDK *R* / et . . . KOD *inter.* (angulus *om.*) *O* / KOD: KCD *FS*; KED *P1* / OKD: CKD *O* / KCD: KPD *R* 88 OCD: OPD *R* / KCD: KPD *R* 90 tria puncta *transp. FP1* / K: H *S* 91 *ante* cum *add.* si transeat per D *R* / OTK: OTH *FP1* 92 ODK: CDK *S*; *inter. L3* / *post* rectis *add.* si transeat supra D eadem est demonstratio *R* / ND: non *S* / dividet: dividit *C1* 94 *post* si *add.* autem *R* 96 C: P *R* / CK: C et K *S*; *corr. ex* C et K *L3*; PK *R* / CO: PO *R* / supra: super *R* / a: ex *O* 97 *post* perpendicularis *add.* erit acutus *FP1SOL3* / C: P *R* / *post* erit *scr. et del.* perpendicularis *C1* / supra: super *R* 98 C[1,2]: P *R* / cadat: cadet *FP1C1*; cadit *R* 99 C: P *R* / secaret: secari *P1* / DC: DP *R* 100 triangulus: triangulum *R* 102 C: O *R* / *post* obtusus *add.* ergo P non cadit inter N et O quia *R* / et *om. R* / erit triangulus *transp. OC1* / triangulus: triangulum *R* 104 quoniam: quod *R* / est *om. FP1* / anguli: an *FP1S* / *ante* KTO *add.* sed *P1* / *post* KTO *add.* sed KTO *F* / sed KTE (105) *om. P1* 105 *post* KTE *add.* est *OR* / ETD *corr. ex* ED *L3* / FTO[1] . . .anguli (106) *om. FP1* 106 *post* ETD *add.* est *S* / *post* anguli *rep.* KTF (105) . . . anguli (106) *S*; *rep. et del.* FTO[1] (105) . . . anguli (106) *L3* 108 trianguli: tri *P1*; *corr. ex* tri *O* 109 TED equalis: DED equales *FP1* 110 ODC: ODP *R* 111 contrapositus: circapositus *S* 112 puncto *corr. ex* p *L3* / TZ *corr. ex* TE *L3* 113 et *inter. a. m. C1* 114 accidet: accidit *C1* / sit *corr. ex* si *a. m. L3* 115 quoniam: quod *R* / DO: TO *P1*

[2.449] Probatio: TO aut est equidistans KD, aut concurrit cum ea. Sit equidistans [FIGURES 5.2.46a, 5.2.46b, p. 605]. Erit angulus ODA equalis angulo TOD, et ita TOD equalis angulo OTF. OD, TF aut sunt equidistantes, aut concurrent.

[2.450] Si equidistantes [FIGURE 5.2.46a, p. 605], cum cadant inter equidistantes, erunt equales. Si vero concurrunt [FIGURE 5.2.46b, p. 605], faciunt triangulum cuius latera equalia, quia respiciunt equales angulos, et FD secat illa latera equidistans basi. Erit ergo proportio unius laterum ad DO sicut alterius ad FT, et ita TF equalis DO.

[2.451] Et hoc dico si concurrant sub KD. Et si concurrant sub TO eadem erit probatio, quia fiet triangulus cuius unum latus TO et alia duo latera equalia, et erit proportio unius ad DO sicut alterius ad TF. Item, angulus TDK equalis angulo DTO, quia DT inter equidistantes. Igitur est equalis angulo DTK, quare DK, TK sunt equalia. Igitur proportio TK ad TF sicut KD ad DO.

[2.452] Si vero TO concurrit cum KD, concurrat ex parte A in puncto P [FIGURE 5.2.46c, p. 605]. Scimus quoniam proportio KT ad TF compacta est ex proportione KT ad TP et TP ad TF. Sed proportio KT ad TP sicut KD ad DP, quoniam DT dividit angulum KTO per equalia. Et proportio TP ad TF sicut DP ad DO, quoniam angulus ODP equalis angulo PTF, et angulus supra P communis. Erit partialis triangulus similis totali. Igitur proportio KT ad TF constat ex proportione KD ad DP et proportione DP ad DO. Sed proportio KD ad DO constat ex eisdem, quare proportio KT ad TF sicut KD ad DO.

[2.453] Si vero TO concurrat cum KD ex parte G [FIGURE 5.2.46d, p. 605], sit concursus L, et a puncto D ducatur equidistans linee KT, que sit DR, concurrens cum TO in puncto R. Igitur angulus KTD equalis

116 probatio *om. R*/*post* TO *add.* enim *R*/equidistans: equidem *F*/*post* KD *scr. et del.* et con *P1*/concurrit: concurrat *FP1SL3C1*; *alter. ex* concurrat *in* concurret *O*/ea: eo *FP1* 117 equidistans: equidem *F*/*post* erit *add.* ergo *R* 118 *ante* ODA *scr. et del.* cum ea/TOD[1] *corr. ex* DTOD *F*/et ita TOD *om. FP1*; *inter. L3*/OTF: OZF *FP1*; *alter. in* ODF *O*/*post* OTF *add.* et *R*/OD TF *inter. L3* 119 concurret: concurrunt *R* 121 equidistantes: equidem *F* 122 faciunt: facient *R* 123 FD *corr. ex* secundus *L3*/equidistans: equidistanter *C1ER*/ergo *inter. a. m. S* 124 *post* unius *scr. et del.* ad *E*/ad[2] *inter. C1*/TF: F *FP1*/DO[2]: TO *E* 125 et hoc (125) . . . ad TF (127) *transposui a fine* 2.445 *ad initium* 2.451/si[1] *inter. L3*/*post* si[1] *add.* linee ille *R*/sub[1]: cum *FP1*/sub[1] . . . concurrant *om. L3*/sub[2] *corr. ex* supra *O*/TO *corr. ex* DO *F* 126 probatio: improbatio C1/triangulus: triangulum *R*/*post* latus *add.* est *R*/TO *corr. ex* DO *F* 127 unius: unus *S*/*post* unius *add.* laterum *R*/ad . . . alterius *mg. O*/DO: TO *P1E*/*post* TF *add.* et ita TF equalis DO *R* 128 *post* equalis *add.* est *OR* (*inter. O*)/quia: quare *FP1*/*post* DT *inter.* est *O* 129 DK: DF *E*/*post* DK *add.* equalis est *R*/sunt equalia *om. R* 131 TO *om. P1* 132 P: L *R*/quoniam: quod *R*/KT: BT *O* 133 TP[1,2,3]: TL *R*/proportio *om. R* 134 *ante* sicut *add.* est *R*/DP: DL *R*/quoniam DT: quare DO *FP1*; *mg. L3*/dividit *rep. P1*/et *corr. ex* ET *O* 135 TP: TL *R*/*post* TP *scr. et del.* similis (137). . . quare (139) (*ante* similis *add.* ad) *S*/DP: DL *R*/*post* angulus *inter.* est *a. m. E*/ODP: ODL *R*/*ante* equalis *add.* est *R*/*post* equalis *inter.* est *O* 136 PTF: per TH *F*; PTH *P1*; PTK *S*; LTF *R*/supra P: super L *R*/partialis . . . similis (137): partiale triangulum simile *R* 138 DP[1,2]: DL *R*/sed . . . DO[2] *mg. a. m. L3* 139 KD: BD *P1* 140 TO: DO *FP1*/KD *corr. ex* AD *L3* 141 L: A *P1*; S *R*/D *om. FP1*/KT *inter. a. m. C1* 142 DR: DZ *FP1*/TO: DO *FP1L3*/angulus *corr. ex* triangulus *L3*/equalis est (143) *transp. R*

est angulo TDR, sed idem est equalis angulo DTO, quare DR est equalis TR. Sed quoniam triangulus LTK similis triangulo LRD, erit proportio DR ad RL sicut KT ad TL, et ita RT ad RL sicut KT ad TL. Sed RT ad RL sicut DK ad DL. Igitur KT ad TL sicut KD ad DL.

[2.454] Sed quoniam angulus FTO equalis est angulo ODA, erit angulus ODL equalis angulo FTL, et angulus supra L communis. Erit triangulus ODL similis triangulo FTL. Igitur TL ad TF sicut DL ad DO, et ita KT ad TL sicut KD ad DL. Et TL ad TF sicut DL ad DO, quare KT ad TF sicut KD ad DO, quod est propositum.

[2.455] Sed quoniam KZ [FIGURE 5.2.46, p. 604] equidistans TF, erit angulus KZE equalis angulo ETF, et ita triangulus KZE similis triangulo ETF, quare proportio KE ad EF sicut KZ ad TF. Sed KE ad EF sicut KT ad TF, propter angulum supra T divisum per equalia. Igitur KZ equalis KT.

[2.456] Verum quoniam KQ est perpendicularis super EZ, erunt omnes eius anguli recti. Sed angulus ETD est acutus, quoniam est medietas anguli. Igitur KQ concurret cum TD. Sit concursus H, et ducatur linea EH, et a puncto E ducatur equidistans KH producta usque ad DH, que sit EC.

[2.457] Et mutetur figura propter intricationem linearum [FIGURE 5.2.46e, p. 606], et fiat circulus transiens per tria puncta C, T, E. Et producatur KD usque in circulum cadens in punctum M, et ducatur MT. Erit angulus TME equalis angulo TCE, quia cadunt in eundem arcum, et angulus TCE equalis angulo CHK. Erit TME equalis angulo CHK.

[2.458] Secetur ab angulo TME angulus equalis angulo DHE, qui sit FMD, et punctus in quo FM secat TC sit I. Palam quoniam triangulus IMD similis est triangulo EDH, quare proportio HD ad DM sicut EH ad IM.

[2.459] Et similiter, triangulus TMD similis triangulo KHD, et proportio KD ad DT sicut HD ad DM, et ita KD ad DT sicut EH ad IM.

143 TDR: DDN *FP1* 144 TR: ZR *E* / quoniam . . . similis: quia triangulum STK simile est *R* / LRD: SRD *R* 145 RL^1: SR *R* / sicut^1 *mg. a. m. C1* / TL^{1,2}: RL *F*; TS *R* / RL^{2,3}: RS *R* / sed . . . RL^3 *mg. a. m. C1* / RL^3 *corr. ex* TL *L3* 146 DL^1 *corr. ex* LDL *P1* / DL^{1,2}: DS *R* / KT: DT *E* / TL: TS *R* / KD: DK *R* 148 ODL: ODS *R* / FTL: FTS *R* / et . . . FTL (149) *om. ER* / angulus *corr. ex* angulo *S* 149 TL: ST *R* / DL: DS *R* / DO: TO *P1S* 150 ita: est *R* / TL^1 *corr. ex* KL *L3* / TL^{1,2}: TS *R* / DL^{1,2}: DS *R* / DO: TO *E* / quare . . . DO (151) *om. FP1* 151 TF: TC *S*; *corr. ex* TC *O* 152 KZ: HZ *FP1* / equidistans: equidistat *C1E* 153 *post* equalis *scr. et del.* e *S* / triangulus: triangulum *R* / similis: simile *R* 154 TF: EZ *P1* / sed . . . TF (155) *mg. a. m.* (KE: FE) *L3* 155 supra: super *R* 156 equalis: EK *S* / *post* equalis *add.* est *R* 157 quoniam: quod *FP1* / super: supra *L3E* 158 eius anguli *transp. L3* / recti *mg. O* / ETD: EDT *S* 159 KC: CQ *E* 160 KH: HK *R* 161 EC: EX *R* 162 figura: figuram *C1* 163 C: X *R* / T *corr. ex* B *O*; Z *E* 164 circulum: triangulum *C1* / cadens *mg. L3* / ducatur: educatur *ER* 165 TME . . . angulo *om. FP1* / TCE: RCE *S*; TXE *R* 166 TCE: RCE *S*; TXE *R* 167 angulus *om. L3ER* 168 punctus: punctum *R* / TC: TX *R* / quoniam triangulus: quod triangulum *R* 169 IMD: MID *O* / similis: simile *R* / *post* DM *scr. et del.* similis *P1* 171 triangulus: triangulum *R* / similis: simile *R*

[2.460] Sed proportio KD ad DT nota, quoniam semper una et eadem permanet, cuicumque punctus reflexionis sit T in arcu EG, quia semper linea TD una, et KD similiter. Linea etiam EH una in quacumque reflexione permanet, et non mutatur eius quantitas, quare linea IM semper erit una, quare punctus F notus et determinatus.

[2.461] Si ergo a tribus punctis arcus BG fieri posset reflexio, esset ducere a puncto F ad circulum TCE tres lineas equales, quia esset proportio KD ad DT sicut EH ad quamlibet illarum. Et patet ex superioribus quod non nisi due equales duci possunt, quare a duobus tantum punctis fiet reflexio, quod est propositum.

[2.462] **[PROPOSITIO 47]** Amplius, datis duobus punctis K, O [FIGURE 5.2.47, p. 607] in diversis dyametris inequaliter distantibus a centro, est invenire punctum reflexionis.

[2.463] Verbi gratia, sumatur linea ZT, et dividatur in puncto E ut sit proportio ZE ad ET sicut KD ad DO. Quoniam KD maior DO, erit ZE maior ET. Dividatur ZT per equalia in puncto Q, et a puncto Q ducatur perpendicularis super ZT, et fiat angulus ETD equalis medietati anguli ODA. Erit quidem acutus. Igitur TD concurret cum perpendiculari.

[2.464] Sit concursus in puncto H, et ducatur linea DEK ut sit proportio KD ad DT sicut KD ad semidyametrum spere. Et angulo quem habemus KDT fiat in speculo angulus equalis KDT. Dico quoniam T est punctus reflexionis, et si predictam probationem replicaveris, manifeste videbis.

[2.465] **[PROPOSITIO 48]** Amplius, sumptis duobus punctis in diversis dyametris, que puncta inequalis sint longitudinis a centro, si fuerint extra circulum, et reflectantur ab aliquo puncto arcus oppositi dyametris, non reflectentur ab alio eiusdem arcus.

173 sed: si *L3* / KD: KC *FP1*; KT *SOC1* / semper: super *FP1E* / semper una *transp.* (una *mg.*) *C1* / *post* semper *add.* et *L3* / et *om. O* 174 cuicumque punctus: quodcumque punctum *R* / EG: OG *S*; *alter. in* BG *L3*; BG *R* 175 *post* TD *add.* est *R* 176 eius quantitas *transp. C1* / IM *inter. O* / semper erit (177) *om. O* 177 punctus: punctum *R* / notus: notum *R* / determinatus: determinatum *R* 178 si ergo: sibi igitur *FP1* / EG: OG *FP1OC1*; *corr. ex* AG *S*; *alter. in* HG *L3*; *corr. ex* KG *a. m. E*; BG *R* / posset: possit *C1* 179 TCE: XTE *R* / *post* equales *add.* quarum cuiuslibet pars interiacens diametrum TX et circumferentiam circuli esset equalis linee IM *R* / quia: quare *FP1* / esset: semper erit *R* 180 DT: TD *FP1* 181 duci *om. ER* 183 *post* punctis *inter.* sit *L3* 186 ZT: et T *S* / et *corr. ex* aut *P1* / ut: et *S* 187 sit *om. FP1* / KD^2 *corr. ex* QD *C1* 188 Q et *mg. a. m. C1* / Q^2 *corr. ex* quod *S* 189 ETD *inter. L3* 190 ODA: EDA *FP1S* 193 angulo: alio *FP1SO*; *corr. ex* angulus *L3* 194 quem: quoniam *FP1SOL3* / equalis *om. L3* / *post* equalis *add.* scilicet *OC1ER* (*inter. E*) / KDT: TBDC *FP1*; CKDT *S* / quoniam: quod *R* 195 punctus: punctum *R* 198 dyametris *mg. F* / a *inter. O* / si *om. S* 199 extra: ex *O* / et *inter. a. m. E* / aliquo: alio *O*; *corr. ex* alio *L3* 200 reflectentur: reflectantur *SL3C1*; *corr. ex* reflectatur *O* / alio *corr. ex* aliquo *P1* / *post* alio *inter.* puncto *L3*

[2.466] Verbi gratia, sint A, B [FIGURE 5.2.48, p. 607] puncta in diversis dyametris extra circulum, G centrum, T punctus reflexionis, et ducantur BT, AT, GT. BT secabit arcum circuli. Sit punctus sectionis Q. AT secabit similiter arcum circuli. Sit punctus sectionis M.

[2.467] Quoniam angulus BTG equalis est angulo ATG, cadunt in arcus circuli equales, quod patebit, producto dyametro TG. Erit ergo arcus QT equalis arcui MT. Si igitur B refertur ad A ab alio puncto, sit illud H, et ducantur linee BH, AH, GH. Secet BH circulum in puncto L, AH in puncto N.

[2.468] Secundum supradictam rationem, erit HL equalis NH. Sed iam habemus quod QT equalis TM, quod est impossibile. Restat ut B non reflectatur ad A a puncto H vel ab alio puncto arcus oppositi dyametri preter quam T.

[2.469] Similiter, si fuerit alterum punctorum in circulo, alterum extra, ab uno tantum puncto arcus poterit reflecti ad aliud.

[2.470] Amplius, si linea ducta ab uno duorum punctorum ad aliud contingat circulum, aut tota sit extra, sumpto quocumque puncto in arcu opposito dyametris, altera linearum a punctis duobus ad illud punctum ductarum tota erit extra circulum. Et sic neuter punctorum ad alium reflectetur ab aliquo puncto illius arcus, et ab uno solo puncto speculi arcus oppositi reflectetur, et ita ab uno solo puncto speculi.

[2.471] **[PROPOSITIO 49]** Si vero linea ducta ab uno puncto ad alium secet circulum, fiat circulus super centrum speculi et illa duo puncta. Circulus ille aut totus erit intra circulum, aut continget ipsum, aut secabit.

[2.472] Sit totus intra, et ducantur due linee a duobus punctis ad aliquod punctum arcus oppositi. Angulus quem facient erit minor

201 puncta: puncto *FP1* 202 *post* G *scr. et del.* est *C1* / punctus: punctum *R* 203 GT: TG *L3ER* / *ante* BT *scr. et del.* T *L3* / secabit: secat *C1* / punctus: punctum *R* / AT *corr. ex* DT *O* 204 *post* secabit *add.* arcum secabit *P1*; *scr. et del.* circulum *C1* / similiter arcum *transp. FP1* / punctus: punctum *R* 205 cadunt: cadent *R* 206 producto dyametro: producta semidiametro *R* / *post* TG *add.* in P *R* 207 QT: QP *R* / MT: MP *R* / *post* B *scr. et del.* ad *E* / refertur: reflectitur *R* / ad A *om. R* / alio *om. P1* 208 illud: illum *FP1* / BH[1]: KH *F* / BH[2] *corr. ex* LH *L3*; LH *E* 209 *post* N *add.* et producatur BG in K *R* 210 *post* secundum *add.* igitur *R* / supradictam: predictam *L3ER* / rationem: probationem *R* / HL: LH *SOC1L3E*; LK *R* / NH: NK *R* 211 QT: QP *R* / TM: PM *R* / est *om. FP1* / *post* impossibile *add.* ut *F*; *scr. et del.* ut *P1* / B: L *FP1SOL3E*; LI *C1* 212 ad *om. S* 213 dyametri: diametris *R* / *post* quam *add.* a *L3C1ER* 214 fuerit: fuerint *FP1L3* 215 tantum puncto *transp. FP1* / poterit: potest *FP1* / aliud: alium *FP1* 216 ducta *corr. ex* data *L3* / duorum punctorum *transp. C1* / ad aliud *om. R* / aliud: alium *FP1* 217 aut *corr. ex* at *F* 218 *post* duobus *add.* punctorum duorum altero *R* / illud: illum *FP1* 219 neuter: neutrum *OR* 220 alium: aliud *C1R* / reflectetur *corr. ex* reflectatur *E* / aliquo: alio *FP1SOE*; *corr. ex* alio *L3* 221 arcus . . . speculi[2] *om. FP1R* / reflectetur: reflectentur *O* 223 alium: aliud *R* / *post* circulum *add.* et *FP1SOL3C1* / super: supra *FP1*; per *R* 224 circulum *alter. in* speculum *O* / continget *corr. ex* contingit *O* / *post* ipsum *add.* intrinsecus *R* 226 et ducantur *corr. ex* educantur *O* 227 aliquod *om. O*

angulo quem unus dyameter facit cum alio ex alia parte centri, et quilibet angulus sic factus super arcum oppositum minor erit illo angulo.

[2.473] Quoniam angulus factus in interiori circulo per lineas a punctis ad arcum eius interiacentem ductas erit equalis illi angulo, quoniam cum angulo dyametrorum supra centrum valet duos rectos. Sed angulus arcus minoris circuli maior angulo arcus speculi.

[2.474] Igitur in arcu circuli non fiet reflexio nisi ab uno puncto, cum iam dictum sit quod non est possibile reflexionem duobus punctis fieri ut sit uterque angulus minor angulo dyametrorum ex alia parte centri.

[2.475] Si vero circulus ille contingat circulum speculi, angulus factus a lineis ab illis punctis ad punctum contactus ductis erit equalis angulo dyametrorum ex alia parte centri, quare ab illo punto contactus non fiet reflexio. Et angulus factus super quodcumque punctum aliud arcus maioris circuli erit minor illo, quare a duobus punctis arcus non fiet reflexio, secundum predicta.

[2.476] Si vero circulus interior secet circulum speculi, duo puncta aut erunt extra circulum; aut intra; aut unus intra, alius extra; aut unus in circulo, alius extra vel intra.

[2.477] Si fuerint extra, vel unus in circulo alius extra, circulus secans non secabit arcum circuli speculi interiacentem dyametris, et ita quilibet angulus factus super arcum illum erit maior angulo dyametrorum ex alia parte centri. Et iam probatum est in precedenti figura quod hec puncta ab uno solo puncto arcus interiacentis poterunt reflecti.

[2.478] Si vero duo puncta fuerint intra, secabit circulus interior arcum interiacentem in duobus punctis, et restabunt ex eo duo arcus ex diversis partibus.

228 unus: una *R* / alio *corr. ex* alia *L3*; alia *R* / alia *om. R* 229 sic: sit *FSOL3C1* 231 interiori: interiore *R* 232 ductas *corr. ex* duas *L3* 233 supra: super *R* / *post* duos *add.* angulos *L3ER* / rectos *om. FP1S*; *inter. L3*; *inter.* r *O* 234 arcus[1] . . . circuli *corr. ex* minoris circuli arcus *C1* / *post* maior *add.* est *R* 235 *post* arcu *add.* speculi vel *E* / circuli: speculi *R* / *post* circuli *add.* speculi *FSO*; *add.* vel speculi *L3C1* (*deinde del.* vel *C1*) / non: ? *inter. P1* / puncto *corr. ex* punctis *P1* 236 *post* reflexionem *add.* a *OL3R* (*inter. OL3*) 237 *post* angulus *add.* constans ex angulo incidentie et reflexionis *R* / minor *corr. ex* minori *L3* 239 circulus: arcus *E* / *post* contingat *add.* intrinsecus *R* / *post* circulum *add.* illum *C1* / speculi: spere *FP1* 240 contactus: contactis *S*; *corr. ex* contactis *O* 242 quodcumque: quocumque *S* / aliud: alium *FP1* 243 arcus[1] *om. R* 246 unus[1,2]: unum *R* / alius: aliud *SR* 247 circulo: circumferentia *R* / alius: aliud *R* 248 fuerint: fuerit *FP1SO* / vel . . . extra[2] *om. R* 249 dyametris: diametros *R* / et . . . centri (251) *om. R* 251 est *om. L3* / precedenti: precedente *R* 252 *post* interiacentis *add.* diametros *R* / poterunt: potuerit *FP1*; poterint *C1* / *post* reflecti *add.* si vero unum fuerit in circumferentia aliud extra circulus secans secabit arcum circuli speculi diametros interiacentem in unico puncto et quilibet angulus factus super arcum illum erit maior angulo diametrorum ex alia parte centri et sic ab uno puncto vel a duobus potest fieri reflexio *R*

[2.479] Si unus punctorum fuerit intra circulum, alius in circulo vel extra, secabit circulus arcum interiacentem in unico puncto, et restabit unus arcus tantum.

[2.480] Si secet in duobus punctis, omnes anguli facti super arcum interiacentem duo puncta sectionis erit maior angulo dyametrorum ex alia parte centri, et ab hoc arcu poterit fieri reflexio forsitan ab uno puncto tantum, forsitan a duobus.

[2.481] Et a duobus arcubus qui restant ex arcu totali, et ex diversis partibus, omnes anguli erunt minores angulo dyametrorum, et tantum ab uno eorum puncto fiet reflexio.

[2.482] Et in hoc situ poterit fieri reflexio a duobus punctis arcus interiacentis dyametros, aut a tribus.

[2.483] Et palam quod ab uno tantum puncto arcus oppositi fiet reflexio, et ita in hoc situ aliquando a tribus, aliquando a quatuor.

[2.484] Si vero secetur arcus interiacens dyametros in uno tantum puncto a maiori circulo, omnes anguli facti in parte illius arcus inclusa minori circulo erunt maiores angulo dyametrorum, et poterit fieri reflexio a duobus punctis illius partis, vel ab uno.

[2.485] Omnes anguli alterius partis arcus interiacentis erunt minores angulo dyametrorum, et ab uno puncto tantum illius partis fiet reflexio, et ita, cum ab uno puncto arcus oppositi semper fiat reflexio in hoc situ, aliquando a tribus, aliquando a quatuor, non a pluribus poterit esse reflexio.

[2.486] Palam ergo quod puncta inequalis longitudinis a centro, aliquando ab uno puncto tantum, aliquando a duobus, aliquando a tribus, aliquando a quatuor, numquam a pluribus, reflectuntur. Cum autem puncta eiusdem longitudinis fuerint, poterit fieri reflexio aut ab uno tantum puncto, aut a duobus, aut a quatuor, numquam a tribus.

[2.487] Ubi ab uno puncto fit reflexio, una apparet ymago; ubi due, due; ubi tres, tres; ubi quatuor, quatuor. Si vero punctus visus et cen-

256 si . . . punctis (259) *om. R* 257 *post* interiacentem *add.* et *FP1* 259 duobus punctis *transp. P1* / *ante* omnes *add.* et *R* 260 erit maior: erunt maiores *R* 261 alia: aliqua *S* / alia parte *transp.* (alia *inter.*) *L3* / poterit: possit *L3E*; posset *R* 262 puncto *om. FP1* / *post* tantum *rep.* forsitan (261) . . . tantum (262) *S* / forsitan: forsan *L3* 263 *post* et[1] *add.* si *R* / *post* arcubus *add.* fiat reflexio *R* / qui restant *transp. C1* / *post* ex[1] *scr. et del.* cu *F* / et[2] *om. O* 264 *post* partibus *add.* et *FP1* / minores *inter. L3* 268 et *om. ER* / *post* palam *add.* etiam *R* 269 tribus *rep. F* / *post* quatuor *add.* punctis fiet reflexio *R* 270 secetur . . . circulo (271): unum punctorum fuerit intra circulum aliud in circumferentia vel extra secabit circulus arcum interiacentem in unico puncto et restabit unus arcus tantum et *R* / interiacens *corr. ex* interiacentis *P1L3* / in: ab *O* 271 a *om. FP1* 272 minori: minore *C1*; a secante *R* / erunt *corr. ex* essent *P1* 273 reflexio . . . punctis: a duobus punctis reflexio *FP1* 274 *post* omnes *add.* vero *R* / arcus *om. L3ER* 275 puncto tantum *transp. R* / partis *om. S* 276 et *om. FP1* / fiat: fiet *L3* 277 situ *om. FP1* / *post* quatuor *scr. et del.* a *P1*; *add.* et *R* / a[3] *om. S* 281 pluribus: qualibet *FP1* / *post* autem *scr. et del.* a *F* 282 *post* reflexio *scr. et del.* ab *F* 283 a[2] *om. O* / *post* numquam *add.* vero *R* 284 ubi[1]: nisi *FP1S* / reflexio *inter. O* / ubi[2]: ibi *R* / due: a duobus *R* 285 ubi[1]: nisi *S* / tres[1]: a tribus *R* / ubi[2]: ibi a *R* / quatuor[2] *om. FP1* / punctus visus: punctum visum *R*

trum visus fuerint in eodem dyametro, fiet reflexio a circulo toto, et locus ymaginis erit centrum visus. Verum, si centrum visus fuerit in centro speculi, nichil videt. Si vero punctus visus fuerit in centro speculi, non videbitur, quoniam forma eius accedet ad speculum super perpendicularem, nec reflecti poterit nisi super perpendicularem.

[2.488] Cum autem centrum visus et punctus visus fuerint in diversis lineis extra centrum, linee ille ad centrum producte secabunt in diversis partibus ex circulo spere duos arcus. Ab uno puncto unius tantum fiet reflexio, ab alio forsitan a tribus. Quod si centrum spere fuerit ex una parte, centrum visus et punctus visus ex una, arcus quem secant dyametri propter oppositionem capitis abscondetur, unde tunc a tribus tantum punctis fiet reflexio. Et si dirigatur in hoc situ visus ad arcum unius reflexionis tantum, abscondetur alius trium, et unica apparebit ymago.

[2.489] Item, si integrum fuerit speculum, non erit ibi perceptio. Oportet igitur ut in eo sit abscisio, et accidet non numquam arcum interiacentem dyametros abscisum esse, et tunc nichil in eo videri, quare raro eveniet quatuor ymagines in hoc speculo comprehendi. Unde si quis hanc pluralitatem ymaginum voluerit videre, disponat visum intra speculum circa ipsum ut modicam partem eius abscondat mole capitis, et totam speculi superficiem visu discurrat.

[2.490] Cum autem aliquid in hoc speculo percipietur duplici visu, si linea reflexionis fuerit equidistans perpendiculari, erit locus ymaginis punctus reflexionis, et cum distent a se puncta reflexionis respectu duorum visuum, apparebunt duobus visibus due ymagines eiusdem puncti. Si vero linea reflexionis non sit equidistans perpendiculari, et punctus visus tantum distet ab uno visu quantum ab alio, vel modica sit differentia, erit locus ymaginis respectu utriusque visus idem, aut diversus, sed modicum distans. Unde aut una apparebit ymago, aut fere una, sicut probatum est in speculis spericis exterioribus.

286 eodem: eadem *R*/toto: tota *FP1* 287 visus[2] *om. FP1SC1E; mg. O; inter. L3* 288 centro[1]: centrum *P1*/punctus visus: punctum visum *R*/*post* punctus *scr. et del.* fuerit *P1*/in: a *O*/*post* centro[2] *scr. et del.* N in eodem diametro fiat reflexio *C1* 289 *post* non *scr. et del.* fi *P1*/videbitur *alter. in* videbit *a. m. E*/accedet: accidet *L3* 290 nec: non *S*/nisi super *om. S* 291 et . . . fuerint *corr. ex* visus fuerint et punctus *S*/punctus visus: punctum visum *R*/visus[2] *om. E*/fuerint: fuerit *O* 292 lineis: locis *E; corr. ex* locis *L3* 294 quatuor: tribus *C1* 295 punctus: punctum *L3ER*/visus: visum *R*/*post* una *add.* parte *C1*/quem: que *FP1* 296 oppositionem: oppositum *P1*/unde: unum *S* 297 tantum punctis *transp. L3* 298 unius *inter. O*/reflexionis *corr. ex* rationis *L3; om. E*/*post* trium *add.* reflexionum *R* 300 item *om. R*/fuerit *om. S*/ibi *om. C1* 1 ut: quod *P1* 2 dyametros: dyametro *FP1S* 5 speculum *corr. ex* circulum *C1*/ipsum *inter. a. m. E* 7 aliquid *om. O*/speculo *om. E*/*post* speculo *inter.* aliquid *O* 8 *post* perpendiculari *scr. et del.* fuerit *F* 9 punctus: punctum *R*/distent: distant *R* 10 duorum: diversorum *C1*/due *inter. L3* 11 *post* puncti *add.* et locus cuiusque imaginis est in puncto sue reflexionis *R*/equidistans: equidem *S*/et *om. FP1* 12 punctus visus: punctum visum *R*/*post* uno *scr. et del.* p *F*/visu: viso *FP1* 14 unde: unum *S*/apparebit: apparet *C1*/aut *corr. ex* ut *F* 15 speculis . . . exterioribus *mg. O*/*post* speculis *scr. et del.* singulis *E*/spericis exterioribus *transp. SO*

[2.491] In speculis columpnaribus concavis, aliquando linea communis est linea recta. Cum superficies reflexionis transit per axem, aliquando linea communis est circulus—cum superficies illa est equidistans basibus—aliquando linea communis est sectio columpnaris. Quando fuerit linea recta, erit locus ymaginis et modus reflexionis sicut in speculis planis. Quando fuerit circulus, erit idem modus qui in concavis spericis. Cum vero fuerit columpnaris sectio, aut erit locus ymaginis ultra speculum, aut citra visum, aut in centro visus, aut inter speculum et visum, aut in ipso speculo, quod sic patebit.

[2.492] **[PROPOSITIO 50]** Sit ABG [FIGURE 5.2.50, p. 608] sectio. Ducatur perpendicularis in hac sectione, que sit DG, quam secundum predicta patet esse dyametrum circuli, et unicam posse esse, cum ab alio puncto sectionis non possit duci perpendicularis super superficiem contingentem. Sumatur aliud punctum, et sit B, et ducatur ab eo in sectione linea perpendicularis super lineam contingentem sectionem in puncto B, que quidem linea, secundum predicta, necessario concurret cum perpendiculari. Concurrat in puncto D, et sumatur B circa punctum G ut angulus BDG sit acutus.

[2.493] Deinde a puncto G ducatur in sectione linea equidistans BD, que sit GH, que quidem cadat intra columpnarem sectionem, quia erit angulus HGD acutus, cum sit equalis GDB. Et a puncto G inter D et H ducatur linea, que necessario concurret cum BD. Concurrat in puncto N, et inter N et G sumatur punctus quicumque, qui sit O. Ultra punctum N sumatur punctum T. Item, a puncto G ducatur supra GH alia linea GZ tamen intra sectionem, que necessario concurret cum BD ex alia parte. Sit concursus E. Ducatur GQ linea ut angulus QGD sit equalis angulo ZGD, et fiat angulus LGD equalis angulo HGD, et angulus MGD equalis angulo NGD.

17 reflexionis: remotionis *E* 18 est[1]: erit *R* 19 equidistans: equidem *FS* / columpnaris: piramidalis *FP1SO* 20 linea recta *transp. FP1* / *post* ymaginis *add.* ultra speculum *C1* 21 circulus: circularis *mg. a. m. L3* 22 concavis spericis *transp. R* / vero: autem *FP1*; ergo *C1* / *post* vero *add.* linea communis *R* / columpnaris: piramidalis *FP1SOE* / erit . . . ymaginis (23): locus ymaginis erit *S* 24 quod: quo *S* 27 esse[1]: etiam *FP1* / unicam: unica *FP1OL3E* 30 lineam: superficiem *FP1* 31 secundum *inter. L3* 32 cum: in *S* / *post* perpendiculari *add.* GD *R* / et *om. L3* / sumatur: sumptum sit *R* 33 ut: et *S* / BDG *corr. ex* BGD *S* 34 in *om. FP1* / equidistans: equidem *S* 35 GH: DH *P1* / cadat: cadit *C1*; cadet *R* / intra: inter *FP1S* / columpnarem: piramidalem *FP1SO*; *corr. ex* piramidem *E* / erit . . . HGD (36): angulus HGD erit *R* 36 *post* angulus *scr. et del.* GG *F* / sit *om. FP1* / GDB: GDH *SE* / *post* H *add.* et *L3* 38 et inter N *om. E* / punctus . . . qui: punctum quodcumque quod *R* / punctum (39) *corr. ex* puncta *S* 39 N . . . T *mg.* (punctum *om.*) *O* / sumatur: supponatur *FP1* / punctum: punctus *FP1* / supra: super *L3* / alia *inter. L3* 41 GQ linea *transp. C1* / equalis *om. L3* 42 angulo[1]: circulus *L3*; *om. ER* / LGD: BGD *S* / et[2]: sed *S*; *om. O* / angulus[2] *corr. ex* angulo *F*

[2.494] Palam quod, si fuerit visus in puncto Z, reflectetur punctus Q ad ipsum a puncto G, et punctus ymaginis E. Et si fuerit visus in puncto H, reflectetur ad ipsum punctus L a puncto G, et erit locus ymaginis G. Si vero fuerit visus in puncto O, reflectetur ad ipsum punctus M, et locus ymaginis N. Si autem fuerit in N, erit locus ymaginis puncti M in centro visus, id est in N. Si autem fuerit in T, erit locus ymaginis inter visum et speculum, quia in N, et ita propositum.

[2.495] Hec quidem intelligenda sunt cum punctus visus non fuerit super perpendicularem cum ipso visu, tunc enim, cum infinite superficies possunt intelligi quarum quelibet ortogonalis super superficiem contingentem speculum, et omnes sint super illam perpendicularem, quedam illarum superficierum efficiet lineam communem lineam rectam, et non fiet reflexio nisi super eandem perpendicularem, et locus ymaginis centrum visus, et non videbitur punctus nisi qui fuerit in superficie visus.

[2.496] Quedam autem illarum superficierum efficit lineam communem circulum, et tunc puncta inter que et visum fuerit centrum circuli poterunt reflecti ad visum singula a duobus punctis circuli, cum a singulis ducantur linee facientes angulum cum superficie contingente quem per equalia dividat perpendicularis ducta ad centrum. Et hoc quidem dico de punctis que sunt in illa perpendiculari, et loca ymaginum erunt in centro circuli. Alia puncta illius perpendicularis non reflectentur ad visum preter punctum quod est in superficie visus, et illud per illam perpendicularem.

[2.497] Cum autem fuerit linea communis sectio columpnaris, non poterunt puncta perpendicularis reflecti ab aliquibus punctis sectionis, cum forma accedens super perpendicularem reflectatur super perpendicularem, et in sectione unica sit perpendicularis, quare per

44 punctus: punctum *R* 45 Q: quasi *FP1* / punctus: punctum *L3R* / *post* ymaginis *add.* est *R* / fuerit visus *transp. R* 46 punctus *om. R* / et . . . G (47) *inter. a. m. L3* / erit locus *corr. ex* locus erit *C1* 47 O *corr. ex* H *S* 48 punctus: punctum *R* / *post* ymaginis *add.* erit *R* / in N: NM *FP1* / *post* N *add.* erit locus ymaginis N si autem fuerit NM *F* 49 id est: scilicet *O* 50 *post* ymaginis *add.* tunc *R* / quia: qua *S* / in N: NM *SP1* / *post* ita *add.* patet *R* 51 *post* quidem *add.* iam dicta *R* / punctus visus: punctum visum *R* 52 visu *corr. ex* visui *F* / enim: et equali *FP1*; et ? *S* / infinite: finite *S* 53 possunt: possint *R* / quarum: quoniam *FP1S* / *post* ortogonalis *add.* sit *R* 54 omnes: omnis *F*; omni *P1* / sint: sit *F*; secent se *ER*; sunt *inter. L3* 55 quedam . . . perpendicularem (56) *mg.* (non: tunc) *O* 56 rectam *inter. L3* / et^1 *om. FP1* / eandem: illam *R* 57 *post* ymaginis *add.* erit *R* / et *inter. O* / punctus: punctum *R* / qui: quod *R* 58 superficie: se *S* 59 *post* autem *scr. et del.* in se visus *S* / efficit: efficiet *R* 60 que: queque *S* 61 poterunt: poterint *C1* / a^2 *inter. O* 62 *post* cum *add.* super *F* 63 quem: que *FP1O* / quem per *mg. a. m. C1* / dividat: dividit *R* / perpendicularis: perpendicularem *E* / hoc: hec *R* 64 quidem *inter. OL3* (*a. m. L3*) / quidem dico *transp. FP1* / dico *om. S*; *corr. ex* ducta *C1* / punctis: predictis *E* 66 reflectentur: reflectuntur *L3* / quod *om. L3* 67 *post* illud *inter.* punctum *L3* / perpendicularem *om. FP1* 68 columpnaris: piramidalis *FP1SOE* 69 ab *inter. O* / *post* aliquibus *add.* aliis *R* 71 et *inter. O* / unica *corr. ex* unicam *L3*; *corr. ex* una *a. m. E*; una *R*

hanc solam perpendicularem fiet reflexio, et solus punctus superficiei visus, et locus ymaginis centrum visus.

[2.498] Si vero fuerit centrum visus in centro circuli, reflectetur portio visus quam secant perpendiculares ducte a centro visus ad circulum a portione simili in circulo quam secant similiter eidem perpendiculares. Cum quelibet linea ducta a centro visus ad circulum sit perpendicularis, fiet reflexio per perpendicularem, et locus ymaginum centrum visus, quod est centrum circuli.

[2.499] Amplius, fiat super punctum A angulus acutus quoque modo, qui sit FAG. Palam quoniam concurret FA cum GZ. Sit concursus in puncto Z, et fiat angulus CAG equalis angulo FAG. Concurret quidem AC cum GQ. Sit concursus in puncto C. Palam quoniam C refertur ad Z a puncto G, et etiam refertur ad Z a puncto A, et non ab alio puncto sectionis, quia non poterit reflecti nisi a termino perpendicularis, et una est in sectione illa perpendicularis, scilicet GA.

[2.500] **[PROPOSITIO 51]** Amplius, sumptis duobus punctis in axe columpne, erit unum reflecti ad aliud ab uno circulo columpne toto, et loca ymaginum erit circulus quidam extra columpnam.

[2.501] Verbi gratia, sit EZ [FIGURE 5.2.51, p. 609] axis, T, H duo puncta sumpta in axe, AG, BD bases columpne. Dividatur TH per equalia in puncto Q, et fiat circulus cuius Q centrum, scilicet LM, qui erit equidistans basibus, eius dyameter LM, latera columpne BLA, DMG. Fiat etiam circulus KC cuius H centrum, CK dyameter, et ducantur linee TL, TM, HL, HM.

[2.502] Palam quoniam quatuor angulorum super Q quilibet est rectus, et TQ equalis QH, et QL equalis QM. Erunt illi trianguli similes, et anguli TLQ, QLH equales; similiter, anguli TMQ, QMH equales. Si

72 perpendicularem *om. FP1* / solus punctus: solum punctum *R* / *post* solus *inter.* videbitur *O* / *post* superficie *scr. et del.* punctus *F* 73 *post* visus *add.* videbitur *FP1R* / *post* ymaginis *add.* erit *R* 74 centrum *om. R* / *post* circuli *scr. et del.* et *L3* 75 quam: quem *C1* 76 portione: proportione *P1* / in *om. R* / quam: quem *SL3C1E* / eidem *corr. ex* eisdem *S*; *corr. ex* idem *C1*; eedem *ER* / perpendiculares: perpendiculari *FP1* / *post* perpendiculares *add.* quia *R* 77 *post* sit *scr. et del.* o *P1* 78 *post* fiet *add.* f *E* / reflexio: ratio *F* / per *om. FP1OE*; *inter. L3*; super *R* / perpendicularem: perpendicularis *P1* / ymaginum: ymaginis *L3ER* / *ante* centrum *add.* erit *R* 80 fiat . . . A: super punctum A fiat *ER* / *post* angulus *scr. et del.* a *P1* / quoque: quocumque *O*; quoquo *L3C1R* 81 quoniam: quod *R* 82 Z *corr. ex* S *L3* 83 quidem: equidem *FP1SOR*; *alter. in* equidem *L3*; equidistans *E* / quoniam: quod *R* 84 refertur[1,2]: reflectetur *R* / etiam: ita *R* / refertur[2]: reflectatur *L3E* / ad[2] . . . A: a puncto A ad Z *R* 85 a termino *om. FP1S* 86 *post* una *add.* sola *OL3* / scilicet *om. FP1* 88 columpne[1]: NE CO *E* / erit: poterit *R* / aliud: alium *FP1* / columpne[2] *corr. ex* columpnari *L3*; columpnali *E* / toto: tote *S* 89 loca ymaginum: locus ymaginis *R* / circulus quidam *transp. FP1* / columpnam *corr. ex* columplam *S* 90 duo *om. R* 91 columpne: columpnale *P1*; *om. L3ER* / dividatur *om. FP1* / equalia (92): qualia *F* 92 scilicet LM: eius diameter *R* / LM: HN *S* / qui erit *om. L3* 93 equidistans: equidem *F* / eius . . . LM *om. R* 94 fiat *corr. ex* fit *L3* / KC: KP *R* / cuius *corr. ex* minus *L3* / CK: PK *R* 96 quoniam: quod *R* 97 QL: QH *FP1* / illi . . . similes: illa triangula similia *R* / similes: similis *FP1* 98 anguli[1]: angulus *L3* / TLQ: DQ *FP1*; *corr. ex* LQ *L3*

igitur fuerit H centrum visus, reflectetur punctus T ad punctum H a puncto L, et similiter a puncto M. Si igitur moveatur triangulus TLH, immoto axe TH, describet punctus L circulum, et semper duo anguli TLQ, QLH manebunt equales, et semper in hoc motu reflectetur T ad H.

[2.503] Producatur autem linea CHK donec concurrat cum linea TL, et sit concursus F. Palam quoniam F erit locus ymaginis, et motu trianguli TLH, movebitur triangulus TFH, et hoc motu punctus F describet circulum extra columpnam. Et totus ille circulus erit locus ymaginum, et hoc est propositum. Idem erit probandi modus, sumptis quibuslibet duobus in axe punctis.

[2.504] **[PROPOSITIO 52]** Amplius, punctorum extra perpendicularem visus sumptorum, quedam unicam habent ymaginem, quedam duas, quedam tres, quedam quatuor, non plures.

[2.505] Verbi gratia, sit A [FIGURE 5.2.52, p. 610] punctus visus extra perpendicularem visus, et fiat superficies transiens per A equidistans basibus speculi. Faciet quidem circulum in columpna. Sit centrum illius circuli H, et sumatur in superficie circuli aliud punctum, quod sit B, et ducantur dyametri AH, BH.

[2.506] Palam ex eis que dicta sunt in speculis spericis concavis quod ab uno puncto arcus quem intercipiunt hii duo dyametri potest A reflecti ad B forsitan a duobus, aut tribus, sed non a pluribus; ab arcu autem opposito non nisi ab uno puncto. Sit igitur quod A refertur ad B a tribus punctis arcus intercisi, et sint puncta illa G, D, E, et ducantur linee AG, HG, BG, AD, HD, BD, AE, HE, BE.

[2.507] Et a puncto A ducantur in eadem superficie tres linee equidistantes tribus dyametris HG, HD, HE, que sunt AK, AF, AN. Cum igitur AK sit equidistans HG, concurret BG cum AK. Concurrat in punc-

99 fuerit *inter. L3* / H^{1,2}: T *R* / punctus T: quidem H *R* 100 triangulus: triangulum *R* 101 immoto: immote *S* / punctus: punctum *L3R* / *post* L *scr. et del.* centrum *E* / semper: super *L3* 102 QLH *mg. F* / T: H *R* 103 H: T *R* 104 CHK: PHK *R* 105 palam . . . F (106) *mg. a. m. C1* / quoniam: quod *R* / et^{2} *corr. ex* in *L3* 106 movebitur: movevebitur *F* / triangulus: triangulum *R* / punctus: punctum *R* 107 columpnam: columpna *S* 108 ymaginum: imaginis *R* / et *om. C1* / hoc: quod *C1* / est *inter. L3* 109 quibuslibet *corr. ex* quibusduobuslibet *F*; quibus *P1* / *ante* duobus *add.* dyameter *P1* / duobus *om. F* / *post* duobus *add.* licet *P1* / in axe punctis: punctis in axe *R* 111 *post* visus *scr. et del.* et fiat superficies transiens per A *C1* 112 *post* quatuor *add.* et *R* 113 punctus visus: punctum visum *R* 114 et *om. O* / equidistans: equidem *F*; *corr. ex* equidem *S* 116 *post* illius *scr. et del.* cum *C1* / et . . . B (117) *inter. a. m. L3* / *post* superficie *scr. et del.* speculi *P1* / aliud: alium *F* 118 eis: hiis *FP1* / quod: quoniam *L3E* 119 puncto *om. FP1SO*; *inter. L3* / arcus: arcu *O* / quem: quam *P1* / hii duo: he due *R* 120 forsitan: forsan *SOC1* / *post* duobus *add.* punctis *R* / *post* aut *add.* a *SOC1* / sed *om. FP1* / autem: aut *P1*; *om. O* 121 non *corr. ex* nec *L3* / ab: a *L3E* / uno puncto *transp. L3E* / A refertur *corr. ex* refertur A *S* / refertur: reflectatur *R* 122 arcus intercisi *transp. R* / et^{2} *om. O* 123 AD *om. E* / AD HD BD: HD BD AD *L3R* (AD *inter. L3*) / BD: DB *C1E* 124 A *inter. a. m. L3* / superficie tres *corr. ex* superficies est *L3* / equidistantes (125): equidem *FL3* 125 sunt: sint *OR* / AK *corr. ex* AE *a. m. E* 126 AK^{2} *corr. ex* AE *a. m. E* / concurrat *corr. ex* concurret *a. m. C1*; concurret *R*

to K. Similiter BD concurret cum AF. Sit concursus in puncto F. Similiter BE cum AN. Sit concursus in puncto N.

[2.508] Deinde a puncto H erigatur axis, que sit HU, et a puncto B perpendicularis super superficiem circuli. Erit quidem equidistans axi, que sit BT. Et sumatur in ea punctum quodcumque, quod sit T, et ducantur tres linee TK, TF, TN, et a tribus punctis G, D, E erigantur tres perpendiculares super superficiem circuli GM, DL, EQ. Erunt quidem equidistantes TB. EQ igitur erit in superficie trianguli TBN. Igitur EQ secabit TN. Secet in puncto Q. DL secet TF in puncto L; GM secet TK in puncto M. Et erunt hee tres perpendiculares linee longitudinis columpne.

[2.509] A puncto Q ducatur equidistans linee NA, que quidem concurret cum axe UH, quoniam erit equidistans EH. Sit concursus in puncto U, et ducatur linea TA, quam secabit QU, quoniam QU ducitur a latere trianguli equidistanter basi. Sit punctus sectionis I, et ducatur linea QA.

[2.510] Palam quoniam angulus BEH equalis est angulo ENA, et angulus HEA equalis angulo EAN, et angulus BEH equalis angulo HEA. Erit angulus EAN equalis angulo ENA, quare EN equalis EA.

[2.511] Et EQ perpendicularis. Erit triangulus QEA equalis triangulo QEN; erit QN equalis QA, et erit angulus QNA equalis angulo QAN. Sed angulus TQI equalis angulo QNA, et angulus IQA equalis angulo QAN. Erit angulus IQT equalis angulo IQA, quare A refertur ad T a puncto columpne quod est Q.

[2.512] Eodem modo probabitur quod refertur A ad T a punctis L, M, et ita a tribus punctis columpne ex eadem parte.

[2.513] Nec potest a pluribus, detur enim aliud. Ducto latere ab illo puncto, cadet in circulum quem habemus, et probabitur quod a puncto casus qui est in circulo poterit reflecti A ad T, replicata probatione, quod est impossibile.

127 K *corr. ex* E *a. m. E* / BD: KD *S* / concurret: concurrat *L3*   129 a[1] *corr. ex* in *L3* / que: qui *R* / HU: HX *R*   130 *post* circuli *add.* que *R* / quidem *om. L3ER* / equidistans: equidem *F* / axi *corr. ex* axis *L3*   132 TN: ON *S*   134 equidistantes: equidem *F* / igitur erit *transp.* (igitur *inter.*) *O* / erit: erunt *SC1R*   135 DL: D vel O *FP1* / L: LG *FP1* / GM: TM *E* / secet[3] *om. FP1SOC1E; inter. L3*   136 et *om. FP1*   138 ducatur: ducantur *FP1; corr. ex* ducantur *L3* / equidistans: equidem *F*   139 UH: XH *R* / EH: EB *C1* / sit: sint *FP1*   140 TA *corr. ex* TH *a. m. E* / quam: quoniam *S* / ducitur: ducatur *P1*   141 *post* trianguli *add.* et linea EQ *R* / equidistanter: equidistans *E;* equidistante *R* / punctus: punctum *R*   143 quoniam: quod *R* / est *om. L3* / ENA *corr. ex* ENAF *L3*   144 HEA[1]: BEA *FP1*   145 EA *corr. ex* EN *C1*   146 triangulus: triangulum *R* / equalis: equale *R*   147 QEN *corr. ex* QAN *a. m. E* / *post* QEN *add.* et *OR* (*inter. O*) / *post* equalis[1] *scr. et del.* ? *O* / QA *corr. ex* QM *L3* / et *inter. O* / angulus *om. R* / QAN . . . angulo (149) *om. S* / *post* QAN *inter.* et *L3*   148 sed: erit *L3* / TQI *corr. ex* IQT *L3* / QNA . . . angulo (149) *mg. a. m. L3*   149 refertur: reflectetur *R*   151 refertur: reflectetur *R* / A *om. O*   152 punctis *om. S; inter. O*   153 aliud: alium *F*   154 in *om. FP1; inter. L3* / probabitur: probatur *SOC1*   155 qui: que *S* / A ad T: AD DT *FP1* / replicata: repetita *R*   156 est *om. SO*

[2.514] Ex arcu opposito circuli poterit reflecti A ad B ab uno puncto. Sit illud Z, et ducatur dyameter HZ, et ei equidistans AC. Et ducatur BZ, que concurrat cum AC in puncto C. Et erigatur perpendicularis OZ, que erit latus et equidistans TB, et ducatur TC, que secabitur a linea OZ. Sit sectio in puncto O. Probabitur modo predicto quod A refertur ad T a puncto O. Et si sumatur ex illa parte alius punctus columpne a quo possit reflecti, per replicationem probationis probabitur quod ab alio puncto circuli quam Z potest reflecti ex parte illa, quod est impossibile.

[2.515] Si ergo A ab uno puncto circuli refertur ad B ex aliqua parte, refertur ab uno columpne ex eadem; si a duobus, a duobus; si a tribus, a tribus; nec potest amplius; ab opposita parte ab uno circuli tantum, et ab uno columpne tantum.

[2.516] Item, TB equidistans UH, nec potest sumi superficies equalis in qua sit T cum UH preter superficiem TBUH. Similiter, non potest sumi superficies in qua sit A cum UH preter superficiem AUH, que est perpendicularis. T igitur non est in eadem superficie perpendiculari cum A, nec in eodem circulo, nec est in axe, quia est in linea ei equidistante. Superficies igitur in qua A refertur ad T est sectio columpnaris.

[2.517] Verum, producta TA ultra T et A ex utraque parte, et sit RP. Cum quatuor sint superficies reflexionis, quia a quatuor punctis, et in qualibet sint duo puncta T, A, erit RP communis quatuor superficiebus reflexionis. Et quelibet harum superficierum secat superficiem contingentem speculum in puncto super suam lineam communem, non super eandem. Linea RP perpendicularis est super unam linearum

157 *post* arcu *add.* vero *R* / opposito *alter. in* oppositio *L3* / opposito circuli *transp. R* / T: H *F*; *alter. ex* Q *in* B *P1*; *alter. ex* C *in* B *L3*; B *C1ER* 158 illud *om. L3* / Z *corr. ex* S *L3* / dyameter *om. P1* / HZ: H *FP1*; *corr. ex* HS *L3* / equidistans: equidem *F*; *corr. ex* equidem *S* / AC: AS *R* 159 BZ: Z *FP1*; *corr. ex* BS *L3* / AC: AO *E*; AS *R* / C: O *O*; S *R* / *post* perpendicularis *add.* que sit *R* 160 OZ *mg. O*; *corr. ex* OI *L3* / equidistans: equidem *F* / quidistans *C1* / TB *corr. ex* TH *E* / et[2] *om. L3* / TC: TS *R* 161 OZ *corr. ex* OS *L3* / sit *om. FP1* / quod: quia *FP1* / A *inter. OL3* 162 refertur: reflectetur *R* / si *inter. L3*; *om. E* / alius *corr. ex* illius *L3* / alius punctus *transp. E*; punctum aliud *R* 163 *post* probabitur *add.* sic *P1* 164 *post* quod[1] *scr. et del.* ab *E* / Z *corr. ex* S *L3* / reflecti *mg. O* 166 puncto *om. SE*; *inter. OL3* / refertur: reflectitur *R* / B: T *SOC1E*; *corr. ex* T *L3* / aliqua: alia *FP1S* 167 refertur: reflectitur *S*; reflectetur *R* / *post* eadem *add.* ad T *R* 168 a tribus *rep. S* / nec: non *L3* / *post* amplius *add.* ab illa parte *R* / *post* opposita *add.* vero *R* / *post* parte *add.* nisi *L3E* (*inter. L3*); *add.* non nisi *R* / *post* uno *add.* nisi puncto (*inter.* puncto) *O*; *add.* puncto *R* / tantum *om. C1* 170 equidistans: equidistat *C1ER* / equalis *om. SO* 171 T: A *S* / *post* UH *scr. et del.* AI *L3* / TBUH ... superficiem *om. S*; *mg. O* 172 sumi superficies *L3ER* / A: E *FP1* 173 T *corr. ex* circuli *a. m. L3* / T igitur *transp. R* 174 equidistante (175) *corr. ex* quidem *L3* 175 refertur: reflectitur *R* / columpnaris *om. SOC1E*; *inter. L3* 176 *post* producta *add.* sit *R* / ex *corr. ex* et *L3* 177 reflexionis *inter. L3*; *om. E* / *post* quia *scr. et del.* a *L3* / a *rep. C1E* / *post* punctis *add.* fit reflexio *R* 178 *post* qualibet *add.* horum *R* / sint: sunt *E* / puncta TA *transp. L3* / T A *inter. OL3* / erit *mg. L3* 179 et *om. O* / contingentem (180): continentem *F* 180 *post* puncto *add.* sue reflexionis *R* 181 *post* eandem *add.* sed diversas *E* / linea *corr. ex* lineam *C1* / *post* linea *add.* ergo *R* / est *inter. O*

quatuor communium, non super duas, esset enim perpendicularis super superficiem contingentem, et ita perveniret ad axem. Sunt igitur diverse perpendiculares a puncto T ad has quatuor lineas communes, nec est nisi una tantum que transeat per A.

[2.518] Et perpendicularis aut erit equidistans linee reflexionis, aut concurret cum ea ultra speculum, vel intra. Si fuerit equidistans, erit locus ymaginis punctus reflexionis, ut probatum est, et cum quatuor sint reflexionis puncta, erunt quatuor ymagines. Si concurrit, cum quatuor sint perpendiculares, quatuor erunt concursus, et quatuor ymagines.

[2.519] Amplius, datis puncto viso et puncto visus, erit invenire punctum reflexionis. Verbi gratia, sit A punctus visus. Fiat superficies secans columpnam equidistans basi transiens per A, et faciet circulum. B aut est in superficie huius circuli, aut non. Si fuerit, inveniemus punctum reflexionis in illo circulo sicut dictum est in sperico concavo. Si non fuerit, ducatur a puncto B perpendicularis super superficiem huius circuli, et replicetur supradicta probatio, et invenietur punctus reflexionis. Duplici autem visu adhibito, una ymago in veritate efficientur due, sed contigue vel admixte, unde videbuntur una.

[2.520] In speculis piramidalibus concavis, linea communis superficiei reflexionis et superficiei speculi aut erit linea longitudinis speculi, aut erit sectio piramidalis. Si fuerit linea longitudinis, erunt loca ymaginum in ipso speculo. Si fuerit sectio piramidalis, erunt loca ymaginum aliquando citra visum, aliquando in visu, aliquando inter visum et speculum, aliquando ultra speculum, sicut ostensum est in speculo columpnari concavo.

[2.521] Amplius, si in perpendiculari ducta a centro visus ad superficiem contingentem piramidem sumatur punctus corporeus inter visum et speculum, non reflectetur forma eius ad visum per perpendicu-

182 enim *corr. ex* IBN *L3* / *post* enim *add.* esset *E* 185 est *corr. ex* esse *mg. F* / nisi *inter. a. m. E* / *post* nisi *add.* perpendicularis *ER* / transeat: transit *L3ER* 186 et: sed *FP1* / erit *inter. a. m. E*; est *R* 187 concurret: concurrit *R* 188 punctus: punctum *R* / et *om. L3* 189 sint: fuerint *FP1* / concurrit: concurrant *L3*; concurrerint *C1*; concurrunt *E*; *corr. ex* concurrat *O* 190 sint: sunt *R* / quatuor . . . concursus: erunt concursus quatuor *R* 193 sit *corr. ex* si *O* / punctus visus: punctum visum *R* / *ante* fiat *inter.* et *O*; *add.* B centrum visus *R* 194 equidistans: equidistanter *R* / et *inter. O* 195 inveniemus; invenimus *L3* / punctum reflexionis (196) *om. FP1* 196 reflexionis: rationis *S* 197 super *om. S* 198 punctus: punctum *R* / reflexionis (199) *om. P1* 199 in . . . due (200): efficietur due in veritate *O* / efficientur; efficietur *FP1* 200 videbuntur: videbitur *R* / *post* una *inter.* ymago *a. m. C1* 201 piramidalibus *corr. ex* pluribus *O* 202 reflexionis et superficiei *mg. a. m. L3* / superficiei: superficie *P1* / aut[1] . . . speculi *transp. post* piramidalis (203) *mg. O* 203 si *mg. O* 204 in . . . ymaginum *mg. a. m. L3* 206 *ante* aliquando *add.* et *L3ER* / speculum *om. O* / columpnari (207): columpnali *E* 209 piramidem: piramidalem *FP1* / punctus corporeus: punctum corporeum *R* / corporeus *inter. a. m. E* 210 per *om. FP1S*

larem, quoniam punctus ille occultabit terminum perpendicularis, et ob hoc non reflectetur ab eo. Si autem nullus fuerit punctus in perpendiculari illa, reflectetur quidem ad visum per hanc perpendicularem punctum visus, quod iterum secat perpendicularis ex eo, et ille solus.

[2.522] Verum, visu existente in hac perpendiculari et in axe, efficietur circulus ad cuius quodlibet punctum linea ducta a visu erit perpendicularis super superficiem contingentem, unde a quolibet puncto illius circuli fieri poterit reflexio ad visum per perpendicularem. Et fiet reflexio partis visus quam secant due perpendiculares maiorem angulum in eo continentes.

[2.523] Si vero inter visum et speculum fuerit axis, non fiet ad ipsum reflexio per perpendicularem nisi puncti eius quem secat perpendicularis.

[2.524] **[PROPOSITIO 53]** Amplius, existente visu et puncto viso in axe, erit reflecti unum ad aliud.

[2.525] Verbi gratia, sit H [FIGURE 5.2.53, p. 611] centrum visus, T punctus visus. Fiat superficies secans piramidem transiens super axis longitudinem, que sit ABGH, AH axis, AB, AG latera piramidis. A puncto T ducatur perpendicularis super lineam AB, que sit TQ, et producatur usque QL. Sit equalis QT. Et a puncto H ducatur linea ad punctum L, que secabit lineam longitudinis que est AB. Secet in puncto B, et a puncto B ducatur equidistans linee TQ, que necessario perveniet ad axem. Perveniat in puncto D, et ducatur linea TB.

[2.526] Palam, cum TQ sit perpendicularis super AB, et TQ equalis QL, erit triangulus BTQ equalis triangulo BQL, et erit angulus QLB equalis angulo QTB. Sed angulus QTB equalis est angulo TBD, et angulus DBH equalis est angulo QLB. Igitur angulus TBD equalis est angulo DBH, et ita T refertur ad H a puncto B, et locus ymaginis L.

211 punctus ille: punctum illud *R* / *post* terminum *add.* illius *E* / *post* perpendicularis *add.* illius *R* 212 ob: ab *SO* / ob hoc *corr. ex* ad hoc *mg. a. m. C1* / *post* hoc *scr. et del.* re *S* / reflectetur: reflectitur *L3* / nullus: nullum *R* / punctus: punctum *R* 214 punctum *alter. in* puncti *L3* / *post* visus *add.* punctum *OL3C1* (*inter. L3; inter. a. m. O*) / iterum *om. SOR* / ille solus; illud solum *R* 215 existente *corr. ex* exigente *a. m. E* 216 cuius quodlibet *corr. ex* cuiuslibet *L3* / ducta *corr. ex* producta *F* / ducta a visu: a visu ducta *L3* 217 illius circuli (218) *transp. C1* 218 *post* reflexio *add.* a *C1* / per *om. FP1E; inter. SL3;* secundum *R* / perpendicularem: perpendiculares *R* 219 due perpendiculares *transp. L3ER* 220 continentes: contingentes *S* 221 *post* ipsum *add.* visum *L3E* (*inter. a. m. L3*) 222 per *om. FP1; inter. SL3;* propter *O* / quem: quod *R* / secat: secant *L3ER* / perpendicularis (223): perpendiculares *L3ER* 225 erit: poterit *R* / aliud: alium *FP1* 227 punctus visus: punctum visum *R* 228 longitudinem *corr. ex* superficiem *L3* / AB AG *transp. FP1* 230 usque: ut *O;* quousque *R* 231 *post* AB *scr. et del.* in puncto *P1* 232 et . . . B *mg. O* 233 TB: DB *FP1* 235 triangulus . . . equalis: BTQ triangulum equale *R* / BQL *corr. ex* QL *L3* 236 est *om. P1* / TBD: QBD *O* 238 DBH *corr. ex* QLB *P1* / refertur: reflectitur *R* / *post* ymaginis *add.* est *R*

[2.527] Igitur, moto triangulo TLH, describet punctus B circulum in piramide, et a quolibet puncto illius circuli reflectetur T ad H. L vero extra circulum describet circulum qui totus erit locus ymaginis puncti T.

[2.528] **[PROPOSITIO 54]** Amplius, sumptis duobus punctis extra perpendicularem visus et extra axem in hoc speculo, scilicet Z,E [FIGURE 5.2.54, p. 612], fiat superficies equidistans basi super Z. Faciet circulum in speculo. E aut erit in hoc circulo, aut in alia superficie.

[2.529] Sit in superficie illius circuli, et ducatur linea EZ. Palam quoniam Z refertur ad E a circulo illo ex una parte aut ab uno puncto, aut a duobus, aut a tribus; ex alia ab uno.

[2.530] Sumatur punctus circuli a quo refertur ad ipsum, et sit H, centrum circuli T. Et ducantur linee ZH, EH. Et dyameter TH dividet quidem angulum illum per equalia, et secabit lineam EZ. Secet in puncto Q, et sit A conus piramidis, AH linea longitudinis.

[2.531] A puncto Q ducatur linea cadens perpendiculariter super lineam AH, que sit QM, que quidem perveniat ad axem, qui est AT. Cadat in ipsum in puncto D, et ducantur linee ZM, EM. A puncto Z ducatur in superficie circuli linea equidistans linee QH, que sit ZL. Concurrat quidem EH cum illa. Sit concursus in puncto L, et a puncto H ducatur perpendicularis super LZ, que sit HC.

[2.532] Deinde in superficie trianguli EMZ ducatur linea equidistans linee QM, que sit ZO. Concurrat EM cum ea in puncto O, et ducatur linea LO. Et a puncto C ducatur equidistans LO, que sit CN, et ducatur linea NM.

[2.533] Palam quoniam angulus EHQ equalis est angulo QHZ et angulo HLZ, et angulus QHZ equalis est angulo HZL. Erit HL equalis

239 triangulo *corr. ex* angulo *C1* / describet *corr. ex* describetur *C1* / punctus: punctum *R* 240 L *om. FP1* 241 circulum¹ *scr. et del. L3*; speculum *R* / circulum² *om. O* / qui *rep. E* / locus . . . puncti *corr. ex* puncti ymaginis locus *E* 243 *post* punctis *add.* et *ER* 244 Z *corr. ex* S *L3* 245 equidistans *corr. ex* equidem *F* / Z *corr. ex* S *L3* 246 E *inter. O* / alia: illa *P1* / *post* superficie *add.* ipsi equidistante *R* 247 EZ *corr. ex* ES *L3* 248 quoniam: quod *R* / Z *corr. ex* S *L3* / refertur: reflectitur *E*; reflectetur *R* / E: A *SOE*; *corr. ex* A *L3* / parte *om. O* 249 *post* alia *add.* vero *R* 250 *post* sumatur *add.* igitur *R* / punctus: punctum *R* / refertur: reflectitur *R* 251 ZH *corr. ex* SH *L3* / EH: EB *S* / dyameter: dyametrum *FP1* 252 quidem: equidem *FP1* / lineam: linea *E* / EZ *corr. ex* ES *L3* 253 et *inter. O* / conus: vertex *R* 254 cadens *om. R* / perpendiculariter: perpendicularis *R* 255 perveniat *alter. in* perveniet *O*; perveniet *R* / AT: AD *R* / AT cadat (256): et cadat *inter. L3* / *post* AT *add.* et *ER*; *scr. et del.* et *C1* 256 in puncto *inter. O* / ZM *corr. ex* SM *L3* / ZM . . . linee (257) *om. FP1* 257 ZL *corr. ex* SL *L3* 258 concurrat: concurret *ER* 259 H *om. P1* / LZ *corr. ex* LS *L3* / HC *corr. ex* HZ *E*; HP *R* 260 trianguli *corr. ex* circuli *E*; *om. R* / EMZ *corr. ex* EMS *L3* 261 linee QM *corr. ex* QM linee *C1* / ZO *corr. ex* SO *L3*; *corr. ex* TZO *C1* / *post* ZO *add.* et *R* / *post* EM *scr. et del.* ducatur linea *S* / *post* et *scr. et del.* L *F* / ducatur *corr. ex* duducatur *S* 262 linea . . . ducatur² *om. O* / et *om. S*; *inter. L3* / C: P *R* / CN: PN *R* 264 quoniam: quod *R* / EHQ: EQH *C1* / QHZ *corr. ex* QHS *L3* 265 HLZ *corr. ex* HLS *L3* / QHZ *corr. ex* QHS *L3* / *post* angulo² *add.* coalterno *ER* / HZL *alter. ex* HLS *in* HLZ *L3* / erit . . . HZ (266) *om. P1* / *post* erit *add.* igitur *R*

HZ, et HC perpendicularis super LZ. Erit triangulus LCH equalis triangulo CHZ, et erit LC equalis CZ.

[2.534] Et CN equidistans OL; erit proportio LC ad CZ sicut ON ad NZ, quare ON equalis NZ. Item, cum OZ sit equidistans QM, erit superficies ZLO equidistans superficiei QMH. Et superficies EOL secat illas duas super lineas communes que quidem erunt equidistantes, scilicet MH, OL, quare HM, CN equidistantes. Et quoniam HC cadit inter LZ, HQ equidistantes, et est perpendicularis super LZ, erit perpendicularis super HQ, quare CH erit contingens circulo.

[2.535] Igitur superficies AHC est superficies contingens piramidem. In hac superficie est CN et NM, et super hanc superficiem est perpendicularis linea DM. Igitur perpendicularis est super lineam NM, quare NM perpendicularis super OZ, et ON equalis NZ. Erit MO equalis MZ, et proportio EM ad MO sicut EM ad MZ.

[2.536] Sed EM ad MO sicut EH ad HL, et EH ad HL sicut EH ad HZ, et EH ad HZ sicut EQ ad QZ. Igitur EM ad MZ sicut EQ ad QZ, quare angulus EMQ equalis angulo QMZ, quare Z refertur ad E a puncto M. Si ergo Z refertur ad E a puncto circuli H, refertur ad ipsum a puncto piramidis M. Et si a duobus circuli, a duobus piramidis; si a tribus, a tribus; si a pluribus, a pluribus. Eodem modo ex alia parte circuli fiet probatio quod ab uno piramidis sicut ab uno circuli.

[2.537] Si vero E non fuerit in circulo equidistante basi transeunte super Z, erit E [FIGURE 5.2.54a, p. 613] quidem supra aut infra. Sit

266 HZ *corr. ex* HS *L3*/HC: HP *R*/*post* perpendicularis *add.* est *R*/super: SF *E*/LZ *corr. ex* LS *L3*/triangulus . . . equalis: triangulum LPH equale *R*   267 triangulo angulo *O*/CHZ: CHT *P1; alter. ex* THS *in* THZ *L3*; CHE *E*; PHZ *R*/LC: LP *R*/CZ *corr. ex* CS *L3*; PZ *R*   268 CN: PN *R*/*post* equidistans *add.* est *R*/LC: LP *R*/CZ *corr. ex* CS *L3*; PZ *R*   269 NZ[1,2] *corr. ex* NS *L3*/OZ *corr. ex* OS *L3*; *corr. ex* CZ *a. m. E*/equidistans *corr. ex* E *O*/*post* QM *add.* et HQ equidistans LZ *R*   270 ZLO *corr. ex* SLO *L3*; ZOL *R*/equidistans: equidem *S*   271 super *om. FP1*/lineas *corr. ex* illas *S*/scilicet (272) *om. C1*   272 OL: LO *R*/CN: N *S*; PN *R*/*ante* equidistantes *add.* sunt *R*/et *scr. et del. L3*/quoniam: quare *L3*/HC *corr. ex* HI *L3*; HP *R*   273 LZ[1]: Z *F*; ZH *P1*/LZ[1,2] *corr. ex* LS *L3*/equidistantes: EQ *C1*/erit . . . HQ (274) *om. R*   274 CH . . . circulo: PH continget circulum *R*   275 igitur: quare *ER*/AHC: AHP *R*/superficies[2] *corr. ex* superficiei *L3*/contingens: contingentes *F*   276 CN: ON *SC1*; PN *R*/et[1] *inter. C1*   277 *post* perpendicularis[1] *scr. et del.* est super *F*; *scr. et del.* super OE et ON equalis NZ *C1*/linea *corr. ex* lineam *F*/*post* igitur *rep.* perpendicularis[1] . . . igitur *S*   278 *post* NM *add.* est *R*/super OZ et *mg. O*/OZ: OE *SL3C1E*/NZ *corr. ex* MS *L3*/erit . . . MZ[1] (279) *inter. a. m. L3*   279 proportio *om. R*/EM *alter. ex* EZN *in* ZN *F*; EZN *P1*/MZ[2] *corr. ex* MS *L3*   280 et . . . HL[2] *inter. a. m. L3*/EH[3]: HL *SC1E*; HF *O*; HE *L3*   281 HZ[1] *corr. ex* HS *L3*/et . . . HZ[2] *mg. O*/HZ[2]: ZH *L3*/QZ[1] *corr. ex* QHZ *F*; *corr. ex* QS *L3*/igitur EM *rep. P1*/MZ *corr. ex* MS *L3*/QZ[2] *corr. ex* Z *mg. F*   282 *post* EMQ *add.* est *L3*/*post* equalis *inter.* est *O*/QMZ *corr. ex* QMS *L3*/Z *corr. ex* S *L3*/refertur: reflectitur *R*   283 M: E *S*; *corr. ex* E *L3*/M . . . puncto[1] *mg. O*/refertur[1]: reflectitur *R*/refertur[2]: reflectetur *R*   284 piramidis[1]: piramidalis *S*/a[1] *inter. O*/*post* piramidis[2] *add.* et *FP1*   286 probatio *corr. ex* proportio *L3*/*post* ab *add.* u *S*/*post* uno *add.* puncto *C1R*/piramidis: piramide *FP1*/*post* circuli *add.* reflexio fiat *R*   287 E *corr. ex* est *F*; est *P1*/equidistante: equidistans *F*; equidistanti *E*   288 Z *corr. ex* S *L3*/*post* Z *scr. et del.* equi *F*/E *om. FP1SER*/aut: vel *L3ER*/aut infra *corr. ex* infra aut *S*/sit *corr. ex* si *a. m. L3*; si *E*

supra, quia utrobique eadem est probatio. Ducatur linea AE donec contingat superficiem illius circuli, et sit punctus contactus H, Q centrum circuli. Palam quoniam H potest reflecti ad Z ab aliquo puncto circuli. Sit illud T, et ducatur dyameter QT. Et linea HZ secabit hunc dyametrum in puncto quod sit N. Et ducatur EZ et linea longitudinis AT.

[2.538] Palam, cum punctus Z sit ex una parte dyametri QT, ex alia E, linea EZ secabit superficiem AQT. Secet in puncto O, et a puncto O ducatur perpendicularis super lineam AT, que sit OC, que necessario cadet super axem. Cadat in puncto D, et ducantur linee EC, ZC. Dico quoniam E refertur ad Z a puncto C.

[2.539] Probatio: ducatur a puncto Z linea equidistans QT, que sit ZF, et producatur linea HT donec concurrat cum illa. Sit concursus in puncto F. Similiter, a puncto Z ducatur equidistans linee OC, que sit ZK, et producatur linea EC donec concurrat cum illa. Sit concursus in puncto K.

[2.540] Palam, cum linea ZF sit equidistans QT, et ZK equidistans OC, erit superficies ZKF equidistans superficiei OCT que est superficies AQT. Et superficies HFK secat has duas superficies super lineas CT, KF. Igitur CT, KF sunt equidistantes.

[2.541] Ducatur a puncto T perpendicularis super lineam ZF, que sit TP. Palam, cum cadat inter duas equidistantes, erit equidistans linee NZ, et ita erit contingens circulo. Igitur superficies ATP contingit piramidem super lineam AT, et linea OC est perpendicularis super hanc superficiem. Superficies igitur ATQ erit ortogonalis super superficiem ATP, et superficies ATP secat duas superficies ATQ, ZKF, que sunt equidistantes. Igitur linee communes sectionum sunt equidistantes, una harum linearum est CT, alia sit PI. Sed iam patet quod CT est equidistans KF. Igitur PI est equidistans KF.

289 est *om. L3*/probatio *corr. ex* proportio *L3*/AE: ab E *FP1*; *corr. ex* A *L3*; a puncto E *E*; a vertice A per punctum E *R*/*ante* donec *scr. et del.* puncto *C1*/contingat (290): secet *R* 290 *ante* superficiem *scr. et del.* super *C1*/punctus: PG *S*; punctum *R*/contactus: sectionis *R* 291 quoniam: quod *R*/Z *corr. ex* S *L3*/ab *corr. ex* et *L3* 292 illud: illum *FP1*/dyameter *corr. ex* dyatmeter *F*/HZ *corr. ex* HS *L3*/hunc: hanc *R* 293 et[1] *om. S*/EZ *corr. ex* ES *L3* 294 *ante* palam *scr. et del.* et *E*/punctus: punctum *R*/Z *corr. ex* S *L3*/*post* QT *add.* et *P1ER* 295 E *om. E*/EZ *corr. ex* ES *L3*/secet *rep. S* 296 OC: OP *R*/necessario cadet (297) *transp.* (cadet *alter. ex* sit *in* cadat) *L3* 297 EC: EP *R*/ZC: SC *L3*; ZP *R* 298 quoniam: quod *C1R*/E: Z *R*/refertur: reflectetur *R*/Z *corr. ex* S *L3*; E *R*/C *corr. ex* Z *S*; P *R* 299 probatio: proportio *L3*; *om. R*/Z *corr. ex* S *L3* 300 ZF *corr. ex* SF *L3*/et *inter. a. m. C1*/concurrat *corr. ex* conconcurrat *L3* 1 Z *corr. ex* S *L3*/OC *corr. ex* AC *E*; OP *R* 2 ZK *corr. ex* ZE *E*/EC: EP *R* 3 K *corr. ex* E *a. m. E*/*post* K *add.* et ducantur linee KF KH *R* 4 linea *corr. ex* lenea *O*/ZK *corr. ex* ZE *a. m. E* 5 OC: OP *R*/*ante* erit *add.* quod *R*/ZKF *corr. ex* ZEF *a. m. E*/superficiei *om. ER*/OCT: OPT *R* 6 *ante* AQT *scr. et del.* super lineas *F*/HFK *corr. ex* HFE *E*; HKF *R* 7 CT[1,2]: PT *R*/KF[1] *corr. ex* EF *a. m. E*/KF[2]: EF *E*/equidistantes: equidem *S*/*post* equidistantes *add.* et *L3* 9 TP: TS *R*/cadat: cadit *C1*/equidistans . . . NZ (10): angulus QTS rectus *R* 10 erit . . . circulo: continget circulum *R*/ATP: ATS *R* 11 et *inter. O*/OC: OP *R* 12 ortogonalis: ortogonaliter *E* 13 ATP[1,2]: ATS *R*/et . . . ATP[2] *inter. a. m. C1*/ZKF *corr. ex* ZEF *a. m. E* 14 sunt: fiunt *P1* 15 CT[1,2]: PT *R*/PI: SI *R*/patet: patuit *R*/est[2] *om. FP1*/est equidistans (16) *transp. R* 16 equidistans[1]: equidem *S*/igitur . . . KF[2] *om. C1*/igitur . . . sed (17) *om. FP1*/PI: SI *R*/KF[2] *corr. ex* EF *a. m. E*

[2.542] Sed planum est quod angulus NTZ equalis est angulo TZF, et angulus HTN equalis angulo TFZ, et TP perpendicularis. Erit FP equalis PZ. Sed proportio FP ad PZ sicut KI ad IZ; erit KI equalis IZ.

[2.543] Ducta autem linea CI, cum superficies ATPI sit ortogonalis super superficiem ZKF, erit CI ortogonalis super ZK, et erit angulus CKZ equalis angulo KZC. Sed angulus ECO equalis est angulo CKZ, et angulus OCZ equalis est angulo CZK, quare angulus ECO equalis est angulo OCZ. Et ita E refertur ad Z a puncto C, quod est propositum.

[2.544] Si autem sumatur aliud punctum in circulo a quo H reflectatur ad Z, probabitur quod ab alio puncto piramidis quam C refertur E ad Z. Et si reflectatur H ad Z a tribus punctis circuli, reflectetur E ad Z a tribus piramidis; si a quatuor, a quatuor.

[2.545] Punctum autem reflexionis a quo E refertur ad Z facile est invenire, invento puncto circuli a quo punctus H refertur ad Z, et erit inventio modo predicto.

[2.546] Si vero dicatur quod a pluribus punctis piramidis quam quatuor possit punctus E reflecti ad Z, per replicationem predicte probationis poterit ostendi quod punctus H refertur ad Z a pluribus punctis circuli quam quatuor, et ubi accidet punctum H reflecti ad Z ab aliquot punctis circuli vel ab uno tantum, accidet punctum E reflecti ad Z a totidem punctis piramidis aut ab uno tantum, et econverso. Quod si dicatur contrarium, poterit improbari predicto modo.

[2.547] Palam ergo quod punctorum quedam habent unicam ymaginem, quedam duas, quedam tres, quedam quatuor, sed non possibile quod plures. Verum adhibito speculo duplici visu eiusdem ymaginis

17 TZF: DZF *FP1; corr. ex* TZP *a. m. E* 18 TFZ *corr. ex* TZF *C1*/TP: TS *R*/FP: FS *R* 19 PZ[1,2]: SZ *R*/FP: SP *SE*; FS *R*/*post* PZ[2] *rep.* sed . . . PZ[2] *P1*/KI[1,2] *corr. ex* EI *a. m. E*/*post* erit *add.* ergo *R*/IZ: ZI *O* 20 CI: PI *R*/cum: circuli *S*/superficies ATPI: super fiat PL *FP1*/ATPI *corr. ex* AQF *E*; ATF *R* 21 ZKF *corr. ex* ZEF *a. m. E*/CI *corr. ex* Q *O*; *corr. ex* QCI *L3*; PI *R*/ZK *corr. ex* ZE *a. m. E*/et *om. O* 22 CKZ[1,2]: PKZ *R*/KZC: KZP *R*/ECO: EPO *R*/est *om. R* 23 OCZ: OPZ *R*/est *om. R*/CZK: CZH *P1*; CKZ *S*; *corr. ex* CZE *E*; PZK *R*/*ante* quare *add.* et angulus CEZ equalis est angulo CZK *S*/ECO: EPO *R* 24 est[1] *om. C1*/OCZ: COZ *S*; OPZ *R*/E *inter. L3*; Z *R*/refertur: reflectitur *R*/Z: E *R*/a puncto C *om. L3*/C: P *R* 25 sumatur: assumatur *L3*/aliud: alium *F*/H: Z *R* 26 reflectatur: refertur *L3*/*post* reflectatur *add.* H *C1*/Z: H *R*/probabitur . . . Z[2] (27) *om. O*/C: P *R* 27 refertur *om. P1*; reflectitur *L3*; reflectetur *R*/E ad Z: Z ad E *R*/*post* Z[1] *scr. et del.* a tribus pyramidis *C1*/H ad Z: Z ad H *R*/reflectetur: reflectitur *L3* 28 E *om. FP1*/E ad Z: Z ad E *R*/*post* tribus *add.* punctis *R* 29 E: Z *R*/*post* E *add.* punctus *O*/refertur: reflectitur *SL3ER*/Z: E *R* 30 punctus H: punctum Z *R*/refertur: reflectitur *L3R*/Z: H *R* 31 inventio *corr. ex* mutatio *C1* 32 dicatur *inter. O*; *rep. C1*/*post* quam *add.* a *C1* 33 possit: posset *O*/punctus E: punctum Z *R*/E *om. FP1*/Z: E *R*/replicationem: reflexionem *FP1SL3C1E*; conversionem *R* 34 probationis: proportione *L3*/punctus H: punctum Z *R*/refertur: reflectitur *L3R*/Z: H *R* 35 *post* circuli *scr. et del.* reflectetur E ad Z *C1*/H: E *C1*; Z *R*/Z: H *R*/ab . . . Z (37) *mg. C1* 36 punctis: punctus *F*/E: Z *R* 37 Z: E *R*/et econverso: aut econtrario *R*/quod *inter. L3* 38 poterit *corr. ex* verum *L3* 39 habent unicam *transp. R* 40 sed *inter. a. m. E*/*post* non *add.* est *O* 41 quod *om. R*/adhibito . . . visu: duplici visu adhibito speculo *R*/ymaginis: ymagines *FP1*

diversa erunt loca, que diversitas, propter sui imperceptibilitatem, non inducit errorem.

42 *post* diversa *scr. et del.* sunt et *E* / sui: suam *R* / imperceptibilitatem *corr. ex* imperceptibilibtatem *F*; *corr. ex* imperceptibilem *a. m. C1*

FIGURES FOR TRANSLATION AND COMMENTARY

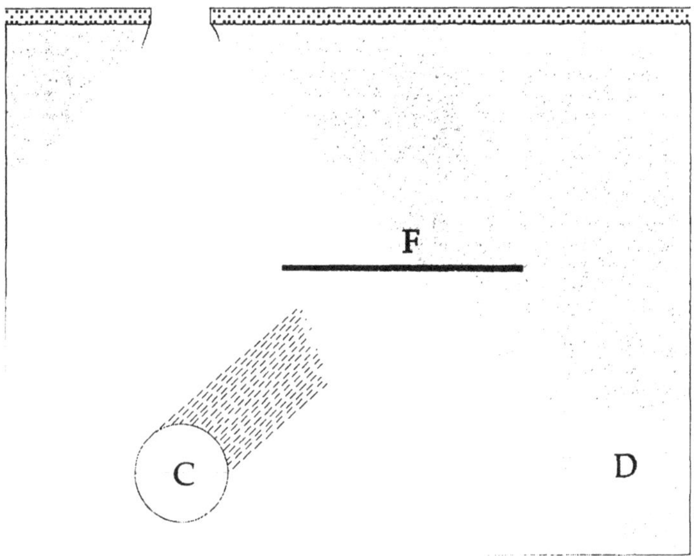

figure 4.2.1

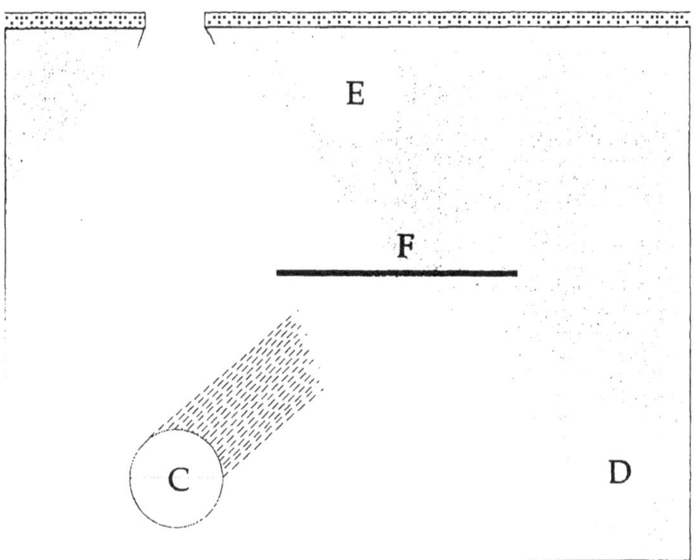

figure 4.2.2

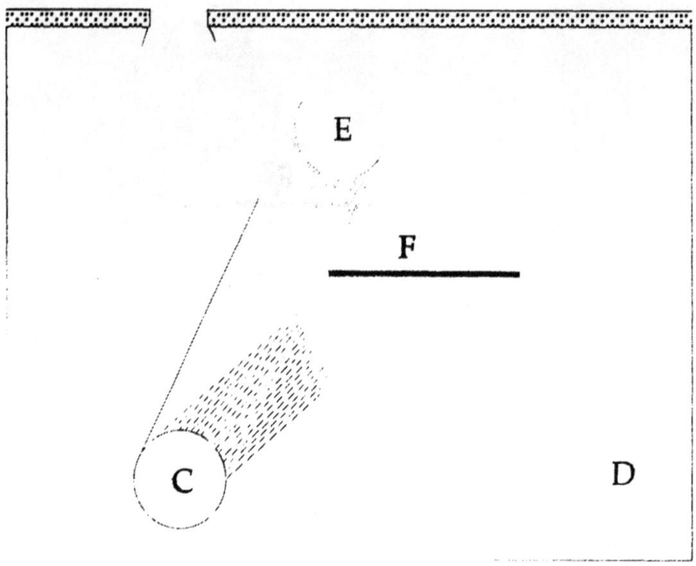

figure 4.2.3

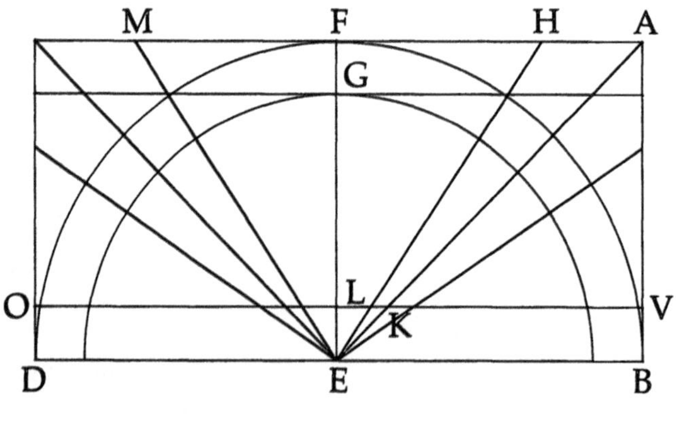

figure 4.3.1

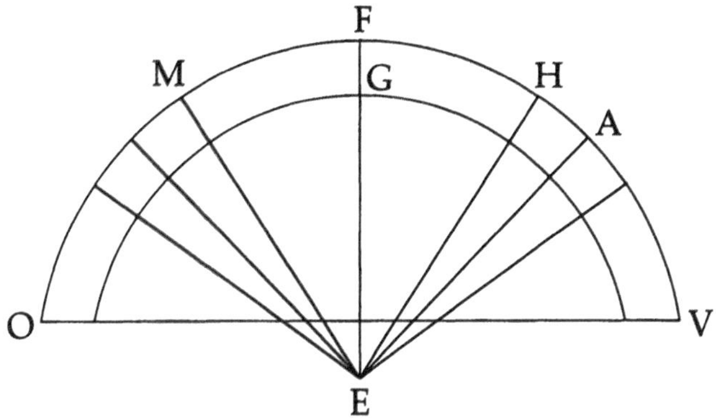

figure 4.3.2

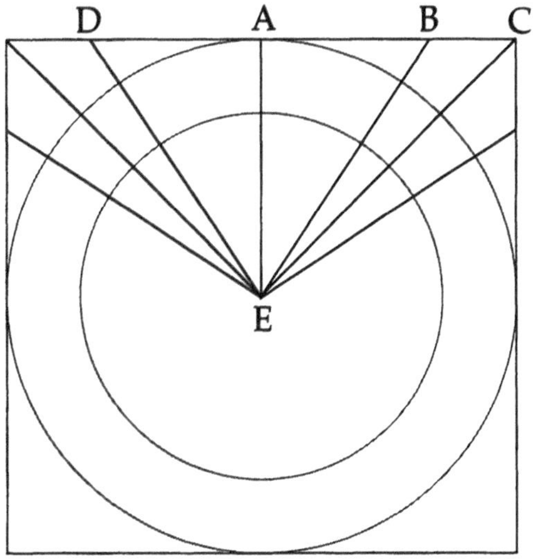

figure 4.3.3

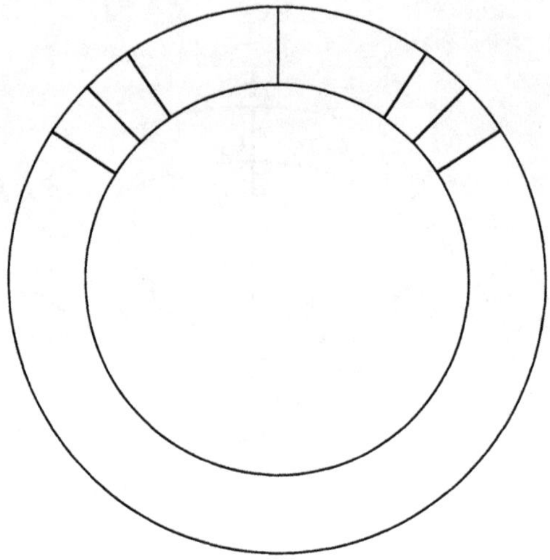

figure 4.3.4

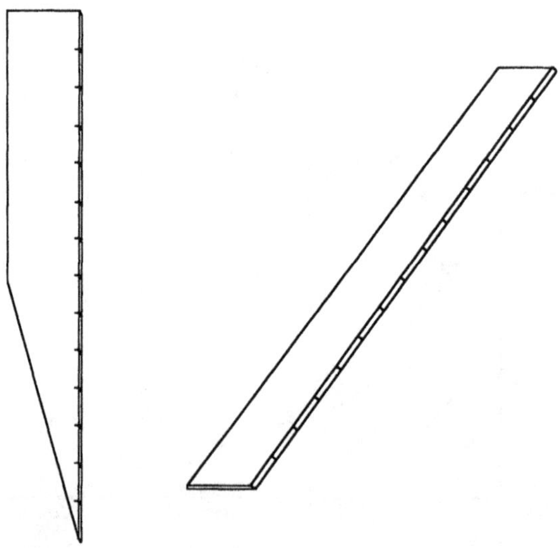

figure 4.3.5

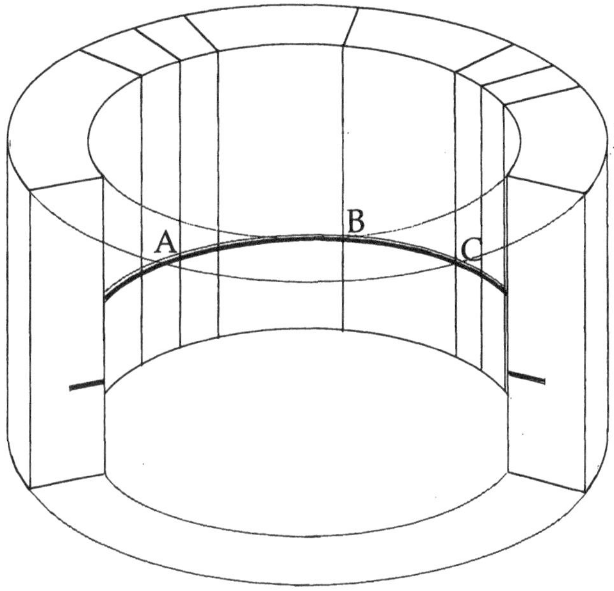

figure 4.3.6

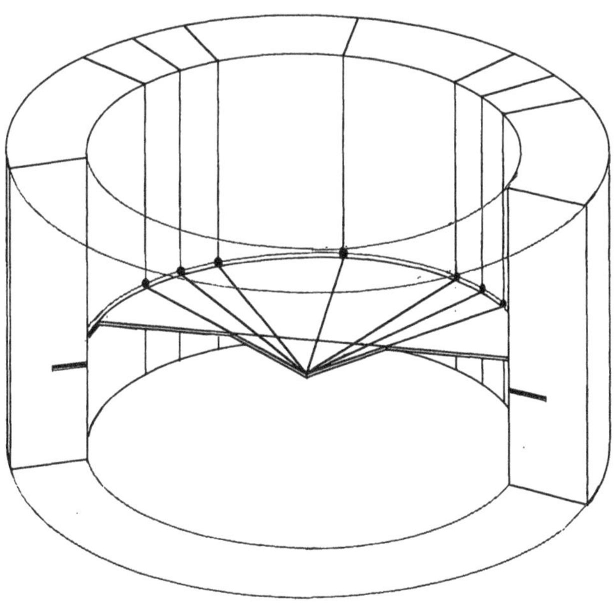

figure 4.3.7

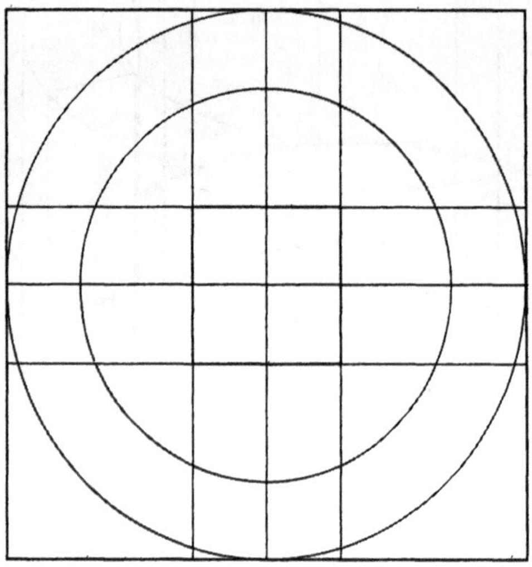

figure 4.3.8

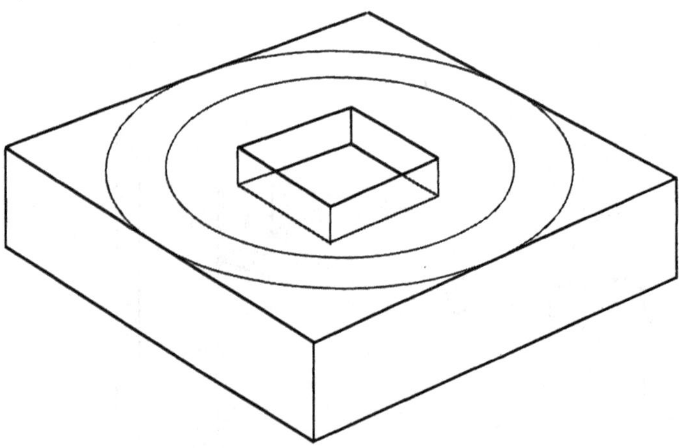

figure 4.3.9

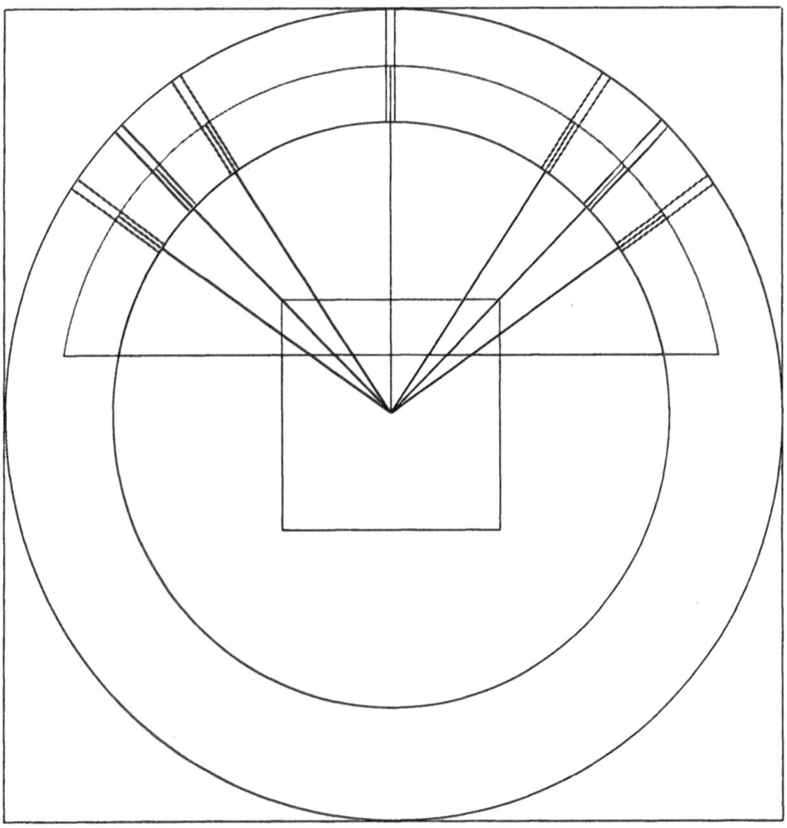

figure 4.3.10

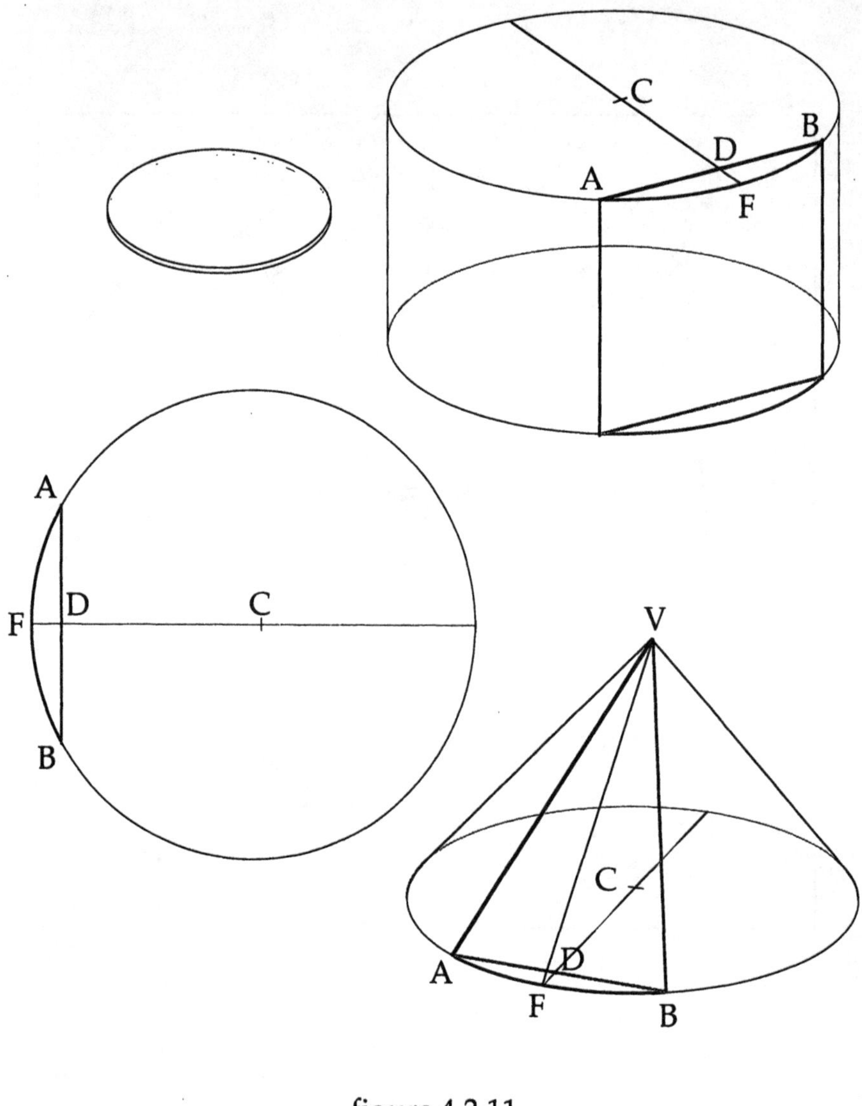

figure 4.3.11

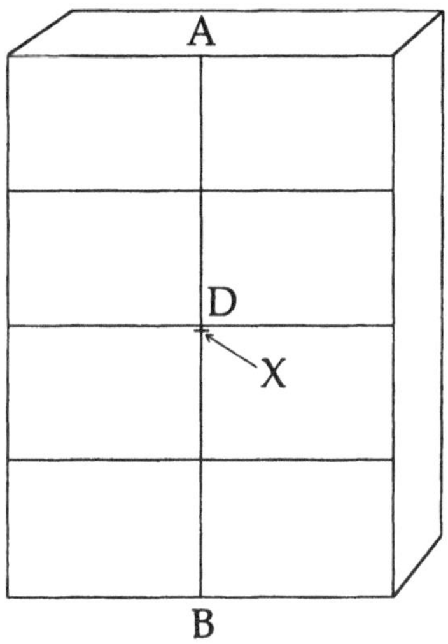

figure 4.3.12

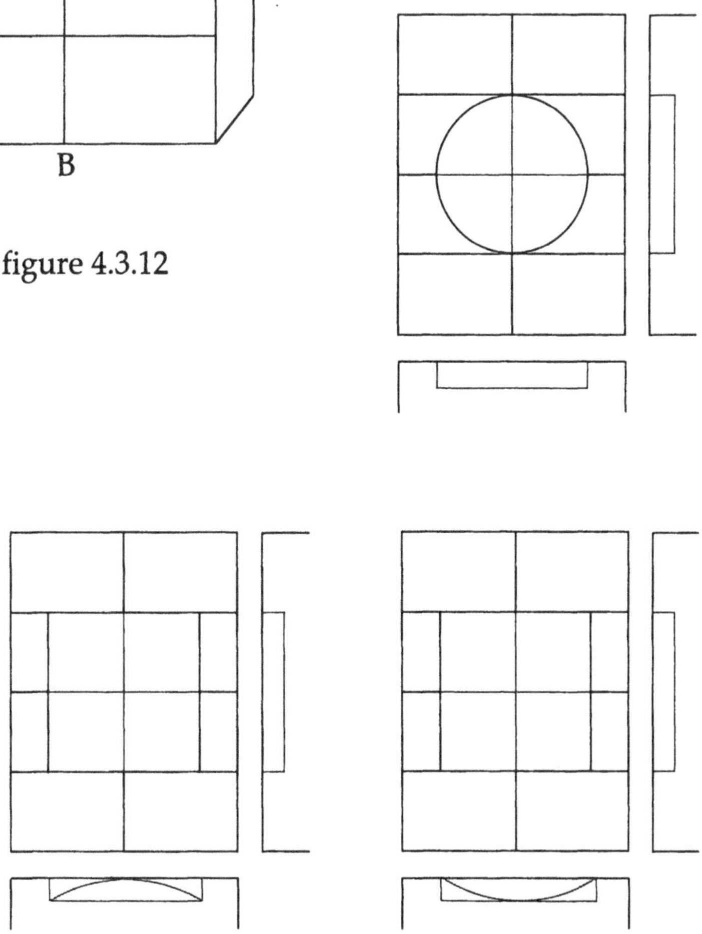

figure 4.3.13

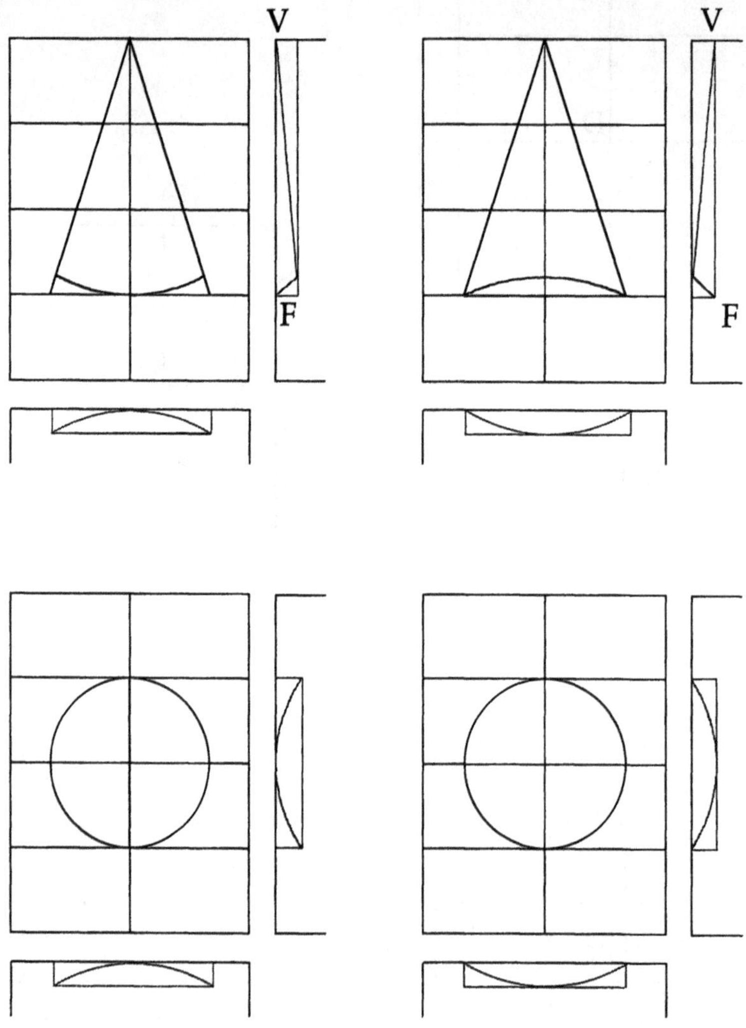

figure 4.3.14

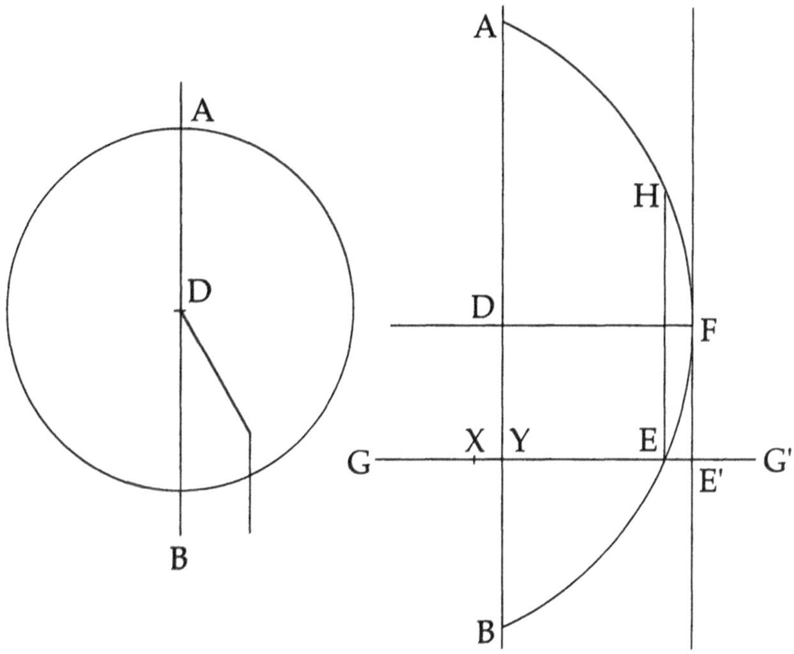

figure 4.3.15

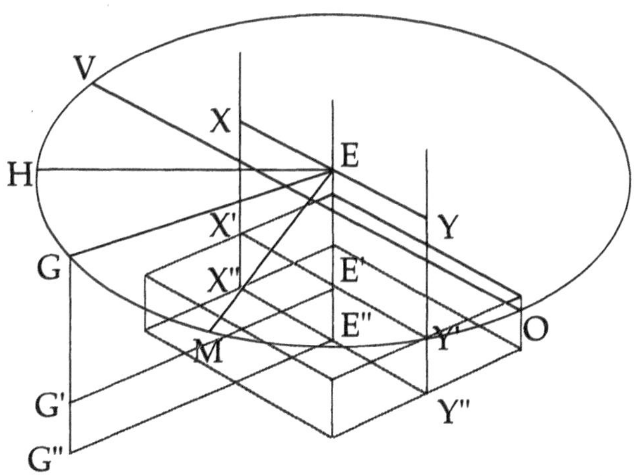

figure 4.3.16

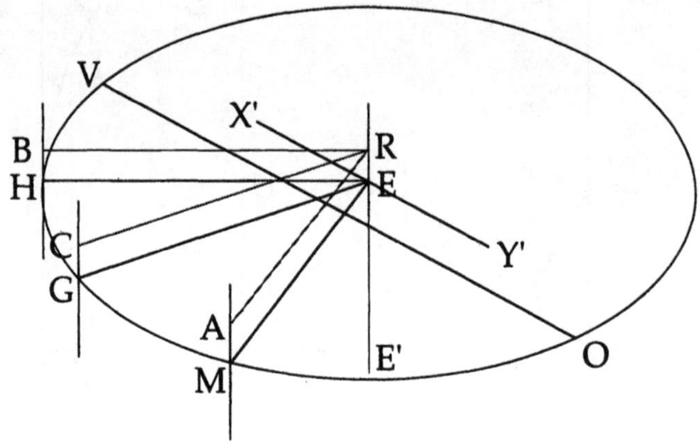

figure 4.3.17

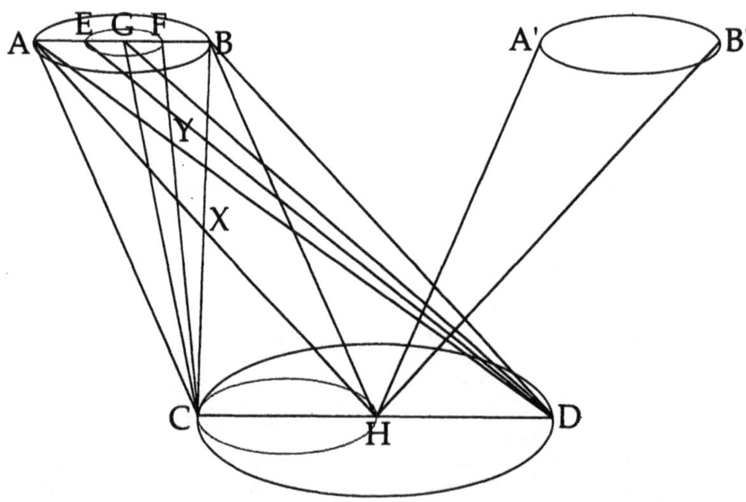

figure 4.3.18

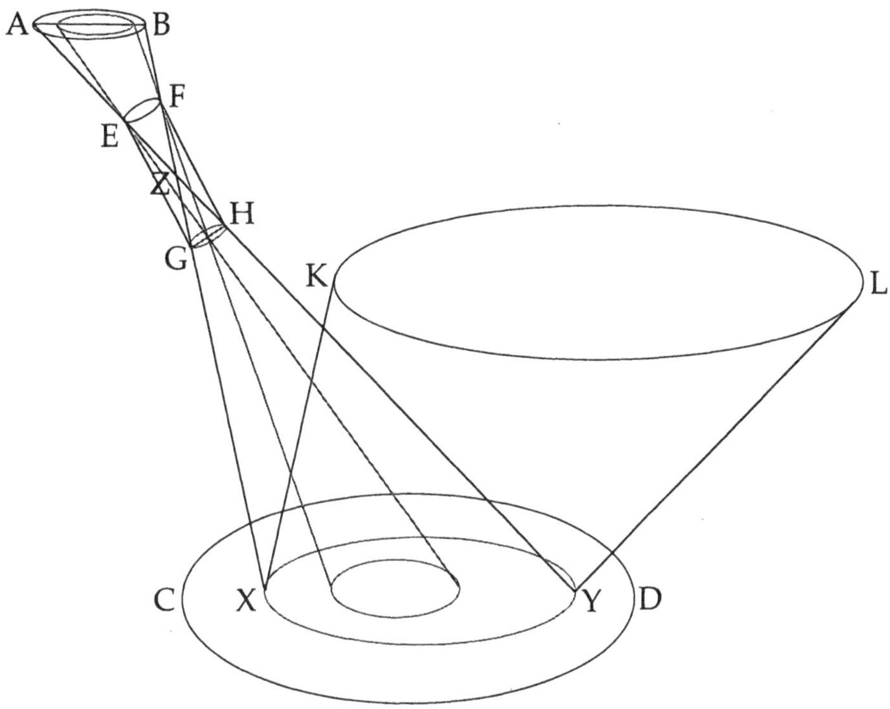

figure 4.3.19

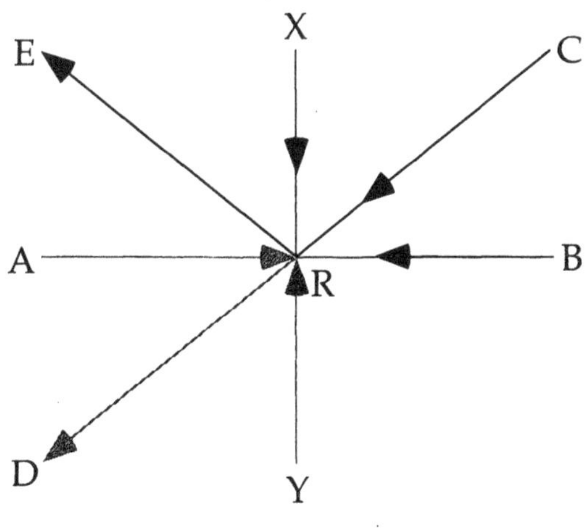

figure 4.3.20

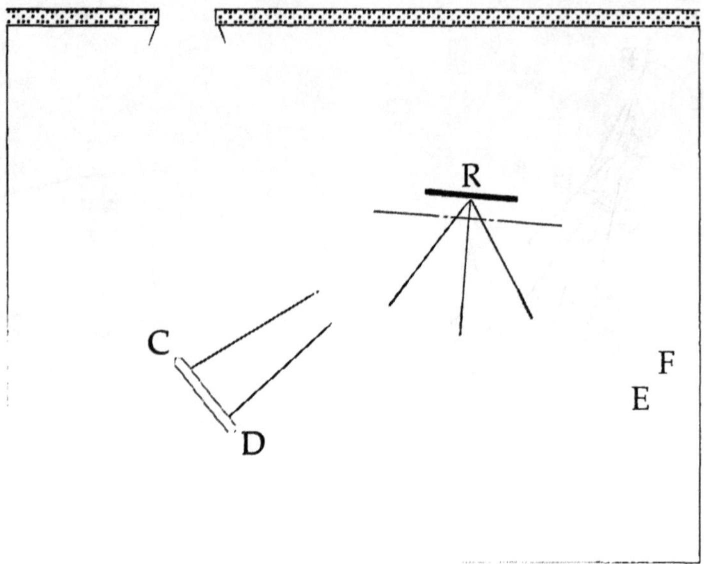

figure 4.3.21

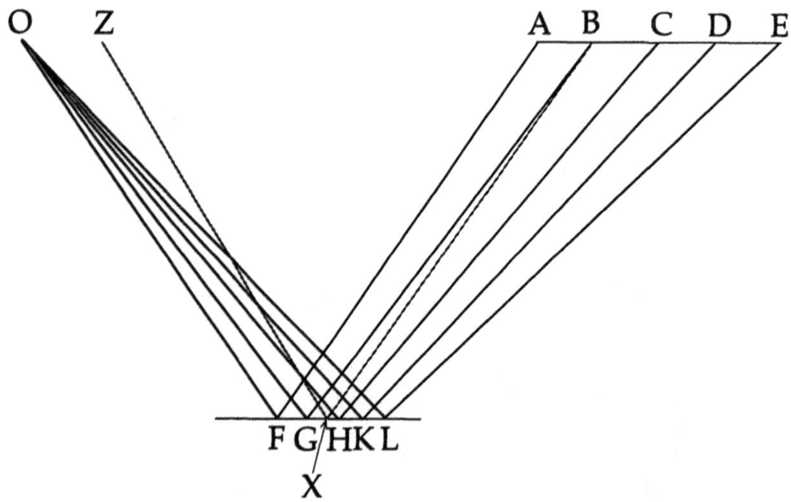

figure 4.5.1

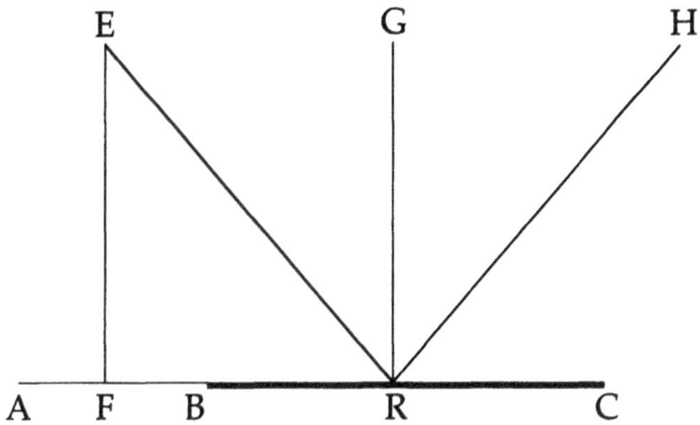

figure 4.5.2

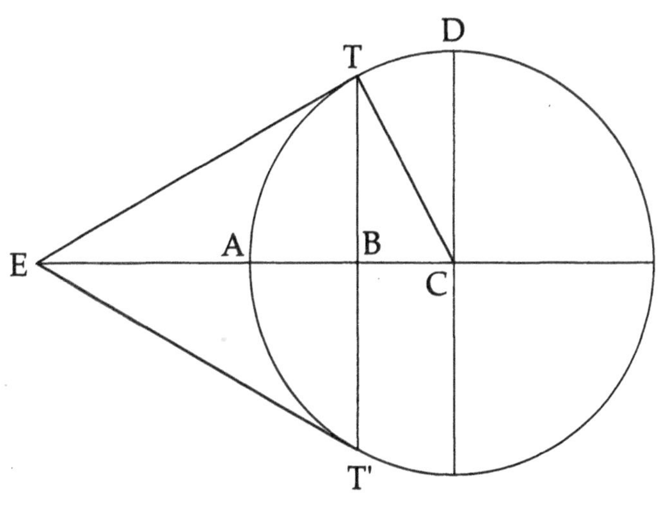

figure 4.5.3

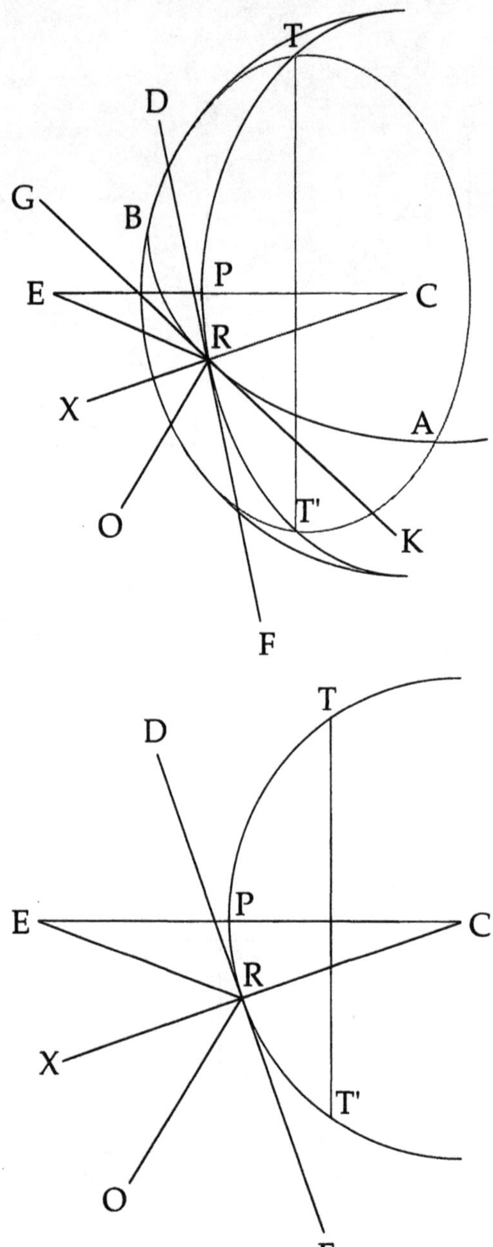

figure 4.5.4

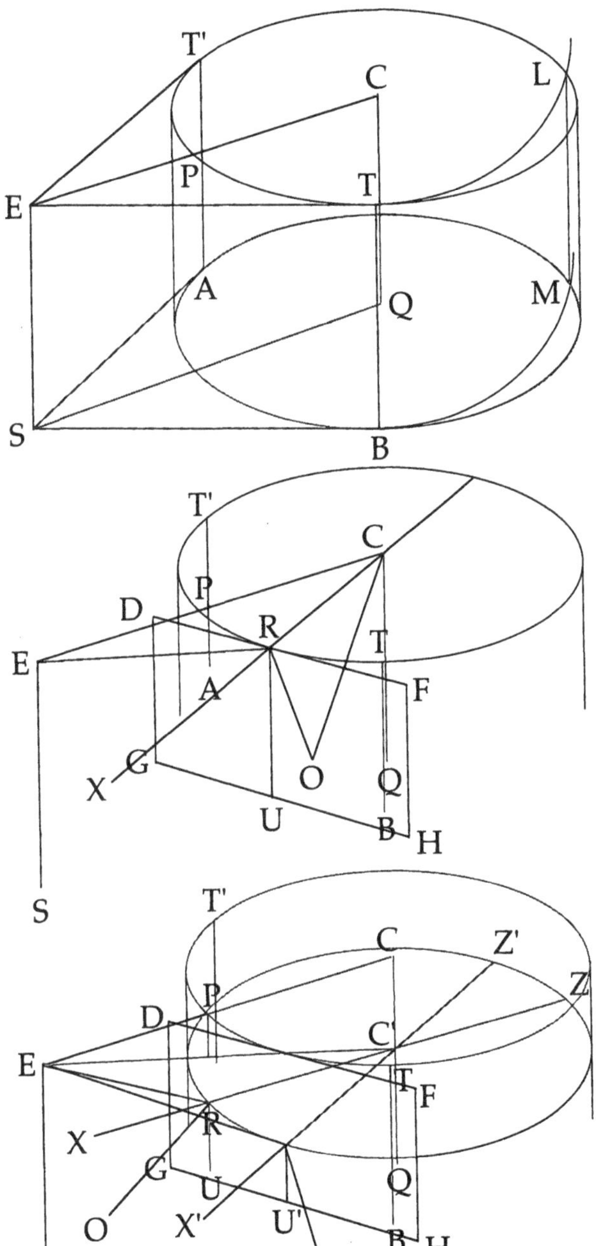
T'
C
L
P
T
E
A
M
Q
S
B
T'
C
P
D
R
T
E
F
A
G
O
Q
X
U
B
H
S
T'
C
Z'
Z
P
D
C'
E
T
F
X
R
G
U
Q
O
X'
U'
B
H
S
O'

figure 4.5.5

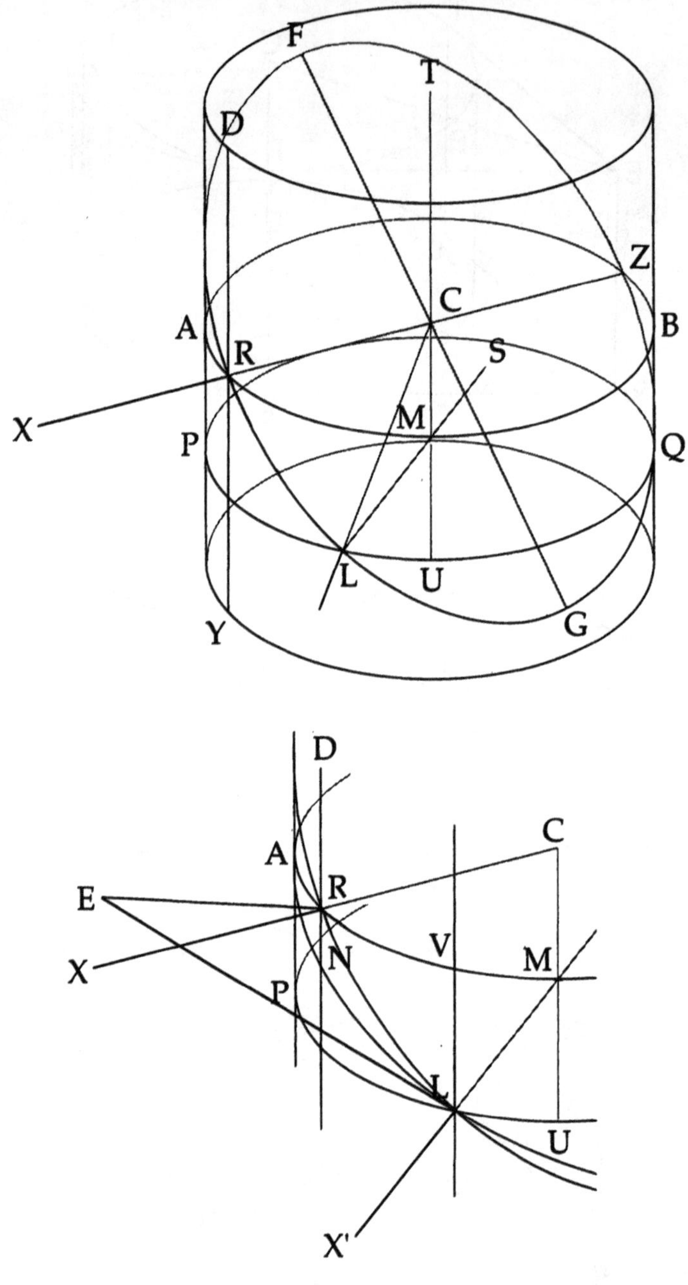

figure 4.5.6

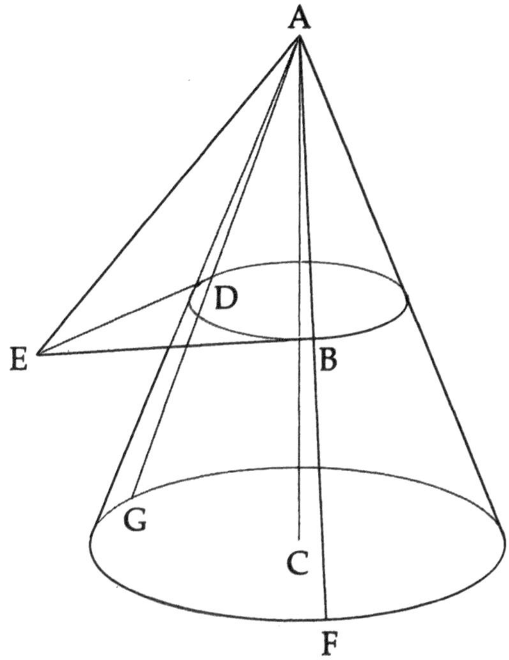

figure 4.5.7

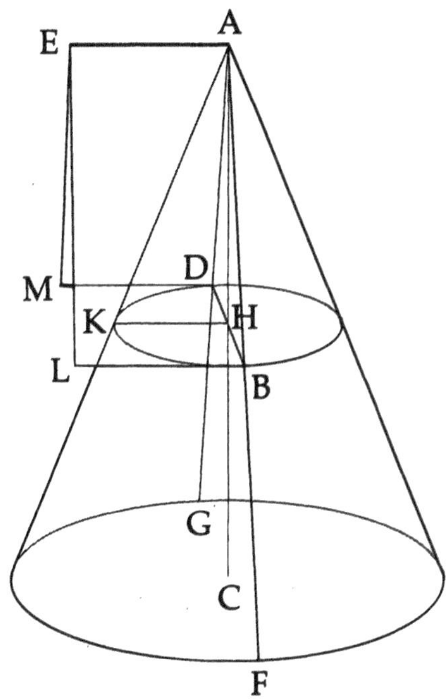

figure 4.5.8

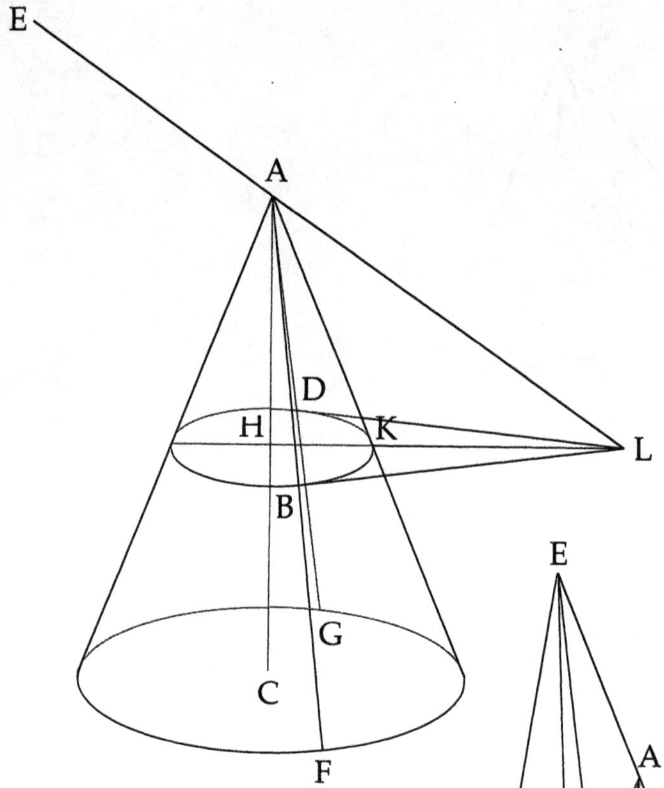

figure 4.5.9

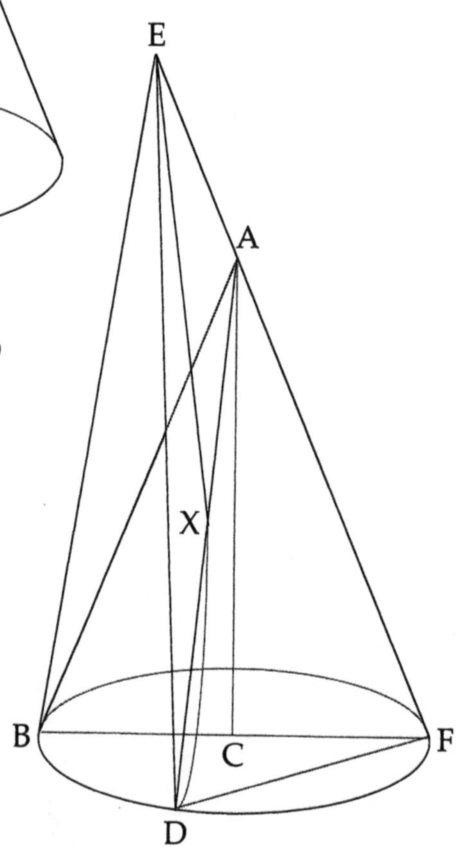

figure 4.5.10

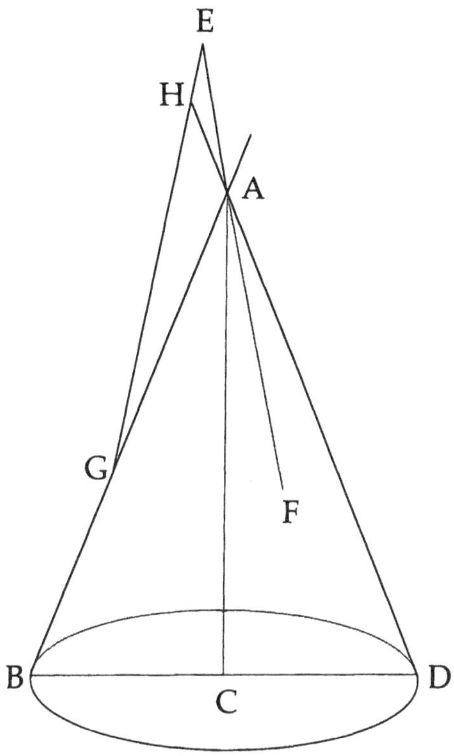

figure 4.5.11

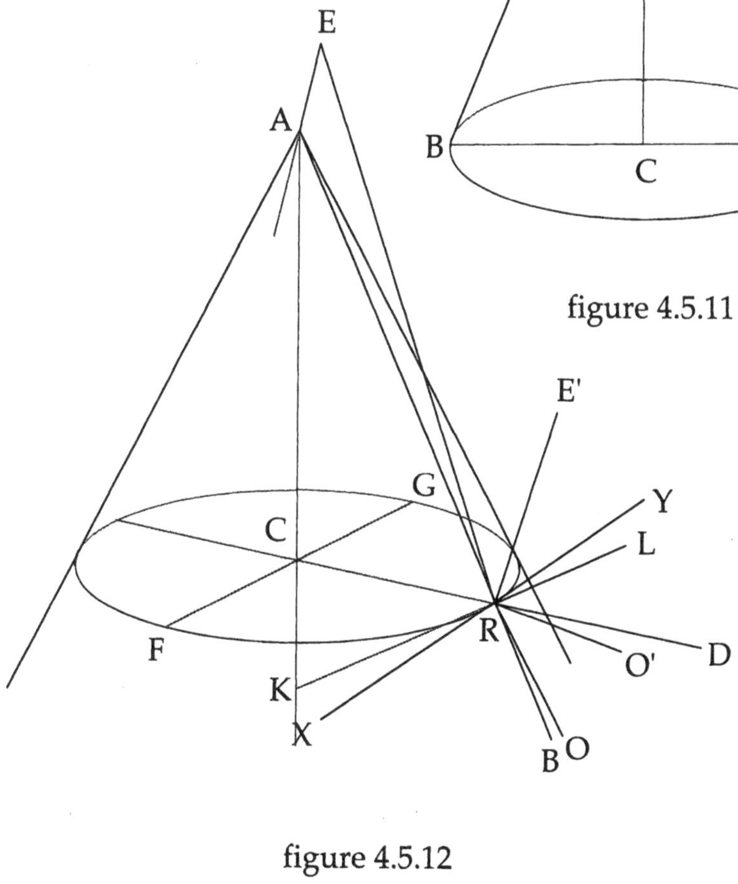

figure 4.5.12

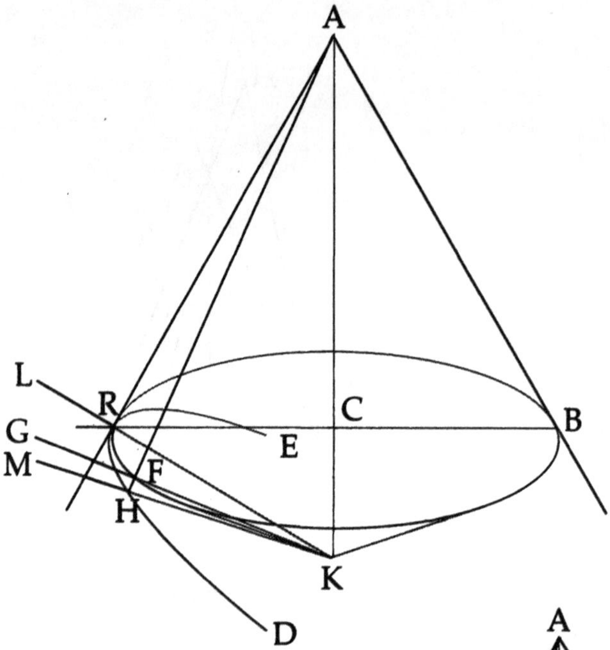

figure 4.5.13

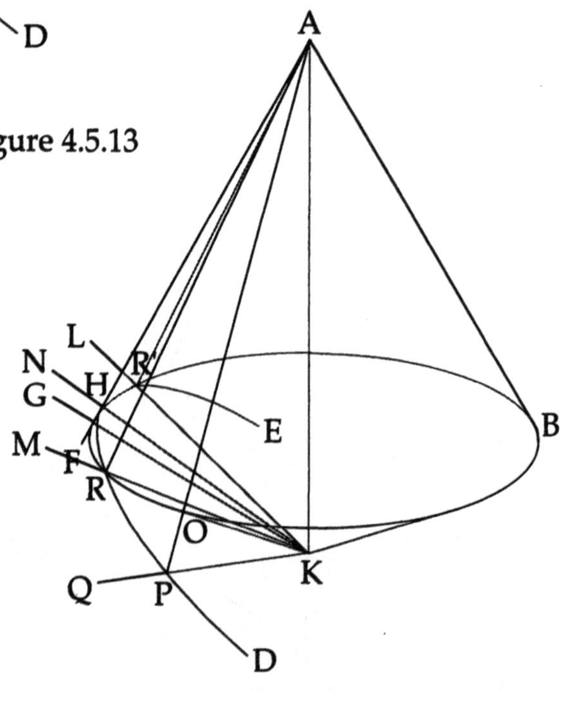

figure 4.5.14

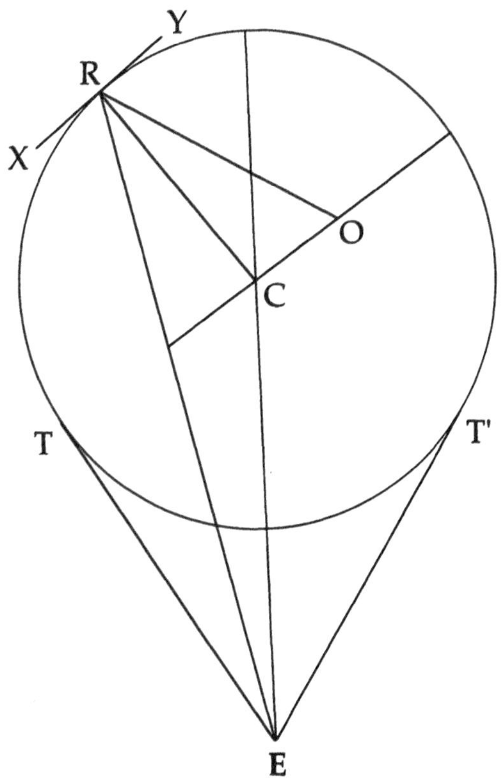

figure 4.5.15

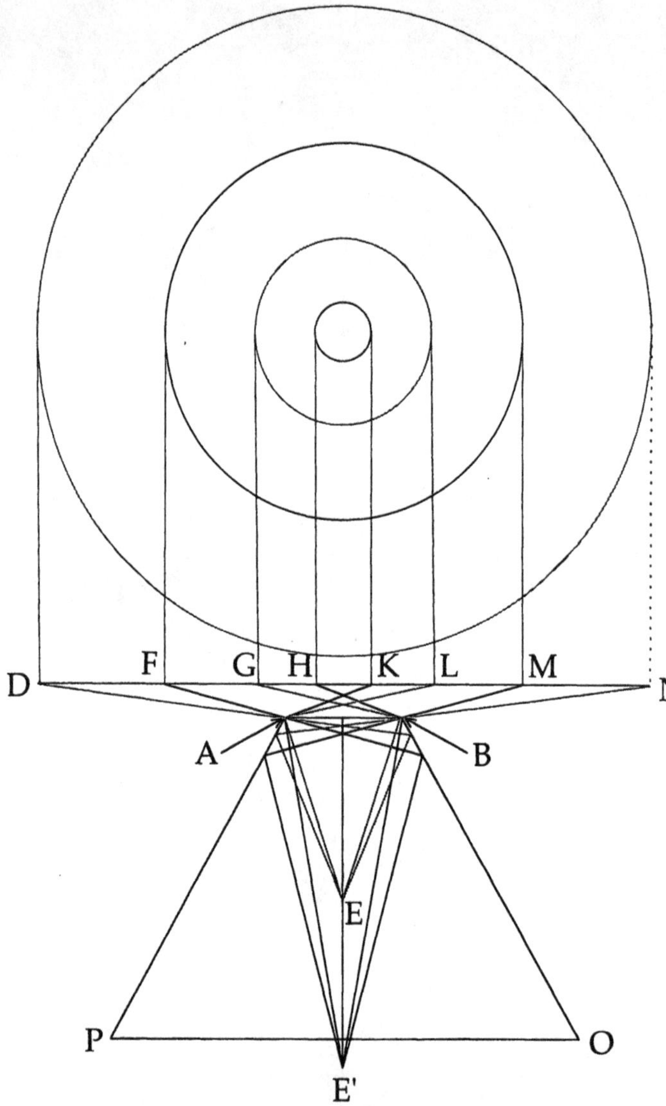

figure 4.5.16

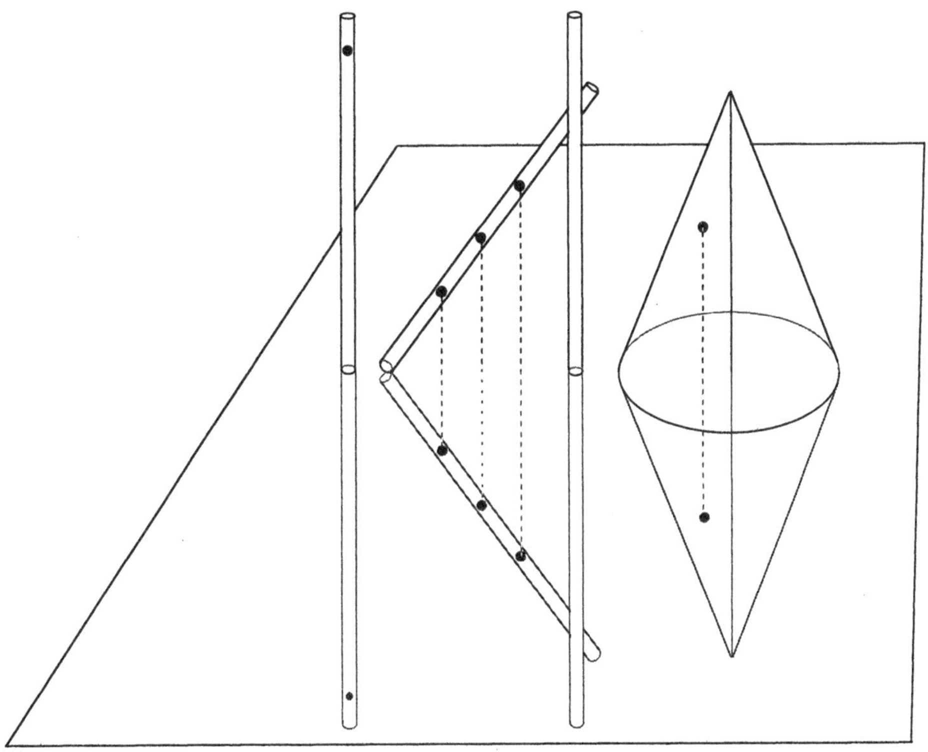

figure 5.1

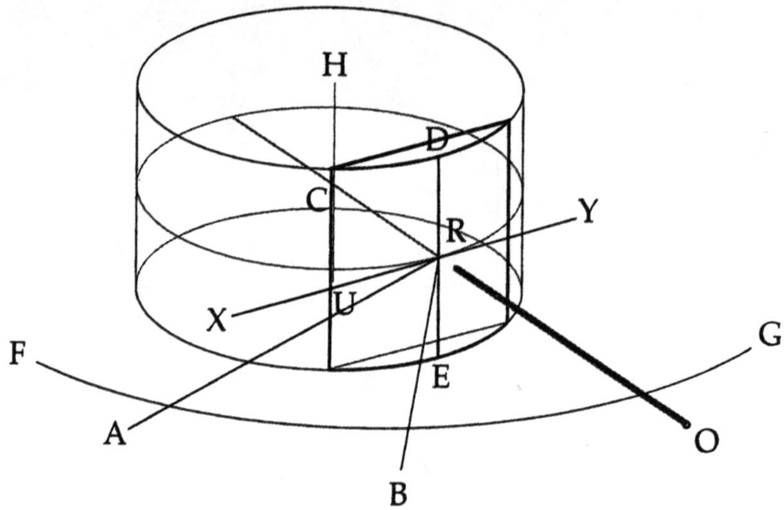

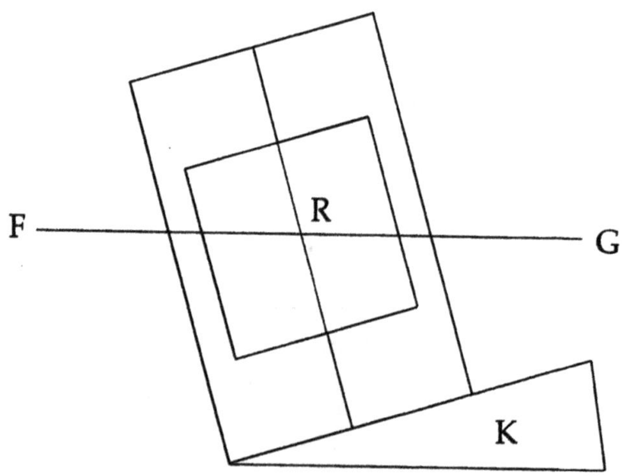

figure 5.2

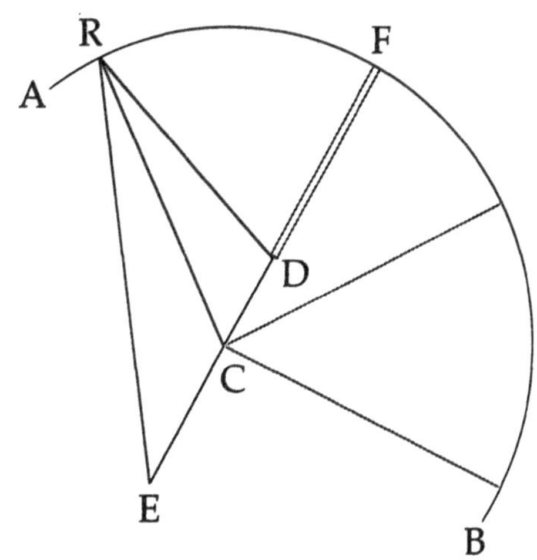

figure 5.3

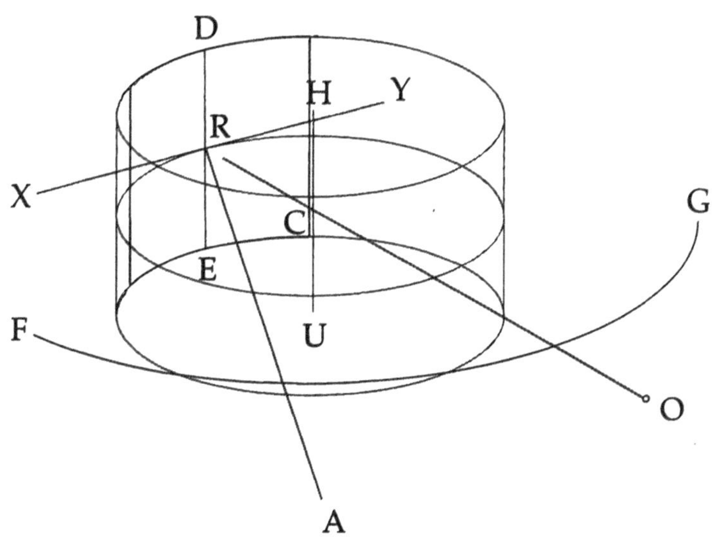

figure 5.4

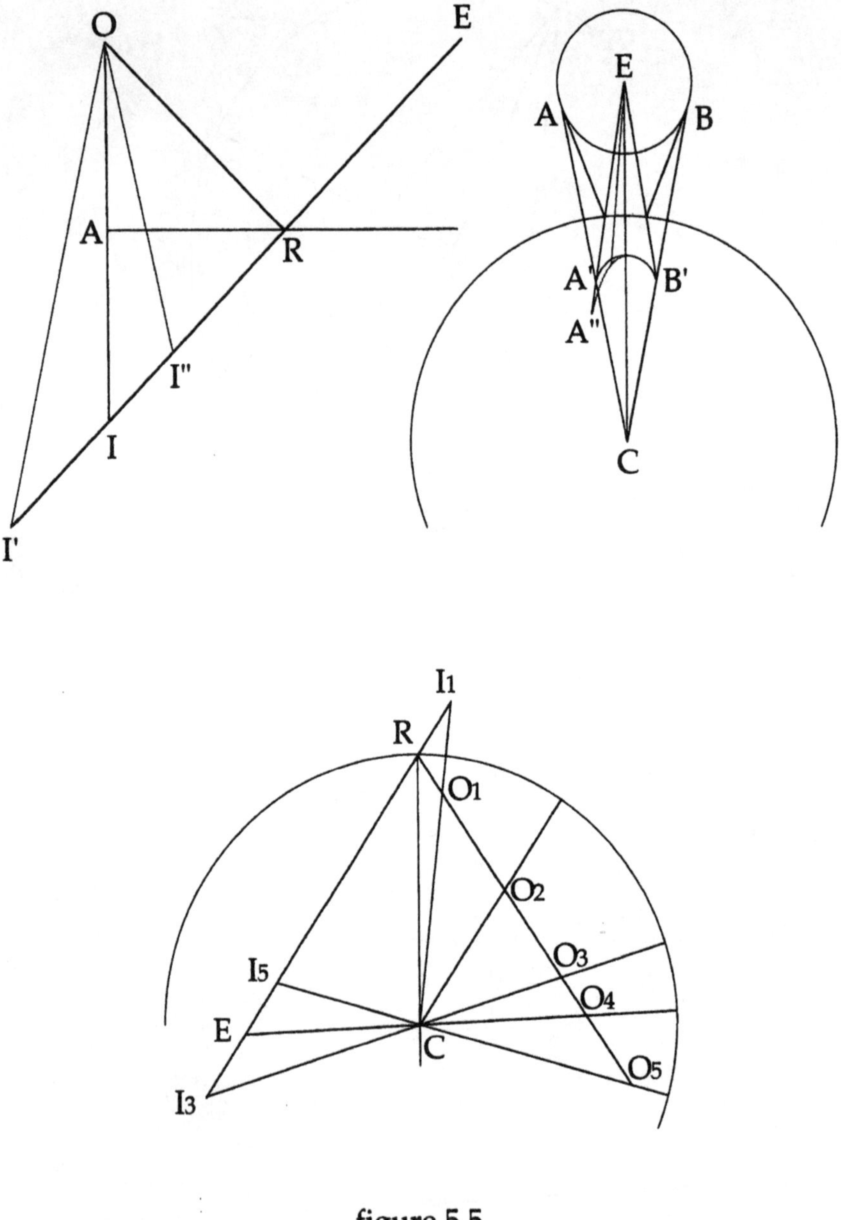

figure 5.5

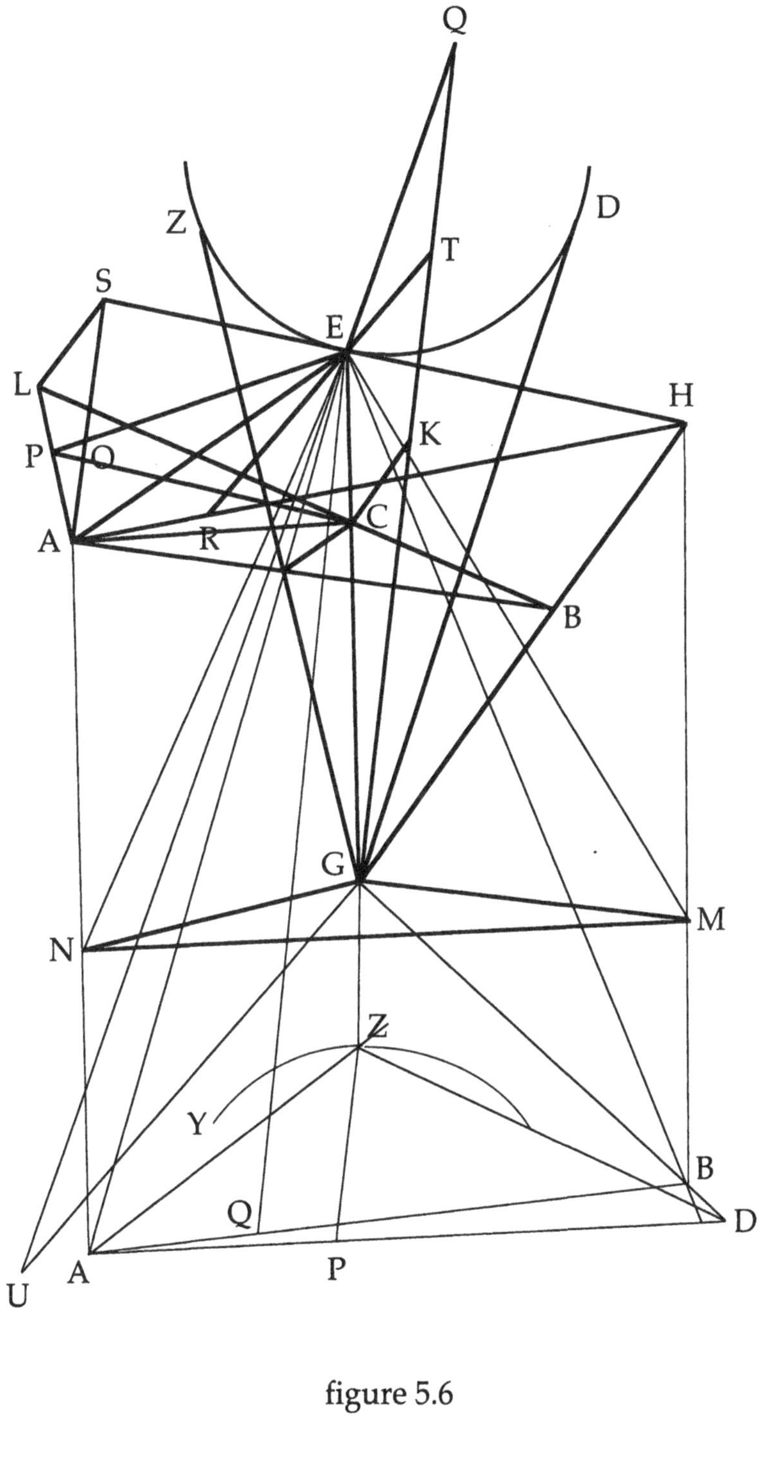

figure 5.6

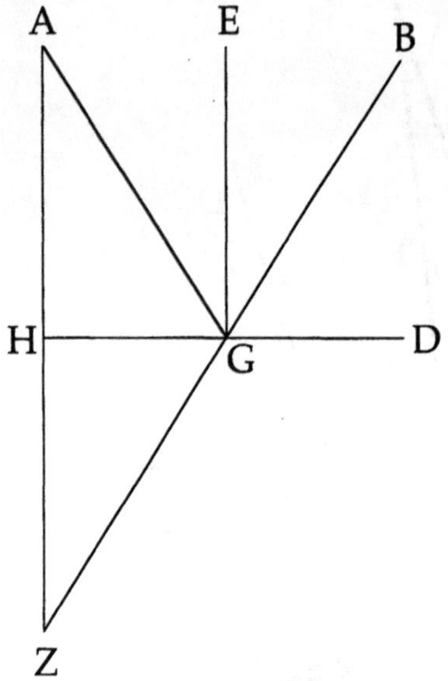

figure 5.2.1

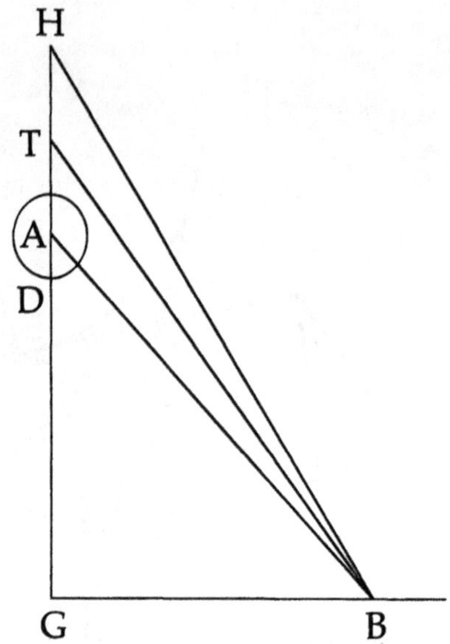

figure 5.2.2

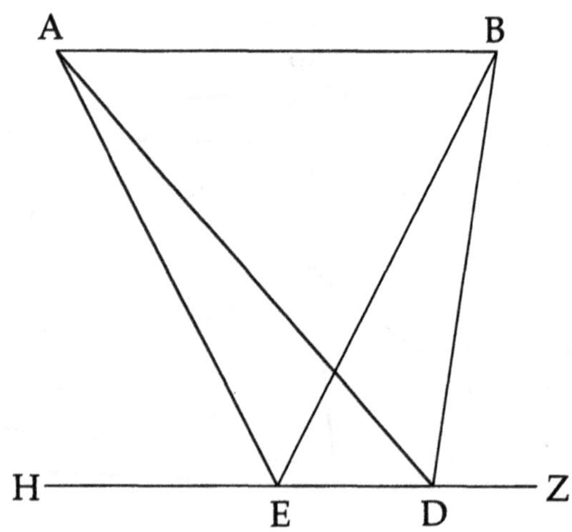

figure 5.2.3

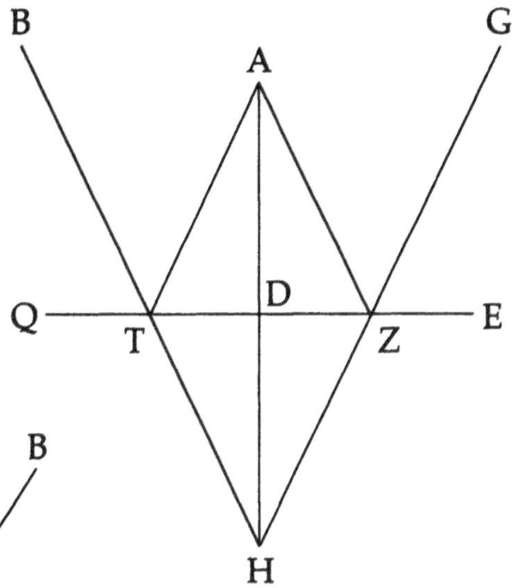

figure 5.2.4

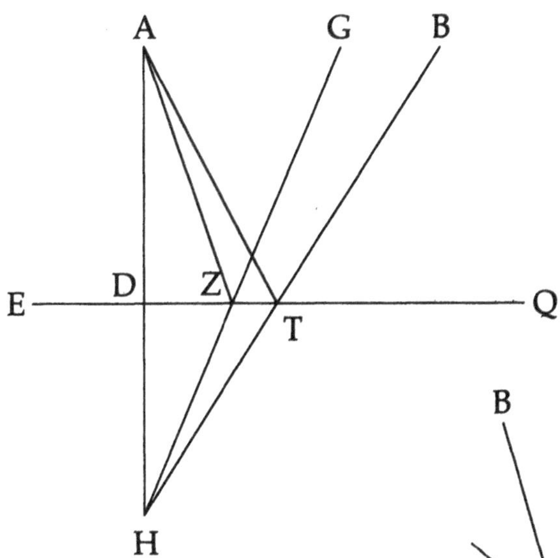

figure 5.2.4a

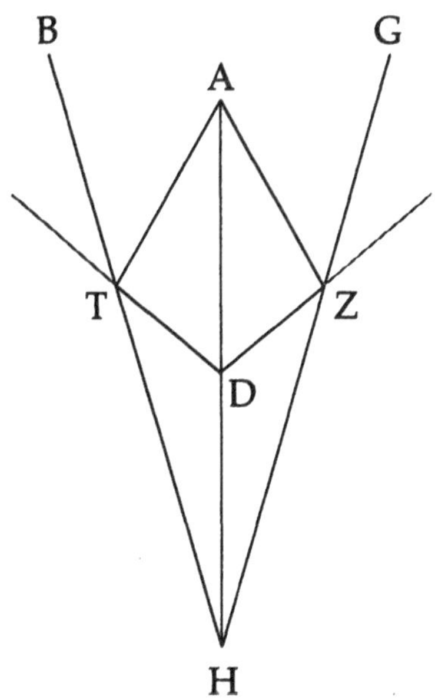

figure 5.2.4b

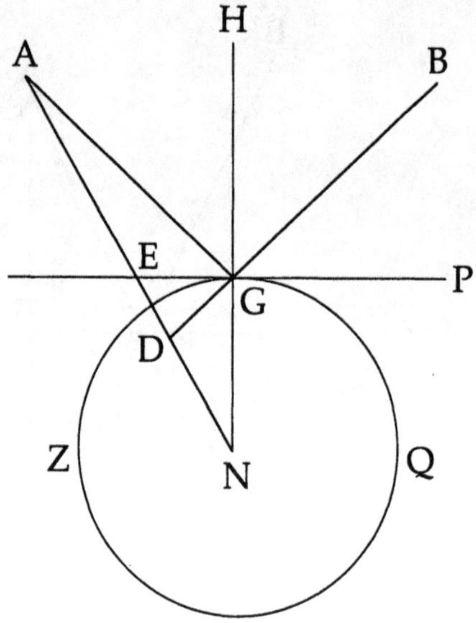

figure 5.2.5

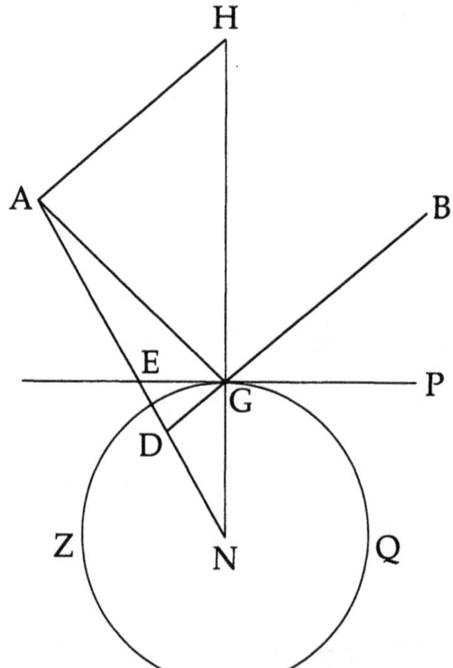

figure 5.2.6

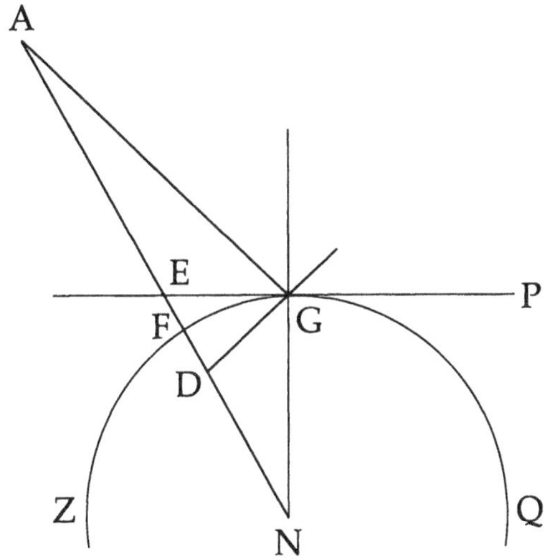

figure 5.2.7

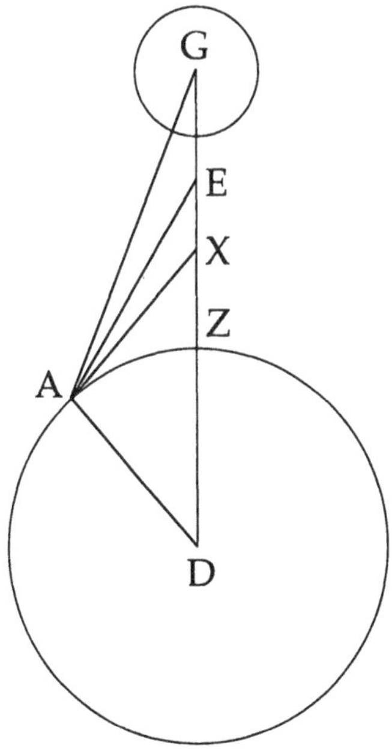

figure 5.2.8

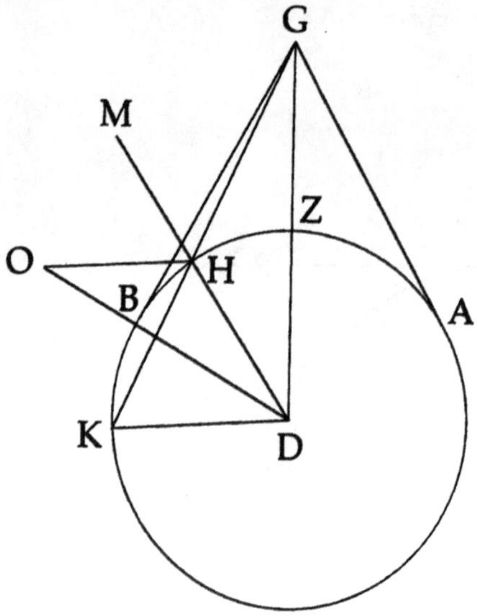

figure 5.2.9

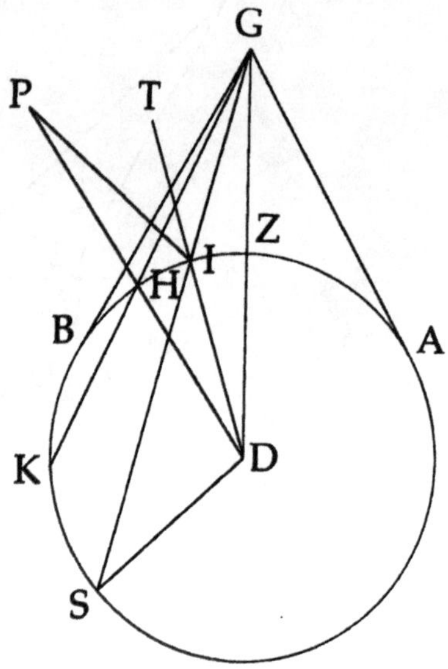

figure 5.2.9a

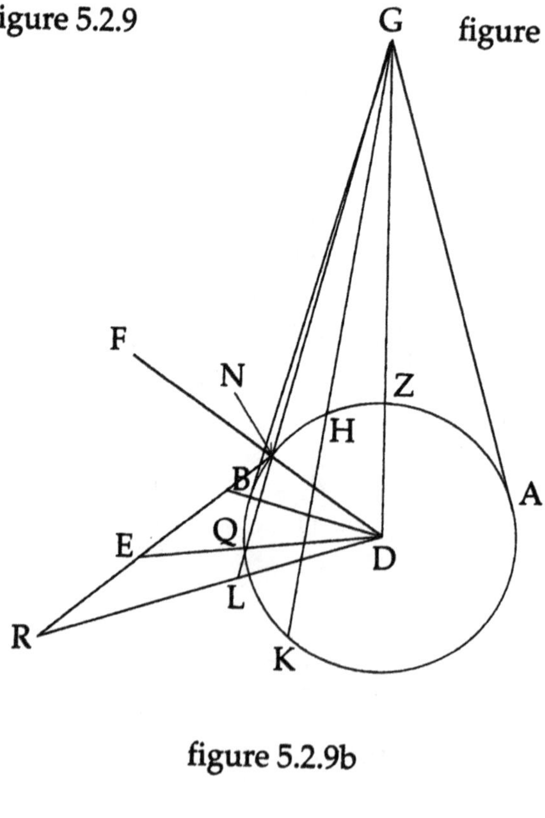

figure 5.2.9b

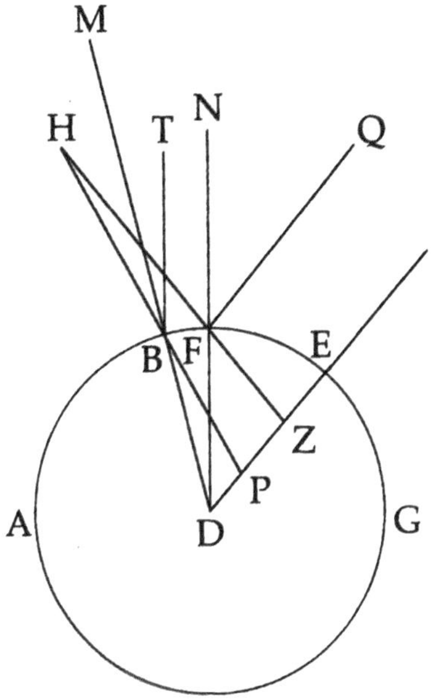

figure 5.2.10

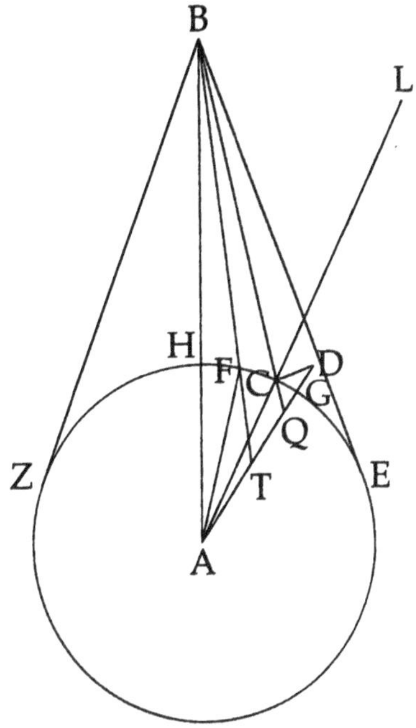

figure 5.2.11

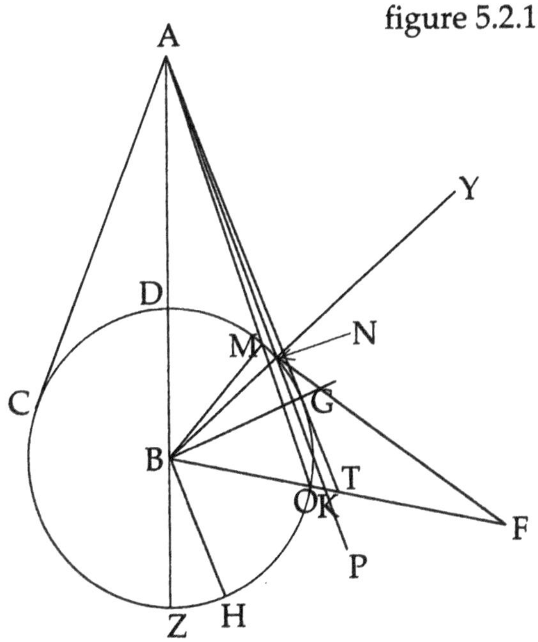

figure 5.2.12

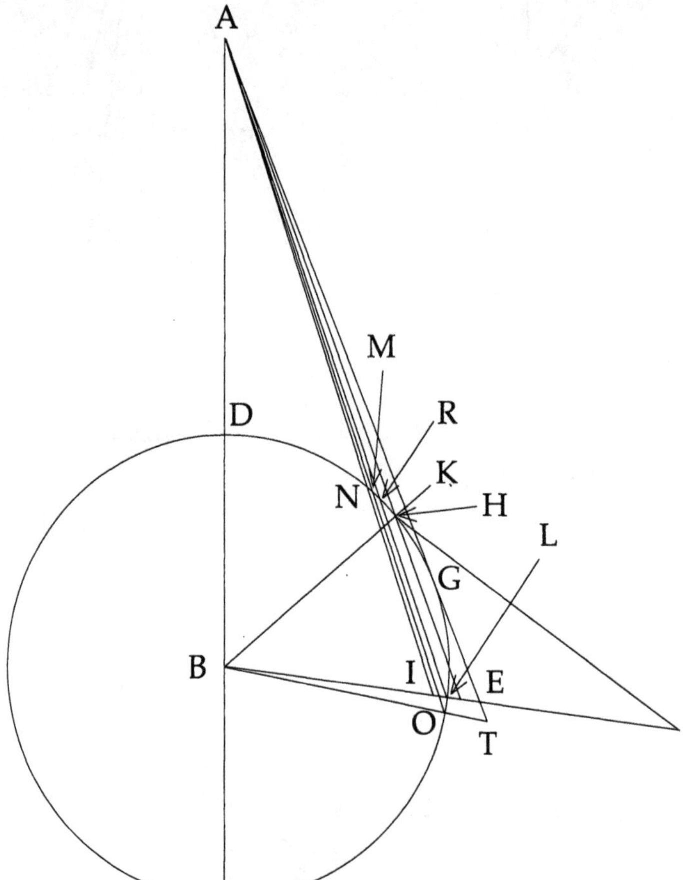

figure 5.2.13

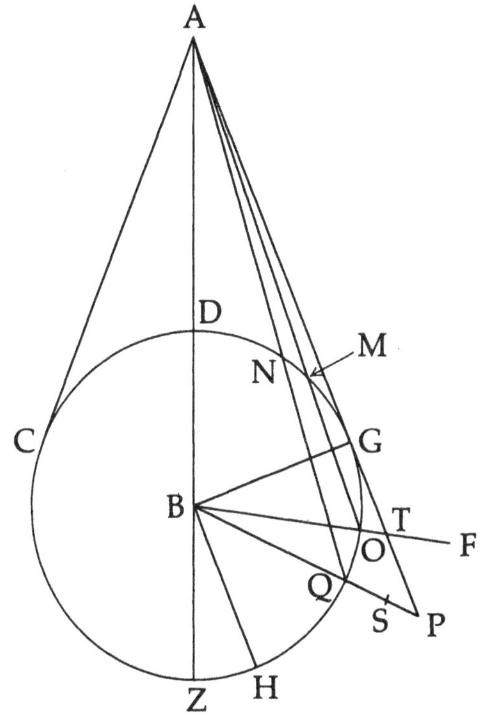

figure 5.2.14

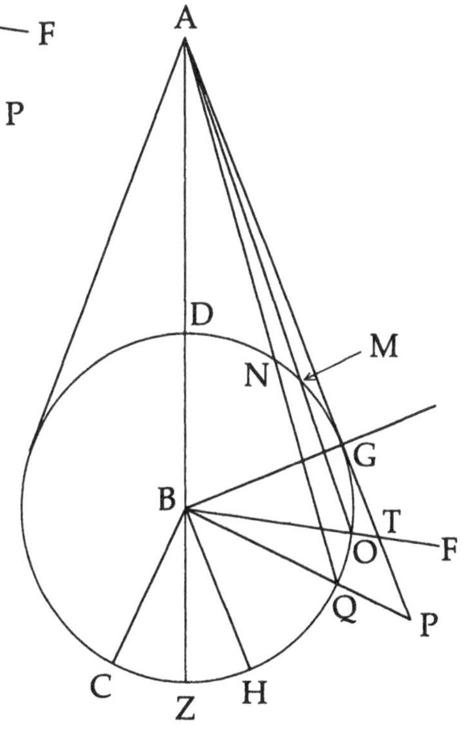

figure 5.2.15

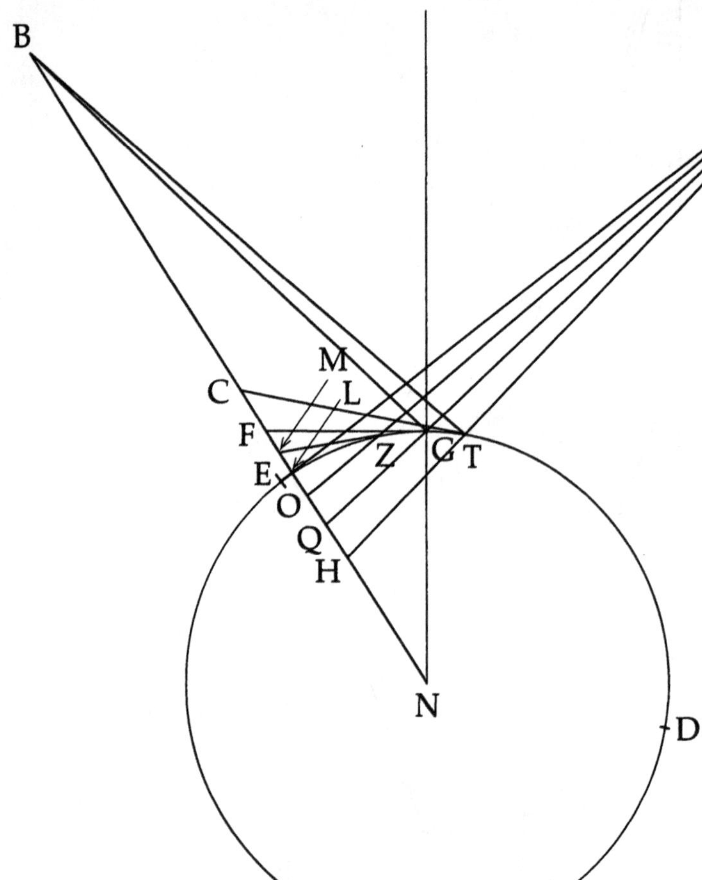

figure 5.2.16

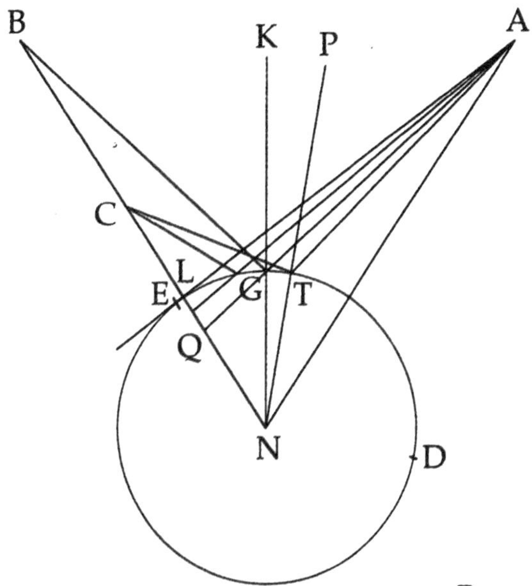

figure 5.2.17

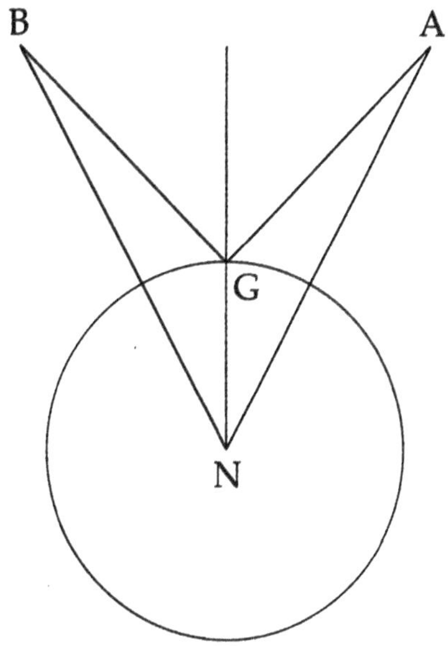

figure 5.2.18

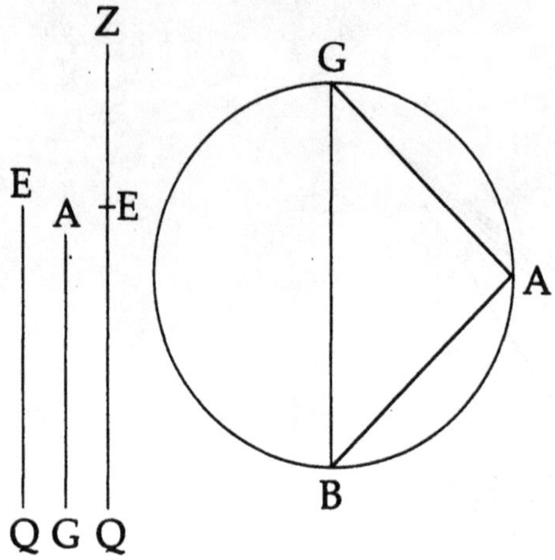

figure 5.2.19

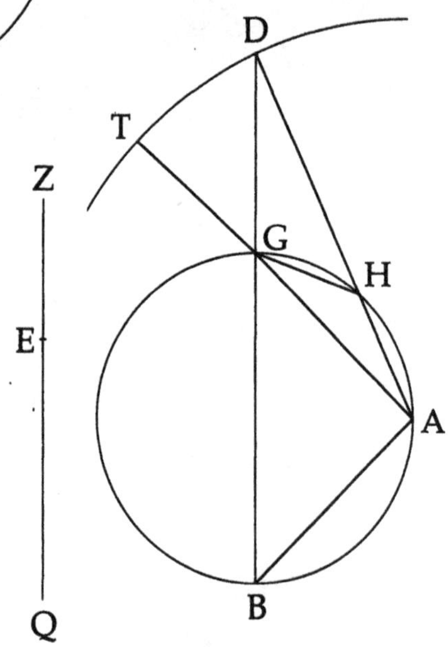

figure 5.2.19a

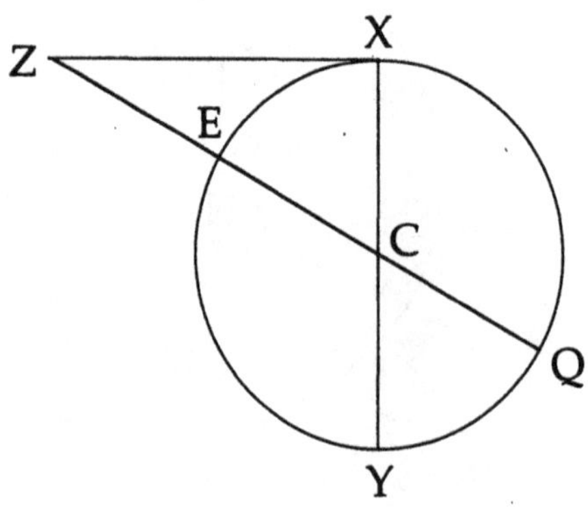

figure 5.2.19b

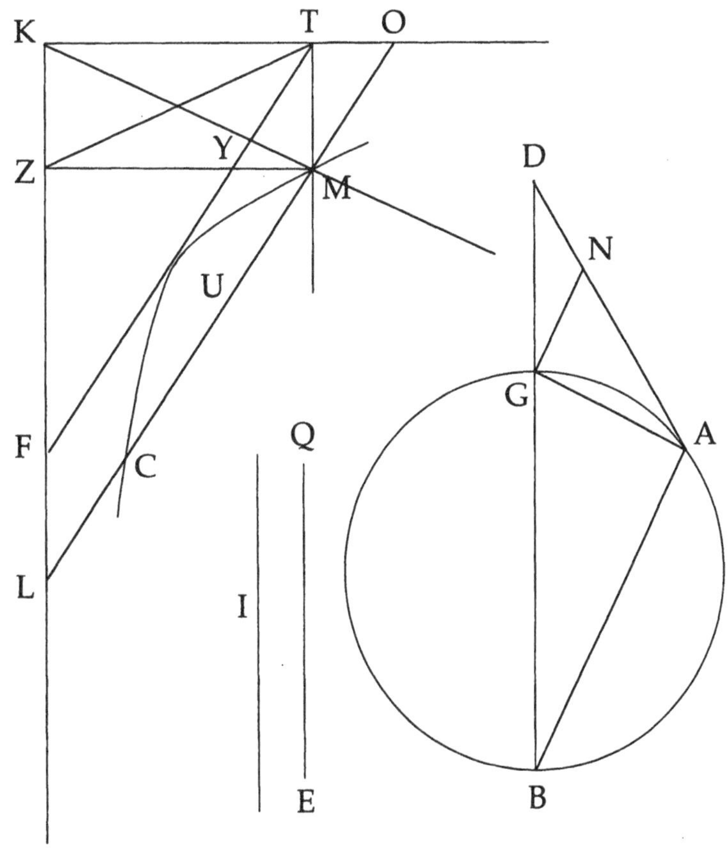

figure 5.2.19c

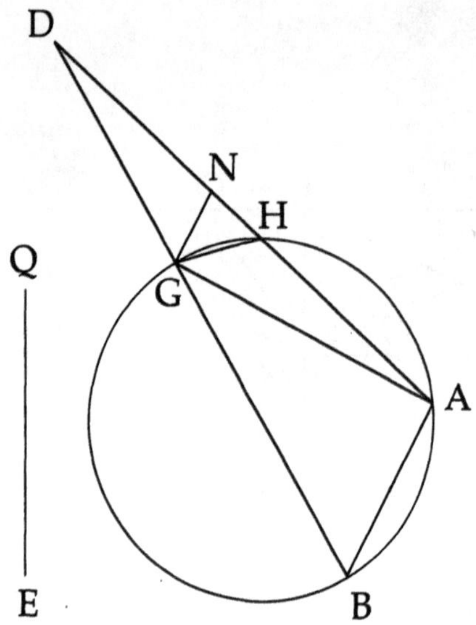

figure 5.2.19d

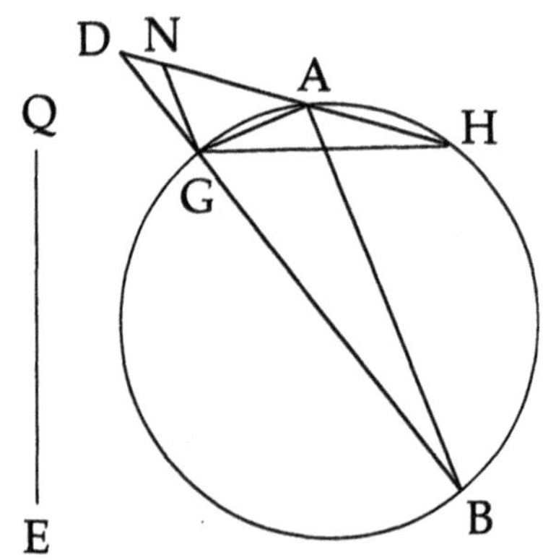

figure 5.2.19e

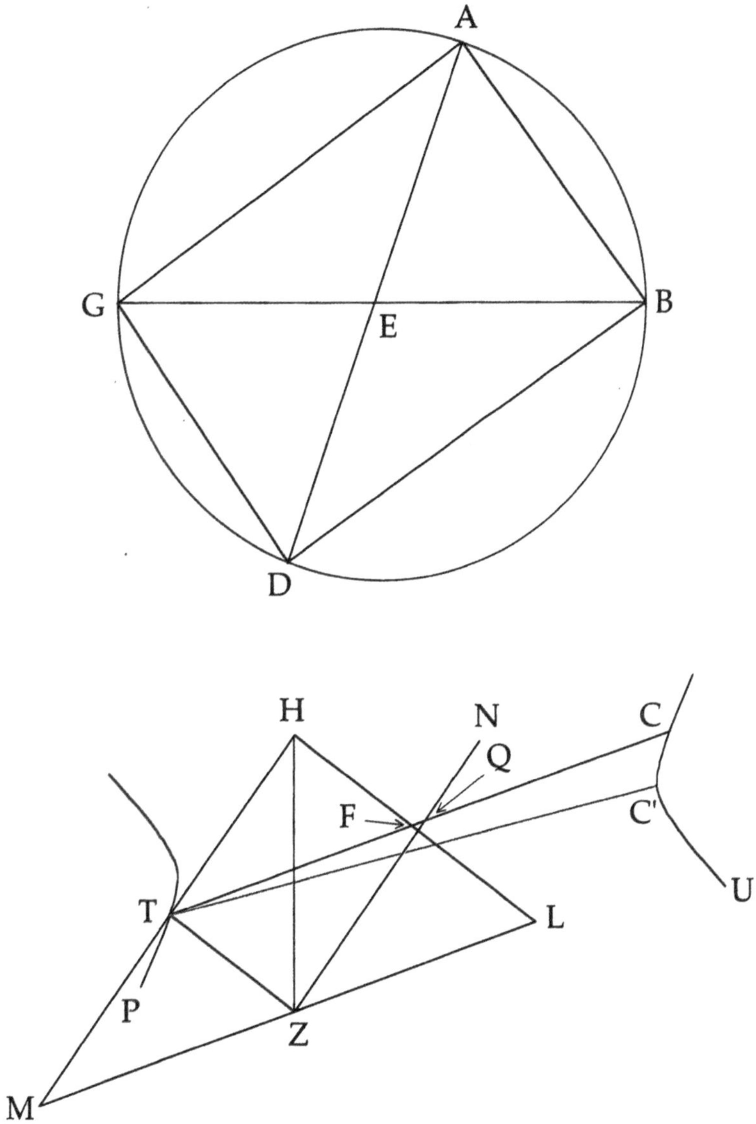

figure 5.2.20

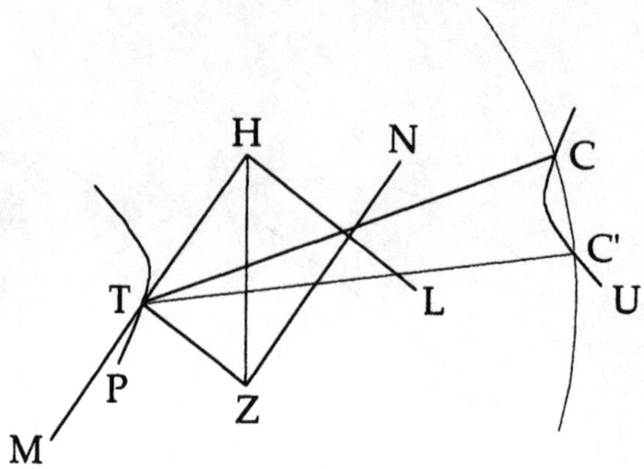

figure 5.2.20a

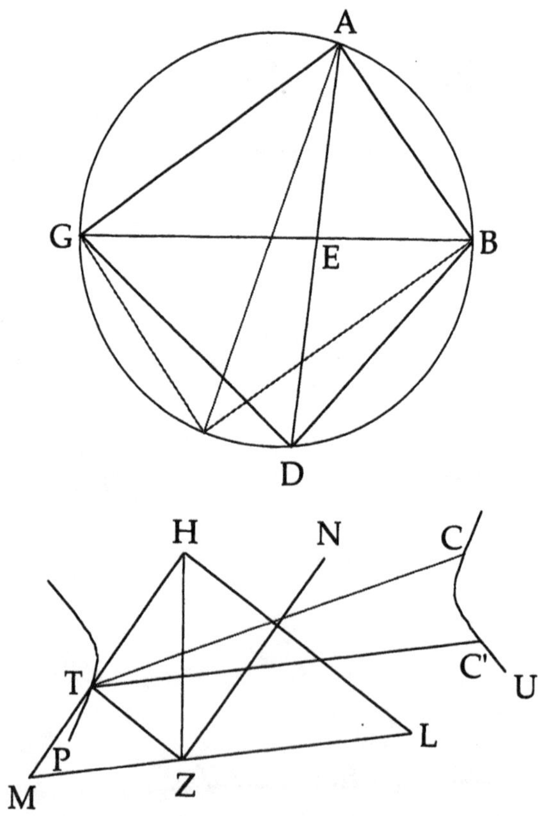

figure 5.2.20b

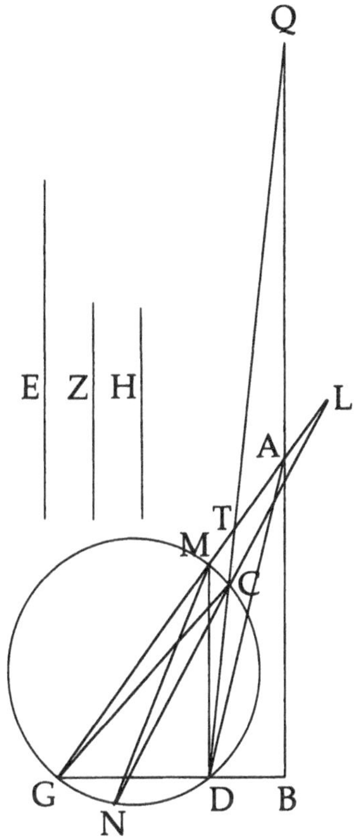

figure 5.2.21

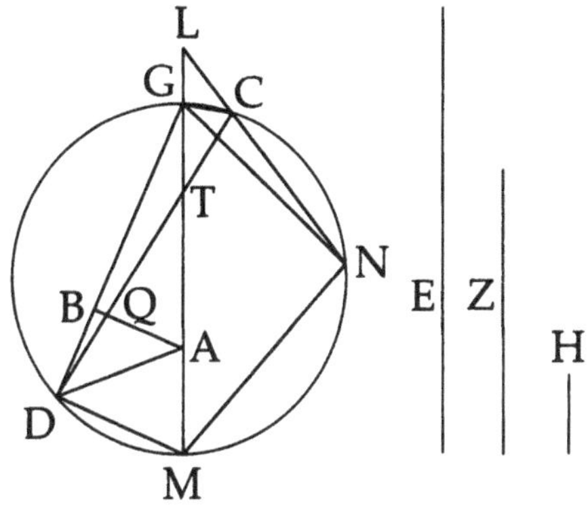

figure 5.2.21a

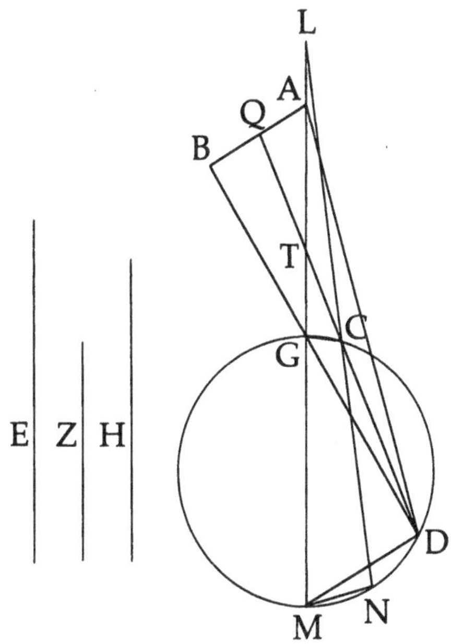

figure 5.2.21b

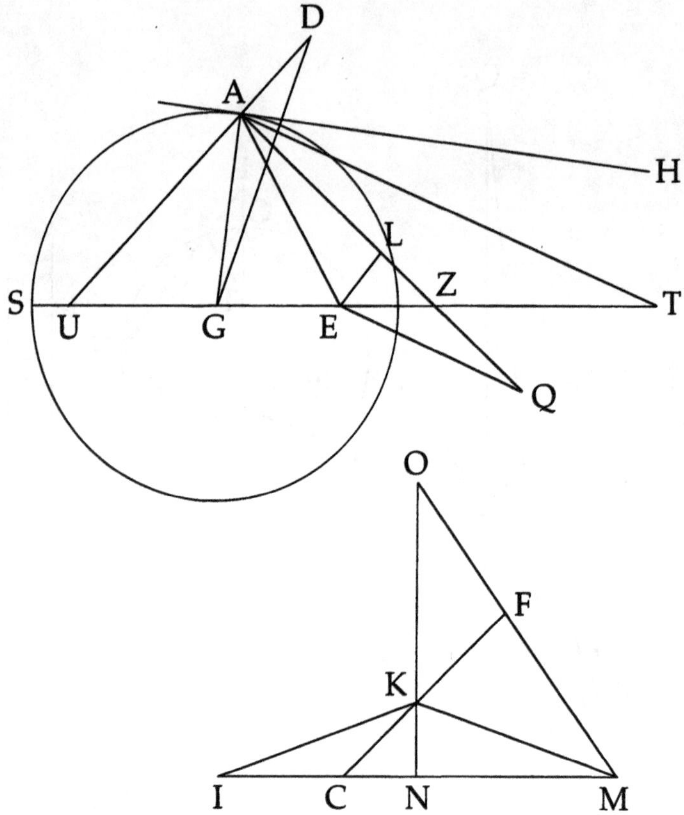

figure 5.2.22

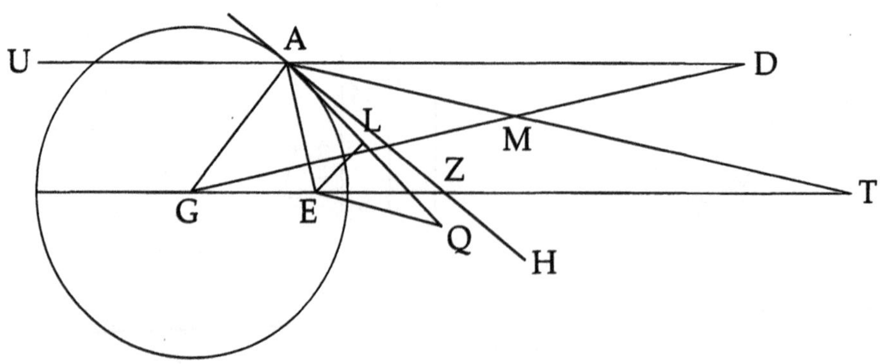

figure 5.2.22a

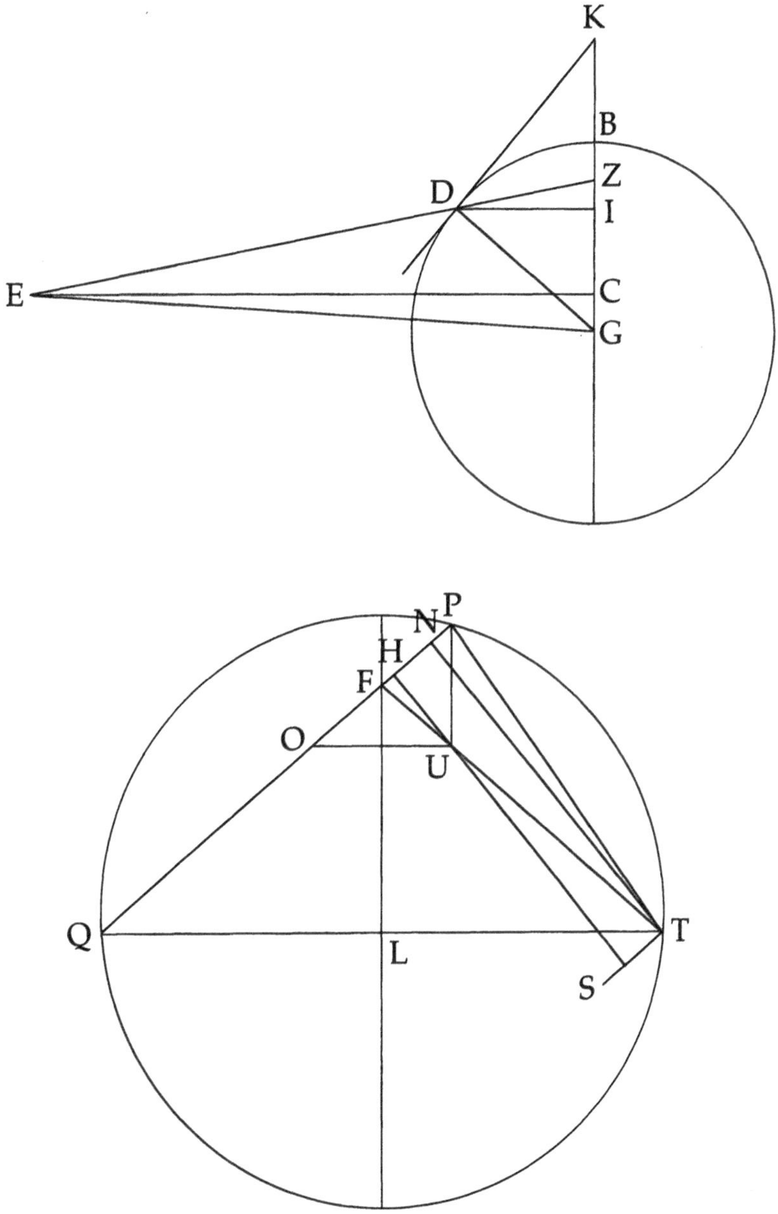

figure 5.2.23

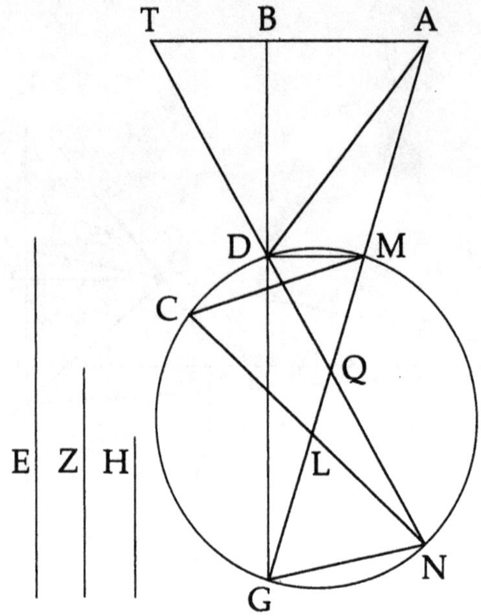

figure 5.2.24

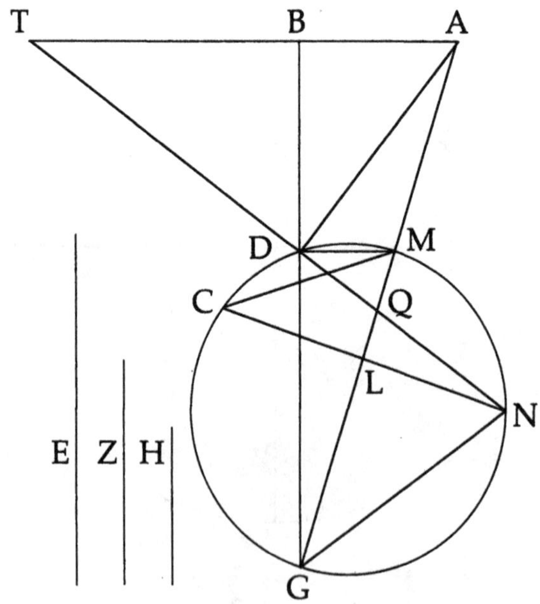

figure 5.2.24a

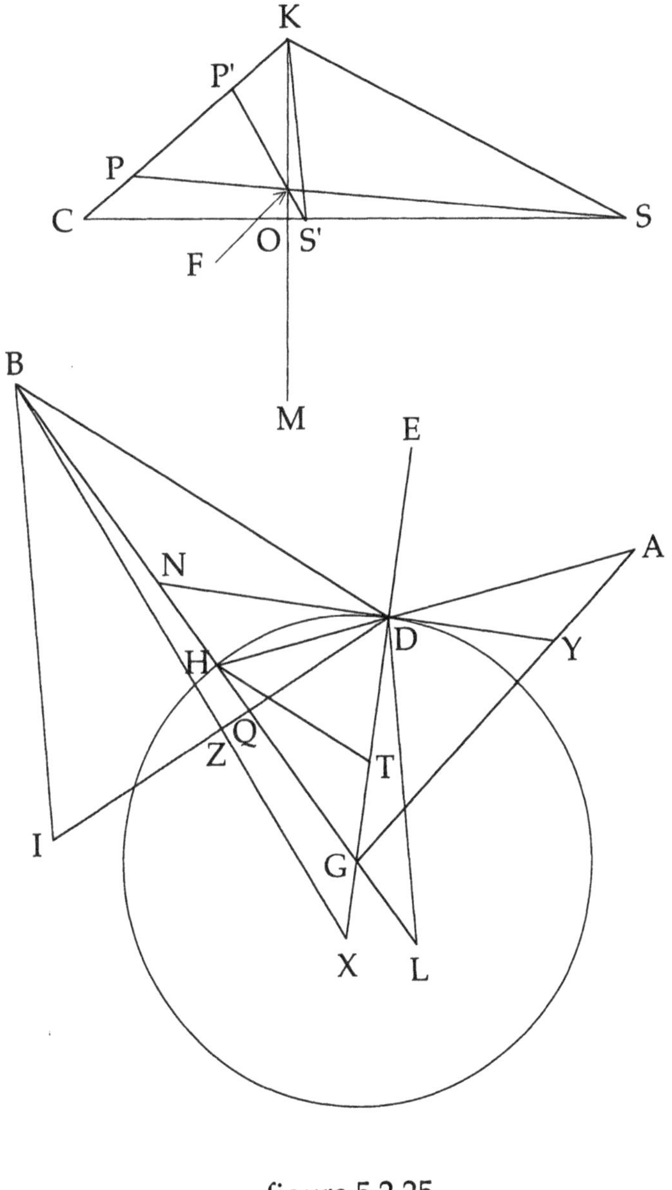

figure 5.2.25

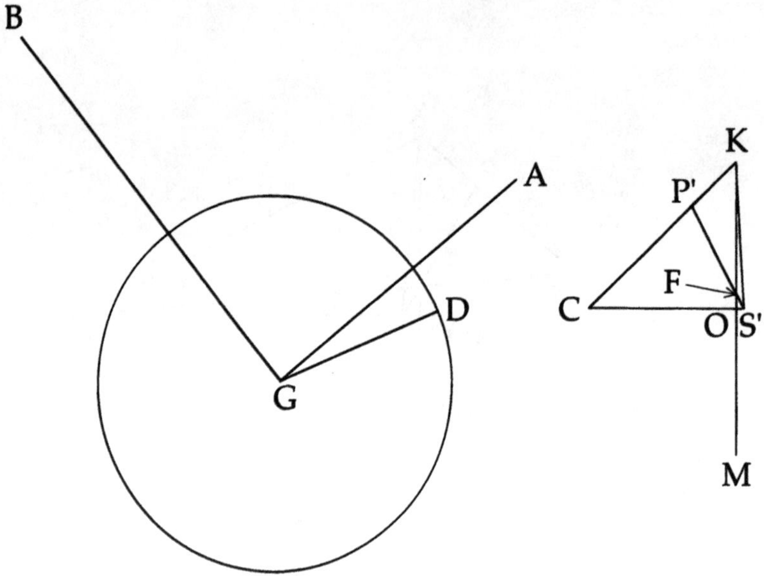

figure 5.2.25a

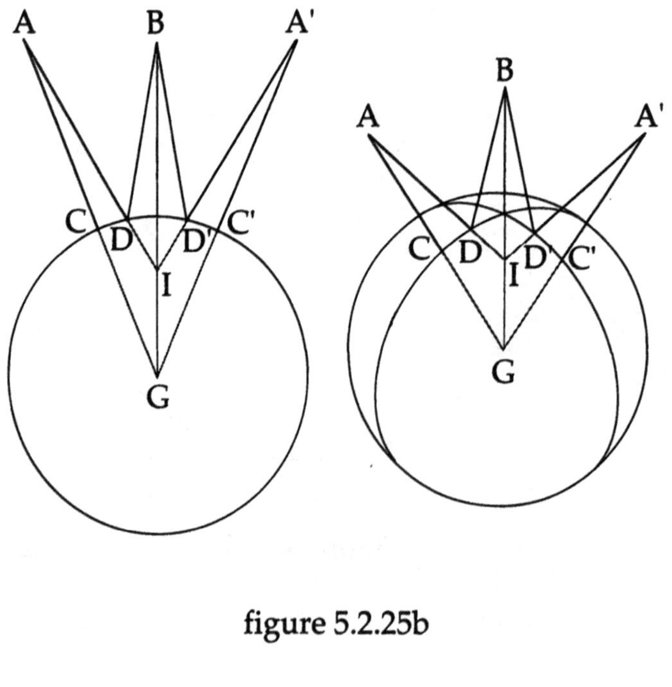

figure 5.2.25b

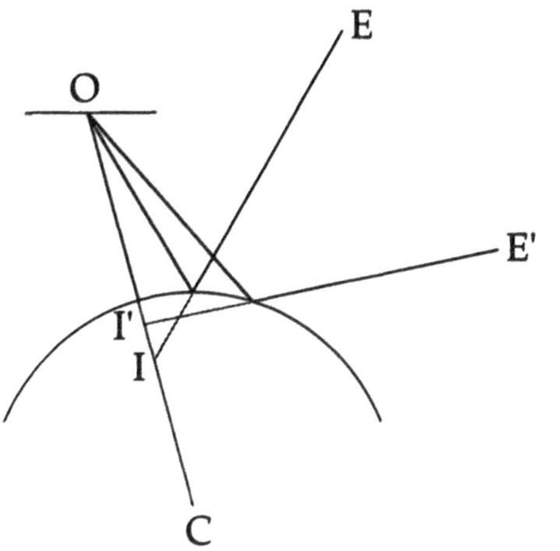

figure 5.2.25c

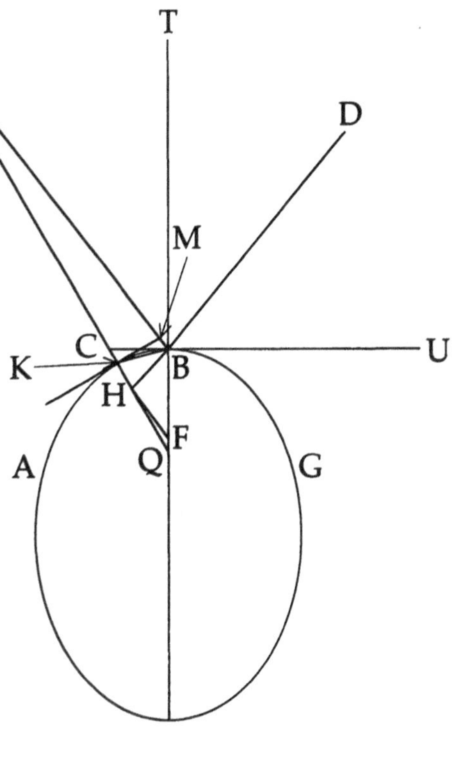

figure 5.2.26

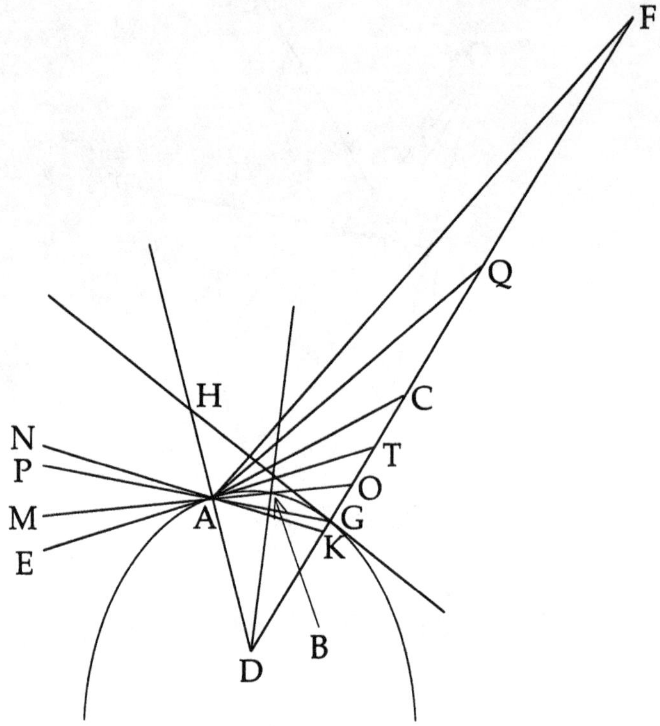

figure 5.2.27

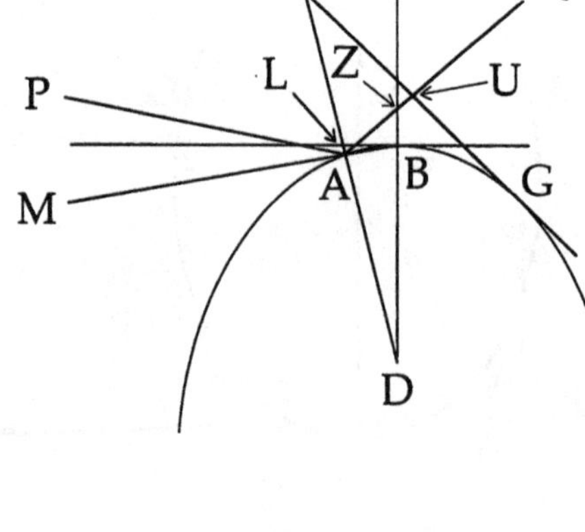

figure 5.2.27a

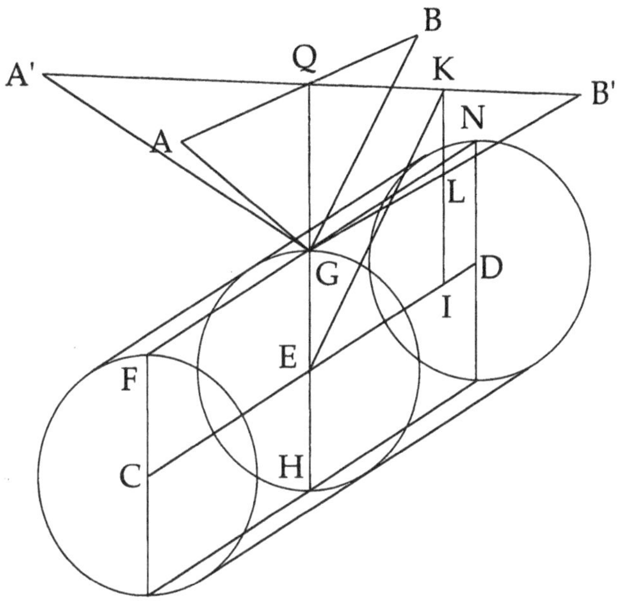

figure 5.2.28

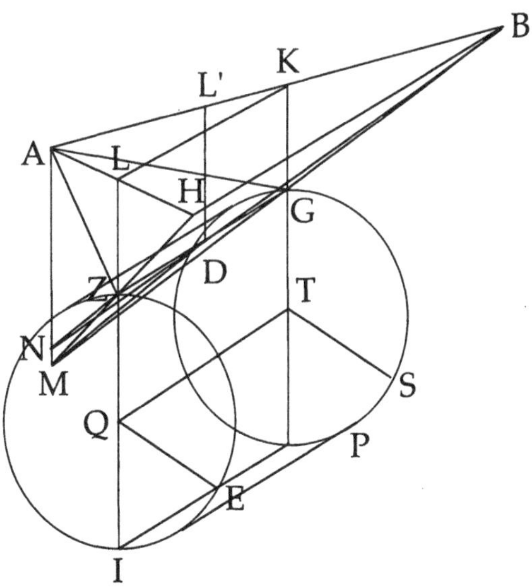

figure 5.2.28a

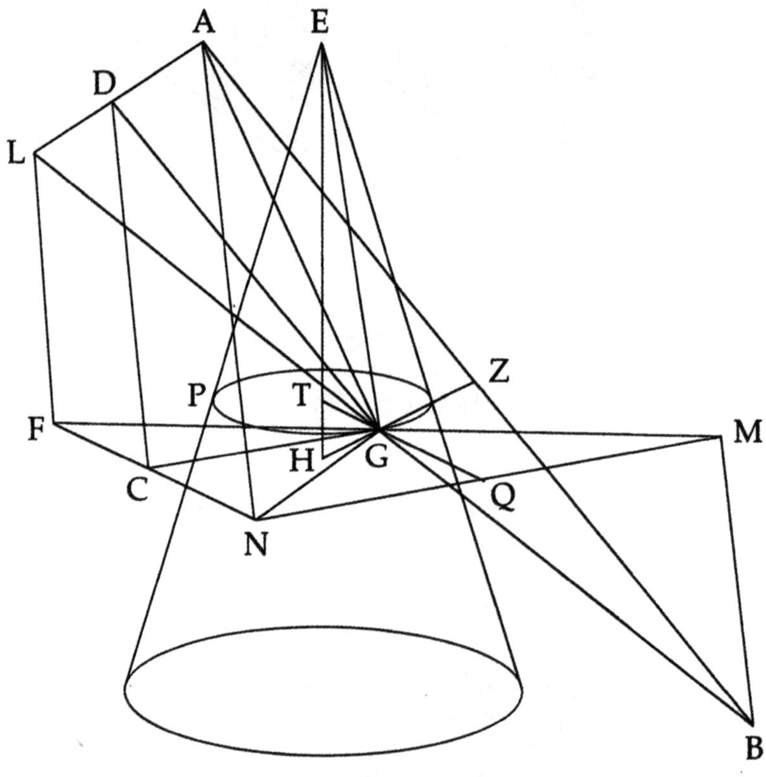

figure 5.2.30

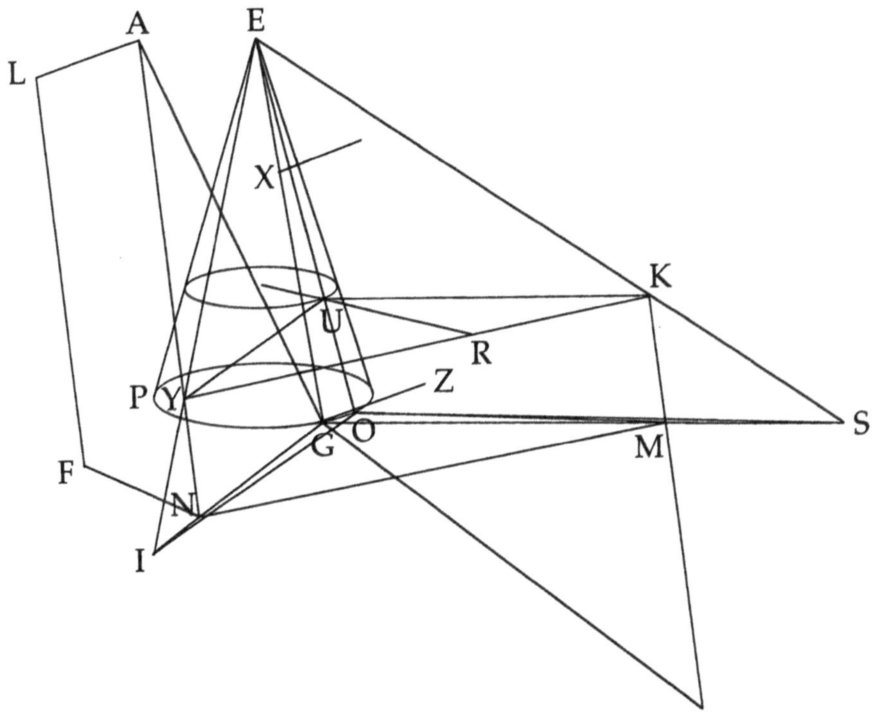

figure 5.2.30a

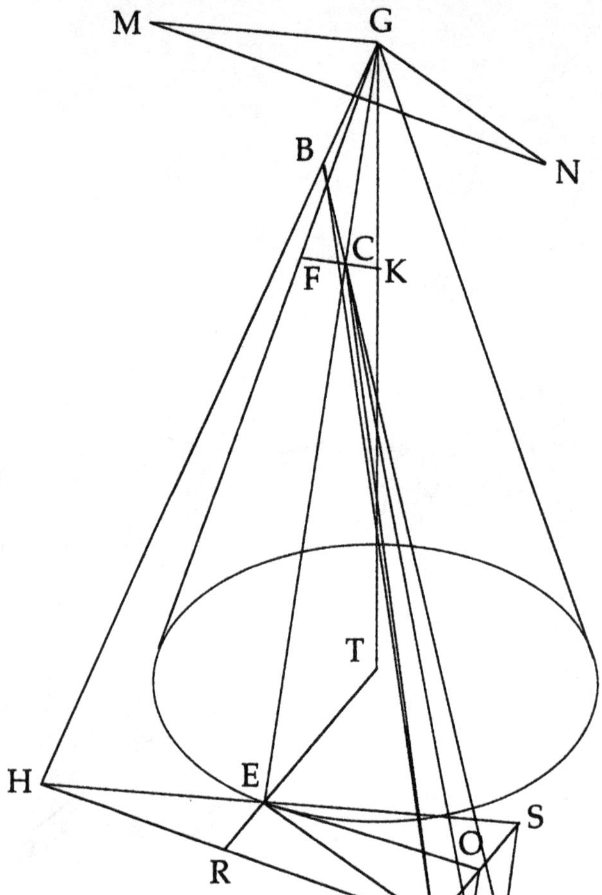

figure 5.2.31

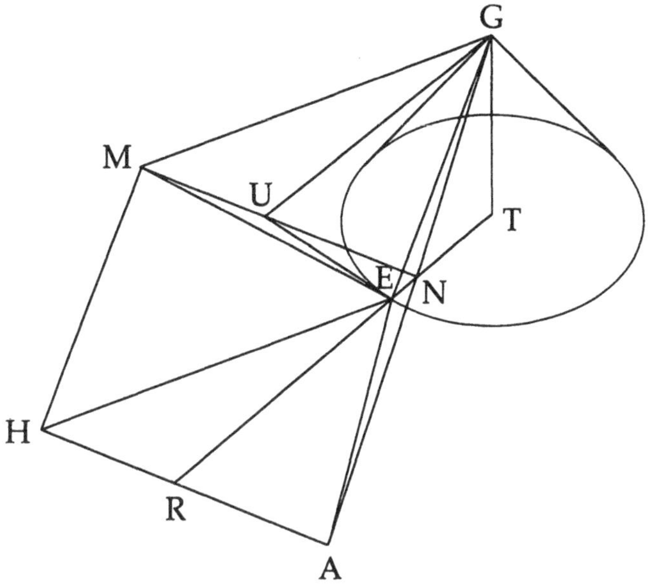

figure 5.2.31a

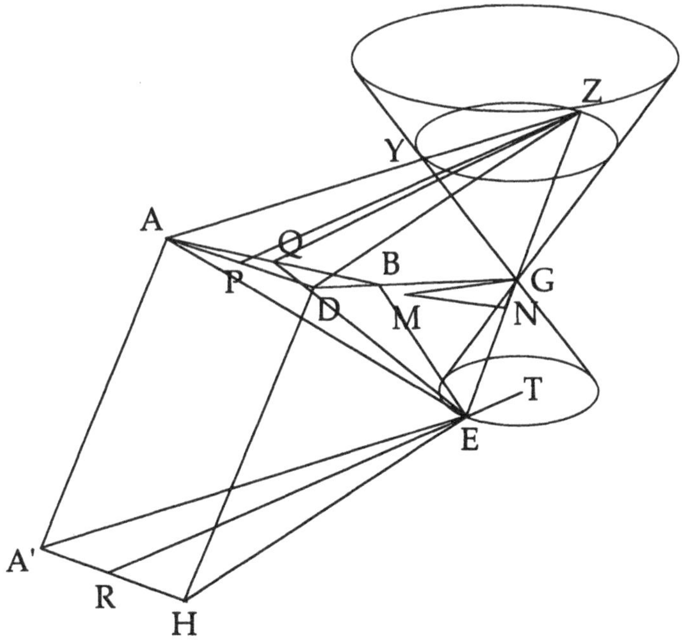

figure 5.2.31b

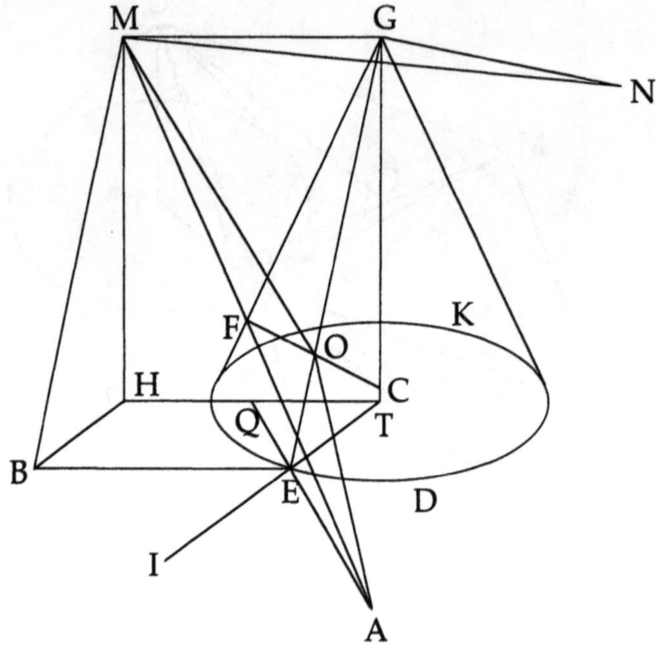

figure 5.2.31c

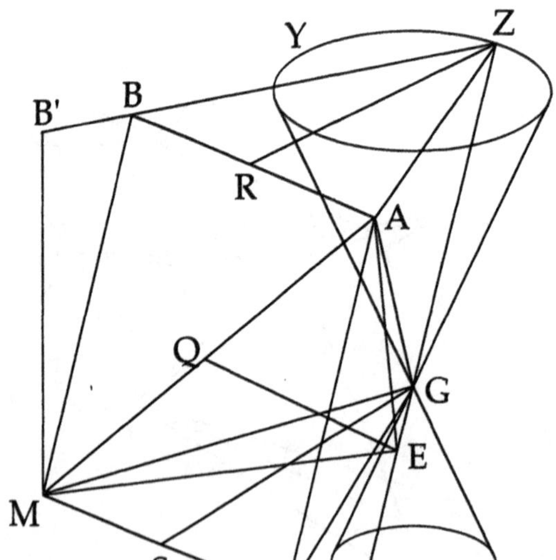

figure 5.2.31d

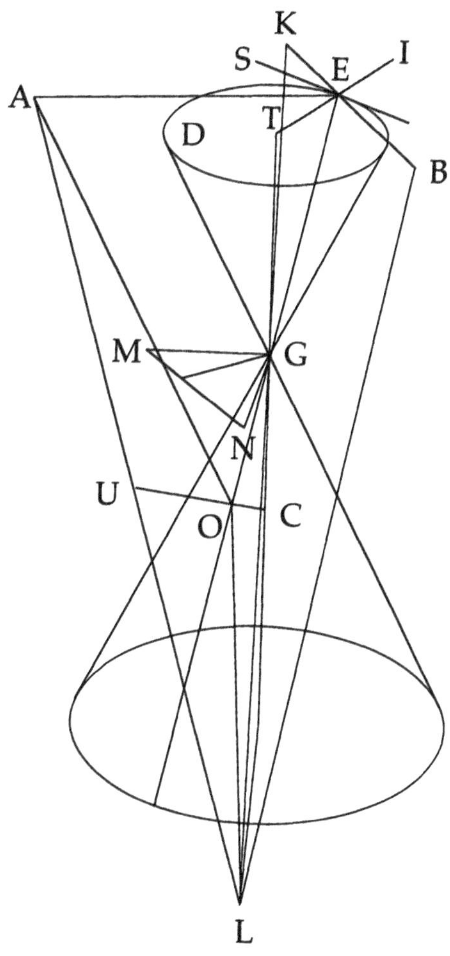

figure 5.2.31e

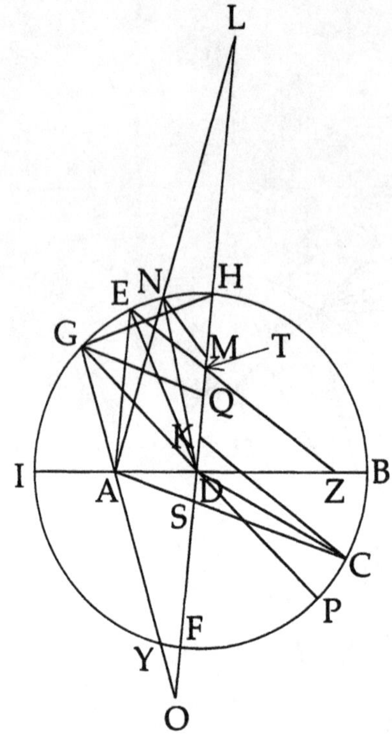

figure 5.2.32

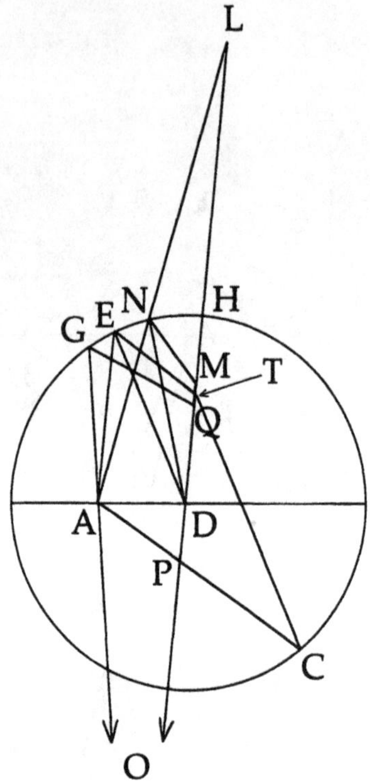

figure 5.2.32a

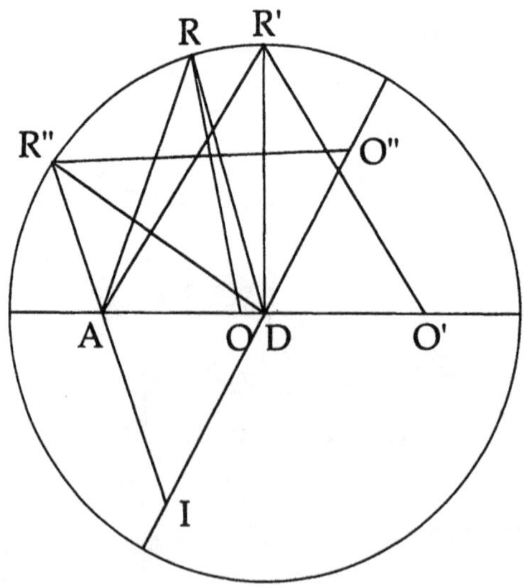

figure 5.2.32b

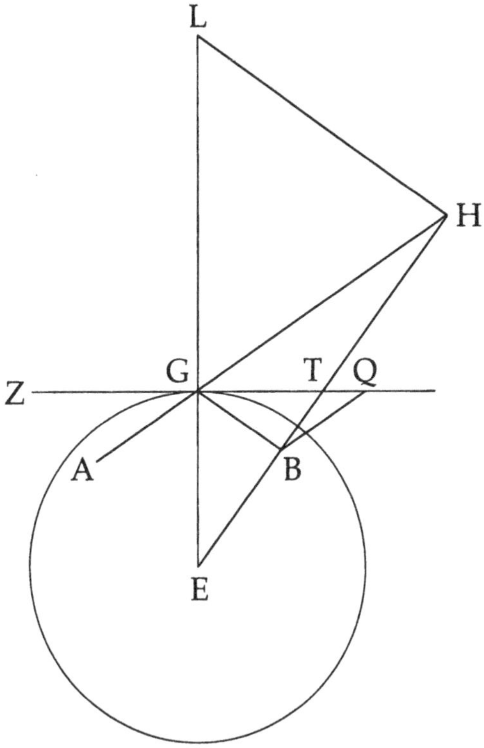

figure 5.2.33

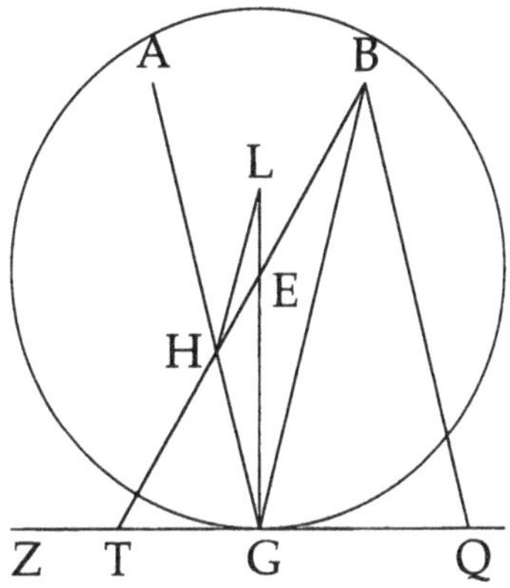

figure 5.2.33a

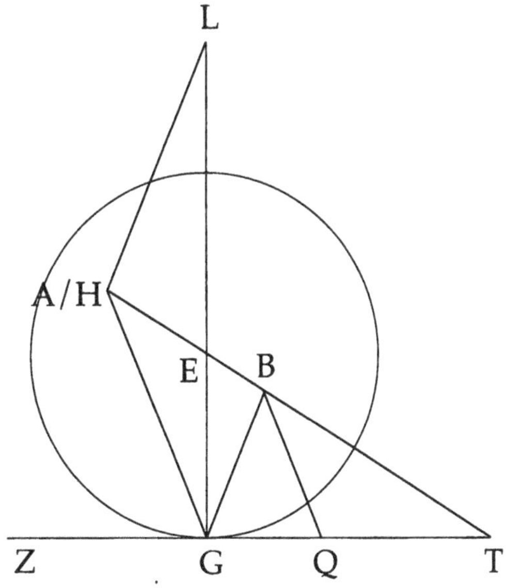

figure 5.2.33b

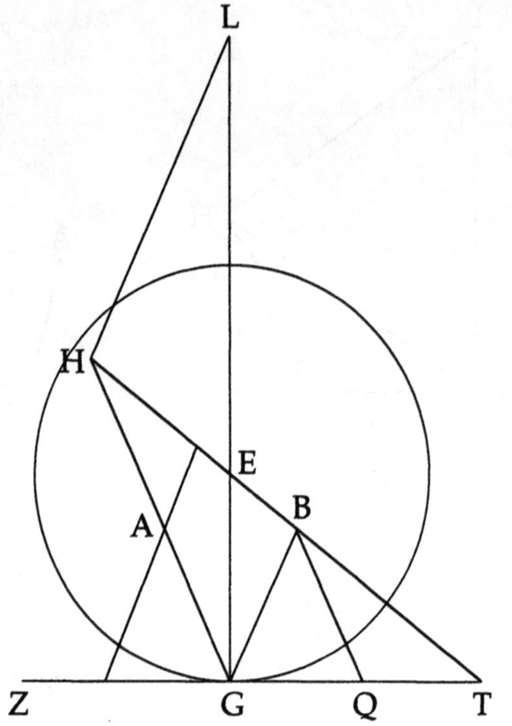

figure 5.2.33c

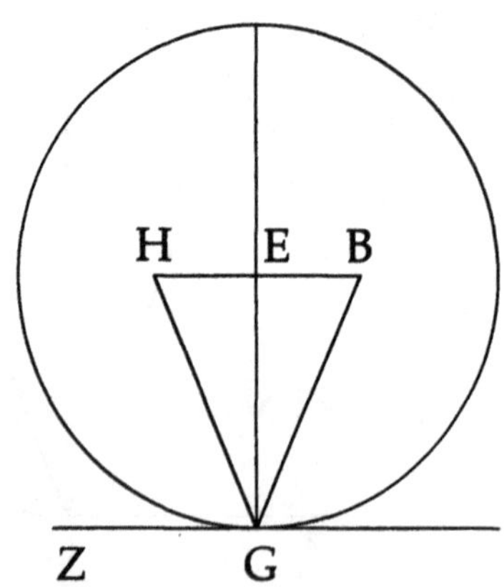

figure 5.2.33d

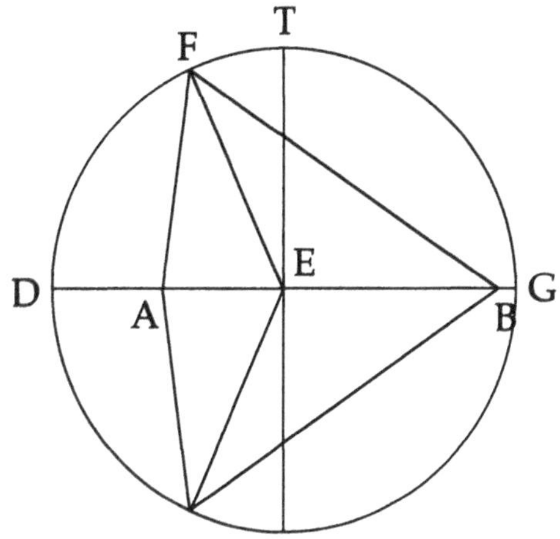

figure 5.2.34

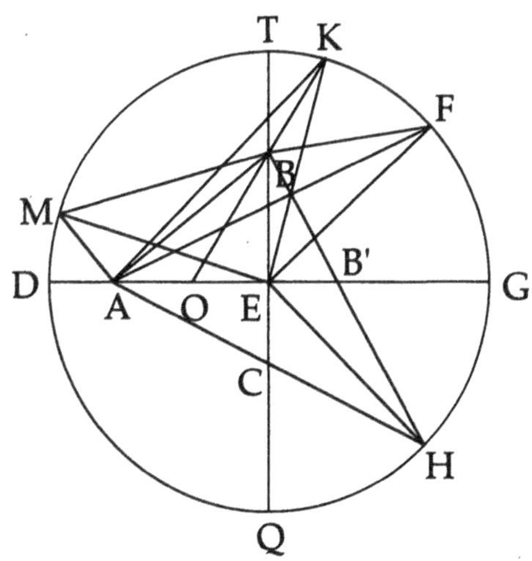

figure 5.2.34a

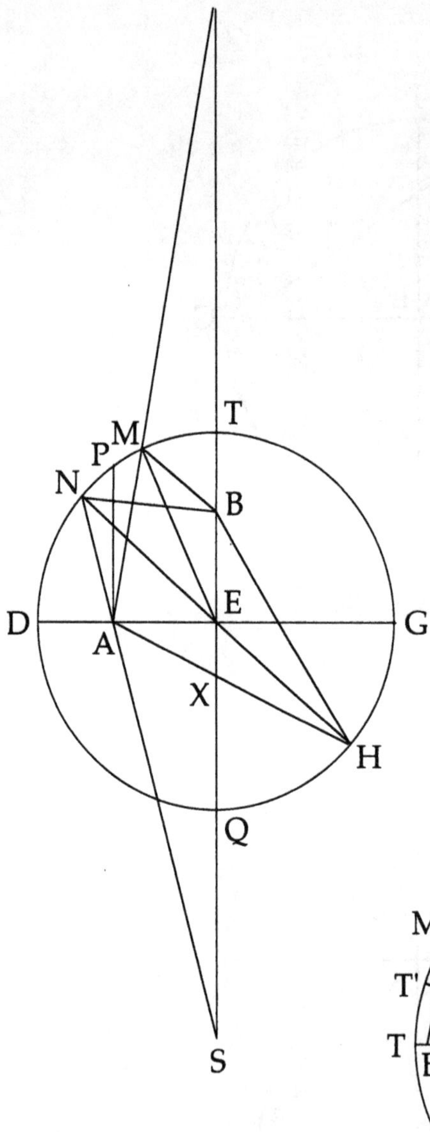

figure 5.2.34b

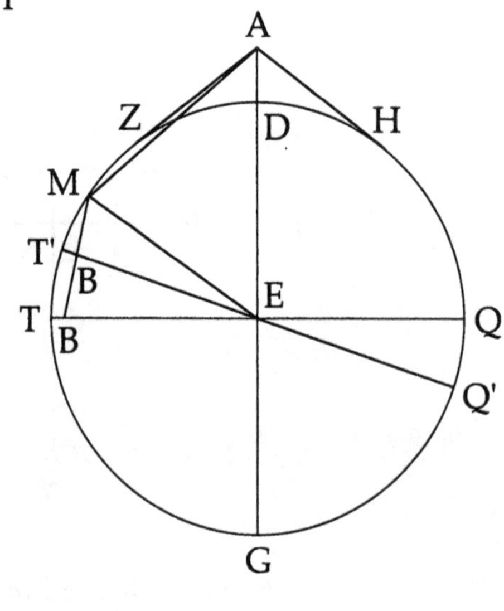

figure 5.2.34c

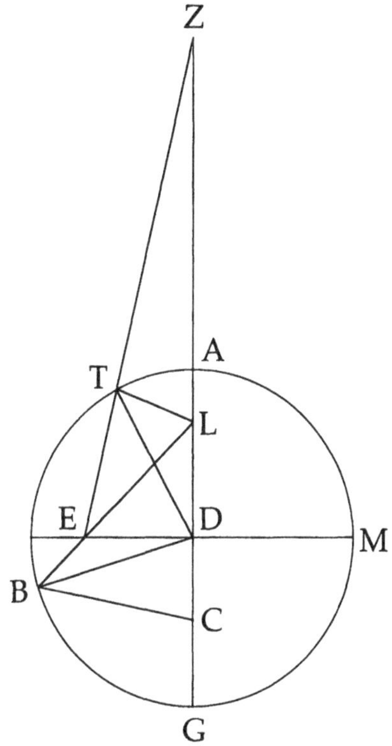

figure 5.2.35

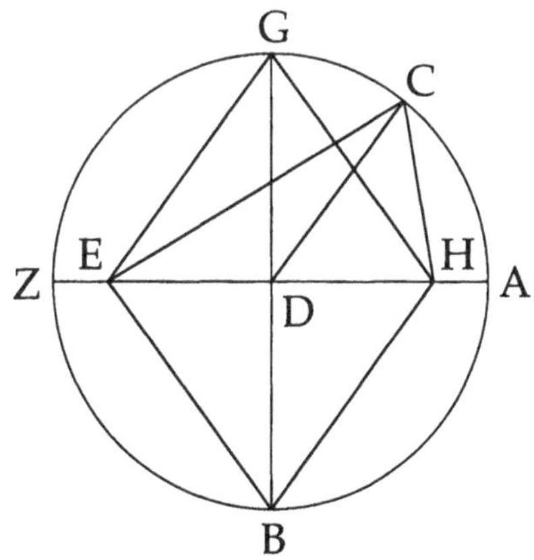

figure 5.2.36

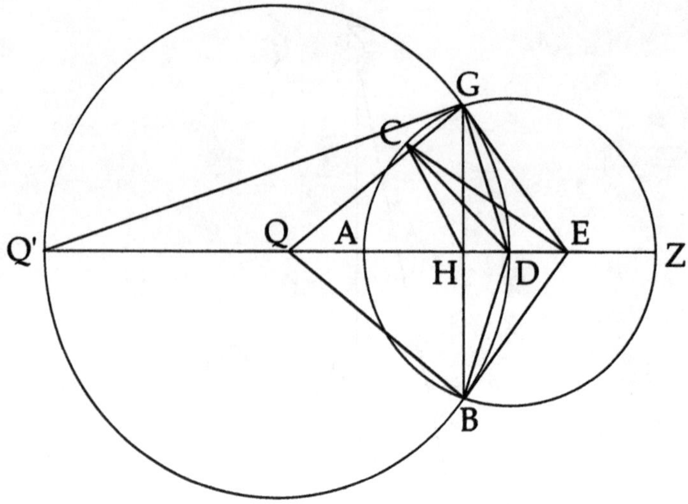

figure 5.2.36a

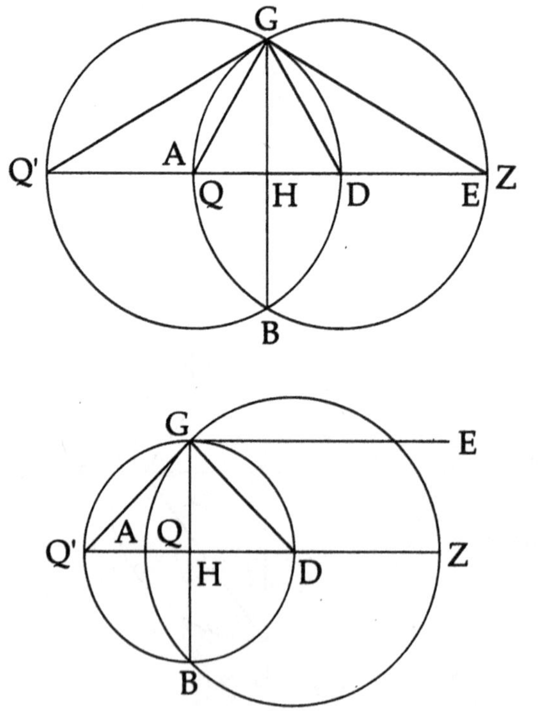

figure 5.2.36b

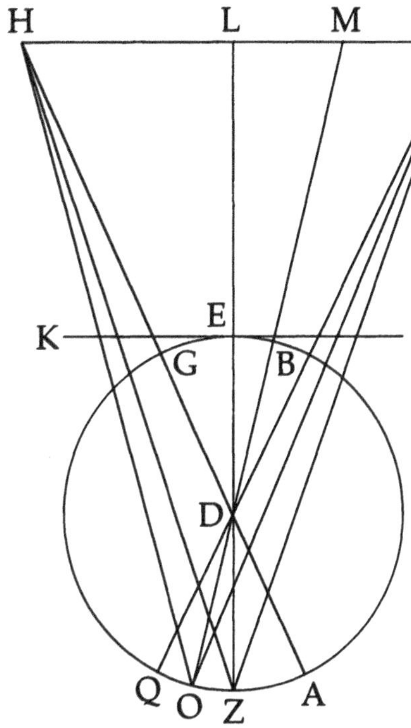

figure 5.2.37

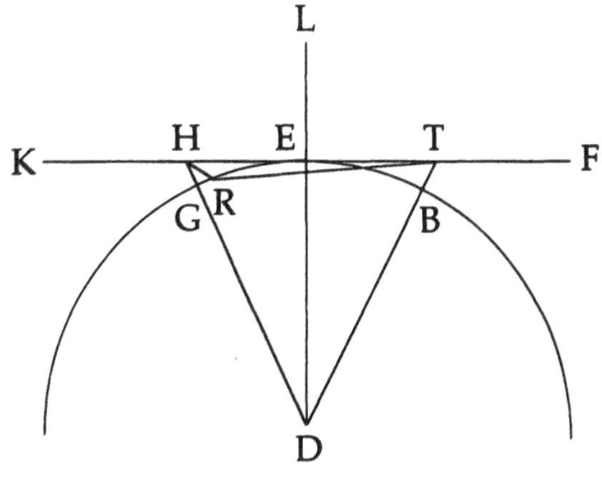

figure 5.2.37a

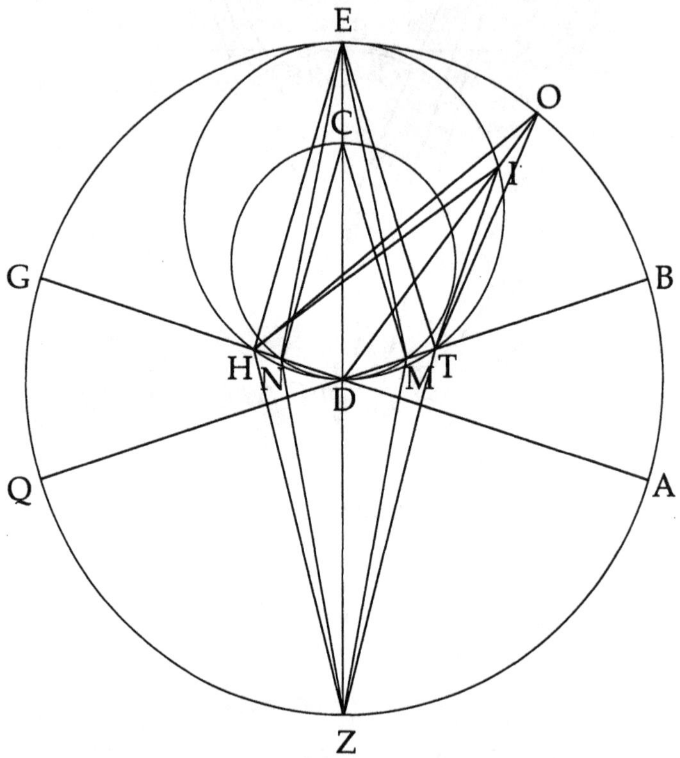

figure 5.2.37b

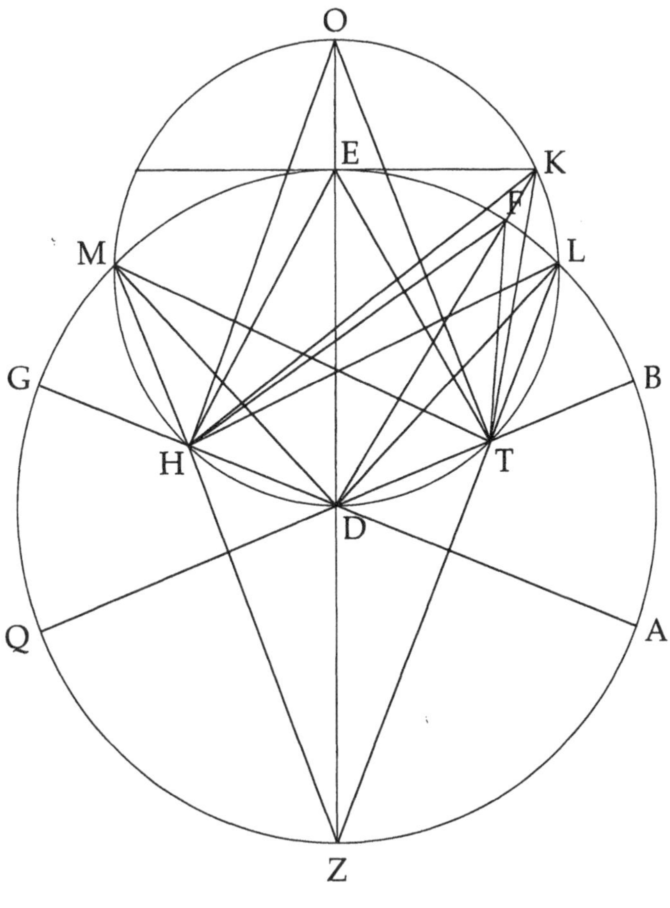

figure 5.2.37c

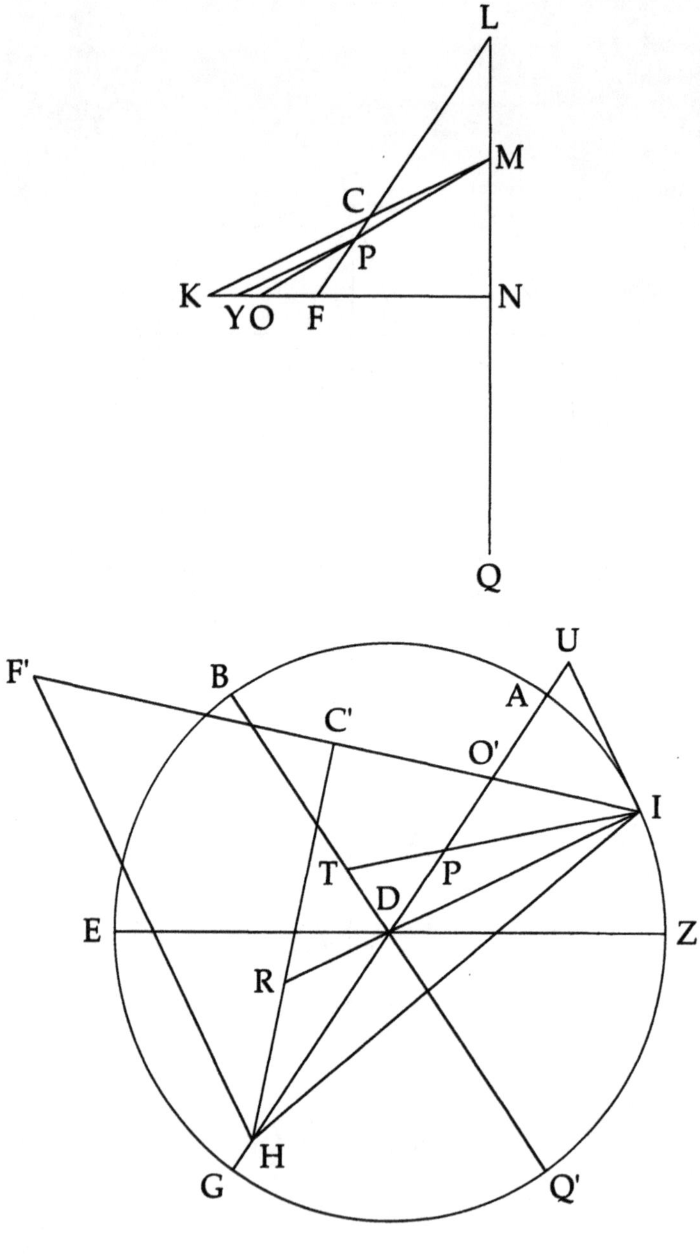

figure 5.2.38

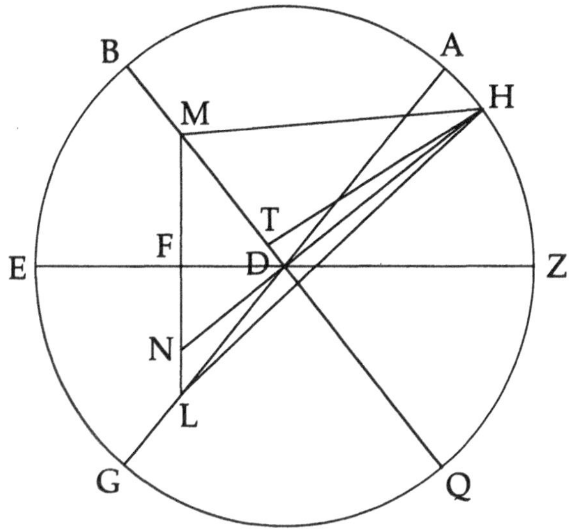

figure 5.2.39

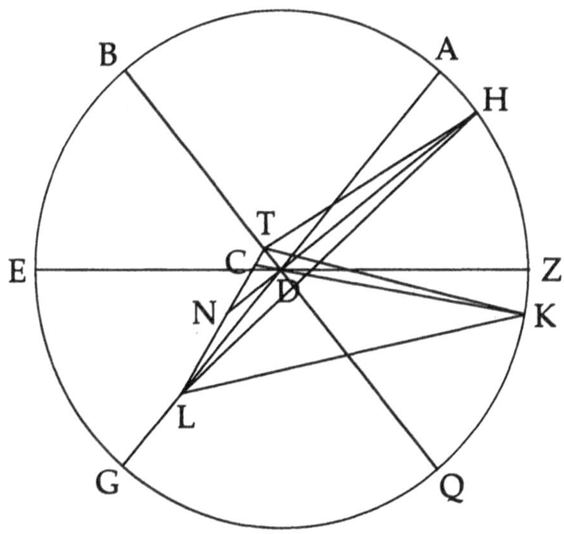

figure 5.2.40

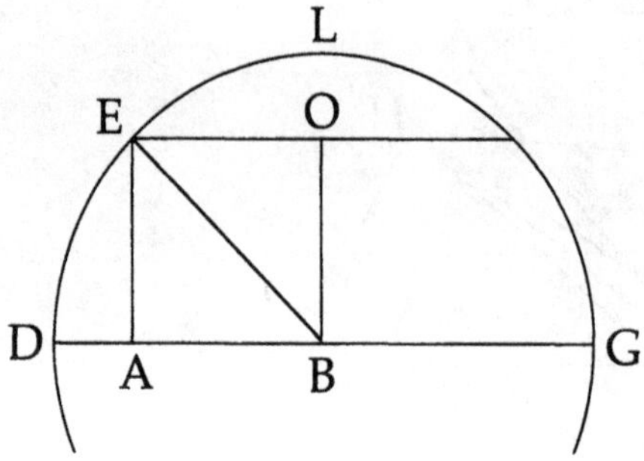

figure 5.2.41

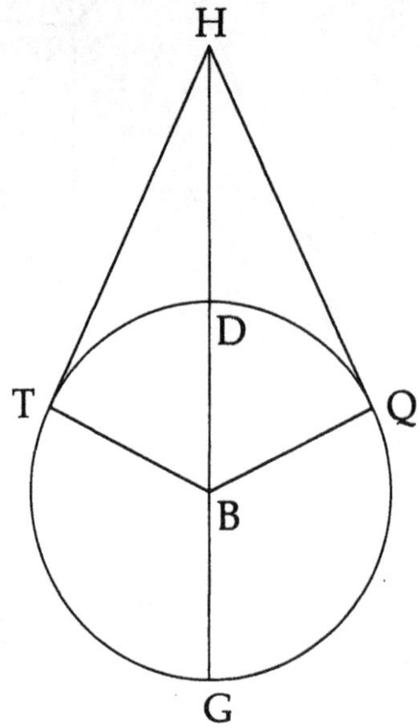

figure 5.2.41a

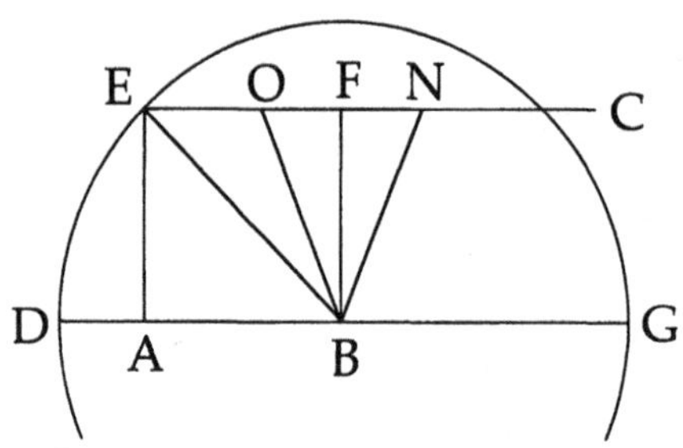

figure 5.2.42

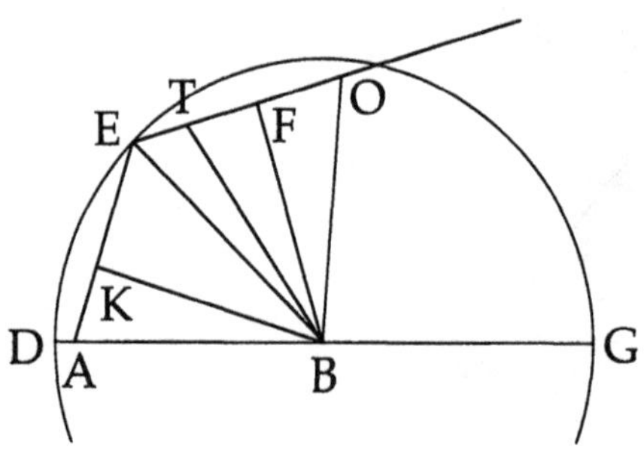

figure 5.2.42a

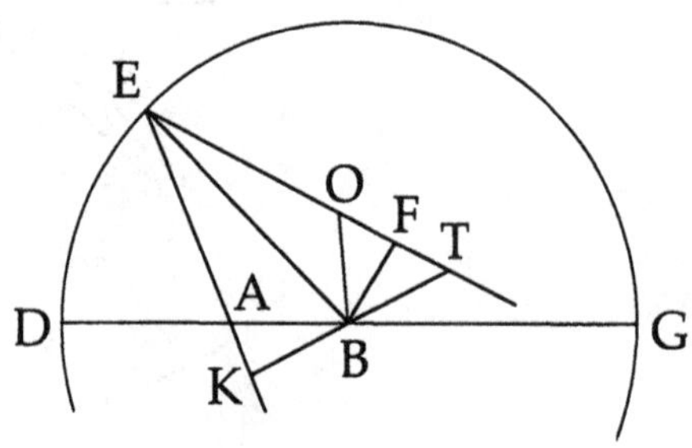

figure 5.2.42b

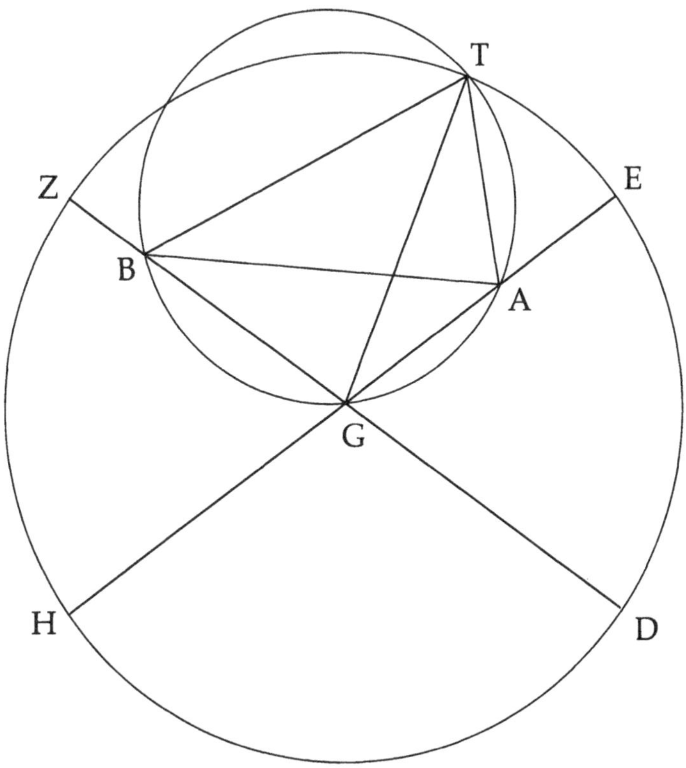

figure 5.2.43

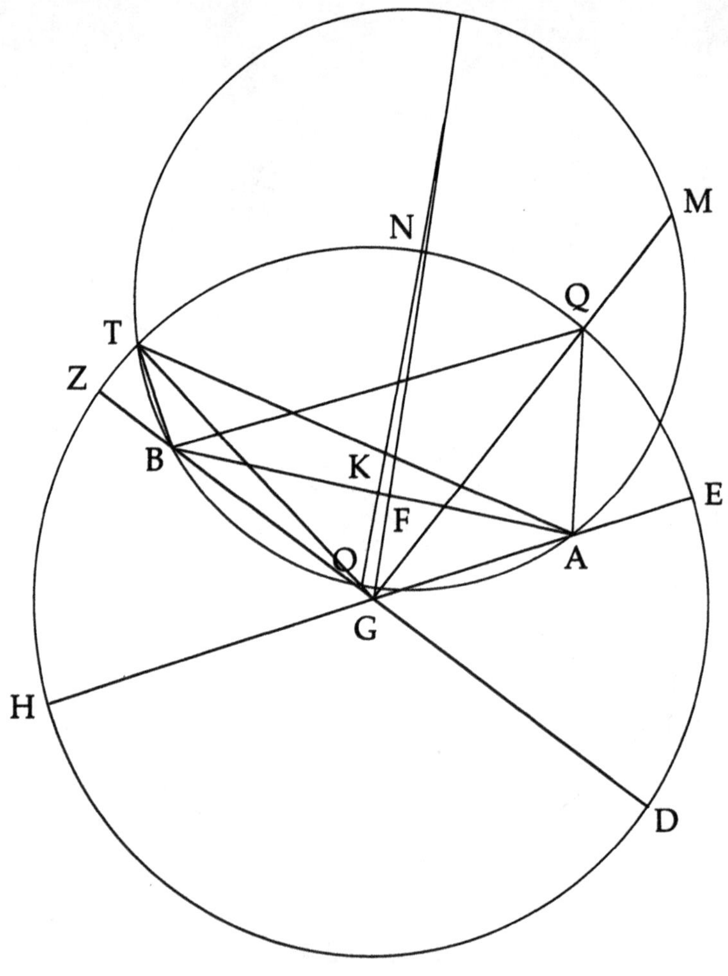

figure 5.2.44

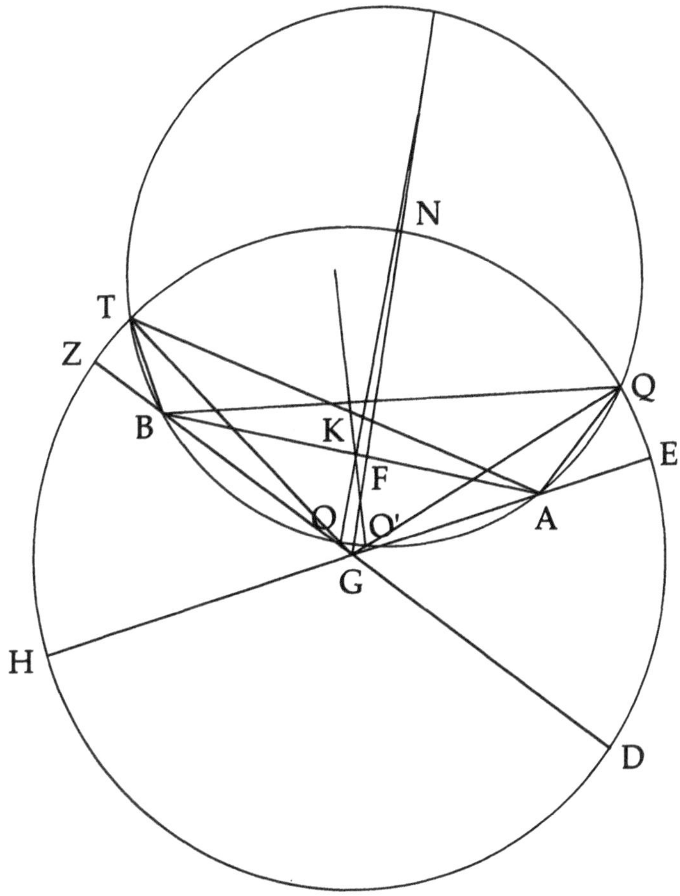

figure 5.2.44a

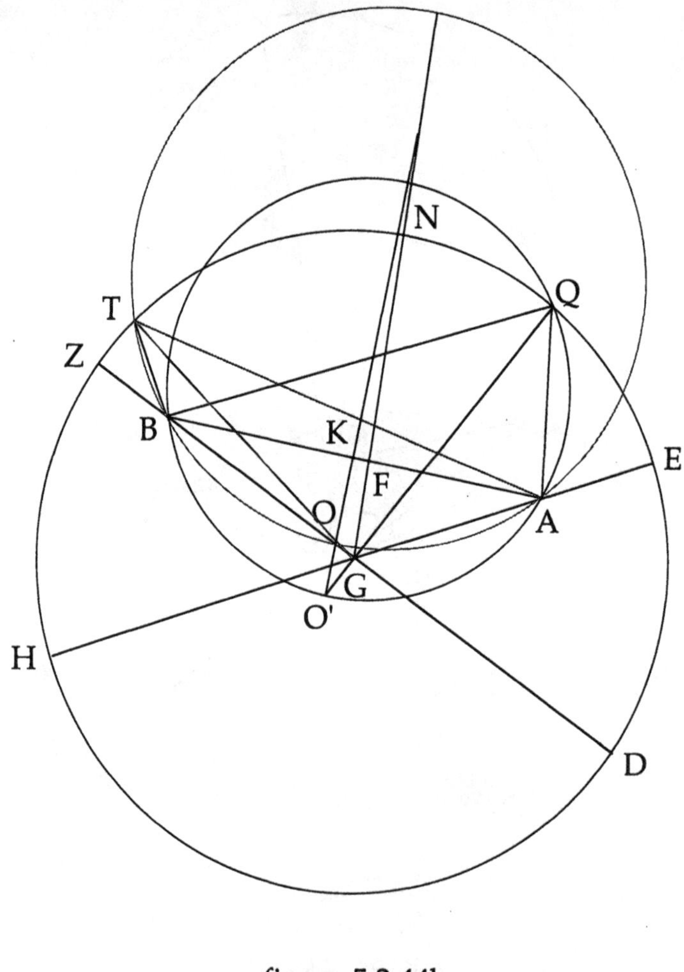

figure 5.2.44b

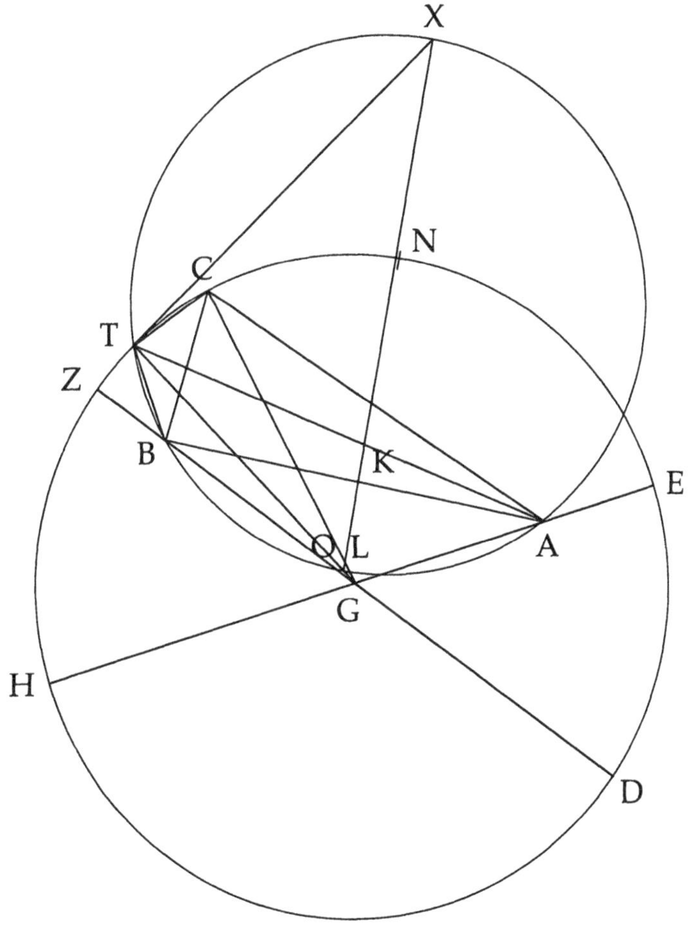

figure 5.2.44c

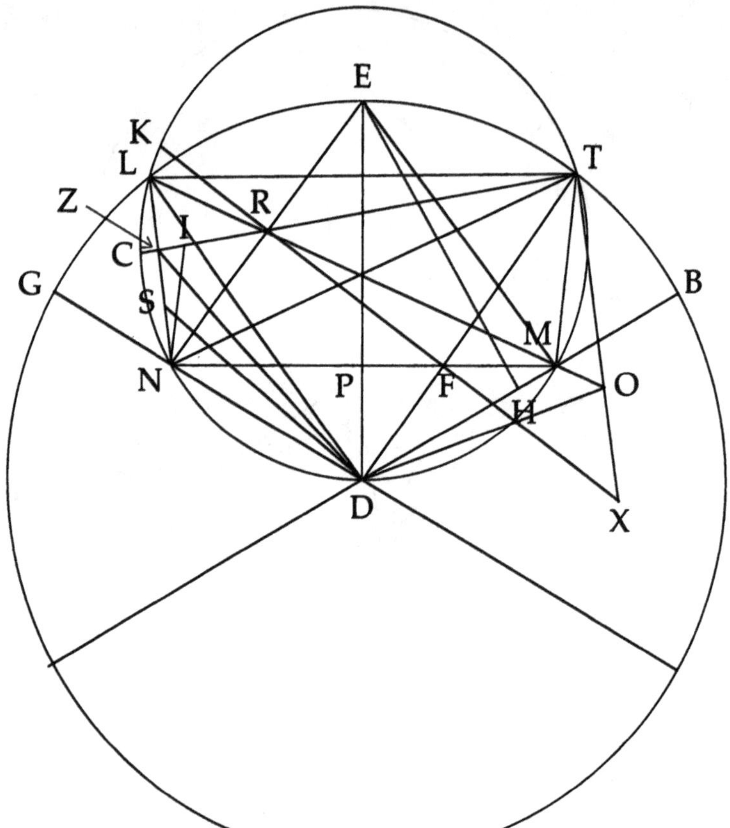

figure 5.2.45

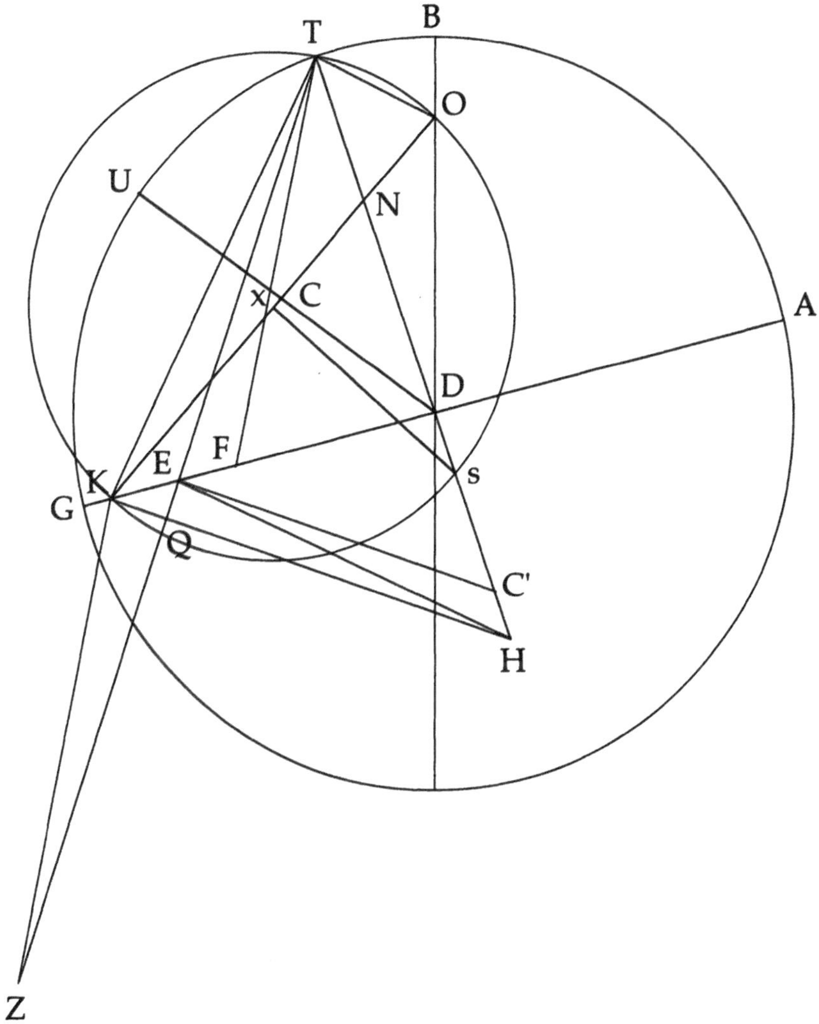

figure 5.2.46

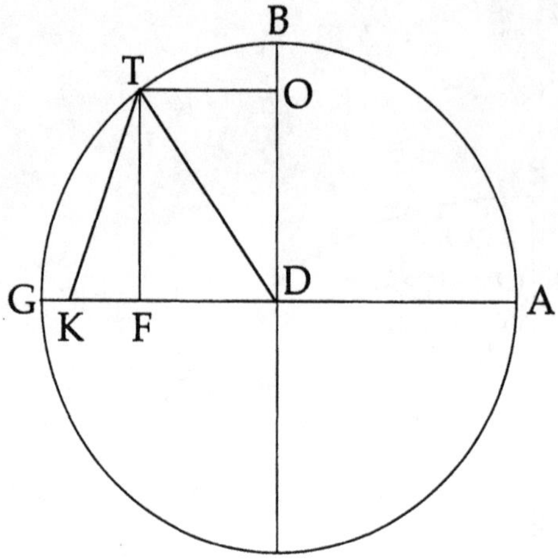

figure 5.2.46a

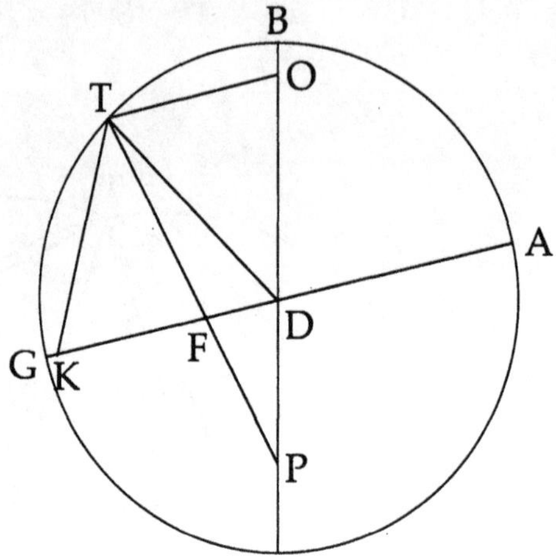

figure 5.2.46b

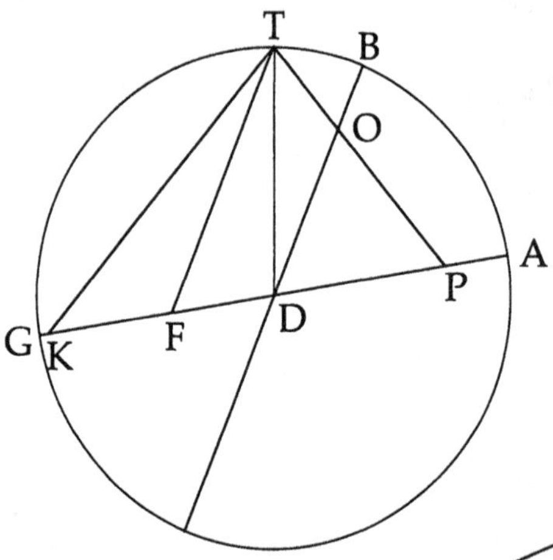

figure 5.2.46c

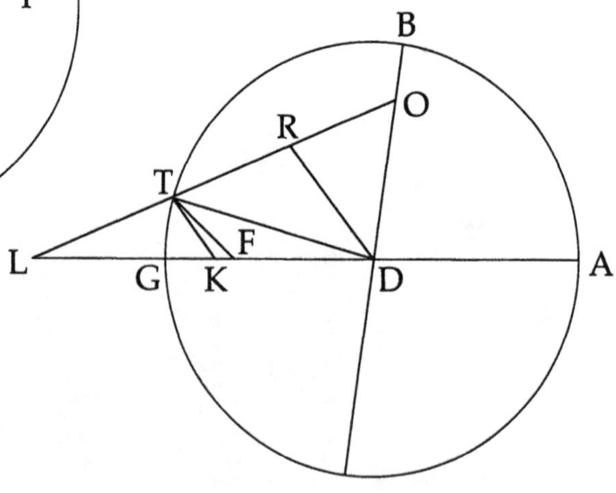

figure 5.2.46d

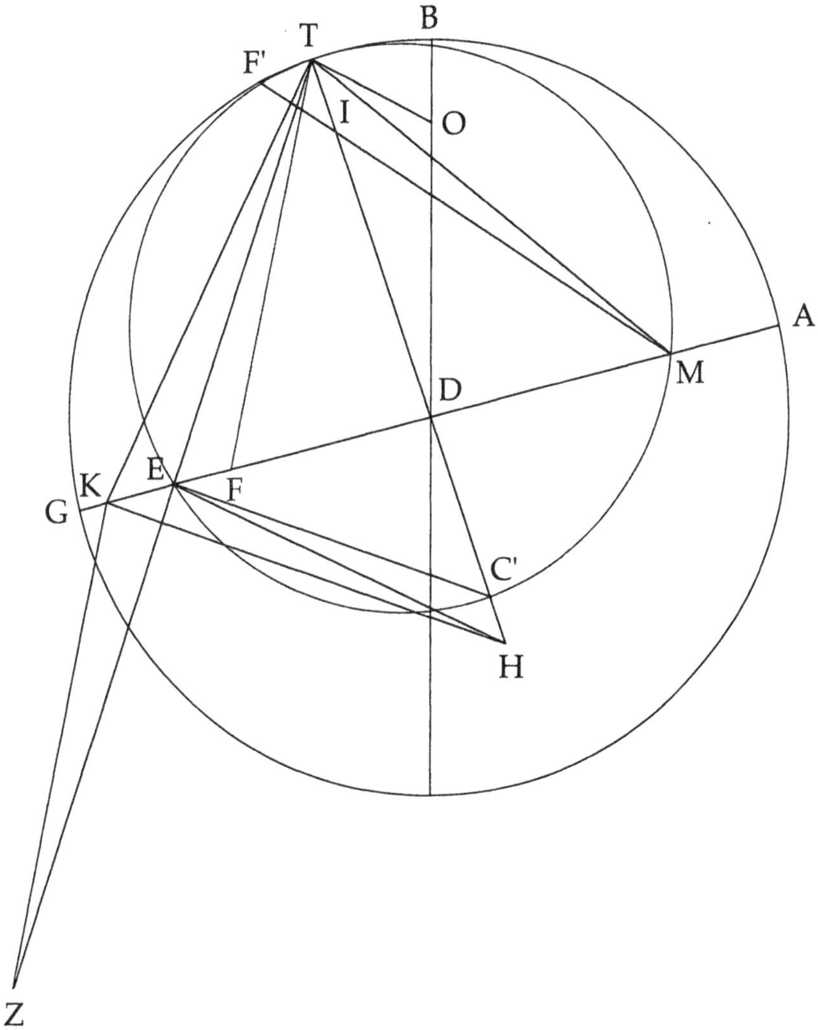

figure 5.2.46e

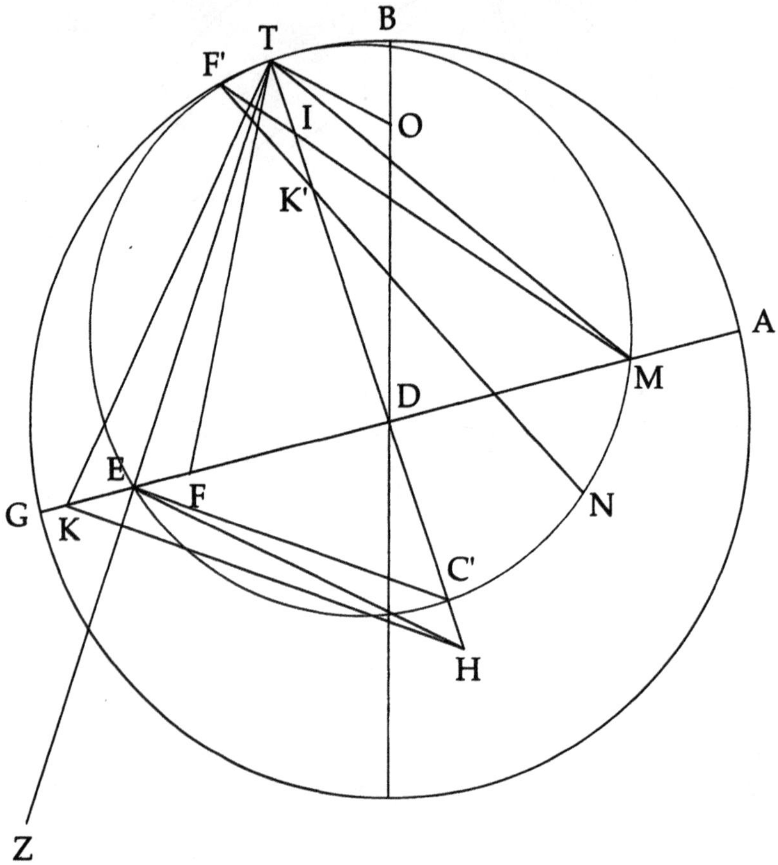

figure 5.2.46f

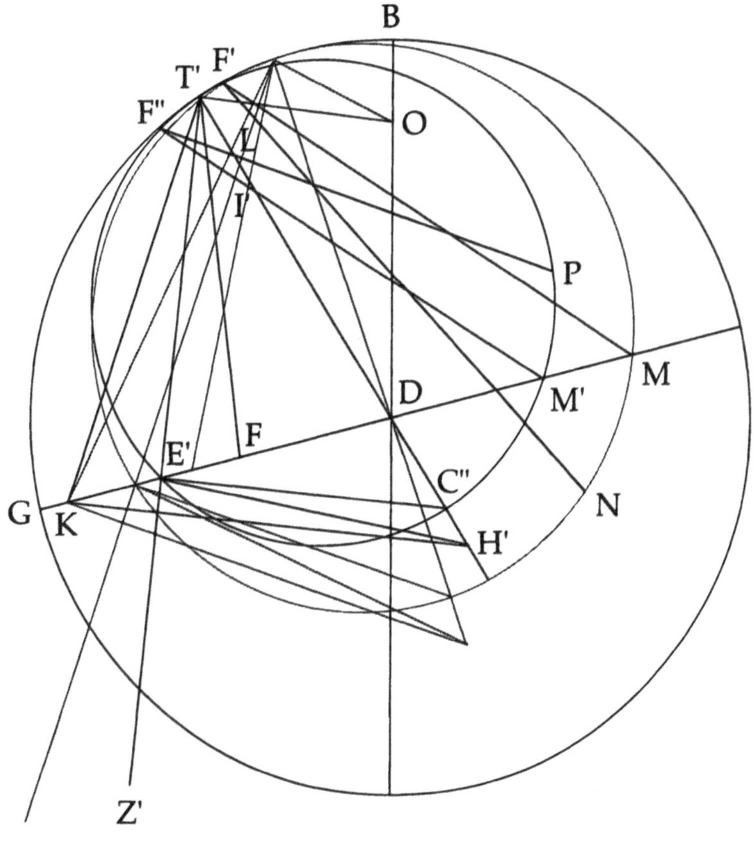

figure 5.2.46g

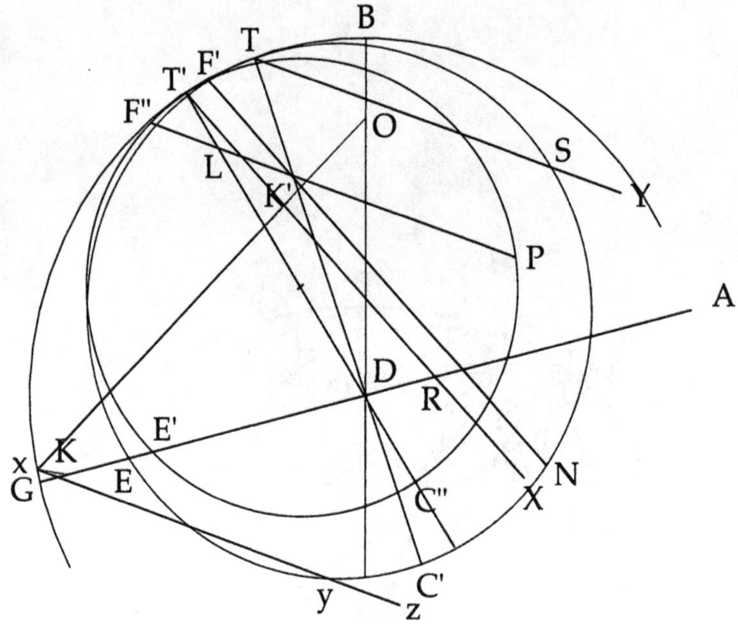

figure 5.2.46h

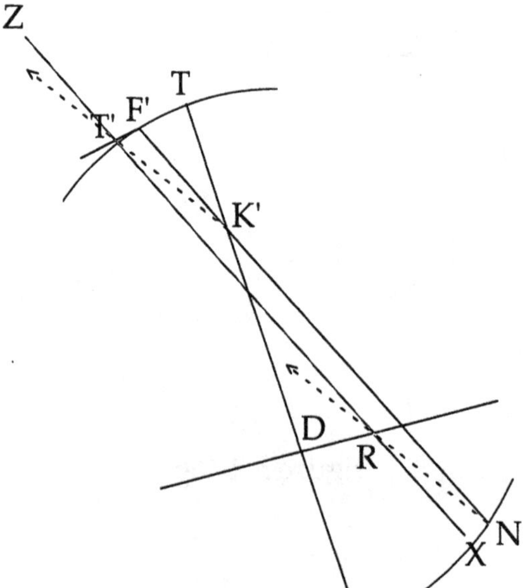

figure 5.2.46k

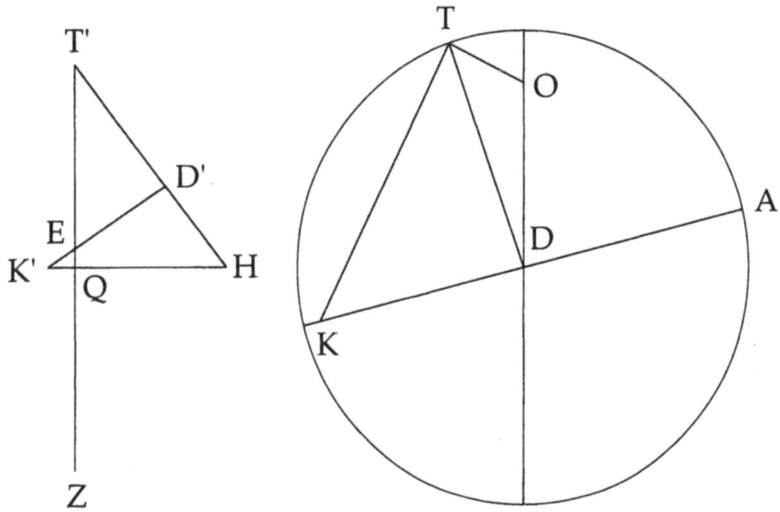

figure 5.2.47

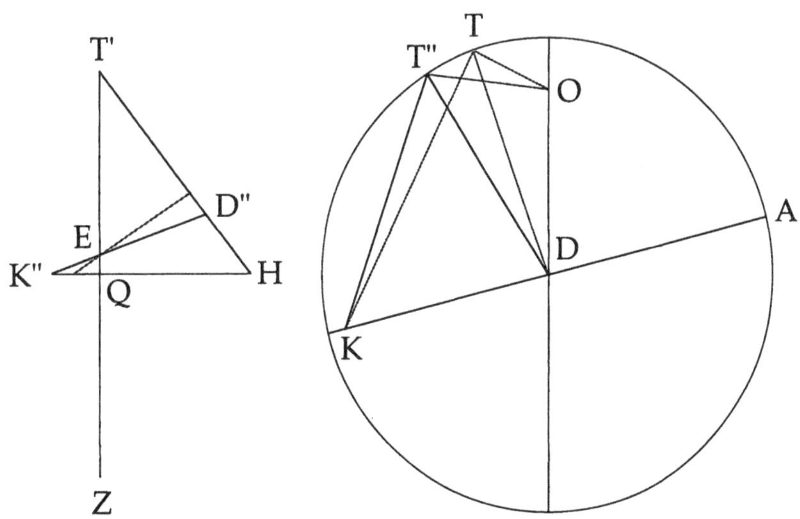

figure 5.2.47a

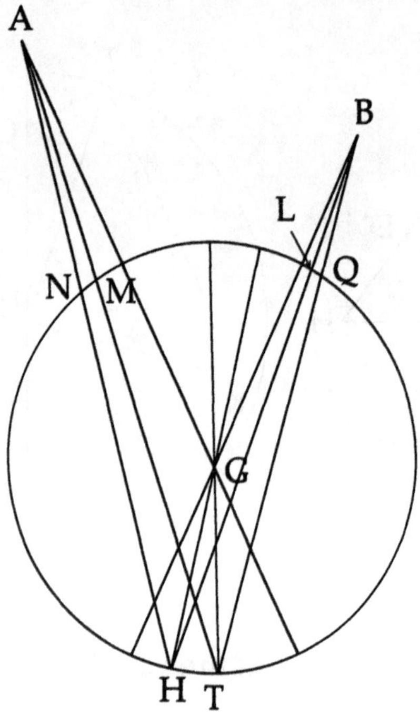

figure 5.2.48

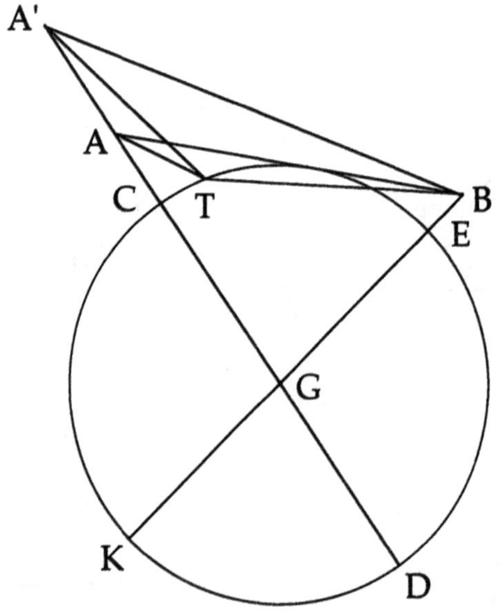

figure 5.2.48a

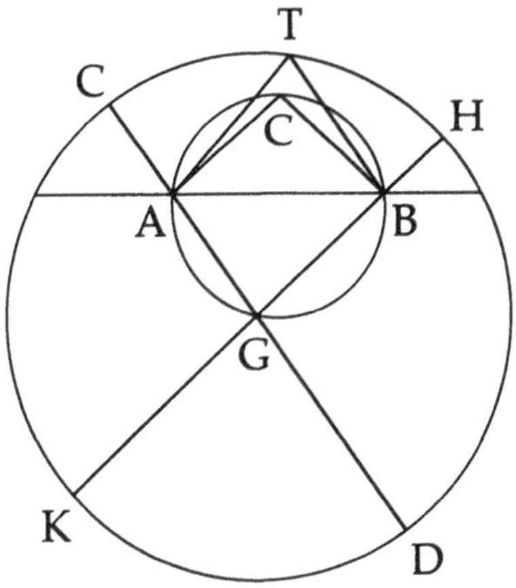

figure 5.2.49

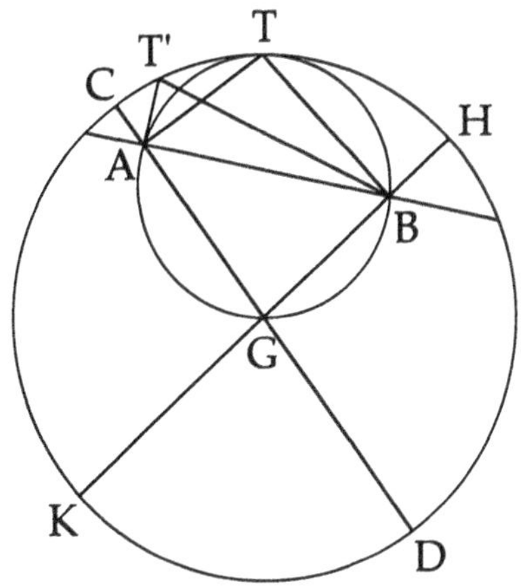

figure 5.2.49a

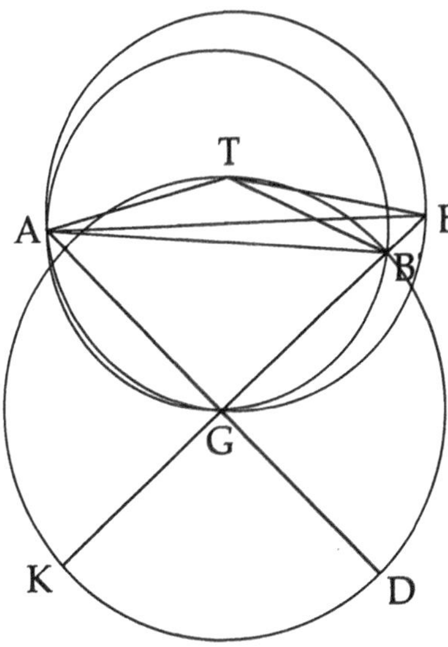

figure 5.2.49b

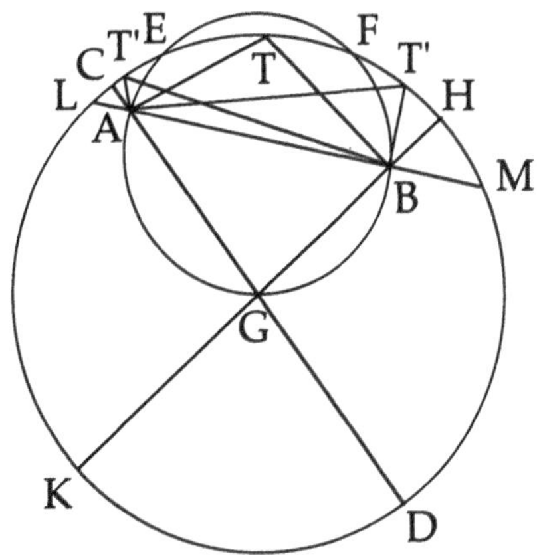

figure 5.2.49c

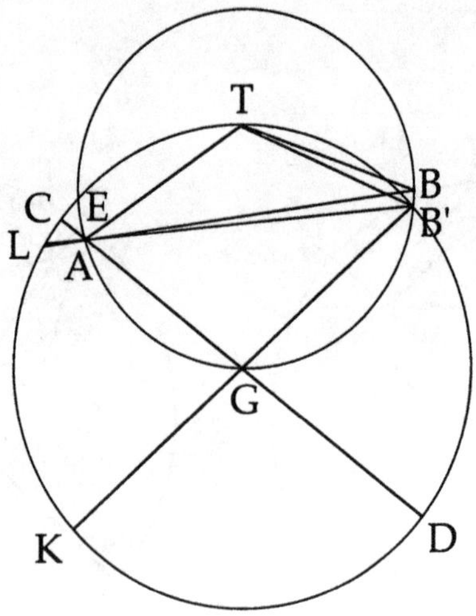

figure 5.2.49d

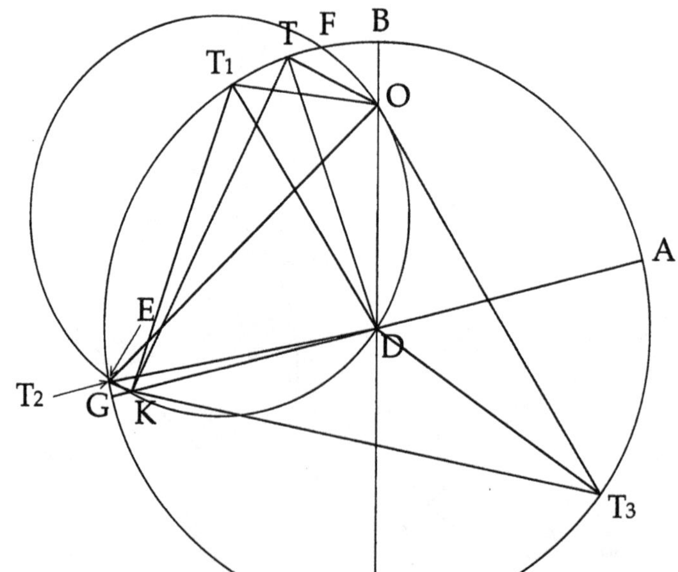

figure 5.2.49e

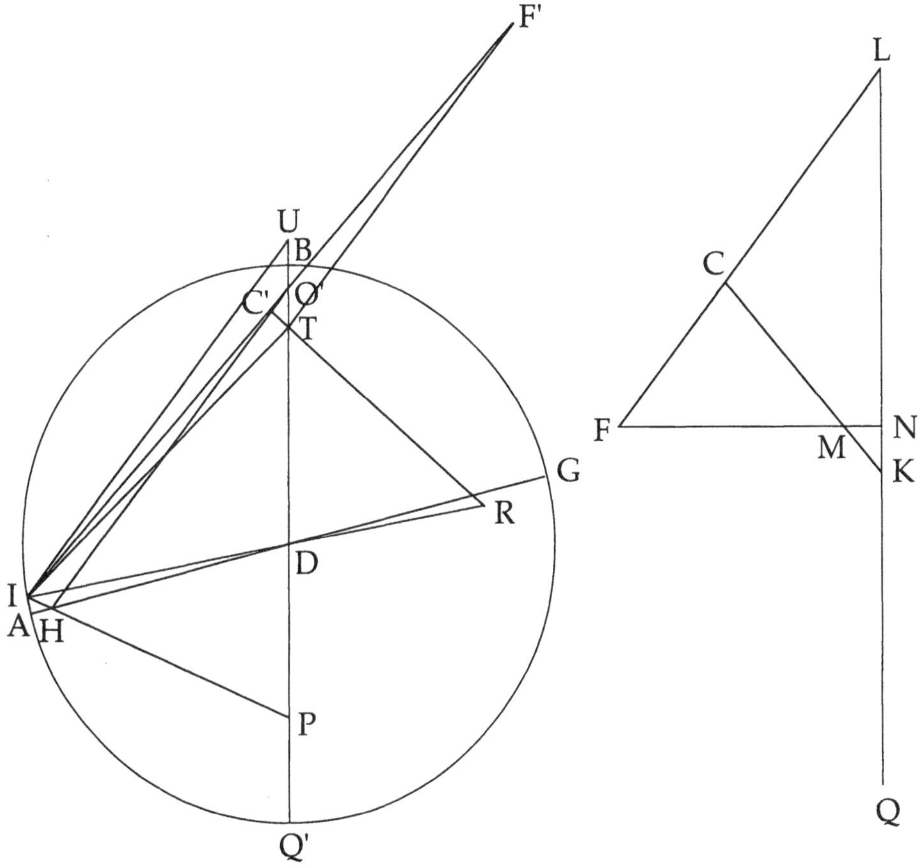

figure 5.2.49f

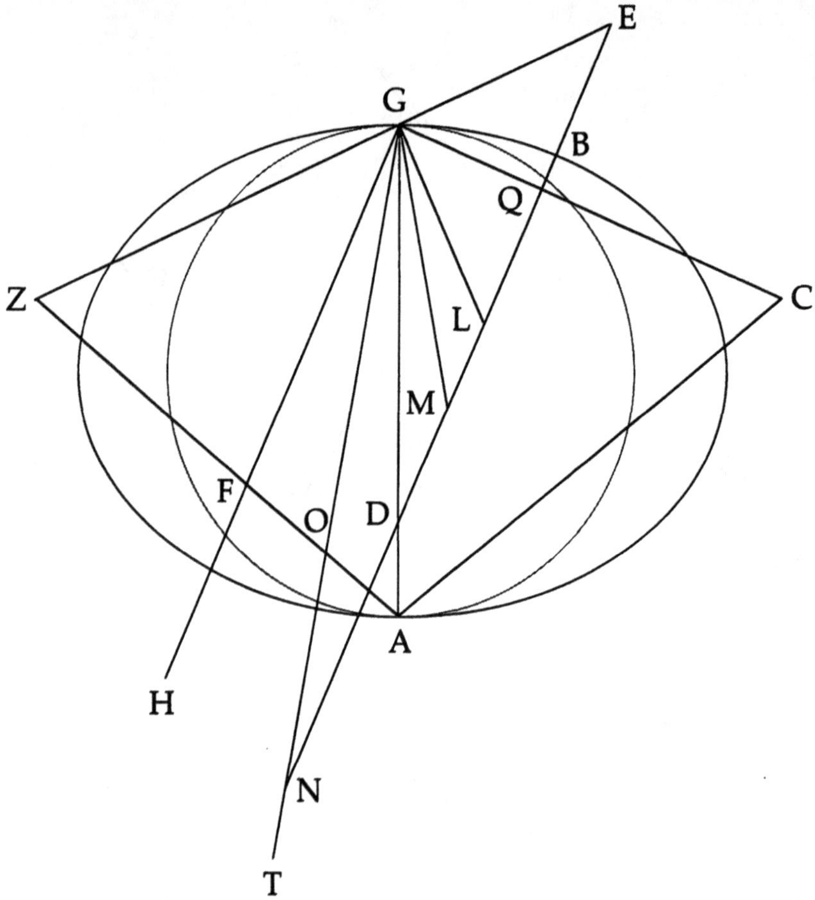

figure 5.2.50

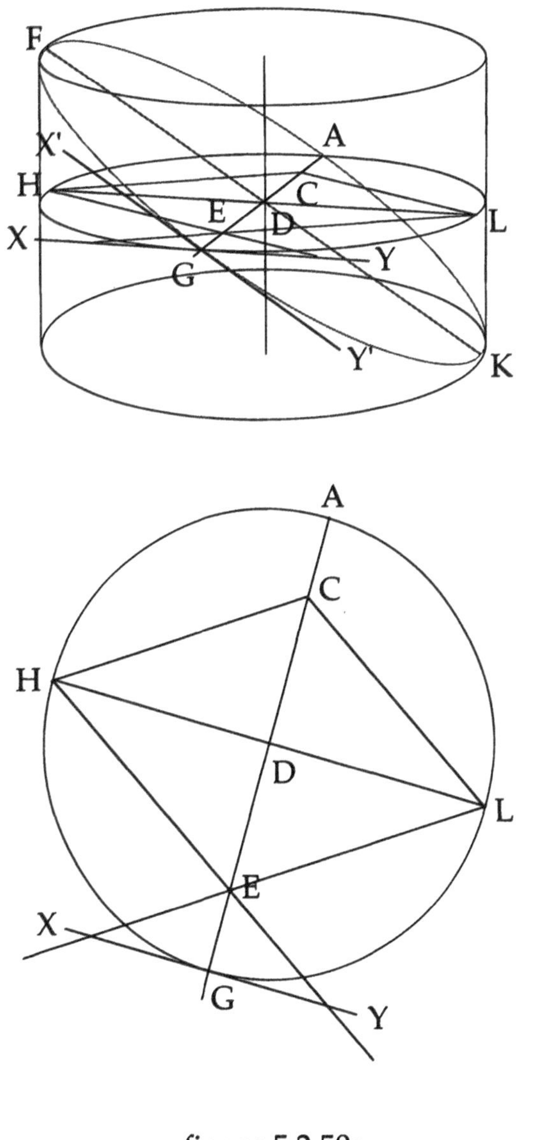

figure 5.2.50a

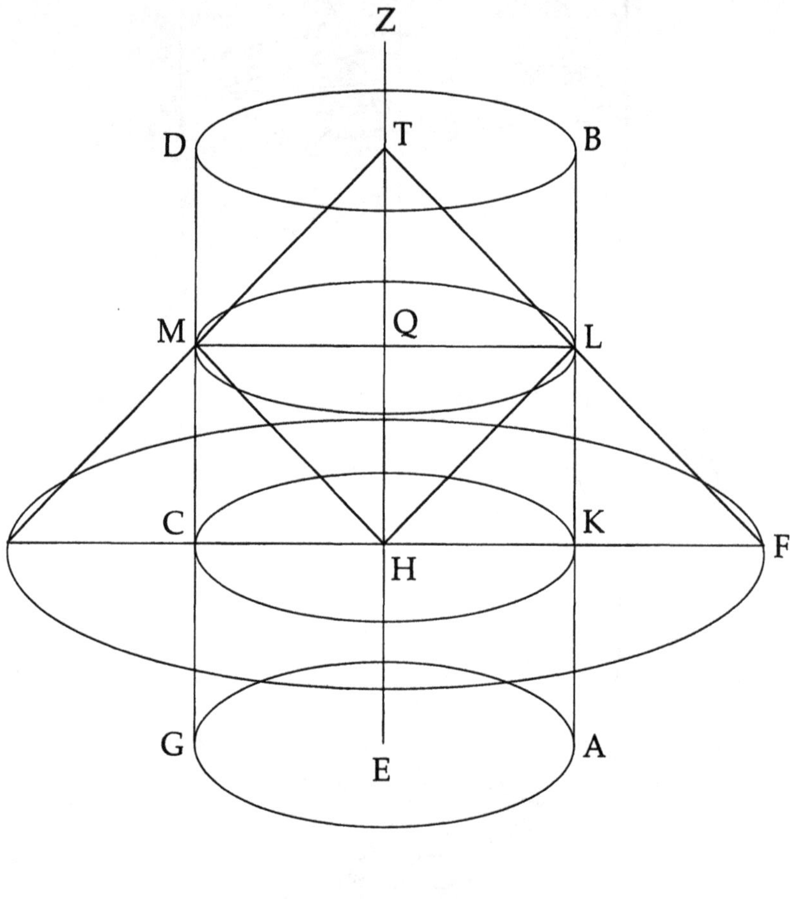

figure 5.2.51

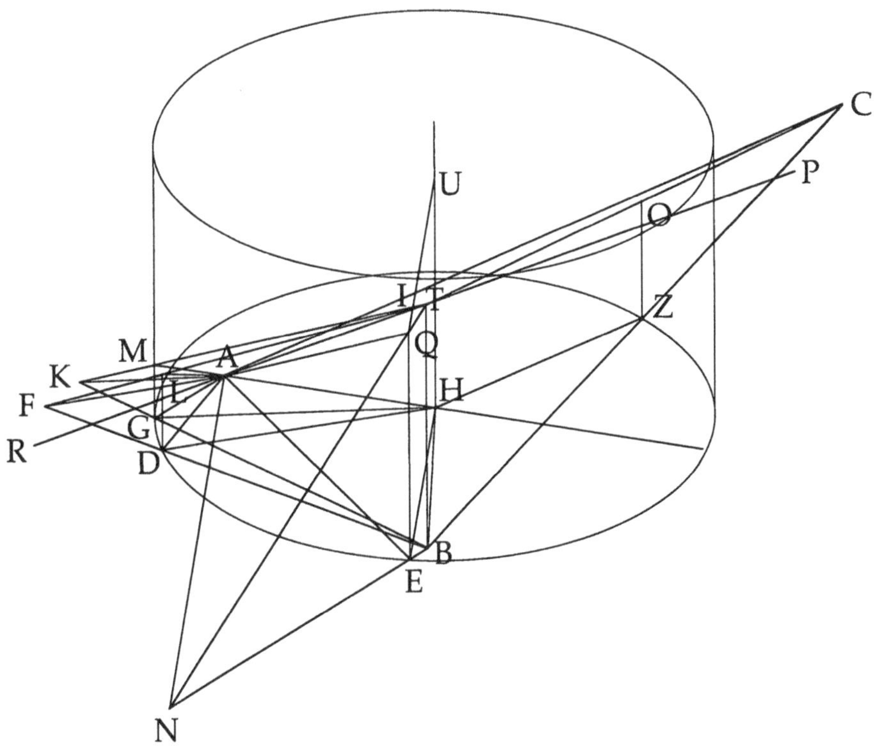

figure 5.2.52

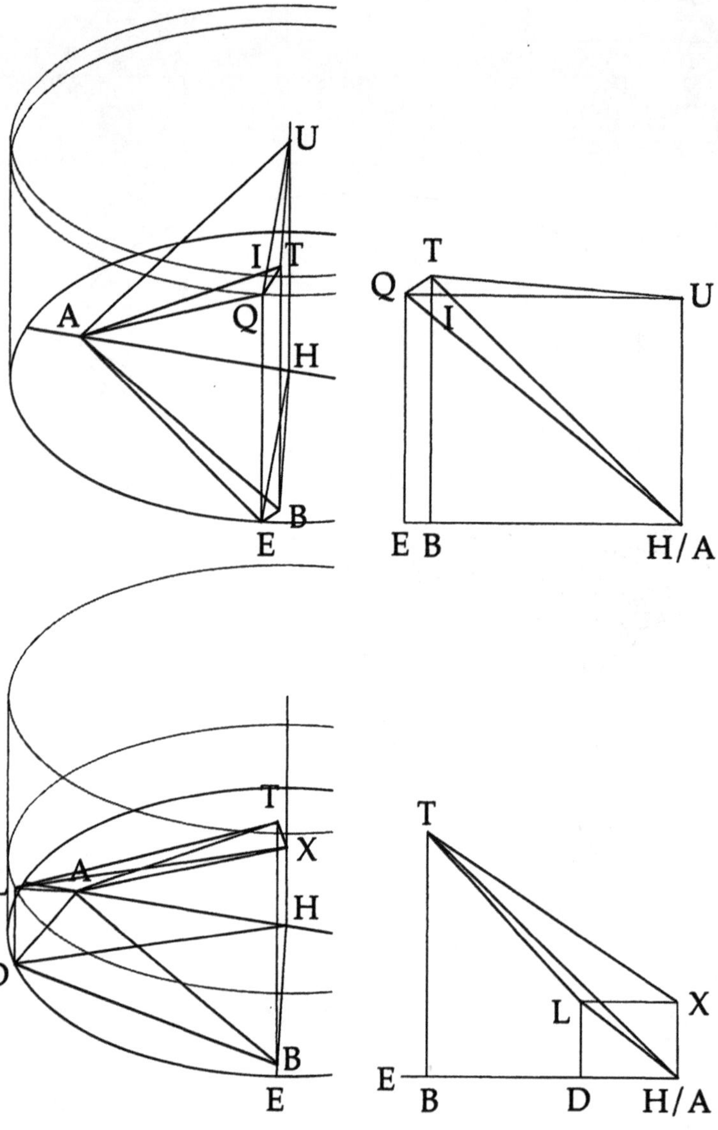

figure 5.2.52a

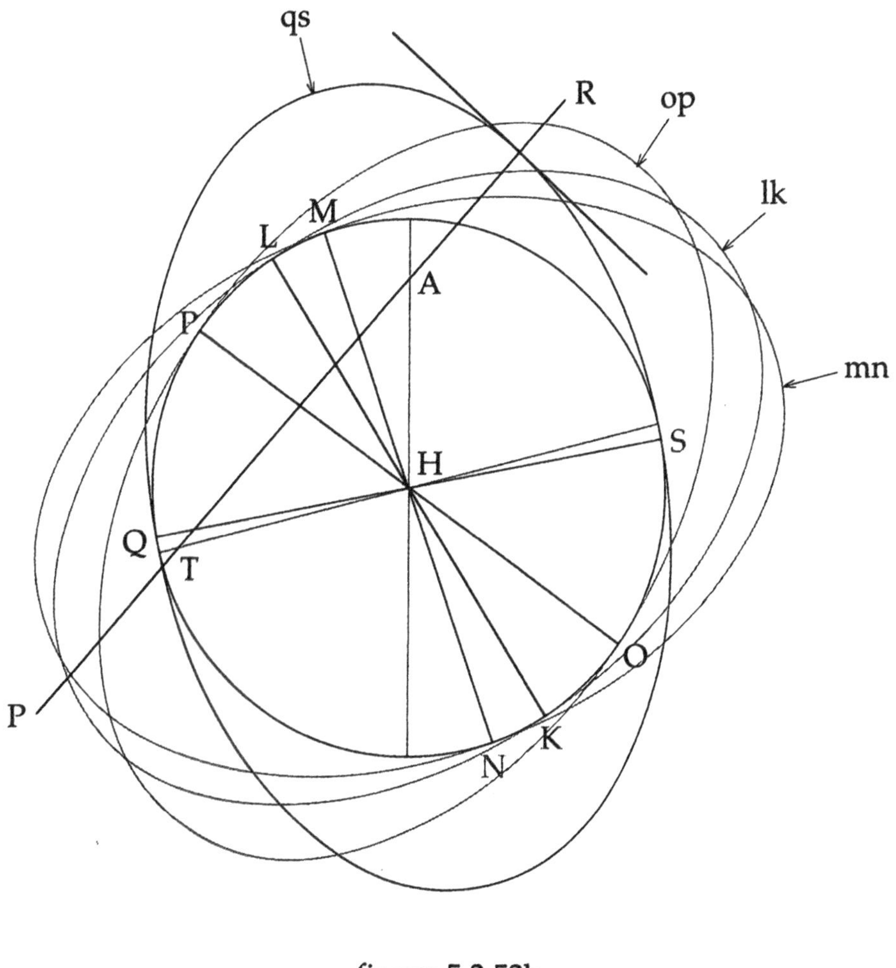

figure 5.2.52b

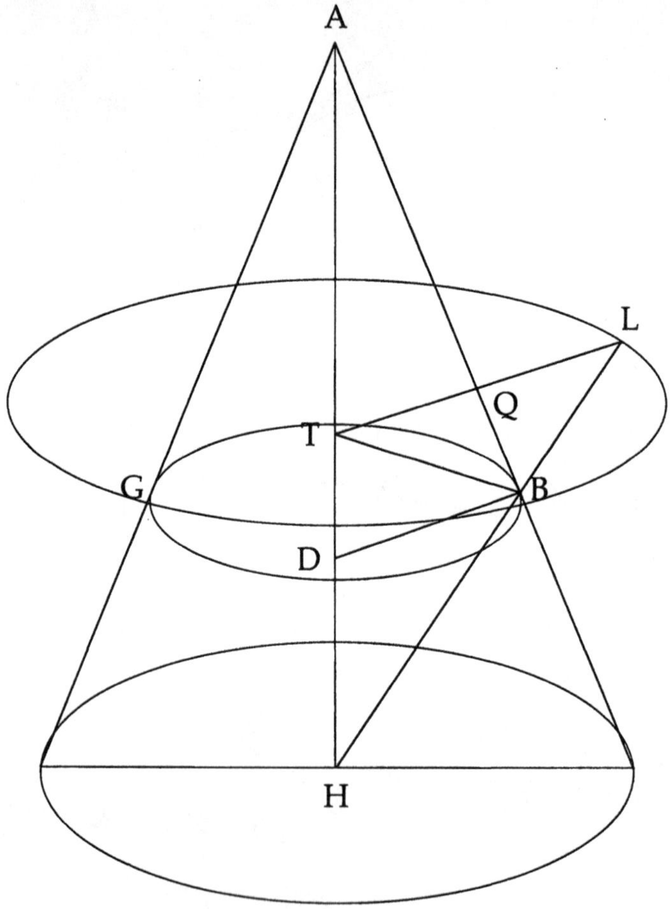

figure 5.2.53

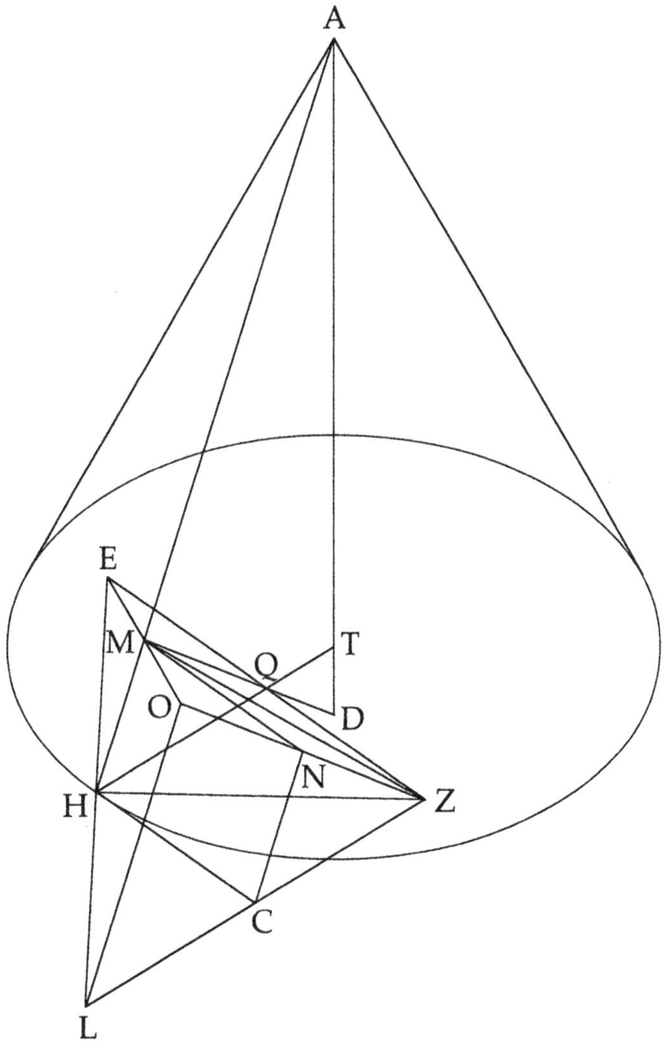

figure 5.2.54

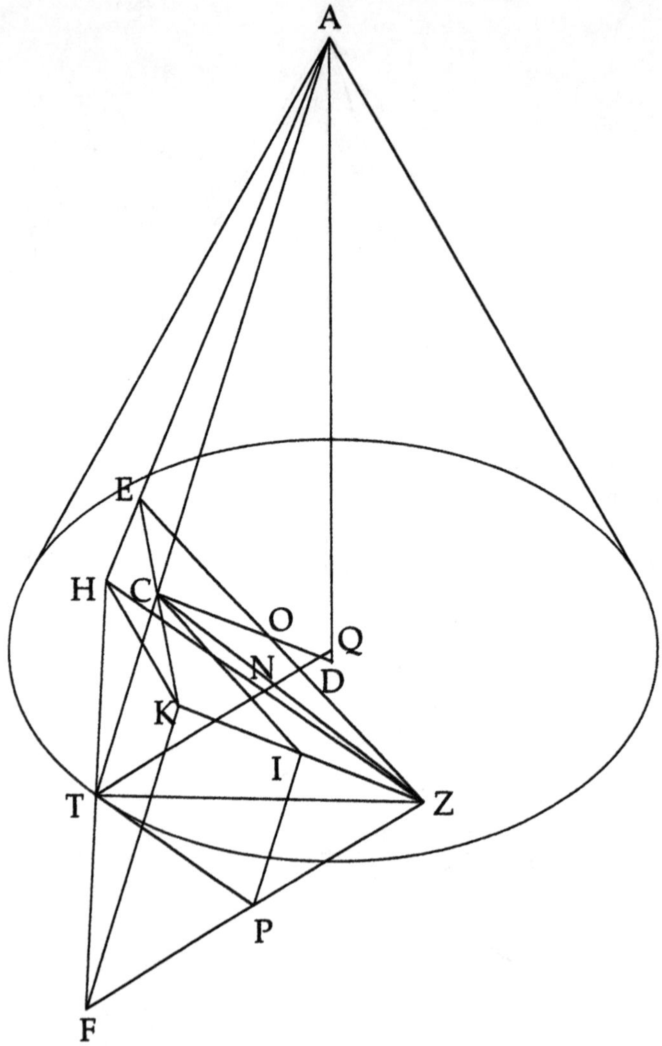

figure 5.2.54a

www.ingramcontent.com/pod-product-compliance
Lightning Source LLC
Chambersburg PA
CBHW081339190326
41458CB00018B/6048